Textbooks in Electrical and Electronic Engineering

Series Editors
G. Lancaster   E. W. Williams

1. **Introduction to fields and circuits**
   GORDON LANCASTER

2. **The CD-ROM and optical disc recording systems**
   E. W. WILLIAMS

3. **Engineering electromagnetism: physical processes and computation**
   P. HAMMOND and J. K. SYKULSKI

4. **Integrated circuit engineering: establishing a foundation**
   L. J. HERBST

Integrated Circuit Engineering
Establishing a Foundation

■

L. J. Herbst
Division of Electronic and Computer Engineering
University of Teesside
Middlesbrough

OXFORD   NEW YORK   TORONTO
OXFORD UNIVERSITY PRESS
1996

Oxford University Press, Walton Street, Oxford OX2 6DP

Oxford New York
Athens Auckland Bangkok Bombay
Calcutta Cape Town Dar es Salaam Delhi
Florence Hong Kong Istanbul Karachi
Kuala Lumpur Madras Madrid Melbourne
Mexico City Nairobi Paris Singapore
Taipei Tokyo Toronto
and associated companies in
Berlin Ibadan

Oxford is a trade mark of Oxford University Press

Published in the United States by
Oxford University Press Inc., New York

© L. J. Herbst, 1996

All rights reserved. No part of this publication may be
reproduced, stored in a retrieval system, or transmitted, in any
form or by any means, without the prior permission in writing of Oxford
University Press. Within the UK, exceptions are allowed in respect of any
fair dealing for the purpose of research or private study, or criticism or
review, as permitted under the Copyright, Designs and Patents Act, 1988, or
in the case of reprographic reproduction in accordance with the terms of
licences issued by the Copyright Licensing Agency. Enquiries concerning
reproduction outside those terms and in other countries should be sent to
the Rights Department, Oxford University Press, at the address above.

This book is sold subject to the condition that it shall not,
by way of trade or otherwise, be lent, re-sold, hired out, or otherwise
circulated without the publisher's prior consent in any form of binding
or cover other than that in which it is published and without a similar
condition including this condition being imposed
on the subsequent purchaser.

A catalogue record for this book is available from the British Library

Library of Congress Cataloging in Publication Data
Data applied for

ISBN 0-19-856279-9 (Hbk.)
ISBN 0-19-856278-0 (Pbk.)

Printed and bound in Great Britain by
The Bath Press, Somerset

To my wife

Preface

Integrated circuits (ICs) occupy a key role in electronics and are to be found in almost every conceivable product of consumer, industrial, and military electronics. They are routinely included in the curriculum of first degrees in electrical/electronic engineering. Many other first degrees in related disciplines like microelectronics and computer engineering also contain substantial material on ICs. The evolution of very-large-scale integration (VLSI) has radically altered the nature of IC engineering. The electronic circuits of a complex system like a computer can now be contained in one or a small number of chips. The nature of VLSI is being brought home to engineers in the form of application-specific ICs (ASICs), which involve customers in their design.

Based on my personal experience of teaching ICs and VLSI on first degrees and on special courses (which I organized) in industry, I perceived the need for a comprehensive text on IC engineering which recognizes the centrality of VLSI, but also embraces lower levels of integration. There is at present a serious gap in such literature. Nothing has been published since 1977 to compare with the comprehensive texts on IC engineering by Hamilton and Howard (1975) and Glaser and Subak-Sharpe (1977). The many books on ICs which have appeared during the last twenty years are generally pitched at a graduate rather than an undergraduate level and tend to be specialized, covering subjects like MOS ICs, VLSI, and IC technology.

This book gives a broad, comprehensive coverage of ICs, embracing fabrication, circuit techniques, and VLSI/ASIC system aspects. Technology is likewise covered in breadth by including bipolar, MOS, and GaAs ICs. It is assumed that readers will have a background in basic transistor operation, analogue and digital circuit practice, and digital design. Alternatively undergraduates may be taught these subjects in parallel with the contents of this book, which is aimed at the second and third years of a British, and the second, third and fourth years of an American degree. The text also meets a vital need for continuing education. Much of the subject matter it contains will probably be new to most engineers and scientists who have graduated some years ago. For that reason this book should prove to be very attractive for career–long learning.

Coming to the details of the contents, the text begins with a general survey (Chapter 1). Device fabrication and packaging (Chapter 2) come next. Chapter 3 covers the formation of transistors, resistors and capacitors, and Chapter 4 outlines device behaviour and modelling. Circuit techniques form the subjects of Chapter 5 (digital) and Chapter 6 (analogue). The last few years have seen a distinct swing towards more analogue and mixed-mode (analogue and digital) ICs, brought about largely by the growth in telecommunications. That situation is reflected in the contents of Chapter 6, the largest chapter in the book, and the inclusion of mixed-mode ASICs in Chapter 10. Chapter 7 on semiconductor memories initiates the VLSI/ASIC sector, which extends to the end of the book. An overview of ASIC design styles in Chapter 8 paves the way for a fuller treatment of ASICs in Chapters 9 to 11. Chapter 9 describes programmable logic devices (PLDs) with a strong emphasis on field programmable gate arrays (FPGAs). Chapter 10 deals with the characteristics and design issues of ASICs, including economics and design for testability, and Chapter 11 covers ASIC design techniques. Finally Chapter 12 looks at submicron scaling and projections of ICs for the immediate future. Electronics abounds with acronyms and jargon, which are defined in a glossary at the end of the text for ease of reference.

The style of presentation departs in a number of ways from the orthodoxy found in the majority of books on ICs. My first and foremost consideration has been to highlight the engineering dimension. Many publications within the last two decades on the teaching of engineering have urged the need for more material on engineering practice and design, and less on analysis (ASEE 1987; Finniston 1980). Inspite of this there is a continuing trend towards a concentration on analysis.

The roles of analysis and synthesis in this book embody some ideas which I advanced at two conferences in the US (Herbst 1987 and 1989). Analysis has been greatly reduced relative to the traditional presentation in many established texts on ICs. Second synthesis has been carefully structured to illustrate current engineering practice. Last synthesis is often placed before analysis, frequently in the form of an overview at the beginning of a chapter. My approach bears some resemblance to the inverted curriculum proposed by Cohen (1987 and 1992). Extensive quotations of IC specifications and performance bring home the capabilities of current ICs, and this information is continually interwoven with the rest of the text, sometimes even within the analytical sections. The profuse quotations of current IC performance are open to the charge that such material will soon become dated. There is a characteristic American saying that if a piece of equipment works it is out of date. However the probable changes in the performance of future ICs, mainly due to continuing miniaturization, are dealt with, and readers should be able to appreciate the capabilities of new products emerging within the next few years without undue difficulty (Mead and Conway 1980).

The accent on engineering design and applications in this text is long overdue. Much of the engineering science and mathematics of an engineering undergraduate curriculum is unnecessary and can be eliminated, making way for material on synthesis, design and applications (Finniston 1980, pp.94–5). Nevertheless a balance must be maintained between analysis and synthesis: the danger of overemphasizing the engineering aspects to the detriment of fundamental concepts is very real.

Muller and Kamins (1986) hold successful engineering to rest on two foundations, physical concepts and technology. Zorpette (1984) warns against excessive emphasis on what he calls the technology of the moment. Everitt (1980) defined the fundamental difference between science and engineering to be the difference between analysis and synthesis, and went on to say that synthesis can only be accomplished after a thorough grounding in analysis. G.M. Trevelyan (1946), the late great English historian, was right when he told us that 'disinterested intellectual curiosity is the life-blood of real civilization'. However hard-pressed engineers may be in industry, where they are often pushed from pillar to post in order to come up with designs for new equipment or to meet critical production schedules, they must preserve an element of detached intellectual curiosity for part of their work. In the light of these comments I have taken care to stress fundamental concepts and to avoid preoccupation with engineering immediacy.

It remains for me to express my gratitude for the help I have received from various sources. Apart from obtaining databooks from semiconductor vendors, I had countless conversations with staff from these establishments to clear up matters on which I felt uncertain, or to obtain additional information. This help was given most willingly without exception and has been of tremendous value. The manuscript has been produced and typeset with the aid of computer typesetting software by Trevor and Rose Atkinson, who have very effectively accommodated special requirements regarding style and have been most helpful throughout. I am indebted to Colin Gregg, who has drawn all the diagrams with the aid of computer graphics, and has in many cases improved on the originals submitted to him. The help and cooperation of the publishers at all stages of the production is gratefully acknowledged.

Last and most I am indebted to my wife for her unstinting support for this undertaking, which made a severe inroad on the time for normal domesticity. It is only her encouragement and understanding which have made this book possible.

Middlesbrough L.J.H.
September 1995

References

ASEE. (1987). *A national action agenda for engineering education. A report of the task force on the national agenda for engineering education.* American Society for Engineering Education (ASEE), USA.

Cohen, B. (1987). The education of the information systems engineer. *Electronics and Power*, **33**, pp.203–5.

Cohen, B. (1992). *The inverted curriculum.* National Economic Development Office, England.

Everitt, W.L. (1980). The phoenix-a challenge to engineering education. Reprinted from the Proceedings of the IRE, 32, September 1944, 509–13. *IEEE Transactions on Education*, **23**, pp.179–83.

Finniston, M. (1980). *Engineering our future. Report of the committee of enquiry into the engineering profession.* Her Majesty's Stationery Office, England.

Glaser, A.B. and Subak-Sharpe, G.E. (1977). *Integrated circuit engineering.* Addison-Wesley, USA.

Hamilton, D.J. and Howard, W.D. (1975). *Basic integrated circuit engineering.* McGraw-Hill, USA.

Herbst, L.J. (1987). *Analysis and synthesis in honours engineering degrees.* Proceedings of the 1987 Frontiers in Education Conference, IEEE and ASEE, Terre Haute, Ind., USA.

Herbst, L.J. (1989). *Placing synthesis before analysis.* Proceedings of the 1989 Frontiers in Education Conference, IEEE and ASEE, Bingham, N.Y., USA.

Mead, C. and Conway, L. (1980). *Introduction to VLSI systems,* p.vi. Addison-Wesley, USA.

Muller, S. and Kamins, T.I. (1986). *Device electronics for integrated circuits,* (2nd edn), p.57. Wiley, USA.

Trevelyan, G.M. (1946). *English social history,* (3rd edn), p.viii. Longmans, England.

Zorpette, G. (1984). EE programs. *IEEE Spectrum,* **21**, (11), 44–50.

Acknowledgements

The following industrial establishments have kindly given permission to quote from their databooks and/or datasheets.

>Actel
>Altera
>Advanced Micro Devices (AMD)
>Analog Devices
>ASM Lithography
>Austria Mikro Systeme International (AMS)
>Burr Brown
>Cadence
>Compass
>Cypress Semiconductor
>Fujitsu
>GEC Plessey Semiconductors (GPS)
>Harris Semiconductor
>Lattice Semiconductor
>Mentor Graphics
>Micro Linear
>Motorola
>National Semiconductor (NS)
>NEC
>Philips Semiconductors
>Synopsys
>Technology Modeling Associates (TMA)
>Texas Instruments (TI)
>VLSI Technology
>Xilinx

Great care has been taken to give due acknowledgement for the use of data, whether in tabulations or in the text. Trademarks have also been listed, especially in the Glossary. Alternatively the introduction of an acronym has been directly linked with its originator, thereby establishing its proprietary nature. The author apologizes if inadvertently the proper acknowledgement has not been made, or if a trademark has been omitted. Such cases and errors in the data should be brought to the notice of the publishers, so that they can be put right in a reprint. The same observation applies to permissions for using information from textbooks.

Philips Semiconductors have requested the insertion of the following disclaimer:

Philips Semiconductors and North American Philips Corporation Products are not designed for use in life support appliances, devices or systems where a malfunction of a Philips Semiconductor and North American Philips Corporation Product can reasonably be expected to result in a personal injury. Philips Semiconductors and North American Phillips Corporation customers using or selling Phillips Semiconductors and North American Phillips Corporation Products for use in such applications do so at their own risk and agree to fully indemnify Philips Semiconductors and North American Philips Corporation for any damages resulting from such improper use or sale.

On a more general note, the author advises that the quotations of data from the establishments listed are only guidelines, but that fuller information, sometimes going beyond what is available in databooks and datasheets, should be consulted for actual designs.

Contents

1 Overview   1
    1.1   IC structure   1
    1.2   Historical perspective   4
    1.3   IC spectrum   8
       1.3.1   VLSI   8
       1.3.2   SSI and MSI   9
       1.3.3   PLDs   10
       1.3.4   Analogue ICs   10
       1.3.5   ASICs   11
       1.3.6   Summary   12
       References   13

2 Fabrication and packaging   15
    2.1   Processing outline   15
    2.2   Processing sequence for transistors   18
    2.3   Wafer preparation   22
    2.4   Deposition and growth   25
       2.4.1   Oxidation   25
       2.4.2   Epitaxy   30
       2.4.3   Diffusion   31
       2.4.4   Ion implantation   34
       2.4.5   Dielectric and polysilicon films   39
       2.4.6   Interconnects   40
    2.5   Lithography   44
       2.5.1   Introduction   44
       2.5.2   Lithographic technology   44
       2.5.3   Resists   50
       2.5.4   Etching   52
    2.6   Packaging and mounting   56
       2.6.1   Introduction   56
       2.6.2   Packages available   57
       2.6.3   Packaging and mounting technologies   59
       2.6.4   Package selection   63
       2.6.5   Multichip modules   65
    2.7   Overview   67
       References   69

3 Component formation   73
    3.1   Introduction   73
    3.2   CMOS   73
    3.3   Silicon bipolar transistors   78

	3.4	BiCMOS		81
	3.5	GaAs transistors		82
	3.6	Resistors and capacitors		84
	3.7	Design rules and device areas		87
		References		91

4 Device behaviour and modelling — 93

4.1	Introduction		93
4.2	Transient response and bandwidth		93
4.3	MOSFET characteristics		96
	4.3.1	DC characteristics	96
	4.3.2	Capacitances	100
	4.3.3	Channel transit time	103
	4.3.4	Hot carriers	103
4.4	GaAs MESFET characteristics		103
4.5	BJT characteristics		105
4.6	The role of SPICE		109
4.7	SPICE models		111
	4.7.1	Introduction	111
	4.7.2	MOSFET models	111
	4.7.3	GaAs MESFET model	112
	4.7.4	BJT model	112
	4.7.5	Junction diode	112
	References		113
	Further reading		114

5 Digital circuits — techniques and performance — 117

5.1	Introduction		117
5.2	Logic Circuits		117
	5.2.1	Design criteria	117
	5.2.2	Overview of logic familes	118
	5.2.3	nMOS	119
	5.2.4	CMOS	121
	5.2.5	ECL/CML	129
	5.2.6	TTL	132
	5.2.7	I^2L	135
	5.2.8	BiCMOS	138
	5.2.9	GaAs	141
5.3	Bistables		145
5.4	Dynamic CMOS		150
5.5	Survey of logic circuits		152
	5.5.1	Overview	152
	5.5.2	Characterization and performance	157
	5.5.3	Comparison of GaAs with Si	172
	Appendix—SPICE parameters		175
	References		177
	Further reading		179

6	**Analogue circuits—techniques and performance**		181
	6.1 Introduction		181
	6.2 Operational amplifiers		181
		6.2.1 Basic concepts	181
		6.2.2 Circuit techniques	186
		6.2.3 Specification and performance	194
	6.3 Voltage comparators		206
	6.4 Voltage references and regulators		209
		6.4.1 Voltage references	209
		6.4.2 Voltage regulators	215
	6.5 Analogue signal processing		224
		6.5.1 Introduction	224
		6.5.2 Analogue filters	226
		6.5.3 ADCs	232
		6.5.4 DACs	242
		6.5.5 Specification and performance	244
		6.5.6 Oversampling sigma–delta converters	249
	References		258
	Further reading		260
7	**Semiconductor memories**		261
	7.1 Introduction		261
	7.2 Organization and operation		263
	7.3 SRAMs		267
	7.4 DRAMs		271
	7.5 ROMs		277
	7.6 Characterization and performance		284
	References		290
	Further reading		291
8	**ASIC design styles**		293
	8.1 Introduction		293
	8.2 Categories		293
	8.3 Gate arrays		294
	8.4 Standard cells		296
	8.5 Cell-based ASICs		297
	8.6 Mixed-mode and analogue ASICs		298
	8.7 PLDs		298
		8.7.1 PLA and PAL	298
		8.7.2 Field programmable gate arrays	300
	8.8 Overview		301
	Further reading		302
9	**ASICs—Programmable logic devices**		303
	9.1 Overview		303
	9.2 PAL-based PLDs		304
		9.2.1 Structures	304

		9.2.2 PAL characteristics	313
	9.3	FPGAs	318
		9.3.1 Introduction	318
		9.3.2 Selected families	319
	9.4	Design outline	332
		References	336
		Further reading	337
10	**ASICs—Characteristics and design issues**		339
	10.1	Introduction	339
	10.2	Design methodology and design tools	342
	10.3	Design for testability	344
	10.4	Economics	356
	10.5	Characteristics and performance	362
		10.5.1 Design styles	362
		10.5.2 Gate arrays	365
		10.5.3 Standard cells	370
		10.5.4 Cell-based ASICs	371
		10.5.5 Mixed-mode and analogue ASICs	372
	10.6	Overview	380
		References	383
		Further reading	384
11	**ASICs—Design techniques**		385
	11.1	Overview	385
	11.2	Design flow and methodology	387
	11.3	Hardware description languages	398
	11.4	Simulation and checking	406
	11.5	Commercial design tools	409
	11.6	FPGA design tools	417
	11.7	Conclusions	419
		References	421
		Further reading	423
12	**Submicron scaling**		425
	12.1	Overview	425
	12.2	MOSFET scaling	431
	12.3	BJT scaling	437
	12.4	Interconnection scaling	439
	12.5	The immediate future	444
		References	449
	Glossary		453
	Index		465

1
Overview

1.1 IC structure

The nature of electronic equipment based on integrated circuits (ICs) is illustrated in Fig. 1.1, the photograph of a motherboard, a printed circuit board (PCB) for a personal computer (PC). The motherboard is the core of the electronics for the computer. It contains numerous ICs of various types and sizes, in addition to a few *discrete* components (resistors, capacitors, crystal oscillators), and sockets for interconnecting the motherboard to other PCBs. The ICs have transistor counts ranging from about 100 to 250 000 per package, and are interconnected to form a highly sophisticated system. The complexity of an IC in this era of very-large-scale integration (VLSI) is evident from the photograph in Fig. 1.2, which shows the surface of a *die* (signifying an unencapsulated chip). That die contains over 100 000 transistors, has 68 bonding pads, and measures about 8×8 mm^2. It is encapsulated in a 68-pin square package not much larger than the die. This section presents a very brief outline of the structure and the processing leading to an IC.

Figure 1.3 shows the profile of a die. The electrodes of all active (transistors and diodes) and passive (resistors and capacitors) components are

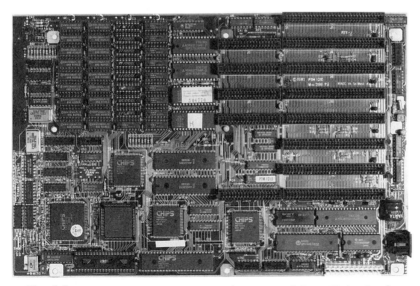

Fig. 1.1 PC motherboard photograph (Courtesy of Opus Technology)

2 Overview

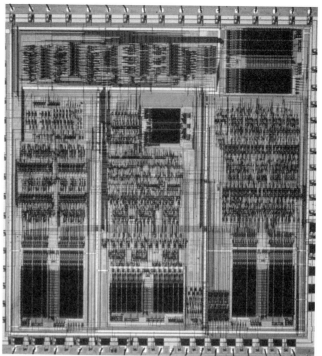

Fig. 1.2 VLSI die photograph (Courtesy of GEC Plessey Semiconductors)

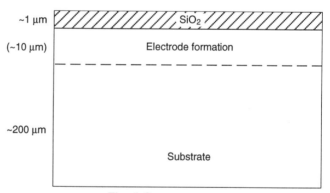

Fig. 1.3 Profile of die

formed (with the exception of the MOSFET gate, which is placed within the SiO$_2$ layer) in the silicon substrate, and are contained within a depth of about 10 μm from its surface. The substrate is typically $\sim$200 to 250 μm deep to give it adequate mechanical strength. A protecting silicon dioxide layer (also called field oxide) covers the substrate.

The interconnections are made by depositing aluminium over the oxide surface after *contact windows* have been opened in the oxide. The aluminium will fill the contact windows and will thereby establish connections to the electrodes. The circuit is formed by etching away the unwanted metal to have the desired interconnect pattern.

Cross sections of a bipolar junction transistor (BJT) and a metal-

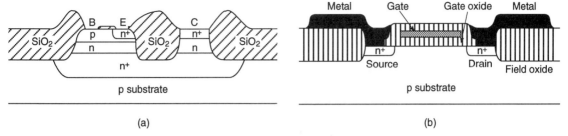

Fig. 1.4 Transistor profiles (a) npn BJT (b) n channel MOSFET

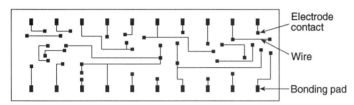

Fig. 1.5 Top view of die

oxide-semiconductor field effect transistor (MOSFET) are sketched in Fig. 1.4. Note that the transistors are isolated from their neighbours—not shown—by means of the oxide layers. A symbolic top view of a die (symbolic because it contains only a very small number of the interconnections) is given in Fig. 1.5, which shows the nature of the interconnects. These are also called *wires*, a term now universally accepted and used freely throughout the book. The peripheral bonding pads serve for the leads connecting the input/output terminations of the die to the pins of the package.

For chips with high component densities, a single metallization layer (also called metal), is insufficient and multilevel interconnects are used instead. An example of three-level metal is sketched in Fig. 1.6. Communication between the various metals is established using vertical aluminium interconnects, known as *vias*. Dielectrics other than silicon dioxide are frequently used for inter-metal isolation and passivation, because they are cheaper and electrically acceptable in these locations. The chip is finally completed by encapsulation of the die.

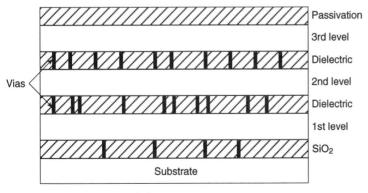

Fig. 1.6 Multilevel interconnect

1.2 Historical perspective

The cornerstone of the integrated circuit is the transistor, which was invented by Bardeen and Brattain, two members of a research group headed by Shockley and Morgan at the Bell Laboratories, in the so-called *magic month* from November 17 to December 16, 1947. The team was attempting to make a device resembling a field effect transistor, when they observed minority carrier injection and pounced on that discovery to invent the germanium point contact transistor. The basic principle behind the action of the the metal-oxide-semiconductor (MOS) field effect transistor, the MOSFET, was proposed much earlier by Lilienfeld (1930) and Heil (1935). Technological limitations prevented a practical realization of that idea. Shockley went on to predict the bipolar junction transistor (BJT) and the junction field effect transistor (JFET), and these two devices materialized in 1951 (Shockley 1976). Germanium soon gave way to silicon, and the first commercial silicon BJTs came on the market in 1954. Another milestone came with the production of the MOSFET in 1960 (Kahn and Atalla 1960).

In the meantime the interconnection of electric circuits was progressing with the development of printed circuit assemblies. The possibility of semiconductor integrated circuits was first advanced by G.W.A. Dummer of the Royal Radar Establishment, Malvern, England in 1952. Addressing the Electronics Components Conference, he said: 'With the advent of the transistor and the work in semiconductors generally, it seems now possible to envisage electronics equipment in a solid block with no connecting wires. The block may consist of layers of insulating, conducting, rectifying, and amplifying materials, the electric functions being connected directly by cutting out areas of the various layers'. Dummer's vision prompted investigations by the US Armed Forces towards such an objective, advanced at first under the name of *molecular electronics*. The breakthrough came in 1958 when Jack Kilby of Texas Instruments, USA, produced the first integrated circuit. That circuit was a phase shift oscillator consisting of a distributed RC network and a flip flop. Another development, of equal importance, was undertaken by Robert Noyce and Gordon Moore of Fairchild Semiconductor, USA, who invented the planar process for passivating the junctions of a silicon bipolar transistor in 1959. Until then, adequate protection of transistor junctions from surface leakage and other effects had proved difficult. The structure of the planar bipolar transistor is shown in Fig. 1.4(a). The salient feature, oxide passivation, holds for discrete and integrated transistors alike, the only difference being that the discrete transistor has a metallization layer for the collecter contact in place of the substrate. Noyce filed his patent for the silicon integrated circuit in July 1959, shortly after the announcement of the integrated circuit by Texas Instruments. The silicon monolithic integrated circuit had been born, with Kilby and Noyce its joint inventors (Kilby 1976).

It was then already evident that computing would become the outstanding growth area of electronics, and the first IC developments were for digital logic. Early design efforts in the analogue field concentrated on the operational amplifier. Once the idea of integrated circuits had taken root, their tremendous potential was soon realized, linked to an awareness that the key to further progress lay in increasing the function density per chip. That increase comes about largely by reducing device dimensions. The processing of ICs advanced rapidly and the exponential increase in transistor count per chip during the two decades 1959–1979 followed fairly closely Gordon Moore's prediction of 1960, in which he forecast that the function density per chip would double every year (Moore 1975).

Data processing, with semiconductor memories at the spearhead, was and remains the driving force behind the ongoing increase in chip function density. Processing power increased until it extended to the system level, where a single chip could perform a complex function. The first microprocessor—by definition a central processing unit (CPU) on a single chip—emerged in 1971. With it, the Intel 8008, came the Intel 1103, a 1 Kb random access memory (RAM), which was an outstanding advance in semiconductor memories at that time. (K is used in preference to k for 1000 in semiconductor electronics.) The Intel 8008 was naturally a very modest CPU by today's standards, but the breakthrough had been made: the move towards very-large-scale integration (VLSI) had begun in earnest. The categorization of chips according to transistor count is shown in Table 1.1.

Table 1.1 IC categories—device count

Designation	Active devices per chip
SSI	<100
MSI	100–1000
LSI	1000–100 000
VLSI	>100 000

SSI: Small-scale integration
MSI: Medium-scale integreation
LSI: Large-scale integration
VLSI: Very-large-scale integration

The ranges in Table 1.1 are broad rather than precise indicators, and have undergone changes with time. For example, the LSI/VLSI boundary, now at 100 000, stood at 10 000 devices per chip in the early years of VLSI. The active device count, to all intents and purposes the transistor count for the great majority of VLSI, is given in preference to an alternative parameter, the equivalent gate count. In VLSI an equivalent gate is usually interpreted to signify four transistors. The term ultra-large-scale integration (ULSI) is sometimes used for chips containing more than one million transistors, but the designation ULSI has so far not caught on.

The arrival of the microprocessor marked the arrival of MOS VLSI. Digital VLSI is now largely MOS; bipolar VLSI is reserved for special applications which demand its superior speed. MOS transistors have a characteristic which is probably as important as their intrinsic low power consumption relative to bipolar transistors. The extremely high, virtually open-circuit input resistance of the MOSFET gate permits *dynamic operation*, in which input power to the circuit is only applied for a fraction of the clock cycle. Logic data remains stored on nodal capacitors during the intervals between activation. The saving of power in this dynamic mode and the increase in chip function density it allows permit the production of components with increased transistor count. The dynamic mode plays a key role in the drive for larger storage in VLSI memories, where dynamic RAMs (DRAMs) have by far the highest storage and had reached 64 Mb in 1991.

The scaling down of active devices and interconnections is the prime thrust of VLSI development. Transistor geometries are specified in terms of the minimum lateral dimension for electrode formation. Various designations in use, design rules, feature size, geometry, line width and structure, all have similar significance. For example a minimum feature size of 5 μm implies that the emitter of the BJT in Fig. 1.4(a), or the channel of the MOSFET in Fig. 1.4(b) have their shortest lateral dimension equal to 5 μm. MOSFET geometries were about 5 μm in the 'early'

Table 1.2 IC evolution.

Event	Date
IC	1958
Silicon planar IC	1959
Digital SSI and MSI	~1961
MOS	1962
CMOS	1963
First linear ICs	1964
MOS memory	1968
Microprocessor	1971
1 Kb memory	1971
64 Kb DRAM	1977
ASICs	~1980
Multichip module	1980
FPGA	1985
64 Mb DRAM	1991

years of 1971–75, reducing to about 3 μm by 1977 and 1.5 μm by 1981. Currently (1995) standard commercial VLSI is on the micron/submicron boundary, with geometries in the range 0.6–1.2 μm. The move towards smaller geometries is in full swing, with structures of about 0.5 μm for the 64 Mb DRAMs, which appeared in 1991, and yet smaller structures in the range 0.25–0.35 μm at an advanced stage of development.

Complementary MOS (CMOS) logic, which was originated in 1963, only consumes power during switching transitions. That stamped it immediately to be a favourite for digital logic, all the more so with the arrival of VLSI, where the maximum transistor density is limited by power capability rather than device geometry. However it took about 15 years to bring CMOS technology to acceptable standards of quality and reliability. Now that has been done, CMOS is without question the preferred technology. By 1990 about 60% of all ICs were CMOS, and that proportion is expected to increase significantly over the next few years. Some milestones of IC evolution are given in Table 1.2. Gallium arsenide (GaAs) devices have an important role. They offer faster performance than either bipolar or MOS transistors in both digital and analogue applications. Their foremost niche is in optoelectronic systems and multi-gigahertz communication systems; these areas are not covered in the text. In terms of volume, GaAs ICs account at present for about one per cent of the total.

Little has been said so far of analogue IC developments. Much smaller in terms of numbers used than digital ICs, they are of course essential and indispensable constituents in all manner of electronic circuits and systems. Amplifier performance has improved over the years with tighter production control and smaller spreads in key parameters. The reduced transistor dimensions have greatly increased amplifier bandwidth. Signal processing, previously carried out in the analogue domain, is now, thanks to the powerful VLSI signal processors, being performed more and more digitally. High-resolution data converters are needed for that purpose and 16-bit resolution is commonplace. The high-frequency response of active filters has been extended by the improved bandwidth of transistors with micron/submicron geometries. MOS technology has permitted the realization of switched capacitor filters in integrated circuit format. Switched capacitor concepts are rapidly gaining ground elsewhere, for example in charge-scaling data converters.

There have been other spin-offs from the shrinkage to micron/submicron structures. The higher bandwidth of amplifiers has already been mentioned. Other analogue elements like comparators and data converters have similarly advanced in speed. New digital logic families in small-scale and medium-scale integration, which came on the market in the years 1983 to 1986, have far superior performance in respect of speed and power consumption than the earlier versions which they replace. The merging of bipolar and CMOS technologies (BiCMOS) on one chip has become economically viable, thanks to advances in processing. On the analogue side BiCMOS leads to circuits which are markedly superior to those obtainable in either bipolar or CMOS technology. In digital ICs, BiCMOS offers a good speed advance over CMOS at the cost of only a modest increase in power consumption.

VLSI chip development has been paralleled by major advances in packaging and mounting. VLSI packages are available with pin counts approaching 500. Many packages are designed for surface mounting on PCBs. Not only does surface mounting save space, it also increases packaging density by permitting chips to be placed on both sides of the PCB. Surface mounting technology (SMT) has existed for over twenty years, but is only now becoming standard practice for VLSI. PCBs, which have conformed to long established standards for line width and plated through-hole dimensions, are changing, with reduced geometries becoming standardized for handling the VLSI chips with their large pin counts. Another step in the drive to reduce equipment space is the multichip module (MCM), employed not only for hybrid ICs (for example a monolithic chip with a thin film resistor bank for a high grade data converter), but also for easing the interconnections of VLSI chips. In a MCM, several chips are placed on a substrate and are interconnected by wires with far smaller dimensions than can be obtained on a PCB. The substrate, like the PCB, is multi-layered. The module is in effect a miniature PCB which is connected to the PCB in the usual way. The MCM made its first commercial appearance in 1980.

The nature of VLSI design is now being brought home to many, probably the great majority of engineers engaged in computer, communication, or instrumentation design, with the arrival of application-specific integrated circuits (ASICs). These ICs are produced to a customer's specification and are designed in close cooperation between the customer and the semiconductor vendor. The economics of the market place have shown that the best and most cost-effective implementation of equipment is frequently achieved by a combination of ASICs and standard (off-the-shelf commercial) ICs, with ASICs often being predominant. That state of affairs is one of the great surprises in integrated circuit development. In this highly competitive field, the route to advancement and the means of survival were hitherto thought to be the large-volume production of standard components. It so happens that equipment frequently requires a mix of chips from SSI to VLSI. This mix can be altered to contain a higher proportion of VLSI by designing special chips to individual specification. Such a move is not only advantageous in terms of space, power consumption and reliability, but has also been proved to be economic. The key is the evolution of pre-fabrication and pre-design techniques, which greatly reduce the development costs of ASICs *vis-à-vis* standard components. ASICs were initiated around 1980. They already account for about 20 per cent of the turnover for digital ICs, and that proportion is set to grow.

An earlier move, originating during the mid-1970s, was the inauguration of programmable logic devices (PLDs), which are—strictly speaking—ASICs. The read-only memory (ROM) comes into that category, but is understandably bracketed together with the other types of semiconductor memories. The ROM architecture was adapted to give two similar structures, the programmable logic array (PLA), followed afterwards by the programmable array logic (PAL). The PLA and PAL architectures permit user-programmed implementation of logic at LSI level, leading to a great reduction of chip count in many cases. These two PLDs are a half-way house towards VLSI. The latest PLD development, and one which is experiencing a meteoric rise, is the field programmable gate array (FPGA), first introduced in 1985. It is a type of ASIC which, unlike other ASICs, is electronically user-programmable, rather like some established types of non-volatile semiconductor memory. That makes development much quicker and cheaper than that of other ASICs. Starting originally at LSI densities, the FPGA is now extending into VLSI territory and is set to become a vital constituent of the ASIC armoury.

Finally we make some observations on the nature of VLSI chip design, which is only practicable with extensive computer aid and supporting design tools. Mead and Conway (1980) paved the way with their seminal approach, replacing the old bottom-up approach with a top-down hierarchical design flow. VLSI chips demand sophisticated design support. In the early stages of VLSI, technology was, and nowadays—to a lesser degree—still is ahead of design. Software design tools and their integration, combined with a hardware platform incorporating a workstation, and a powerful database, are the ingredients for VLSI chip design. VLSI progress depends on matching the technological advances with adequate electronic design automation (EDA).

1.3 IC spectrum

In this survey, we take a glimpse at the IC spectrum paying attention to the following:

(i) the function density of the chips;
(ii) the two silicon technologies, bipolar and CMOS;
(iii) digital, analogue, and mixed-signal chips;
(iv) standard and application specific ICs.

1.3.1 VLSI

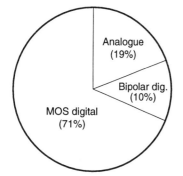

Fig. 1.7 IC consumption (1992)

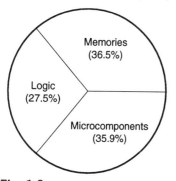

Fig. 1.8 IC MOS consumption (1992)

Figure 1.7 shows the global IC consumption in terms of three categories: digital MOS, digital bipolar, and analogue. BiCMOS is included with MOS; the analogue ICs are bipolar, CMOS, or BiCMOS. Figure 1.8 contains a breakdown of MOS consumption into the categories of memories, and microcomponents (virtually all VLSI), and also logic. The global consumption of ICs totalled $49 B (B = billion) in 1991, is expected to be $68 B in 1994, and should reach $100 B in the year 2000. The predominance of MOS is immediately apparent. That technology is now virtually all CMOS, which is replacing the nMOS technology which went before. The hegemony of digital ICs and VLSI is equally evident. The goals of system design, reliability, ease of manufacture and servicing, and economy are all served by minimizing the package count. A simple example will be given in support of the cost reduction when going from SSI to VLSI.

Consider a digital system with a complexity equivalent to about 20 000 gates, assembled entirely with SSI chips having an average gate count of 12. The total number of chips will be 1667, and using moderately sized PCBs each capable of holding 50 ICs, this will require 33 boards. Let C_{SSI} and C_O be the production and overhead costs per chip respectively; C_O includes items like subrack, rack and backplane assembly, power supplies, system design, checkout and service cost over a specified period. Realistically C_{SSI} might be $0.20, and C_O about twenty times C_{SSI}. The total system cost C_T for a chip count N is given by

$$C_T = N(C_{SSI} + C_O) \qquad (1.1)$$

That leads to a total cost of $7001. Now postulate that a single VLSI chip can replace all the 1667 SSI packages. Assessed purely in terms of gate count, such a step is easily feasible. Putting the cost of the VLSI chip at $30, we add an overhead for the PCB and system expenditure. There is now only one PCB in place of the 33 for the SSI implementation. Assume an overhead cost, based on the figures used before, of $210 ($50 \times 21 \times 0.2$). The system cost then comes to $240, showing that the VLSI implementation leads to a cost saving of 97 per cent! The approach just taken is, it has to be admitted, simplistic. The overhead cost of using a VLSI chip cannot be simply equated with the cost of a PCB for the SSI solution. It is also highly unlikely that a single standard VLSI package will be able to fulfil the required function. In most cases, VLSI has to be combined with chips at lower levels (SSI to LSI) in order to fulfil the desired function. Inspite of that, the cost saving by going from an all-SSI to a predominantly VLSI implementation is typically from 95 to 99 per cent.

The leading standard VLSI chips are memories and microcomponents, the latter embracing microprocessors and their support chips, single-chip microcomputers and the like. Another substantial VLSI category caters for signal processing. There is an increasing plethora of VLSI with ASICs very much on the increase.

1.3.2 SSI and MSI

The ICs considered so far have been digital and we continue in that vein with a look at SSI and MSI. SSI contains logic circuits at the most basic level. Both bipolar and MOS logic families are available. Bipolar families are represented by the saturated transistor-transistor logic (TTL) and the non-saturated, faster emitter-coupled logic (ECL). Each of these families exists in at least two versions, in which speed is traded against power. The device shrinkage of VLSI to micron and submicron dimensions afforded the opportunity to produce improved ranges in TTL, ECL, and CMOS. An early CMOS family has been supplemented with two ranges which attain the speeds of the fastest TTL logic with reduced power consumption. Replacement needs for the huge volume of TTL in existence guarantee the continuation of that logic for sometime yet. A typical selection of products for a standard logic family is listed in Table 1.3. The characteristics of the standard logic families are governed by the perceived needs of the market. Vendors do not simply go for producing logic with, say, the highest possible speed regardless of cost and power consumption. Another important factor is the constraint of external chip interconnects on system speed. The capacity of the fastest ECL family is only realized with a very careful layout of the PCB interconnections with signal paths kept down to 2–3 cm, preferably less. Remember that a significant proportion of applications do not require the highest speeds obtainable with a given technology, and are catered for by the slower ranges with reduced power consumption.

Table 1.3 Typical products of logic family

Function	Classification
NAND/NOR gates AND/OR/ EXCLUSIVE-OR gates	SSI
Inverters/buffers	SSI
Buffer/line driver	SSI
Flip-flops, latches	SSI/MSI
Shift registers	MSI
Bus transceivers	MSI
Arithmetic circuits	MSI
Counters	MSI
Multivibrators	MSI
Analogue and digital multiplexers	MSI
Decoders/encoders	MSI

1.3.3 PLDs

The most important niche of LSI consists of programmable logic devices (PLDs), which are available in bipolar and CMOS technologies. Within the PAL/PLA family, the PAL tends to be the dominant partner. There are numerous applications which do not justify going to VLSI, and where PALs or PLAs, combined with some SSI and MSI, are a very acceptable cost-effective solution. In other applications, VLSI can be advantageously supplemented with PLA and PAL chips, plus some SSI and MSI.

PALs and PLAs are, with only few exceptions, user-progammable. That was not always the case; the PLA, when it was first introduced, was mask-programmable. The user-programmable PLA which subsequently followed was designated field-programmable logic array (FPLA). The letter F tends to be omitted nowadays: the designation PLA or PAL applies to user-programmable devices.

The PLA and PAL are, in the strict sense of the term, ASICs (so, for that matter, are ROMs). They are however not bracketed together with the custom and semicustom ASICs outlined later in Section 1.3.5, because of their structural and programming affinities with ROMs. Another PLD, the field-programmable gate array (FPGA) falls however four-square into the ASIC camp and is mentioned again later in this chapter.

1.3.4 Analogue ICs

The earliest analogue ICs were operational amplifiers, and the operational amplifier (op amp) remains one of the analogue cornerstones. There is a vast range of op amps, with either bipolar or CMOS technology, or with BJT, CMOS, and JFET combinations. High input impedance and low power consumption are the attractions of the JFET and CMOS inclusions. The majority of commercial op amps still have only moderate bandwidths up to a few MHz. Op amps with higher bandwidths, sometimes extending to several hundred MHz, are available.

Additionally semiconductor vendors supply many other types of amplifier. Many of these are for consumer electronics. The consumer products consist largely of audio, radio, and video circuits. Small-signal preamplifiers, and power amplifiers which require some external components, make up the audio section. Radio ICs include AM and FM detectors and stereo decoders. In the video coverage there are amplifiers with bandwidths exceeding 100 MHz, timing ICs (including phase-locked loops), modulators/demodulators, IF amplifiers and TV stereo decoders. This excursion into consumer electronics, not covered in this text, has been undertaken because analogue, unlike digital ICs have their biggest outlet in consumer electronics, witness Table 1.4. Compiled from information available late in 1991, it expresses a distribution which is estimated to hold, with only minor changes, until 1994 and probably beyond. A distribution of analogue IC technologies, drawn up at the same time, is shown in Table 1.5. It highlights the preponderance of bipolar technology, which however is expected to undergo a significant decline at the expense of CMOS.

The increase in digital signal processing has raised the demand for data converters with higher resolution, but 8-bit and 10-bit converters

Table 1.4 Analogue IC consumption by product

Category	%
Consumer products	40
Amplifiers	12
Data converters	9
Voltage regulators	9
Interface	8
Comparators	2
Other	20

are adequate for many applications. Hybrid ICs are much in evidence at the higher resolutions. The hybrid combination usually takes the form of a monolithic chip combined with a laser-trimmed thin-film resistor network.

CMOS is being used more and more in analogue ICs. Switched capacitor circuit techniques permit the IC implementation of analogue filters. The capability of MOS technology to provide a good switch, an op amp, and accurate capacitors has made it the preferred technology for switched capacitor filters, which allow the combination of analogue and digital functions on the same chip. Switched capacitor techniques have given rise to numerous active filters fulfilling lowpass, bandpass, and notch functions.

Table 1.4 points to the large volume of voltage regulators, which cater for a wide range of highly stabilized output voltages. They include CMOS dc/dc switched capacitor voltage converters. Voltage references are a related important category.

Table 1.5 Analogue IC consumption by technology

Technology	Percentage	
	1990	1994
Bipolar	90	75
CMOS	9	18
BiCMOS	1	3

1.3.5 ASICs

An ASIC is user-specified, and the term user-specified integrated circuit (USIC) was in limited use when ASICs emerged in the late 1970s. It has now virtually disappeared, and the designation ASIC is universally accepted. ASICs are largely, although not entirely, at VLSI level. They have to be economically justified, apart from offering the advantages of system compaction and increased reliability.

The economic practicality can be brought home by taking up the example given in Section 1.3.1, where a single VLSI chip replaced 1667 SSI chips. The assumption that a single standard VLSI chip could effect such a replacement was challenged there and then, and will now be elaborated. The number of chips for a specific system is minimized by using ASICs. The question is: can such a move possibly be economic? Fortunately it can. ASIC development (non-recurrent engineering—NRE) costs vary according to the nature of the circuit, but typically the system under consideration might be built with four ASICs at a NRE cost of $32 000. That cost can be recovered by a charge on the development contract, or by pricing the ASIC chips to recover that outlay. Let us postulate an annual demand of 2500 sets of ASICs over a period of three years. A total of 30 000 ASICs will have to be produced and the NRE expenditure will be covered by adding $1.07 to the production cost, likely to be between $5 and $15 for each chip. This example shows that NRE costs are readily recovered with an ASIC solution, giving a tremendous saving over an all-SSI solution.

In practice the saving will be less than indicated in the foregoing example, because an implementation nearly always demands chips at various levels i.e. it will not be entirely VLSI, whether standard or ASIC. For smaller production quantities, the NRE component inevitably raises system cost. There are cases, for example in military equipment, where space is at a premium, and where the relatively high cost on account of the small production quantities required is acceptable.

Custom (full-custom) and the two categories of semi-custom ICs, gate arrays and standard cells, are mask-programmed. A full set of masks is needed for custom and standard cell ASICs, interconnect masks only are required for gate arrays. A newcomer to the ASIC family has emerged, the FPGA, which, because it is user-programmable, offers the prospect of quick development at greatly reduced cost. That makes it attractive for proving a design. Quantity production is however more expensive, because the FPGA is user- and not mask-programmed.

An ASIC, although originated by a single customer, can be potentially of more general interest. Mass market applications like car electronics and washing machine control come to mind. An interesting derivative of the ASIC has emerged in the form of application-specific standard products (ASSPs). Vendors are producing standard chips with prescribed specific functions, ASSPs, when they sense a large potential market.

1.3.6 Summary

The major development continues to be device shrinkage in order to raise chip function density, spearheaded by semiconductor memories, microprocessors, and ASICs. The role of ASICs, already very important, has been enhanced by the FPGA, which complements the mask-programmed custom and semicustom circuits. Small- and medium-scale integration is not witnessing the prominent developments of VLSI, nor is it static. The CMOS logic families have now the potential to replace TTL; ECL remains (also within ASICs) for the fastest applications. The combination of bipolar and CMOS technologies is making progress. The processing for that has been feasible for many years. The issue is the marketing of products whose superior speed justifies the extra cost. The greatest impact so far of digital BiCMOS has been on semiconductor memories and ASICs, but BiCMOS is now extending to SSI and MSI. CMOS technology is being adapted for analogue applications, leading to ICs which implement analogue systems with LSI densities.

The combination of digital and analogue circuits on the same chip is receiving much attention. Mixed-mode operation of this nature is not new and has already been mentioned in this chapter. What is new is the scale of the circuits. Analogue-digital VLSI is being developed where the major digital is accompanied by a minor but nevertheless substantial analogue section. That pushes up the expense, because of the technology, whether CMOS or BiCMOS, and the test costs.

The centrality of VLSI should not be allowed to overshadow the ICs at the lower levels. One design objective is to maximise on VLSI, but in practice most equipments will still contain some chips at lower levels. Also smaller equipment can be designed efficiently without resort to VLSI.

In terms of speed GaAs comes first, bipolar technology second, and CMOS last. The differentials have narrowed over the years: for identical geometries, the speed advantage of one technology over its neighbour in the above ranking is a factor of around two to three. The final choice for a given application will be a weighing of speed, power, cost, and reliability. In terms of consumption CMOS comes first, bipolar technology second, and GaAs is reserved for special applications demanding ultra fast performance beyond the reach of silicon.

References

Heil, O. (1935). British patent, 439–457.

Kahn, D. and Atalla, M. M. (1960). Silicon-silicon dioxide field induced surface devices. *IRE solid-state device research conference*, Carnegie Institute of Technology, Pittsburgh, USA.

Kilby, D. J. S. (1976). Invention of the integrated circuit. *IEEE Transactions on Electron Devices*, **23**, 148–54.

Lilienfeld, J. E. (1930). U.S. patent 1745 175.

Mead, C. and Conway, L. (1980). *Introduction to VLSI systems*. Addison-Wesley, USA.

Moore, G.E. (1975). VLSI: some fundamental challenges. *IEEE Spectrum*, **16**, (4), 30–4.

Shockley, W. (1976). The path to the conception of the junction transistor. *IEEE Transactions on Electron Devices*, **23**, 597–620.

2
Fabrication and packaging

2.1 Processing outline

Fabrication of a silicon integrated circuit begins with the production of a circular doped silicon wafer, covered with a layer of silicon dioxide, simply called oxide. A photosensitive emulsion, known by the generic term of resist, is deposited over the oxide and the stage is set for component formation. An image is produced on the resist by a lithographic technique which defines the lateral dimensions of the layers to be formed during that particular processing step.

The fabrication of the chip is a process of producing a large number of identical dies simultaneously. A photgraph of a wafer containing about 100 dies is shown in Fig. 2.1. The wafer is then dissected into these dies, which are tested and packaged. The key to fabrication is the formation of p- and n-layers superimposed on one another. The principle is explained in this section by illustrating the sequence in forming one such layer, followed by examples of bipolar and CMOS formation in the next section. These examples highlight various features of device processing, and set the scene for a fuller description of die fabrication, followed by accounts of packaging and mounting.

Let us consider the formation of a p-layer within an n-region, which

Fig. 2.1 Wafer photograph (Courtesy of GEC Plessey Semiconductors)

has been formed in a previous process (Fig. 2.2(a)). Development of the resist removes it from the imaged areas; the rest remains masked (Fig. 2.2(b)). The oxide left exposed is etched away, leaving the substrate prepared for formation of the p layer (Fig. 2.2(c)). Only one opening *(window)* is shown in Fig. 2.2; in practice there will of course be many. The p layer is produced by diffusion or ion implantation. In diffusion the wafer is exposed to a dopant containing, in this case, a p-type impurity. The oxide now fulfils its second role: it is a barrier which is impervious to diffusion or implantation. The first role, it will be recalled, is passivation of the substrate surface. The oxide will however only exercise this masking role if it is adequately thick, and this is the case for field oxide with its typical thickness of about 1 μm. Diffusion is inevitably lateral and vertical. The horizontal and vertical dimensions of the diffused layer—see Fig. 2.2(d)—are similar; x_d typically equals 0.7 y_d. In ion implantation, a beam containing p impurity atoms scans the wafer, and the layer may be formed either adjacent to the silicon-oxide boundary (Fig. 2.2(e)) or at a deeper level (Fig. 2.2(f)). The positioning depends on the energy of the beam, which is governed by the accelerator potential of the ion source. An implanted layer has a near-uniform doping profile with a negligible lateral spread, whereas the impurity profile in a diffused layer is graded with a maximum concentration at the oxide boundary. The action just described constitutes a mask level.

It is a necessary condition that the impurity concentration of the diffused or implanted layer exceeds that existing prior to its formation. This principle of compensation permits the formation of very shallow layers, so much so that the diffused base junction depth (meaning the vertical depth y_d) in Fig. 2.2(d) of fast IC switching transistors already had submicron dimensions, typically 0.7 μm, in the late 1960s.

The lateral dimensions of a layer are determined by the lithography, the vertical dimensions by diffusion or implantation. The vertical dimensions of all layers formed simultaneously in a processing step will be equal, but their areas may differ in accordance with the imaging of the wafer. One mask level can serve for forming parts of active and passive devices with areas differing in both shape and size. In bipolar ICs, the level which has just been described would be used for base and resistor formation, possibly for transistors with different dimensions. In digital ICs transistor geometries are more likely to be uniform, but variations may still occur, for example for the buffer input/output transistors of a VLSI chip.

Processing continues with preparation for the next mask level. The resist over the oxide (Figs. 2.2(d) to (f)) is etched away, the oxide is regrown thereby covering the windows, the wafer is covered with a new layer of resist, and a new image is projected on it for the next mask level. This, in a follow-on from the p layer formation just described, could be the production of n^+ emitters and collector contacts for bipolar transistors.

After the formation of the electrodes has been completed, windows, *contact cuts*, are opened in the oxide for depositing metal (usually aluminium) which connects the electrodes to the surface. Further mask levels establish the layers for the interconnections. The number of these steps depends on the interconnect levels, which are tending to increase

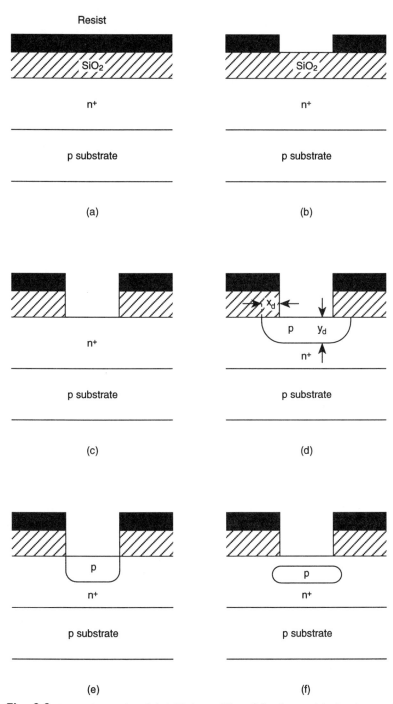

Fig. 2.2 Layer formation (a) initial condition (b) after resist development (c) after oxide etch (d) after diffusion (e) after implantation (f) alternative implantation formation

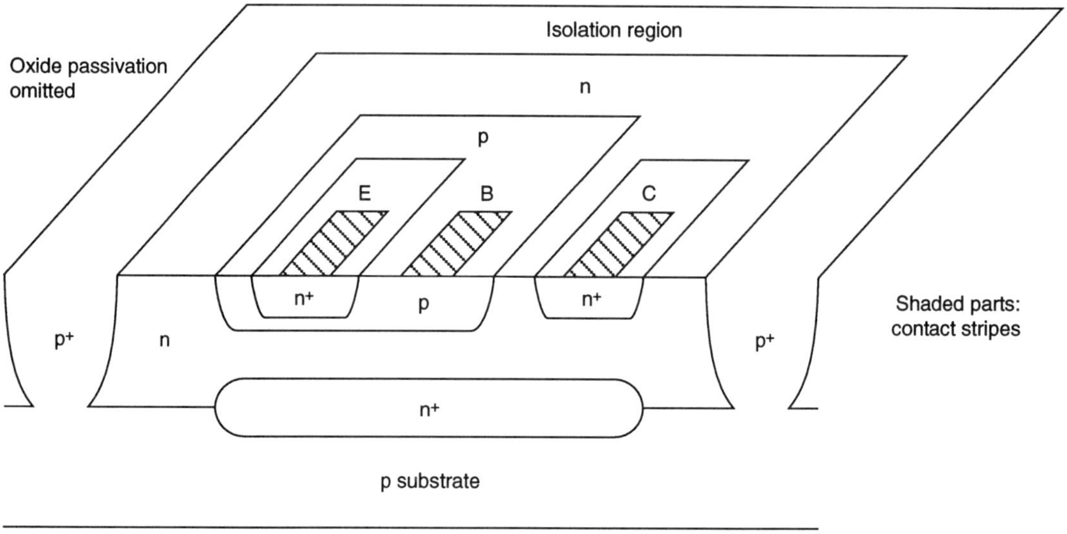

Fig. 2.3 Profile of npn transistor

with VLSI developments; three interconnect levels are shown in Fig. 1.6 of Chapter 1. The final mask level serves for bonding the die pads to the package pins.

2.2 Processing sequence for transistors

Two processing sequences are described in this section, one for a bipolar transistor and one for a CMOS structure. The bipolar transistor is junction-isolated. In this mode the substrate is connected to the lowest supply voltage of the circuit, and all transistors are isolated from their neighbours by a zero—or reverse—biased junction diode. In VLSI junction isolation has largely given way to oxide isolation, but junction-isolated transistors still exist, especially in SSI and MSI. Also the processing sequence for junction isolation is readily adapted for oxide isolation, which has already featured in the transistors of Fig. 1.4.

The two processing sequences given here are basic in that the transistor structures of VLSI, and the processing steps they require have become more complex and numerous over the years. All the same the tabulations bring home the salient issues in transistor formation.

The structure of the n-p-n bipolar transistor, processed in accordance with Table 2.1, is shown in Fig. 2.3 and the processing sequence is illustrated in Fig. 2.4, A set of masks for the transistor is sketched in Fig. 2.5. The processing of the npn transistor starts with a p-type wafer covered with thin oxide and resist. Thin oxide, also called thinox, has a thickness of 0.1 μm or less in contrast with field (thick) oxide which has a depth of about 1 μm.

The steps of removing the resist by development, etching the oxide to open a window, stripping the resist by etching, regrowing the oxide, and applying a new resist for the next cycle occur in most levels. Level 1 is however an exception. There the diffusion of the buried collector region is followed by exposing the entire wafer for an unmasked growth of the epitaxial layer, which has a uniform impurity concentration. Levels 2 to 4 are diffusion steps. The interconnections on top of the oxide are initiated,

Table 2.1 Bipolar processing sequence

Mask level	Formation	Processing step
1	Buried n^+ layer	Develop resist n^+ diffusion Strip resist Etch oxide
Unmasked level	Epitaxial n layer	Grow epitaxial n layer Grow oxide Apply resist
2	Isolation p^+ layer diffusion	Develop resist Etch oxide p^+ diffusion Strip resist Grow oxide Apply resist
3	p base	Develop resist Etch oxide Strip resist p diffusion Grow oxide Apply resist
4	n^+ emitter n^+ collector contact	Develop resist Etch oxide Strip resist n^+ diffusion Grow oxide Apply resist
5	Contact cuts	Develop resist Etch oxide Strip resist Deposit metal Apply resist
6	Interconnections	Develop resist Etch aluminium Strip photoresist Deposit passivation Apply resist
7	Pad contact cuts	Develop resist Etch passivation Bond die pads to package contacts Strip resist Assemble and package

20 Fabrication and packaging

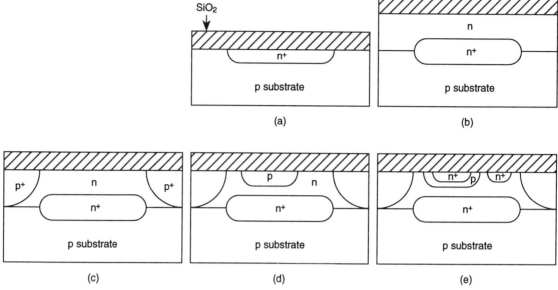

Fig. 2.4 Formation sequence—npn transistor (a) buried n^+ layer (b) epitaxial n layer (c) p^+ isolation regions (d) p base region (e) n^+ emitter and collector contact regions

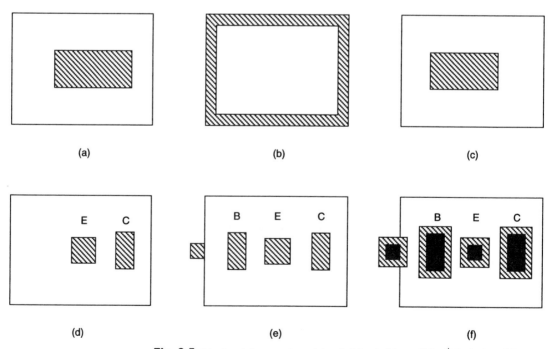

Fig. 2.5 Mask set for npn transistor (a) buried layer (b) p^+ isolation (c) base (d) n^+ emitter and collector contacts (e) contact windows (f) metallization

following the contact cuts of level 5, with the deposition of metal over the entire wafer. They are completed in level 6, and finally the die pads are connected to the package contacts in level 7. For multilevel interconnect layers (see Fig. 1.6) the sequence of levels 5 and 6 will be repeated.

Two points in Fig. 2.4 call for explanation; the movement of the buried layer, and the diffusion of the n^+ pocket for the collector contact. The change in the original position of the buried layer in Fig. 2.4(a) is caused by the high temperature of the diffusions which take place after its deposition. All diffused layers are affected by subsequent high-temperature processes; the effect will naturally be most severe where the number of such cycles is greatest. The n^+ pocket for the collector contact ensures a good ohmic metal-semiconductor contact with a low voltage drop.

A typical p-well CMOS process, which starts with a n-type wafer, is listed in Table 2.2, and the device structure is shown in Fig. 2.6. The information on processing steps has been reduced relative to Table 2.1 by omitting the repetitive steps of oxide etch and growth, and applying, developing, and the etching resist.

Dimensions have been given in Table 2.2 against mask levels 1 and

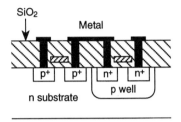

Fig. 2.6 p-well CMOS structure

Table 2.2 p-well CMOS processing sequence

Mask level	Formation	Processing step
1 p-tub mask	p-well	Deep p diffusion 4–5 μm
2 Thin oxide ('thinox') mask	Thin oxide	Growth of thin oxide ~0.05 μm
3 Polysilicon mask	Polysilicon gates, interconnect of gates	Deposition of polysilicon
4 p^+ mask (positive)	p-MOSFET source and drain Polysilicon gates doped	p^+ implantation
5 p^+ mask (negative)	n-MOSFET source and drain Polysilicon gates doped	n^+ implantation
6 Contact mask	Contact cuts	Deposit metal
7 Metal mask	Interconnections	Etch metal
8 Pad mask	Die-pad package contacts	Bond die pads to package contacts

22 Fabrication and packaging

Table 2.3 Units of length

Unit	Ångstrom	Micron	Nanometre	Mil
Ångstrom (Å)	—	10^{-4}	0.1	3.94×10^{-6}
Micron (μm)	10^4	—	10^3	0.0394
Nanometre (nm)	10	10^{-3}	—	3.94×10^{-5}
Mil (0.001 in)	2.54×10^5	25.4	2.54×10^4	—

2 to indicate orders of magnitude. The thin oxide permits penetration by the ion beam to implant the source and drain layers. Implantation is generally preferred to diffusion for VLSI structures. The self-aligned silicon gate process is standard practice in CMOS. Self alignment comes about because the polysilicon gates, formed in level 3, are in effect masks for the source and drain implantations. These implantations also dope the gates, thereby giving them a low resistance, an important condition for proper functioning. Equally important the doping gives a low resistance for the polysilicon interconnect of the two gates. Levels 4 and 5 have complementary masks. Only one image is required to form the p-MOSFET (positive mask) and n-MOSFET (negative mask) electrodes.

The npn transistor has a vertical, the MOSFET a lateral structure. The speed of a transistor is greatly influenced by the transit time of carriers in the base or the channel. Minimum lateral dimensions are governed by lithography and have always been greater than the vertical dimensions achievable by implantation or diffusion. The minimum base width has, until recently, been very much smaller than the minimum channel length, and this is one of the reasons why bipolar has been much faster than CMOS logic. The gap is closing, but it still remains. IC dimensions are expressed in four units, three of them metric. The unit used more than any other is the micron, but dimensions below one micron are frequently quoted in Ångstrom. With the continuing shrinkage into the deep submicron region, increasing use is made of the nanometre. Curiously the inch retains a place in ICs, almost entirely for expressing chip edge lengths and chip areas. The relation between these parameters is shown in Table 2.3.

2.3 Wafer preparation

The wafer consists of high-grade silicon, which has been transformed into a crystal structure by a chemical process. Silicon is an ideal semiconductor. Being the second most abundant material in the earth's crust, it is plentiful and cheap, and silicon semiconductors can work at temperatures up to 200 °C. Silicon also has an oxide (SiO_2) which has excellent qualities in respect of passivation, and for masking against diffusion and implantation; these characteristics have already been mentioned. The basic properties of silicon are shown in Table 2.4.

Preparation of the wafer starts with growing the crystal in the form of an ingot. First electronic-grade silicon (EGS), a polycrystalline silicon material, is produced by purifying raw silica, either mined beach sand

Table 2.4 Properties of silicon at 300 K

Atomic density (atoms cm^3)	4.96×10^{22}
Atomic number	14
Atomic weight	28.09
Breakdown field (V/cm)	$\sim 3 \times 10^5$
Crystal structure	diamond
Energy gap (eV)	1.115
Temperature coefficient of energy gap (eV/°C)	-2.3×10^{-4}
Melting point (°C)	1417
Electron drift mobility μ_n (cm^2/V-s)	1350
Hole drift mobility μ_p (cm^2/V-s)	480
Refractive index	3.42
Intrinsic carrier concentration n_i (carriers/cm^3)	1.45×10^{10}
Thermal conductivity (W/cm-°C)	1.57
Intrinsic resistivity (ohm-cm)	2.3×10^5
Thermal diffusivity (cm^2/s)	0.9
Minority carrier life time (s)	2.5×10^3
Specific heat (J/g-°C)	0.7

or chunks of rock deposit. The primary method of crystal growth is the Czochralski (CZ) method (Brice 1986, Rea 1981). An EGS block is heated in a fused silica crucible. A seed crystal is then lowered into the molten materials allowing the crystal ingot to form on the seed by solidification. The crystal is rotated slowly during growth, thereby stirring the melt and averaging out temperature gradients, which would lead to inhomogeneous solidification.

The silicon ingot is shaped into a wafer by machining, chemical, and polishing operations. The right-circulator ingot is shaped by removing the seed (which initiated the crystal growth) and tang ends. The standard wafer size i.e. diameter has increased over the years, and now stands at 6 in (150 mm). An ingot for that size of wafer weighs about 60 kg, and is grown in lengths of up to 2 m. Silicon crystals are grown in either $\langle 111 \rangle$ or $\langle 100 \rangle$ orientation. The dopants, typically boron for p-, and phosphorus for n- resistivity, are added at the commencement of crystal growth in either solution or powder form. Although the movements of crystal and crucible help to distribute the dopant, the final ingot will have a slightly non-uniform impurity profile which, however, is acceptable.

The ingot is grown a little oversized and is ground to the precise diameter required by a lathe-like rotating diamond cutting tool. Further operations grind at least one, generally two *flats* along the length of the ingot. These flats become features of each wafer. The position of the larger, the *major* or *primary* flat is relative to a specific crystal direction, and is located by an X-ray technique. The flat serves for mechanical alignment of the wafer in automatic processing, and also for orienting the ICs on the wafer relative to the crystal. The smaller secondary flats identify the orientation ($\langle 100 \rangle$ or $\langle 111 \rangle$), and the conductivity (p or n) of the wafer. The position of the flats in accordance with standards laid down by the Semiconductor Equipment and Materials Institute (SEMI)

Table 2.5 SEMI specification for 150 mm (6 in) wafer

Dimension	Magnitude Min	Max
Diameter (mm)	149	151
Thickness, centre point (μm)	650	700
Primary flat length (μm)	55	60
Secondary flat length (μm)	35	40
Bow (μm)		60
Total thickness variation (μm)		50

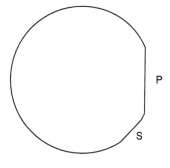

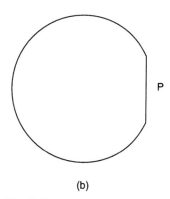

Fig. 2.7 Identifying flats on $\langle 111 \rangle$ silicon wafer (a) n resistivity (b) p resistivity

is shown in Fig. 2.7, and a typical SEMI specification for a 6 in (150 mm) wafer is given in Table 2.5, where the term *bow* signifies wafer curvature.

The slicing is carried out to produce wafers with the correct crystal orientation, which is determined by making X-ray measurements on an initial batch of a few cuts. Wafers of $\langle 100 \rangle$ orientation are cut precisely for that condition in an 'on orientation'. Wafers with a $\langle 111 \rangle$ orientation are cut 'off orientation' with an offset angle of about 3°, in order to minimize pattern shifts of epitaxial layers, a phenomenon which is explained in Section 2.4.2.

Once the flats have been ground, the ingot is sliced into wafers. Slicing is an important operation: the vital features in wafer production are to maintain a flat plane and the desired surface orientation. The ingot is sliced with a circular cutting blade kept in tension on the outer and having the cutting edge on the inner diameter. This mode, *inner diameter cutting*, minimizes the *kerf loss* (kerf, in this context, signifying the slit made by cutting) and gives the best planarity. Even then the kerf loss is severe: it amounts to between 30 and 40 per cent of the ingot.

Three further operations, lapping, etching, and polishing leave the wafer in its final shape. The thickness variation of the wafer after slicing is up to 50 μm. A mechanical two-sided abrasion with lapping plates, removing between 10 and 50 μm of substrate material, achieves a planarity to within about 2 μm. However lapping causes surface damage, which is cured by an etching process, removing from 10 to 30 μm of wafer from both sides. Finally the wafer is polished to eliminate microcracks and debris, a step which removes a further 10 to 35 μm of wafer surface and at the same time flattens the wafer to give the best planarity obtainable. The wafer, after cleaning, is ready for the formation of the dies.

The three steps of lapping, etching, and polishing reduce the wafer thickness by 40 to 145 μm. That, taking typical figures, will give a substrate of around 560 μm, which is much larger than the magnitude of 200 μm, shown in Fig. 1.3. Table 2.5 is a guide, and semiconductor vendors are free to decide on the precise dimensions themselves. In practice the processed 6-inch wafers are typically 250 μm to 500 μm thick. The ongoing increase in wafer size places an additional mechanical strain on the wafer in the mounting for processing and die formation. The thicker substrates are being adopted, because they make the larger wafers easier to handle and reduce breakages.

The thicker wafers can also stand better the thermal stresses of epitaxy, oxidation, and diffusion. These stresses, which occur repeatedly during the formation of the dies, must not exceed the critical stress the wafer can stand without irreversible deformation. If they do, warping will take place and dislocations will form. High temperature processes are usually carried out by ramping the furnace containing the rack which holds the wafers from a lower to the operating temperature and by a similar ramp down at the end of the process. That technique is often reinforced by a slow insertion and withdrawal of the rack from the furnace. Both these steps reduce thermal stress.

The various procedures of wafer processing are aimed at producing the highest possible quality and the best flatness of the substrate for imaging. The image is registered on the surface of the resist, which, it will be remembered, is deposited on the oxide covering the wafer. The flatness of the resist will be a direct reflection of the wafer's planarity. The continuing shrinkage to deep submicron dimensions is making increasing demands for ultra-flat surfaces.

2.4 Deposition and growth

2.4.1 Oxidation

Oxidation, the production of silicon dioxide by thermal growth, has several uses in silicon technology, and is a key process in IC fabrication. Silicon dioxide fulfils the following functions.

(i) It provides surface passivation.
(ii) It is a barrier, masking against diffusion or an implant of an impurity dopant into the substrate.
(iii) It can be used to isolate a transistor from its neighbours. This applies to bipolar and MOS transistors (and also to passive components).
(iv) It is a component, namely the gate oxide, of the MOSFET.
(v) It is one form of the dielectric isolation between multi-level interconnect layers, and for the final die passivation. (Other dielectrics with similar composition can be used for that purpose.)

Various techniques for forming oxide are available; this section covers wet and dry thermal oxidation. The chemical reactions for these growths are

$$Si + O_2 \rightarrow SiO_2 \quad (2.1)$$

and

$$Si + 2H_2O \rightarrow SiO_2 + 2H_2 \quad (2.2)$$

Dry oxidation, (equation 2.1), is a reaction of silicon with oxygen, wet oxidation, (equation 2.2), a reaction of silicon with water vapour or steam. In each case the oxidation demands the existence of a silicon surface, on which it is formed. Thermal growth feeds on silicon and oxygen, which has to diffuse through the oxide to the Si–SiO_2 boundary, where the reaction takes place. When the growth is completed, about 46 per cent (the exact value depends on the processing and lies between 44 and 46 per cent) of the oxide is within the original substrate, and

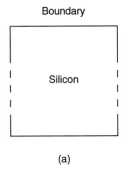

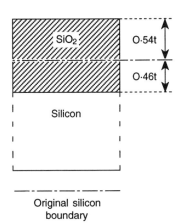

Fig. 2.8 Thermal oxide formation (a) profile prior to formation (b) profile after formation

54 per cent above it (Fig. 2.8): the silicon boundary has been depressed from its original position. The temperature of the furnace lies in the range 950 to 1250 °C and is kept constant to about ±1/2 °C. Growth time, pressure, and oxygen gas (or steam) concentration are very closely controlled with the result that the oxide layer is extremely uniform in depth and composition.

The growth rate is fastest at the beginning and then slows down, because the oxygen atoms for the reaction have to diffuse through the oxide in order to reach the Si-SiO$_2$ interface. At a given temperature wet oxidation is always substantially faster than dry oxidation. An oxide layer 1 μm thick, which is representative of the passivating field oxide covering the substrate, takes 2 hours 30 min. to grow with dry, and 1 hour 30 min. to grow with wet oxidation (Elliott 1989). The time-advantage of the wet process is offset by a higher impurity content of the oxide. MOS ICs demand a very pure oxide for reliable performance and closely controlled characteristics, and dry oxidation is preferred because of this. Dry and wet oxidations are usually performed at a pressure of 1 atm (760 Torr). (The effective value for wet oxidation is about 640 Torr due to the presence of steam.)

Oxidation is, like epitaxy and diffusion, a high-temperature process. It takes place several times in the sequence of die formation, witness the outlines in Section 2.2. One undesirable result is a redistribution of layers from their original locations. Oxidation at a lower temperature will reduce that effect at the expense of increased growth time. The temperature can be lowered by increasing the pressure; an increase by 1 atm permits a reduction of the order 20 °C for no change in growth time. High-pressure oxidation with pressures up to 25 atm is being used successfully at temperatures in the range of 700 to 900 °C.

The silicon dioxide layer deviates from the ideal charge neutrality expected in a dielectric. It contains three charges, a fixed oxide charge Q_f, usually positive, a mobile ionic charge Q_m, and an oxide-trapped charge Q_{ot}. In addition there is an interface-trapped charge Q_{it} at the Si-SiO$_2$ interface (Deal 1980). A low temperature hydrogen anneal (450 °C) reduces Q_{bt}, $\sim 10^{15}$ cm^{-2}, to a negligible level, $\sim 10^{10}$ cm^{-2} or even less. Q_f is restricted to within 30 Å of the Si-SiO$_2$ interface and cannot be changed (Nicollian and Brews 1982). Q_m, also usually positive, is attributed to alkali ions like sodium and potassium. These ions are mobile under the influence of an electric field. Q_m and Q_{ot} are both of the order 10^{10} cm^{-2} to 10^{12} cm^{-2}, values which are vanishingly small compared with the atomic density of silicon, but potentially significant in relation to the impurity concentration density in a MOSFET channel. Q_{ot} can be positive or negative, arising from either holes or electrons trapped in the oxide. It is due to defects in the oxide and may be induced by ionizing radiation, avalanche injection, or high currents in the oxide. The charge density of Q_{ot} is in the range 10^9 cm^{-2} to 10^{13} cm^{-2}. It can be either eliminated or greatly reduced by annealing. Snow et al. (1965) were first to identify that alkali ions, which are linked with Q_m, were the main cause of instability in oxide passivation. Where necessary, dry oxidation is performed using a dry O$_2$-HCL agent, with HCL being a small proportion, five per cent or less. The amount of mobile alkali ions, and hence Q_m, is thereby greatly reduced.

The issue of oxide charge is of very great importance. Charge inversion in the MOSFET channel caused MOS ICs to have extremely variable characteristics and led to reliability problems, especially when operating at high temperatures or voltages, in the early years of MOS development. Only tremendous efforts in the decade 1970 to 1980 improved oxidation to the point where it was acceptable for MOS ICs.

Thin oxides with thicknesses in the range 20 to 200 Å are increasingly in demand for submicron VLSI. The quality and reliability of such oxides call for special attention in their production. The growth rate has to be much slower for thin oxide in order to obtain reproducible uniformity. Growth rate reduces with lower temperature and pressure. Oxides between 30 and 300 Å thick have been grown mainly at pressures of 0.25 to 2.0 Torr (Adams and Chang et al. 1980). In another technique, a thin oxide is produced by an initial growth at 1000 °C, using an O_2-HCL agent, followed by a passivating heat treatment in N_2, O_2, and HCL at 1150 °C to achieve the desired thickness (Hashimoto et al. 1980). Yet another technique combines high pressure (10 atm) with low temperature (750 °C) for growing an oxide layer 300 Å thick in 30 minutes (Hirayama et al. 1982). Whatever technique is adopted, ultra-pure chemicals and processing conditions are absolutely essential to produce thin oxides of the required quality.

This section concludes with a description of the two dielectric isolation techniques in current use. These are oxide and trench isolation, and they hold for both bipolar and MOS transistors alike. In bipolar transistors, oxide isolation (Fig. 1.4(a)) has virtually replaced junction isolation (Fig. 2.4). Compared with junction isolation, oxide isolation greatly reduces the total device area by cutting down substantially on the silicon estate taken up by the isolation region. Equally important, it greatly reduces electrode capacitances. Trench isolation has similar advantages.

In oxide isolation, oxidation is localized in contrast to the unmasked growth mechanisms encountered so far. The local oxidation of silicon (LOCOS) process, an established standard technique, begins with the wafer in the condition of Fig. 2.9(a), where windows A and B represent the profile of the isolation area shown in Fig. 2.9(b). The active (transistor) area remains covered with oxide and nitride layers. These layers have been etched away in the windows for the oxidation which now follows. Silicon nitride (Si_3N_4) is impervious to oxidation and masks the active area. The nitride is deposited on a thin layer of stress relief oxide, whose name gives a clue to its purpose. The deposition of a film on silicon leads to a tensile stress which may damage the wafer by causing warpage, film cracking, or defect formation in the silicon. The technology of oxide deposition on silicon has been mastered adequately to contain that stress. The same cannot be said of nitride. The danger of stress damage which may arise from a deposition of nitride directly on the silicon is averted by the interposed thin layer of stress relief oxide. Typical thicknesses are 200 Å and 1500 Å for the oxide and nitride layers respectively.

There will be inevitably some diffusion of the oxidizing gas through the thin oxide, causing lateral growth and lifting the nitride film. Instead of the ideal profile of Fig. 2.9(c), the structure will have the shape of Fig. 2.9(d). The resulting formation of the field oxide is aptly designated

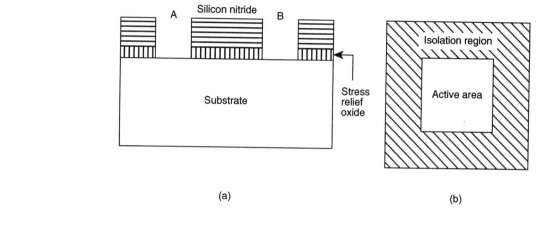

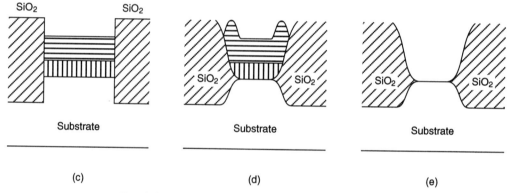

Fig. 2.9 LOCOS isolation (a) before oxidation (b) top plan (c) ideal profile after oxidation (d) actual profile after oxidation; (e) actual final profile

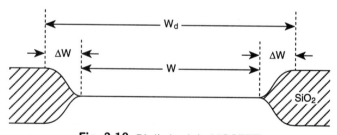

Fig. 2.10 Bird's beak in MOSFET

bird's beak on account of its shape. The bird's beak in a completed transistor structure is evident in Fig. 1.4(a). The bird's beak leads to a loss in silicon estate by causing a serious encroachment on the active area. It also impairs the accurate control of electrode dimensions.

Figure 2.10 illustrates the effect of bird's beaking for a nMOSFET. The conclusions apply equally to a CMOS structure. The channel length L, the spacing between source and drain, remains unaffected although the bird's beak will invade the outer edges of the source and drain layers.

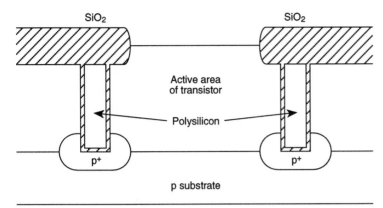

Fig. 2.11 Trench isolation

The channel width is reduced from the *drawn* mask-defined value W_d to W, where

$$W = W_d - 2\Delta W \quad (2.3)$$

The value of ΔW depends on processing details and is typically 0.8 μm for a field oxide thickness t_{fox} of 1 μm, and for the stress relief oxide and nitride thicknesses already given. Advances in the LOCOS technique make it possible to reduce the encroachment of the bird's beak at a cost of considerably increased processing complexity (Uyemura 1988).

The LOCOS isolation is semi-recessed. A fully recessed isolation with deeper penetration into the substrate is possible by etching the silicon of the isolation region prior to oxidation. The depth of etch can be arranged to give a highly planar substrate surface after oxidation in line with the geometry of thermal growth explained with the aid of Fig. 2.8. However subsequent processing steps tend to negate the advantages of fully-recessed isolation, and semi-recession of LOCOS remains the norm.

Trench isolation, a more recent form of dielectric isolation which is rapidly gaining ground, virtually eliminates the loss in silicon estate caused by the bird's beak. The isolation region is in the shape of a deep trench, which surrounds the active area and is cut by an anisotropic reactive ion etch (RIE). Figure 2.9(b) serves for a starting point in the explanation of the trenching sequence. The isolation area therein is the surface of the trench. The trench is 'cut' by a plasma etch. Advances in plasma etch technology have made it possible to form very accurate trenches with a minimum width of 1 μm, maintained closely over a depth of up to 20 μm. That permits deep penetration into the substrate.

The trench is etched with a triple layer sandwich of oxide, nitride, and stress relief oxide covering the active region. A p^+ channel-stop implant through the trench follows. Next comes stripping of the oxide and a selective thermal oxidation, which lines the floor and the walls of the trench with a thin oxide layer. The trench is subsequently filled with polysilicon and then etched back to level the surface. Finally the nitride and stress relief oxide layers are removed, and the wafer is oxidized, thereby capping the surface of the trench with oxide. A simplified profile of the completed trench is shown in Fig. 2.11. The transistor structure has been omitted; the sketch is only intended to illustrate the nature of the trench isolation, which is applied to bipolar and MOS structures.

Trench isolation is very economical in area. There will be some bird's beaking on account of the oxidation beneath the nitride during the lining of the trench with silicon dioxide, but the encroachment is negligibly small. A variant of trench isolation provides a double lining, in which the silicon dioxide is covered by a thin layer of nitride. An oxide infill is preferable to a polysilicon infill; it has a smaller capacitance. One important consideration here is the capacitance between an interconnect which passes over the trench and the trench polysilicon. However an oxide infill is difficult to achieve (although work is going on to make it practical), because there are severe difficulties caused by void formation and other problems. The deep penetration of the trench into the substrate allows the formation of a capacitor with a large vertical area, thereby saving silicon estate relative to a traditional capacitor with horizontal layers. A DRAM single-transistor storage cell requires a capacitance with a minimum demand on chip area, and trench capacitors for that purpose are described in Chapter 7. All in all, trench is more complex than LOCOS isolation, and has problems not brought out in this basic explanation. At present it is the best technique of dielectric isolation for high levels of integration

2.4.2 Epitaxy

In epitaxy—the meaning of the word is 'arranged upon'—a monocrystalline film is formed on top of a monocrystalline (in our case silicon) surface. Epitaxy is a crystalline growth process, in which the foundation layer is the seed crystal. The epitaxial layer may be either p- or n- doped, or intrinsic. The polarity and concentration of the dopant is not restricted by the foundation layer, which may be doped or intrinsic silicon. The silicon process is homoepitaxial: the epitaxial layer and the silicon layer on which it is formed have identical crystal structures. In cases where layer and substrate differ, for example in GaAs technology, the process is heteroepitaxy. There must however be a strong affinity between such layers for epitaxial growth to take place.

Historically silicon epitaxy was developed for the discrete npn planar transistor, whose structure was adopted for ICs. The n-doped epitaxial collector layer (see Fig. 2.3) satisfied electrical performance criteria which could not have been met by other forms of processing at that time. The use of epitaxy in bipolar and CMOS ICs is now widespread. In CMOS it generally forms a lightly doped layer on a heavily doped substrate of the same type (n on n^+, or p on p^+). That structure greatly helps to reduce the likelihood of latch-up, a dangerous and potentially destructive phenomenon in CMOS. Epitaxial growth is not confined to a single layer. Multiple epitaxial layers are feasible, and are indeed found in various applications, for example microwave p-i-n diodes.

The two main epitaxial processes are chemical vapour deposition (CVD), the preferred technique, and molecular beam epitaxy (MBE), which is reserved for special applications. In CVD the film is formed on the surface of the substrate by heat-induced (thermal) decomposition and/or the reaction of gaseous compounds. Epitaxial growth from the gaseous vapour phase is called vapour-phase epitaxy (VPE), and the process will now be described.

The CVD growth takes place in a quartz reaction chamber containing a graphite susceptor which supports the wafer. The four compounds for the reaction are silicon tetrachloride (SiCl$_4$) trichlorosilicane (SiHCl$_3$), dichlorosilicane (SiH$_2$Cl$_2$) and silane (SiH$_4$), with the first-mentioned enjoying the widest industrial use. The temperature for the SiCl$_4$ reaction is between 1150 and 1250 °C; slightly lower temperatures apply to the other vapours. The film grows typically at 1 μm/min, with a decidedly lower rate for silane. Hammond (1978) has surveyed the relative merits of these agents. Doping is achieved by adding the hydrides of arsenic, boron, and phosphorus. These are arsine (AsH$_3$), diborane (B$_2$H$_6$) and phosphine (PH$_3$). The susceptor is usually heated by RF induction, but there is a trend towards the more uniform induction heating obtained with quartz halogen lamps. The doping capability for epitaxial layers lies in the range 10^{12} to 10^{22} atoms cm^{-3}; the majority of requirements are within 10^{14} to 10^{17} atoms cm^{-3}.

MBE is based on evaporation. The substrate is in contact with a solution containing the material in liquid form. The film is evaporated and deposited one layer at a time (hence the name 'molecular'), and film thickness can be controlled to much finer levels than in CVD. The relatively lower temperature for MBE—the range is 600 to 900 °C—is equally conducive to finer widths. The process is performed under ultra-high vacuum (UHV) conditions of 10^{-8} to 10^{-11} Torr. The throughput is slow on account of the low growth rates which range from 0.01 to 0.3 μm/min (Bean 1981). MBE is an expensive process and its use is largely confined to special applications which call for very thin layers, for example silicon-on-insulator (SOI) and silicon-on-sapphire (SOS) structures, and GaAs technology.

The first steps in the processing sequence for the npn bipolar transistor, described in Section 2.2, are the diffusion of the n$^+$ buried layer and the epitaxial growth of the n layer. During the epitaxial formation impurities from the n$^+$ layer move upwards. This *autodoping*, caused by the *out-diffusion*, disturbs the uniform doping of the epitaxial layer, limiting both its thickness and doping level for acceptable composition. The buried layer experiences a lateral shift in ⟨111⟩ oriented wafers during epitaxial growth. The shift is minimized by cutting ⟨111⟩ wafers at an offset angle of about 3° (see Section 2.3), and by the positioning of the area for the epitaxial layer.

2.4.3 Diffusion

Diffusion is historically one of the oldest techniques of forming a doped layer in the substrate. It is a masked high-temperature process in which the conductivity of the substrate is changed over a depth determined by the diffusion. Diffusion is carried out at a high temperature (900 to 1200 °C) with the wafer exposed to a diffusant containing the desired impurity. The imperviousness of silicon dioxide to diffusant vapours may be judged from the diffusivities of dopants in oxide and silicon. Typical values for boron are 7×10^{-6} cm^{2}s^{-1} and 0.76 cm^{2}s^{-1} in silicon dioxide and silicon respectively, with a similar situation for the other impurities in use. The wafer is placed in a quartz boat positioned within the quartz diffusion tube; the same equipment is used for oxidation. The quartz tube holds a stack of vertically mounted wafers, which are diffused simultaneously.

The dopant can be in solid, liquid, or gaseous form; all of these obey the same laws of diffusion. Boron is the standard preferred p impurity; antimony, arsenic and phosphorus are the n dopants. Arsenic is preferred for n^+ source-drain and n^+ emitter diffusions, because its lower diffusivity allows the formation of shallow layers. With solid-source dopants, the carrier gas (nitrogen or oxygen) flows over a platinum source boat (within the diffusion tube), containing the dopant in powder form. Alternatively solid state wafer-shaped dopants are interposed between the wafers to be diffused. With a liquid dopant, the gas is bubbled over a liquid source in a heated bath and re-enters the furnace in a doped state. Lastly a gaseous dopant is introduced directly into the furnace. Gaseous diffusion is a simple process, but it poses problems of toxicity and chemical stability.

Diffusion takes place in one of two modes, single-step or two-step. The first of these is *predisposition* (or *constant-source*) diffusion, in which a gas containing the desired dopant passes over the wafer. The impurity concentration over the surface of the wafer is maintained constant. Its value at the surface of the substrate has an upper limit, the *solid solubility*, which silicon can absorb. Solid solubility is a function of impurity and temperature. Its values at 1100 °C are 3.3×10^{20} cm^{-3} and 1.21×10^{22} cm^{-3} for boron and phosphorus respectively.

The second cycle of the two-step sequence is the *drive-in* (or *limited-source*) diffusion, in which the dopant is removed. The impurity charge in the substrate is now redistributed. The effect will be to lower the impurity concentration at the surface, and to even it out within the substrate. Drive-in diffusion is often combined with a regrowth of the oxide over the wafer by introducing steam or oxygen during that phase. Depending on the required profile of the diffused layer, the drive-in phase either follows on from predeposition or is carried out with the wafer moved to another furnace at a different temperature. An alternative to predeposition is ion implantation of the dopant into the substrate. This method is being preferred for VLSI, because it avoids a high temperature cycle, that of the first-phase diffusion.

Diffusion is covered by Fick's laws, which lead to the following solutions.

For predeposition

$$C(x,t) = C_s \operatorname{erfc}\left(\frac{x}{2\sqrt{Dt}}\right) \tag{2.4}$$

where C = impurity concentration at a vertical distance x from the substrate surface.
C_s = impurity concentration at substrate surface
D = diffusivity (diffusion constant)
erfc = complementary error function
t = time

For drive-in

$$C(x,t) = C_s \exp\left(\frac{-x^2}{4Dt}\right) \tag{2.5}$$

with C_s, now a function of time, given by

$$C_s = \frac{Q_o}{\sqrt{\pi Dt}} \tag{2.6}$$

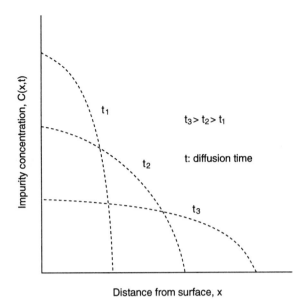

Fig. 2.12 Gaussian diffusion profile

Q_o being the fixed dopant charge.

The impurity profile resulting from drive-in diffusion is contained in Fig. 2.12, which shows that the impurity concentration decreases at the surface, is evened out, and penetrates deeper into the substrate with increasing diffusion time. The area under each of the three curves in Fig. 2.12 equals Q_o of eqn (2.6). Occasionally a layer is formed by pre-deposition diffusion only, but the two-step cycle is the norm.

The *metallurgical junction depth* X_j at which the dopant concentration equals C_B, the background substrate concentration, is obtained from eqns (2.5) and (2.6).

$$X_j = \frac{2\sqrt{Dt}}{\text{erfc}\,(C_B/C_s)} \qquad (2.7)$$

for single-step diffusion

$$X_j = 2\sqrt{Dt}\sqrt{\ln\left(\frac{C_s}{C_B}\right)} \qquad (2.8)$$

for two-step diffusion.

Diffusion is not only vertical, but also lateral. In broad terms, the lateral spread and the depth of the diffused layer are comparable. The extent of lateral diffusion has been analysed by Kennedy and O'Brien (1965), and their results remain an accepted guide. Typically x_d in Fig. 2.2(d) is ~0.7 to 0.9 y_d. The loss in silicon estate due to the lateral spread was not serious in the pre-VLSI era. A diffused base layer for a fast transistor in the mid-1970s might have had a mask opening of 60×40 μm^2 and a depth of 2 μm. The lateral diffusion of about $2 \times 0.7 = 1.4$ μm would have been of little consequence. Consider now a modern VLSI structure with a mask opening of 2×2 μm^2 for a diffused layer 1 μm deep. This, allowing for spreads at each edge, will demand a total substrate surface area of $(2 + 2 \times 0.7)^2$ μm^2 = 11.56 μm^2. The lateral spread has nearly trebled the silicon estate, and this loss is a vital factor in the preference of implantation over diffusion for VLSI. Another feature of diffusion, the non-uniform impurity concentration, may be an advantage or a disadvantage, depending on the function of the layer within the IC structure.

Diffusion is not confined to doping crystal silicon, but is also an important process for doping polysilicon. The resistance of intrinsic polysilicon, which is used to form MOSFET gates, their interconnections, and other interconnections in CMOS, has to be lowered for proper operation. In VLSI, implantation is however preferred to diffusion for that purpose.

2.4.4 Ion implantation

Ion implantation has become the established method for introducing impurities for layer formation in the substrate. It is preferred over diffusion mainly for the following reasons:

(i) the impurity concentration of the implanted regions is highly uniform, typically to within one per cent, over the wafer. A similar degree of uniformity is maintained from wafer to wafer;

(ii) the lateral spread is very small;

(iii) the layer can be formed anywhere within the substrate.

Ion implantation, generally masked, is a low-temperature process. The second phase of the cycle, the anneal, is not. The entire wafer—sometimes only a selected part of it—is exposed to a beam which injects the dopant ions into the unmasked sections of the substrate. The depth of penetrations is governed largely by the energy of the incident beam, the doping concentration by the beam current. Masking is accomplished by the field oxide, typically 1 μm thick, or by a somewhat thicker layer of resist. On the other hand implantation is sometimes made through thin oxide.

The major constituents of an ion implanter are the ion source, a bending analyzer magnet, an acceleration tube, focussing plates, electrostatic XY scanners, and a target chamber for holding the wafer. The ion plasma initially contains the desired ions together with many others. The ions required are obtained by the analyzing magnet, which selects the ions with the desired charge to mass ratio the others impinge on the magnet walls. The XY scanner arranges the sweep of the beam over the wafer, which is placed in the target chamber. Alternatively the sweep is mechanical, and there are hybrid arrangements in which the sweep is partly electrical and partly mechanical. The implanter is operated in a vacuum at a pressure of about 10^{-6} Torr to minimize ion scattering caused by collisions of ions and gas molecules. Accelerator voltages range from a few kV to 200 kV for medium-energy, and up to 2 MV for high-energy implanters. The implantation dosage extends from about 10^{10} to 10^{17} atoms cm^{-2}.

The ionic depth of penetration into the substrate is governed by the implantation mechanism. Ionic bombardment dislodges atoms from the crystal lattice and will generally turn the substrate adjacent to the surface into amorphous silicon to a depth up to several thousand Å. The two prime factors to be assessed are the depth of ionic penetration, and the inevitable spread of the ions, which determines the impurity profile of the implanted layer.

The ions lose energy in the substrate by elastic collisions with nuclei and by inelastic collisions with electrons. The energy loss in the nuclear collisions is, because these are elastic, transferred to the target atoms, displacing them from their lattice sites. The loss mechanism is determined by the accelerator voltage E_o and the atomic number and mass of the ion (Z_1, M_1) and the substrate (Z_2, M_2). Nuclear stopping predominates at low energies and high Z_1, electron stopping at high energies and low Z_1.

Assuming an amorphous region with a Gaussian implant distribution, the range R, the total distance in Å from the substrate surface traversed by the ion is given by (Glaser and Subak-Sharpe 1979)

$$R = \frac{0.7\sqrt{Z_1^{\frac{2}{3}} + Z_2^{\frac{2}{3}}}}{Z_1 Z_2} \frac{M_1 + M_2}{M_2} E_o \qquad (2.9)$$

for nuclear stopping, and

$$R \simeq 20\sqrt{E_o} \qquad (2.10)$$

for electronic stopping.

The ion path in the substrate is not a straight line, because the ions move laterally and vertically before they come to a stop. The range of more importance is R_p, the projected range. R_p is the vertical penetration of the ions, reckoned from the substrate surface. There is no accepted formula relating R and R_p; R is evidently the greater of the two. However R and R_p have similar magnitudes and eqns (2.9) and (2.10) are useful approximate guides for relative projected ranges obtainable with different dopants and beam energies.

Although a completely amorphous region has been assumed, the region covered by the implant is in practice still crytalline in parts. The incoming ion beam, if not aligned with the crystal plane, will miss a good many atoms. Ions will penetrate much further than predicted according to the postulated random orientation, travelling through these channels either undisturbed, or by glancing collisions with atoms of the channel walls. Channelling, because it leads to an extension, not accurately predictable, of implantation depth should be avoided. This is largely achieved by tilting the target wafer 7 ° off the normal in line with the incident beam. It then behaves like an amorphous surface, and this method of reducing the channelling effect is normally adopted.

The impurity profile, centred on R_p (Fig. 2.13),is given by

$$N(x) = N_p \exp\left(\frac{-(x - R_p)^2}{2\Delta R_p^2}\right) \qquad (2.11)$$

where $N(x)$ = concentration at a vertical distance x from the substrate surface.

ΔR_p, the standard deviation of R_p, is called the straggle. It follows from eqn (2.11) that $N(x)$ equals $0.6 N_p$ for $x = R_p \pm \Delta R_p$, and $0.13 N_p$

36 Fabrication and packaging

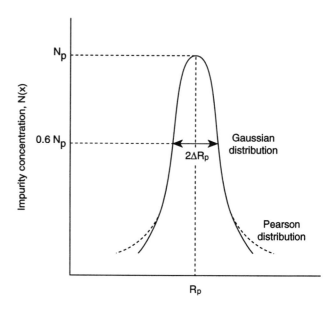

Fig. 2.13 Implantation distribution

for $x = R_p \pm 2\Delta R_p$. If the impurity dose Q resides entirely within the substrate, it is defined for Fig. 2.13 by

$$Q = \int_0^\infty N(x)dx \qquad (2.12)$$

and is related to N_p by

$$N_p = \frac{Q}{\sqrt{2\pi}\Delta R_p} \qquad (2.13)$$

The actual impurity distribution differs from the representation expressed by eqn (2.11) for a number of reasons. Nevertheless the *Gaussian* distribution gives fairly accurate profiles in the peak region $x = R_p \pm \Delta R_p$. A more accurate result is obtained within the Pearson distribution. The main effect this has on the impurity profile is shown in Fig. 2.13. It is relatively easy to determine N_p to better than one per cent by dosimetry. The beam current is measured by arranging for a good electrical contact between the wafer and the target holder. Integrating the beam current over the time of implant gives N_p to better than one per cent. Typical values of R_p and ΔR_p are shown in Table 2.6. The relative straggle $\Delta R_p/R_p$ is governed primarily by M_2/M_1.

Implantation is accompanied by a lateral scatter, giving rise to a lateral straggle, denoted by $\Delta R\bot$. The two straggles are sometimes distinguished by calling ΔR_p normal, and $\Delta R\bot$ transverse. In practice they are virtually identical for antimony and arsenic, and very nearly so for phosphorus. The difference is more marked for boron, but even there it is slight: $\Delta R\bot$ and ΔR_p are about 0.10 and 0.07 μm respectively at 100 keV. The similar values for normal and transverse straggle might conjure up the spectacle of significant lateral spread in implantation,

Table 2.6 R_P and ΔR_p for implants in silicon
Copyright ©1974 IEEE

Energy (keV)	20	40	100	200
Boron				
R_P (μm)	0.064	1.250	2.800	4.850
ΔR_P (μm)	0.028	0.040	0.070	0.092
Phosphorus				
R_P (μm)	0.025	0.047	0.120	0.250
ΔR_P (μm)	0.012	0.021	0.047	0.078
Arsenic				
R_P (μm)			0.057	0.110
ΔR_P (μm)			0.020	0.037

From Lee and Meyer (1974)

paralleling lateral diffusion. Fortunately this is not the case, because the straggle is much smaller than R_p. It also has to be remembered that a thick implanted layer is obtained by several implants at different accelerator voltages. Such a layer is flat-topped with ΔR_p and $\Delta R\perp$ determining the impurity gradients at the ends.

The annealing phase of the implantation cycle restores the amorphous region to a crystal. Second it activates the dopant atoms by placing them in substitutional locations within the crystal. In a thermal anneal the wafer is heated to between 800 and 1000 °C for about 30 minutes. Annealing increases the spread in distribution and decreases the peak concentration N_p. The high-temperature process will also cause some movement of previously formed layers. A good approximation for the change in straggle caused by thermal annealing is

$$\Delta R'_p = \sqrt{(\Delta R_p)^2 + 2D(T_a)t_a} \tag{2.14}$$

where $\Delta R'_p$ = straggle after anneal
T_a = Temperature of anneal
$D(T_a)$ = diffusivity of dopant at T_a
t_a = duration of anneal

Annealing demands careful choice of time and temperature. The restoration of the amorphous section to a crystal is a process of solid phase epitaxy (SPE), which requires a lower temperature than the activation of the dopant atoms. There is also the danger of a clash between SPE and local diffusion, leading to the formation of polysilicon. Frequently annealing is a two-step process with a low-temperature phase leading to a regrowth of the crystal, followed by a high-temperature phase which activates the dopant. Thermal is being replaced by rapid anneal techniques, with laser annealing becoming established practice. In that mode a laser beam scans the wafer, causing local melting and recrystallization. The instantaneous high-temperature process of very short duration does not disturb previous formations. Alternative high-temperature anneals use tungsten-halogen lamps or graphite plates to heat the wafer. In all cases there is only a minimal disturbance of the impurity concentration.

Two types of implants deserve special mention. MOSFETs often contain shallow layers less than 1000 Å deep for threshold voltage control or other purposes. An interesting technique for forming these is a low-energy arsenic implant, which produces a completely amorphous layer, annealed at a relatively low temperature of about 600 °C to give an abrupt junction. An entirely different process entails implanting a film of silicide or polysilicon on top of the substrate, and following this with a drive-in diffusion. The dopant diffuses much faster in the film than in the silicon, and will leave the film with a low resistivity to fulfil its intended role of conductor whilst at the same time forming a shallow layer in the substrate.

The other formation singled out is a layer deep within the substrate, or within a well (like the p-well in Fig. 2.6), which also extends deep into the substrate. Special attention has to be paid to the thickness of the field oxide, which must present a barrier to implantation. The problem arises because implantation into oxide sets up a pattern very similar to implantation into silicon; Fig. 2.13 applies. An acceptable stipulation is that the impurity level at the SiO_2-Si interface, arising from implant penetration of the oxide, is less than 10 per cent of the background impurity concentration N_b at the substrate surface, i.e.

$$N(x_o) < 0.1 N_b \quad (2.15)$$

where x_o is the depth of field oxide reckoned from the surface.

Applying eqn (2.11) to the inequality of eqn (2.15) leads to a minimum oxide thickness given by

$$x_o = R_p + \Delta R_p \sqrt{2 \ln(10 N_p / N_b)} \quad (2.16)$$

Numerical substitutions give $x_o = (R_p + 3.7 \Delta R_p)$ and $x_o = (R_p + 5.3 \Delta R_p)$ for (N_p/N_b) ratios of 10^2 and 10^5 respectively. These results show the extent to which the oxide thickness must exceed R_p plus a multiple of the straggle in accordance with eqn (2.16), and call for a justification of the assertion that the field oxide is a barrier to implantation. The typical thickness of field oxide, about 1 μm, is adequate to make it an effective mask for implants leading to layers up to about 0.5 μm thick immediately below the substrate surface. Implantations of a tub like the p-well in the CMOS structure of Fig. 2.6, or of a buried layer, extend 2 to 3 μm into the substrate and demand a similar but higher thickness of field oxide. The same kind of stipulation applies to a photoresist coating, which should have about twice the thickness of the field oxide. Oxides and resists with such thicknesses have a number of disadvantages, and heavy metals like tungsten are being used in their place; a thin layer of such metals presents an adequate barrier to implantation.

Summing up, ion implantation is now in universal use for bipolar and MOS ICs. MOS applications include the following:

(i) source and drain layers;
(ii) channel layers;

(iii) CMOS-well formation;
(iv) self-aligned gates;
(v) buried layers;
(vi) buried insulators;
(vii) polysilicon interconnect doping.

2.4.5 Dielectric and polysilicon films

The electrodes formed by diffusion, epitaxy, or implantation have all been within the substrate, with the exception of MOSFET gates and their interconnections. Dielectric and polysilicon films are formed above the substrate. The techniques for depositing these thin films, which have thicknesses up to 1 μm, are chemical vapour deposition (CVD and its variants) or physical. The physical techniques are evaporation, sputtering, and spin-coating, which is used for depositing resist. Metallization and silicides, the interconnect materials, are covered in a separate section because of the special functions they perform.

The dielectric films most widely used are intrinsic and doped silicon dioxide, and silicon nitride. Silicon dioxide cannot be thermally grown once metallization has been deposited and the interconnections have been formed on the substrate surface. The necessary condition for thermal growth, the existence of a clear silicon surface, has been removed. Silicon dioxide films are demanded for passivation of the substrate, and occasionally for increasing the thickness of the thermal field oxide.

Insulation between multilevel metallization, and the final die passivation (see Fig. 1.6) could be achieved with silicon dioxide, but oxide-related dielectrics are used instead. These consist of oxide doped with phosphorus, arsenic, or boron. The two established dielectrics are phosphorus-doped SiO_2 (known by the names of phosphosilicate glass, or PSG) and SiO_2 doped with both phosphorus and boron (borophosphosilicate glass, BP-glass, BPSG). These dielectrics have the advantages of inhibiting diffusion of sodium impurities and creating a smooth surface topography for depositing the metal of the next layer. The surfaces of the higher metal layers are undulating; the reasons for that shape can be deduced from the transistor profiles shown in Fig. 1.4. P-glass and BP-glass are formed by low temperature (300–400 °C) low pressure CVD (LPCVD). High-temperature cycling makes the glass turn viscous and reflow, smoothing out the uneven surface. The temperatures for this are about 1000 to 1100 °C for P-glass, and—advantageous, because it is smaller—700 °C for BP-glass. Such temperatures can only be tolerated if refractory metals, which have a high melting point, are used for metallization. Much lower temperatures are mandatory for aluminium metallization. The eutectic point of a silicon-aluminium contact is 577 °C, and an upper limit of around 450 °C is adhered to in practice for depositions on aluminium.

The chemical processes for depositing thin films of dielectric or polysilicon are atmospheric pressure CVD (APCVD), low pressure CVD (LPCVD), and plasma-enhanced (PECVD). The method adopted is governed by the use of the layer. Intrinsic silicon dioxide is formed by APCVD. LPCVD is carried out at pressures between 30 to 250 Pm (255 mTorr to 1.87 Torr). It leads to films which are relatively free of contamination and which provide better adherence over the undulations of the surface. LPCVD has a lower deposition rate than APCVD, but an equivalent throughput of wafers is achieved by special mounting which allows a large batch of wafers to be processed simultaneously. PECVD is a low temperature process favoured for silicon nitride deposition on metals.

P-glass or BP-glass are acceptable for the final passivation of the die, but another dielectric, over-glass, in which a layer of silicon nitride is deposited on a silicon dioxide layer covering the top metal level, has become established. Its outstanding feature is the imperviousness of silicon nitride to the diffusion of sodium and the ingression of moisture.

The over-glass passivation is one role of silicon nitride. Its masking for selective recessed oxidation, described in Section 2.4.1, is another. Composite oxide-nitride layers are also finding use in very thin gate insulation. The standard deposition for nitride is LPCVD at temperatures between 700 and 900 °C. PECVD is being pursued for nitride and dioxide. It is a low temperature process (250–350 °C) and permits deposition over aluminium.

Polysilicon films form MOSFET gates, interconnects, conductors, high-value resistors, and ohmic contacts for shallow junctions. Other uses are as a diffusion source for shallow junctions and to fill in trenches for dielectric isolation (see Section 2.4.1). Deposition is usually by LPCVD in the reaction arising from depositing silane between 550 and 650 °C in a LPCVD reactor. The film is deposited undoped, and is subsequently doped by diffusion or implantation to lower its resistivity. Ion implantation of this nature is part of the sequence for the CMOS structure described in Section 2.2. Doping by diffusion is superior in respect of achieving a low resistivity in the range 0.001 to 0.01 ohm-cm, about ten times below the range obtainable with ion implantation.

2.4.6 Interconnects

Interconnects fulfil a variety of functions. These are:

(i) contacts between the first layer of metallization and the substrate electrodes;
(ii) gate interconnects;
(iii) conductors (interconnects) of the metallization layers/metal levels;
(iv) via contacts between the levels of multi-level metallization

Physical vapour deposition (PVD) is preferred for interconnects, although CVD is still used on occasions. The interconnect is patterned by etching, which is described in the next section because of its importance in realizing the window dimensions prescribed by lithography.

The two PVD techniques are evaporation and sputtering. Evaporation is a process carried out in a vacuum, in which a source is heated until it evaporates and falls on the wafer, condensing on its cooler surface. The pressure of the vacuum chamber is of the order 10^{-6} to 10^{-7} Torr. In sputter deposition the target is bombarded with energetic ions (usually argon) which knock out metal atoms. These land on the wafer and its surroundings; travel is not straight like in evaporation. The wafer is usually biased in order to reject secondary electrons. Sputtering, like evaporation, takes place in a vacuum, but at a much higher pressure (5 to 50 mTorr).

The characteristics and materials for interconnects vary to satisfy the appropriate electrical requirements. A low ohmic resistance is a dominant criterion, which arises from the nature of the wire lengths, in VLSI more than elsewhere. These lengths become significantly high, even at moderate sub-VLSI function densities. Next-neighbour interconnections are limited to a small proportion of the total. The average wire length in VLSI is typically five per cent of the *chip edge* (which signifies the length and breadth of a square die). The wire resistance should be as small as possible to avoid IR losses, and to minimize the wire RC time constant (per unit length) in order to obtain the fastest possible performance. In this respect aluminium is one of the best metals with a resistivity around 3 $\mu\Omega$-cm. However its low melting point (660 °C) restricts processes subsequent to its deposition to less than 500 °C. Consequently it can only be deposited after all high-temperature processes have been completed.

The contact resistances of the metal-substrate junctions should be ohmic and have low values. In this respect an aluminium-p type silicon junction is a good ohmic contact with an acceptably low resistance for impurity concentrations of 10^{16} atoms cm^{-3} or higher. The same cannot be said of an aluminium-n type silicon junction, which constitutes a Schottky diode. (Remember that aluminium is a p-type metal.) The 'diode' effect is constrained to a sufficiently small voltage drop across the contact (a few mV) if the n silicon is heavily doped. That is the reason for the n$^+$ collector contacts in Figs. 1.4(a) and 2.3.

Thin films are patterned in one of two ways. The traditional etch formation, shown in Fig. 2.14(a), has been used for all the patterning described so far. An unmasked deposition of the film is followed by a deposit of the masking resist. The unmasked film is then removed and an etch of the resist follows, leaving the patterned film. The lift off technique for producing an identical pattern is illustrated in Fig.2.14(b). The sequence starts with depositing the masking resist. The mask for that purpose is complementary in area to the mask deposited in Fig. 2.14(a). The principle of complementarity is similar to that of mask levels 4 and 5 in Table 2.2. Next comes the deposition of the film over the entire wafer. Step coverage in PVD, without taking special measures to improve it, is relatively poor, with distinct gaps at the resist corners. This deficiency is used to advantage, because the final step consists of dissolving the resist, thereby removing the part of the film deposited on it and leaving the required pattern.

After deposition and patterning, aluminium is annealed at a temperature around 450 °C to ensure good contact. Silicon diffuses into the aluminium film during that process and causes the aluminium to extend in the form of spikes into the electrode. The significance of that penetration depends on the depths of the aluminium spike and the electrode. It can lead to a short by virtue of the spike penetrating through a shallow layer. The danger of such an occurrence is prevented by the addition of a small amount of silicon (1–2 per cent) to the aluminium. Alternatively a thin film of polysilicon is deposited between the aluminium and the substrate. In both cases, the aluminium will feed on the extra supply of silicon during the anneal without the creation of spikes.

42 Fabrication and packaging

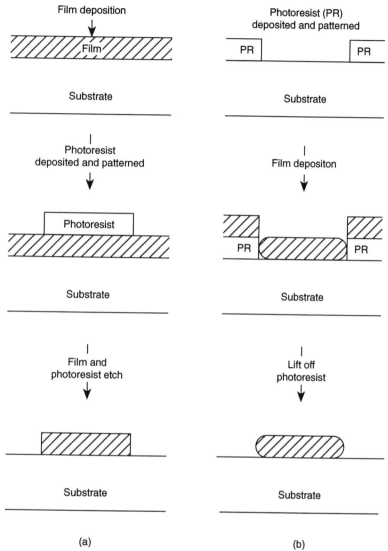

Fig. 2.14 Thick film patterning (a) traditional etch (b) lift off

Conduction in metals is subject to *electromigration*, the transport of atoms by momentum exchange with conducting electrons. At high current densities or elevated temperatures metal atoms move towards the positive, and voids towards the negative end of the wire. Electromigration can ultimately lead to an open circuit, i.e. a break in the wire. Another related danger is a short circuit between adjacent wires, caused by a deposition of dislodged atoms building up on the wire surface. The danger levels for failure are current densities of about 10^5 A/cm^2 and temperatures in excess of 100 °C. The mean time to failure (MTTF) is expressed by

$$\text{MTTF} \propto J^{-n} \exp\left(\frac{\phi}{kT}\right) \quad (2.17)$$

Table 2.7 Properties of various metallizations

	Metals			Silicides			
	Al	Mo	W	MoSi$_2$	TaSi$_2$	TiSi$_2$	WSi$_2$
T$_m$ (°C)	660	2620	3410	1980	~2200	1540	2165
ρ ($\mu\Omega$-cm)	2.5–3.0	6–15	6–15	40–100	38–50	13–16	30–70
Stable on Si up to (°C)	~250	400	600	>1000	≥1000	≥950	≥1000

T$_m$ = melting point ρ = resistivity, range of typical thin film values.

From Sze, S.M. (1988) VLSI Technology (2nd edn), p.383. (Courtesy of McGraw-Hill, Inc.)

where J is the current density, and ϕ the diffusion activation (typically 0.4–0.5 eV for aluminium). The exponent n lies between 1 and 3 (generally it is close to 2).

Electromigration reverses with reversal of current flow. The reversibility is, in terms of quantification, not exact; it depends on the energy of the voids created in the first place. All the same it ensures that electromigration is negligible for the many wires where the net current arising largely from charging and discharging of nodal capacitances is zero. Electromigration needs to be considered when there is a net average dc current, and this current will be largest in the wires distributing the power supply. These wires are dimensioned accordingly to give a large MTTF. Efforts are continuing to reduce electromigration, and a great improvement is obtained with aluminium-silicon-copper alloys in which small proportions of silicon (1–2 per cent) and copper (4–5 per cent) prevent spiking and reduce electromigration to a negligible level.

The limitations of aluminium have led to the alternatives of either refractory metals or silicides, compounds of refractory metals with silicon. These are acceptable in respect of temperature and electromigration, and also have adequately low resistivities. Much ongoing research and development continues into interconnect metallization. Two key issues, which have already been mentioned, are step coverage and contact resistance. Step coverage becomes highly important in the contact cuts of inter-metal dielectric layers and the oxide covering the substrate.

The properties of various metals and silicides are listed in Table 2.7. The refractory metals most widely used are tantalum (Ta), titanium (Ti), molybolenum (Mo), and tungsten (W). Silicides are especially suited for MOS gate interconnects. Polysilicon on its own is the natural choice for the gate; a polysilicon gate structure is described in Section 2.2. Unfortunately polysilicon, even when doped, is unsuitable for interconnects because of its high resistance. Its resistivity after doping is typically 1500 Ω-cm, which is one or two magnitudes above the resistivities of the metallizations listed in Table 2.7. Such a value is unacceptably high—except for very short local next-neighbour interconnects—in view of the declared objectives to minimize wire resistance and RC time constants. The MOS gate structure increasingly adopted consists of a polysilicon-silicide sandwich with the polysilicon on top of the gate oxide. The silicide extends beyond the gate to form the interconnect. That composition, called *polycide*, gives good electrical characteristics with a well defined threshold voltage, and a low interconnect resistance.

Surveying briefly the usage of the various metallization and polysilicon, aluminium is still heavily used for contacts to the substrate. It is also the top-level metallization in multilevel VLSI. There its low resistivity minimizes IR losses where the dc currents in the wires tend to be largest. Metallization selectively formed on silicon is aluminium, tungsten, or a silicide. A wide variety of materials is in use for the intermediate metal levels and for the via contacts between them. Special attention has to be paid to step coverage and contact resistance in order to achieve good contact connections.

2.5 Lithography

2.5.1 Introduction

The lateral dimensions of the electrodes and metal interconnections are prescribed by the areas of the windows (contact cuts) opened for their formation. The processing sequence for opening the windows has already been outlined in Section 2.1. Lithography has three constituents, the deposit of the resist, the imaging of the wafer (to be precise, of the resist covering the wafer), and the etching of the oxide.

Lithography is very much the key to processing. Tremendous advances in lithography have spearheaded the shrinkage of device dimensions, which has made VLSI possible and is still in full swing. To emphasize lithography is not to deny the importance of, and developments in, resist processing and etching. Indeed with the ongoing thrust towards deep submicron VLSI, we are witnessing a symbiosis of etching and lithography. Nevertheless advances in submicron structures are likely to be governed more by lithography than anything else.

2.5.2 Lithographic technology

IC lithography started with optical (UV) beams and has largely remained so. It is complemented by electron beam (e-beam) techniques, which are preferred for making masks and reticles and also have a niche in direct imaging of the wafer. Optical lithography has proved to be amazingly resilient, and currently achieves structures in standard production down to 0.5 μm, with pilot production of 0.35 μm, and an anticipated capability of 0.25 μm geometries in the near future.

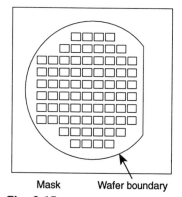

Fig. 2.15 Multiple image formation

The function of lithography is to produce a multiple image on the resist covering and wafer. Figure 2.15 shows such an image; the outlines only are given. The mask for each level will have a multiple image with identical boundaries but different patterns. The various methods of achieving the image pattern of Fig. 2.15 are illustrated in Fig. 2.16. The imaging source, either a mask or a reticle, is usually a glass plate coated with patterned chromium. The distinction between these two is that the mask contains the multiple images with the actual size for printing on the wafer, whereas a reticle contains a single image only, enlarged

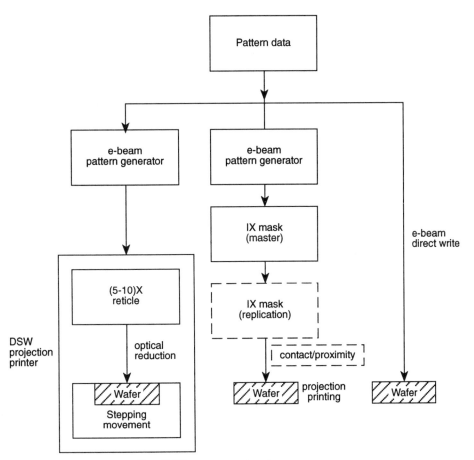

Fig. 2.16 Imaging of wafer

typically 5 X or 10 X the actual size. The process of depositing resist, exposing it to the imaging pattern, selectively removing part of it after exposure, and etching to obtain the patterned chromium layer is identical to the sequence described in Section 2.1. One problem, the high reflectivity of chromium, is resolved by forming a double antireflective layer on the substrate, with chromium (Cr, ~600 Å) on the glass and chrome oxide (Cr_2O_3, ~200 Å) on the chromium.

The masks and reticle in Fig. 2.16 are produced with e-beam optics, the customary method for VLSI. The pattern data base contains the digitized information for the image. The pattern generator processes that information and produces the e-beam scan which sweeps over the mask or the reticle. The e-beam is focused to give a spot with a diameter about one-fifth of the minimum linewidth i.e. 0.2 μm for a 1 μm structure. It is *flashed* (turned on) for each position; with a displacement between adjacent steps of about one-half the diameter. Inevitably it takes much longer to produce a mask than a reticle, because of the multiple image pattern for the former.

The optical alternative, shown in Fig. 2.17, is similar in principle. A UV beam of light takes the place of the e-beam in Fig. 2.16. The pattern generator processes the information obtained from the pattern

46 Fabrication and packaging

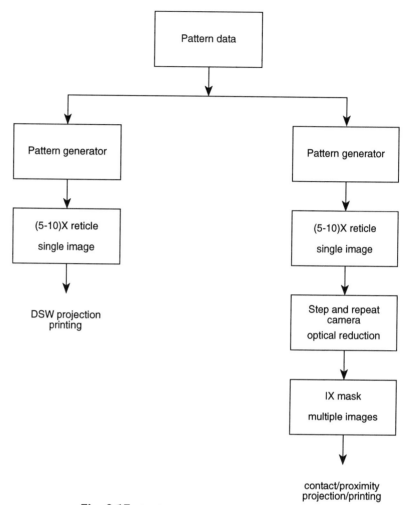

Fig. 2.17 Optical mask and reticle generation

data bank, and contains an optics assembly and a positioning stage. The image is built up by optically flashing minute rectangular blocks unto the reticle. The number of such flashes and the time taken would be prohibitively large for printing the multiple image pattern on a 1 X mask. For that reason such a mask is produced by making a 10 X reticle first, and then printing multiple images on the mask with a step-and-repeat camera, which contains the reticle and is stepped interferometrically over the wafer. The higher speed of flashing an e-beam leads to much faster mask fabrication, typically 2 hours compared with 20 hours or more for optical generation. That, added to the better resolution obtainable with an e-beam, establishes its superiority over optical lithography for mask and reticle production.

Returning to Fig. 2.16, the imaging techniques are:

(i) transfer from a reticle;

(ii) transfer from a mask;

(iii) direct e-beam writing.

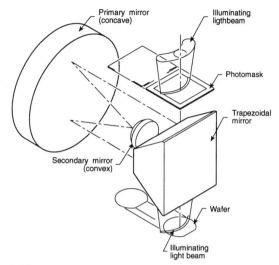
Fig. 2.18 Principle of projection printing (Courtesy of Canon)

The oldest method of printing the image is to place the mask in contact with the wafer. This preserves the intrinsic resolution of the mask. Resolution, interpreted in the context of lithographic capability, is the minimum feature size which can be achieved in production. Unfortunately the mask is likely to be contaminated with resist from the wafer, an abrasive process which impairs the registration accuracy and severely limits the life of a mask. This disadvantage is overcome by proximity printing, in which the mask is held about 20 to 40 μm above the wafer, thereby avoiding contamination by contact. Proximity printing is however accompanied by other problems. Precise alignment is not easy, and the small separation carries with it the danger of accidental contact; it also leads to some loss in resolution on account of diffraction. Contact and proximity printing have virtually been replaced by projection printing, in which wafer and mask are spaced well apart.

The optical arrangement for 1:1 projection printing is the reflection system outlined in Fig. 2.18. A light beam scans a small segment of the mask and the reflected light illuminates the corresponding segment of the wafer. Scanning comes about by perfectly synchronized movements of mask and wafer. The exposure of only a small segment at a time is absolutely essential in order to minimize optical distortion, and to obtain a resolution equal to that for contact printing. A projection printer is a highly sophisticated item of equipment whose performance is largely determined by the quality of its optics.

e-beam lithography has already featured in Fig. 2.16 for the production of the mask or the reticle. Another of its uses is for direct writing of the multiple image on the wafer. e-beam lithography has a higher resolution than optical lithography, but in the case of forming a multiple image on the wafer, it suffers from a low throughput compared with the other methods shown in Fig. 2.16. There are however instances where the slow throughput is acceptable. One such case is VLSI pilot production. Take for example ASICs. Production of a single wafer by e-beam lithography will yield several hundred chips, a sufficient quantity for initial evaluation, at very much less cost than production by other lithographic means.

The best possible resolution is obtained with direct-step-on-wafer (DSW) projection, alternatively called step-and-repeat projection, or step-and-repeat printing. The equipment for that purpose is a DSW wafer stepper, sometimes simply called wafer stepper, or step-and-repeat projection printer. The technique combines a stepping movement with an optical reduction for printing multiple images on the wafer. The step-and-repeat sequences for the DSW projection in Fig. 2.16 and the 1X mask generation in Fig. 2.17 are very similar, but differ significantly in two respects. An understanding of these differences is vital for appreciating the advance in resolution obtainable with DSW projection. The differences are:

(i) the multiple image on the 1X mask in Fig. 2.17 is formed by stepping the reticle and the camera containing it; the mask remains stationary. In DSW projection, it is the wafer which is stepped to obtain a multiple image on it; the reticle and the optical apparatus remain stationary;

(ii) in DSW projection, each die on the wafer is identified by a mark, and the wafer is stepped to obtain precise alignment between die and reticle. Additional highly sophisticated alignment procedures level the wafer i.e. they obtain the best possible parallelism between it and the reticle. It is this die-by-die alignment, combined with wafer levelling, which results in the superior resolution of DSW projection.

The five–or tenfold reduction of the reticle image leads to a corresponding reduction in errors of the wafer image arising from optical distortion, line deviations, and some other shortcomings of 1:1 projection. The improved resolution of DSW projection is obtained at the expense of greatly reduced throughput.

IC fabrication involves typically 5–12 mask levels. That number is tending to increase with advances in VLSI developments. High numbers of mask levels are not confined to VLSI. They occur for example in recent analogue SSI, which contains entirely separate npn and pnp transistors unlike the SSI of old, which made do with pnp transistors obtained from adaptations of the npn structure at the cost of very poor gain and bandwidth. More will be said on that subject in the next chapter. BiCMOS is another example where the mix of bipolar and MOS technologies leads to a higher number of mask levels not only in digital VLSI, but also in analogue sub-VLSI chips. The superposition of masks (or reticles) causes overlay errors, which determine the overall lithographic performance and the production yield.

The customary practice adopted is a *mix-and-match* lithography, in which DSW stepper projection is reserved for the more critical, and projection 1:1 printing for the less critical mask levels. Indeed submicron VLSI lithography is sometimes a blend of DSW projection, scanner 1:1 projection, and e-beam writing. Another factor is the use of multiple machines, for example of several DSW steppers in which mask levels are divided among the steppers, in order to expedite production. The overall resolution in such cases involves not just a single machine overlay, but also alignment errors arising from machine-to-machine overlay. The

Table 2.8 Selected specifications of ASM lithography PAS5500 Wafer Steppers

Family member	PAS 5500/20	PAS 5500/80	PAS 5500/90
Resolution (μm)	0.70	0.50	0.35
Wavelength (nm)	365[1]	365[1]	248[2]
Lens NA	0.40	0.48	~0.50
Lens DoF (μm)	1.6	1.2	0.7
Throughput (W/h)			
6 in wafer	>75	>88	>80
8 in wafer	>50	>64	>60
Alignment accuracy (nm) (99.7%)	<15	<15	<15
Overlay error (nm) Longterm (99.7%)	≤115	≤85	≤70

[1] i-line   (Courtesy of ASM Lithography)
[2] KrF excimer laser
NA = Numerical aperture
DoF = Depth of focus

overlay error is expressed by the standard deviation σ_t related to the standard deviations σ_i of individual errors in accordance with

$$\sigma_t = \sqrt{\sum_i (\sigma_i)^2} \tag{2.18}$$

Assuming a Gaussian error distribution the proportion of errors which fall in the range $\pm n\sigma_t$ is given by $erf(n\sigma_t)$, where erf is the error function. For n=3, $erf(n\sigma_t) = 0.997$, and the overlay error is usually expressed by $3\sigma_t$. Alternatively the same value is quoted with a bracketed qualification (99.7 per cent) against the error parameter (Table 2.8). That specification is used for individual and global errors. The overlay budget is made up of the individual error contributions. Its composition depends on the lithographical strategy adopted, and on the error data supplied for the various machines. Typical individual contributions are: image errors, machine-to-machine overlay error, reticle overlay error, single machine overlay error, etc. Representative values of DSW stepper performance at the cutting edge of technology are given in Table 2.8. which lists some other parameters in addition to overlay errors.

Optical lithography continues to be the unquestioned technology well into the submicron region. Its capability will now be examined with the aid of expressions for resolution (R) and depth of focus (DoF). These are

$$R = \frac{K_1 \lambda}{NA} \tag{2.19}$$

$$DoF = \frac{K_2 \lambda}{(NA)^2} \tag{2.20}$$

$$NA = \frac{1}{2f} \tag{2.21}$$

K_1, a constant for a given system, is a contrast value linked to the composition of the resist. It is typically 0.8 for lithographies of about 1 μm, with a lower theoretical limit of 0.5. λ is the wavelength of the light source, and NA, the numerical aperture of the lens, is defined by eqn (2.21) where f is the ratio focal length/effective diameter. K_2 is again a constant for a given system, also with a theoretical limit of 0.5. That value is frequently assumed in the literature.

The UV light sources and the appropriate lenses for them usually operate at the g-, h-, and i-lines of the mercury spectrum, corresponding to wavelengths of 436 nm, 405 nm, and 365 nm respectively. The two regions for lower wavelengths are mid and deep UV. Mid UV spans a range from 380 down to 240 nm, deep UV applies to shorter wavelengths. Resolution is not simply a matter of going to shorter wavelengths. DoF is a complementary key parameter, vitally important for production control. Wafer stepper registration, overlay accuracy, and wafer topography have to be allowed for. The DoF value should be large enough to cover the spreads in these quantities. Another factor is the availability of a suitable resist for the wavelength in use. Recent improvements are aimed at changing the values of K_1 and NA in eqns (2.19) and (2.20), whilst retaining the i-line wavelength, which is supported by wafer steppers and resists with established high quality and characteristics. Predictions point to an extension of optical lithography to 0.25 μm (Flores and Kirkpatrick 1991, and Hohn 1993). Deep UV lithography, for example with ArF excimer lasers (193 nm), may push the limit down further.

Ultimately X-ray lithography, with its nanodimension capability arising from its ultrashort wavelength (0.5 to 5 Å) will be needed. At the moment it is beset by problems in the fabrication of masks, the development of acceptable resists, and the design of uniform, stable X-ray sources at economic cost. It also has to demonstrate that it is capable of cost-effectiveness by achieving the throughput of optical lithography. The economics of the market place continue to encourage the stretching of optical lithography to its absolute limit. When X-ray lithography eventually becomes economically viable, it will most likely be confined to submicron structures beyond the reach of deep UV lithography.

2.5.3 Resists

The steps leading to the opening of windows for electrode formation occur in the following sequence. The resist is applied over the wafer, heated in a *prebake* (or *softbake*), and exposed to a beam of light (or electrons etc.) and patterned by masking in the manner described in Section 2.5.2. Next the resist is developed; the part remaining protects the underlying layer, the remainder of which is exposed. The resist subsequently undergoes a *postbake* (or *hardbake*), which is similar to the prebake. The final steps are the etch of the exposed layer (oxide, silicon nitride, or polysilicon).

The wafer surface is first cleaned and prepared for bonding by covering it with a resist promoter. In this technique the *promoter* is applied in liquid form by spin coating: the wafer is held in a vacuum chuck and is spun at high speed. The application of the resist, also by spin coating, comes next. Its thickness t is given by

$$t = \frac{k}{(rpm)^{\frac{1}{2}}} \quad (2.22)$$

where k is a constant, which is a function of the resist's composition and viscosity, and rpm is the speed of revolution. Speeds are several thousand rpm and spinning takes up to 1 min in order to obtain highly uniform thickness. The softbake of the wafer consists of heating it to a temperature of ∼85 to 95 °C for typically 10 to 30 min, and serves a twofold purpose. It strengthens the adhesion of the resist to the wafer, and makes the resist photosensitive by removing the solvent.

The exposure of the resist is closely linked to its composition. Resists can be divided into two broad categories, negative and positive. The same categorization applies to e-beam and X-ray resists. Photoresists are, like films for ordinary photography, organic compounds whose solubility is affected by exposure to light. A negative photoresist is composed of a polymer, a sensitizer, and a solvent. It polymerizes on exposure and the exposed portion is thereby hardened. The developer solvent removes the unexposed section of the resist (Fig. 2.19(c)). A positive resist consists of resin, solvent, and a photoactive compound, which gives the resist develop-inhibiting properties. The photoactive compound chemically decomposes on exposure, making the resist readily soluble and removable by a developer. Now it is exposed to light which defines the window (Fig. 2.19(d)). Negative preceded positive photoresist, which has almost entirely replaced it for the following reasons:

(i) the exposed part of the negative resist swells during development. The resulting distortion limits resolution to between 2 and 3 times the thickness of the resist. This effect is absent in positive resist;

(ii) Negative resists have less contrast, and contrast is a very important parameter in determing resolution. The higher the contrast, the better are the edge and pattern resolutions;

(iii) A short exposure time is desirable for a high throughput in DSW projection. Exposure time can be traded off against the intensity of the light source, provided the resist possesses *reciprocity*, i.e. a near-constant value of the light intensity exposure time product over a range of high intensities. Positive resists unlike negative resists, have adequate reciprocity.

Examples of widely used positive resists are Shipley 1350J, MP-2400 and HPR 206. These have a peak sensitivity in the 300 nm to 400 nm range. For shorter wavelengths there are resists like polybutene-1-sulfone (PBS), which has a good sensitivity for $\lambda < 200$ nm, and poly-methyl methacrylate (PMMA). These two resists are also extensively used for e-beam lithography, where PMMA has an outstanding resolution (<0.1 μm).

The accuracy with which the image on the resist is replicated into a developed pattern depends critically on exposure optimization. The prime factors are resist composition and thickness, conditions of the softbake, line width requirements and tolerances, and the composition

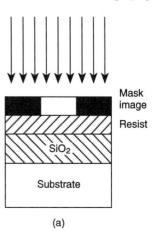
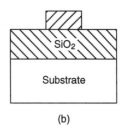
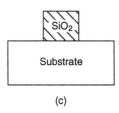
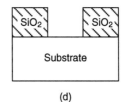

Fig. 2.19 Window profiles (a) exposure (b) after development—negative resist (c) after etch—negative resist (d) after etch–positive resist

of the developer. Another vital issue, the process of etching, has to be established at this stage. Although etching comes after development, it has a bearing on exposure optimization.

The continuing shrinkage of structures to submicron dimensions has not been accompanied by a corresponding reduction in the thickness of the field oxide, which remains at about one micron. The ratio of oxide depth to window area has therefore become larger and the vertical resolution of the etch has assumed more importance. That is one of the reasons for preferring positive over negative resist. The resolution of resist improves with thinner layers. Lithography for submicron geometries demands thin layers in the range 0.1 to 0.3 μm. Such layers, especially in the case of e-beam resist, are prone to *etch invasion*. The function of the resist, to protect the layer beneath it from the etch which removes the unprotected part of that layer, is thereby impaired. The problem is greatly alleviated with a multilayer resist, in which a thick bottom layer, composed of a process-resistant polymer, presents a truly planar surface for a much thinner coating, which serves for imaging. In some cases a third layer of either SiO_2 (optolithography) or a conducting material (e-beam lithography) is interposed to harmonize between the conflicting chemistries of the top and bottom layers. The improvement resulting from multilayer resist is inevitably obtained at the cost of increased complexity.

The last step prior to etching is a postbake, typically for 30 min at 120 °C. Its purpose is to increase the etch resistance of the resist, to strengthen its bond to the underlying layer, and to remove any solvents still left.

2.5.4 Etching

The final step in the opening of the windows is the etch of the exposed dielectric or polysilicon. Etching is also the means of forming the interconnects of the various mask levels, the vias, and lastly the patterning of the opaque sections—usually chrome—of the masks.

The chief requirements to be met are accurate penetration i.e. removal by the etchant, minimum removal of the masking resist, and a high throughput. Wet etching was the universal practice until the early 1970s. Nitric and hydrofluoric acids are the wet etch agents which attack and remove the layer to be etched. Wet etching has the advantage of a high throughput, because a large number of wafers can be immersed in the etchant simultaneously. Its main limitation is its isotropic nature.

Wet etching, rather like diffusion, penetrates both vertically and horizontally, leading to the profile shown in Fig. 2.20(b). The isotropic dissolution is a chemical reaction. The lateral etch, like lateral diffusion, which is explained in Section 2.1, undercuts the resist and widens the layer to be formed in the substrate. For geometries of $\sim$3 μm or larger, that effect can be absorbed without causing a serious change in the intended pattern. For smaller geometries this is no longer the case, and wet etching has given way to plasma-assisted reactive ion etching (RIE) largely for that reason. It still finds uses, for example in cleaning the processing quartzware, furnace tubes etc. It is also used for the etching of chromium masks, where it has some advantages over the dry plasma etch. e-beam resist is less susceptible to attack by a wet than by a plasma etch. Furthermore a wet etch is less demanding on adjustments of the resist image in order to accommodate other variations in processing. Lastly dry etching is far more susceptible to defects arising from particulates.

A vital parameter in etching is the selectivity S, defined by

$$S = \frac{E_e}{E_o} \qquad (2.23)$$

where E_e is the etch rate of the layer being etched, and E_o the etch rate of another layer involved in the process. There are two such layers, the resist and the underlying layer, the substrate. The higher the selectivity, the better will be the quality of the etch. High selectivity minimizes the lateral removal of resist (not shown in Fig. 2.20(b) to (d)), and the removal of substrate material by an overetch, which is a continuation of the etch beyond the dioxide/substrate interface. Layers are usually overetched to allow for variability in thickness and etch rate, and in other process parameters. The etch rate in wet etching is also highly temperature-dependent. Selectivities of 15 or higher are aimed at in practice.

Dry plasma etching, including the RIE mode, has now largely replaced wet etching. The basic apparatus for plasma etching, a parallel-plate reactor, is sketched in Fig. 2.21(a). An rf source, which typically operates at 13.56 MHz, a frequency internationally agreed for industrial processes, produces a plasma discharge in a reaction chamber held at a pressure in the range ~3 to 150 mTorr. The composition of the plasma is determined by the incoming gases, which contain one or more types of halogen atoms. The plasma is a collection of electrons, positive and negative ions and molecules, produced by an interaction between the introduced gases and the AC field. (DC plasma excitation is also possible.) There are only very few negative ions, so that the numbers of electrons and positive ions are very nearly equal. In the vicinity of the electrodes the high electron mobility and the strong field combine to deplete the electron population and weaken the plasma to the point where the discharge glow cannot be observed. A *sheath* (also called *dark space*) is formed near each electrode. The sheath is a region depleted of free electrons. A substantial voltage is developed across the shield between the end of the plasma and the electrode, and this voltage plays an important part in RIE. Silicon and dielectric layers react with fluorine or chlorine to produce halogenated layers. The reactions which desorb the layers being etched are purely chemical and, for that reason, isotropic. A wide variety of inlet gases is used to suit various types of layer, and a representative list is given in Table 2.9.

Plasma etching takes place with the wafers placed on the grounded electrode, and with the rf supply connected to the other electrode via an impedance matching network, see Fig. 2.21(b). The two common arrangements for plasma etching are the planar reactor of Fig. 2.21, and the barrel reactor. The planar reactor operates at a relatively high gas pressure (~0.2 to 0.5 Torr) and with a potential difference ~100 to 200 V between the electrodes and the plasma. A barrel reactor houses a bank of wafers loaded vertically, and operates at about twice the pressure and with about one tenth of the plasma-electrode potential difference. It is

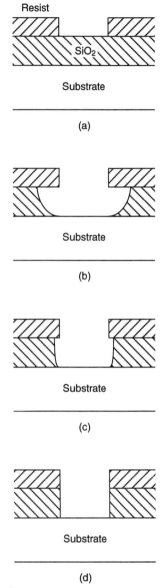

Fig. 2.20 Isotropic and anistrotropic profiles (a) before etching (b) isotropic etch (c) near anistrotropic etch (d) ideal anistrotropic etch

54 Fabrication and packaging

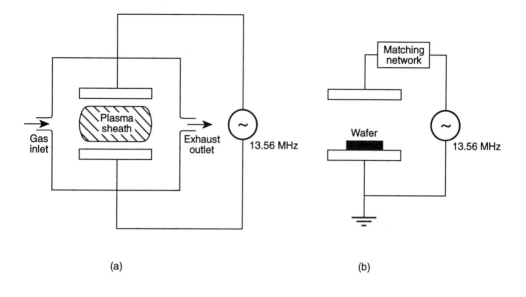

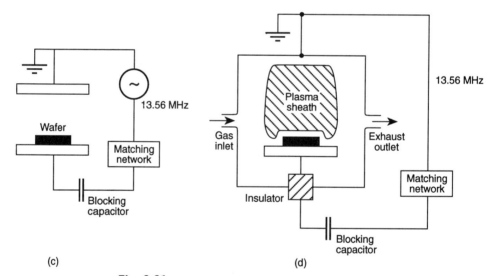

Fig. 2.21 Reactor structure (a) outline (b) electrode arrangement—plasma etch (c) electrode arrangement—RIE (d) outline—RIE

mainly used to strip the resist after etching; O_2 plasma stripping is the standard method for resist removal. Although the plasma etch does not achieve anisotropy, it scores on other grounds over wet etching. It has better wafer-to-wafer reproducibility, the end point of the etch can be accurately determined by spectroscopy or interferometry, and control of the gas flow permits very accurate control of the etch rate. It also allows better adhesion between phosphosilicate glass, a dielectric favoured over oxide in multilevel isolation (Section 2.4.5) and photoresist.

Anisotropic etching, essential for the precise and small dimensions of submicron structures, is achieved with RIE. The technique is a mod-

Table 2.9 Typical gases used in plasma etching

Layer material to be etched	Source gases
Silicon Dioxide (SiO_2)	CF_4, CF_4O_2, CCl_2F_2, $SiCl_4$,
Silicon nitride (Si_3N_4)	CF_4, CF_4O_2
Silicon (Si)	CF_4, CCl_4, F_2, CCl_2F_2
Silicides	CCl_4, $CFCl_3$, $CF_4 + O_2$, HBr
Aluminium (Al) and alloys	CCl_4, $CCl_4 + Ar$

ification of the plasma process. The electrode arrangement for RIE is sketched in Fig. 2.21(c). The electrode driven by the rf supply now holds the wafer, the other electrode is grounded. An amplified diagram of the structure, Fig. 2.21(d), shows that the grounded electrode, the conducting wall of the chamber, has a far larger area than the electrode holding the wafer. The plasma-substrate potential difference is in the range 100 to 500 V, leading to a highly directive bombardment of the substrate by reactive ions. There is a build-up of negative charge on the electrode holding the wafer, in effect on the wafer surface. Electrons are attracted to it during the positive cycle of the AC supply, but relatively few positive ions will reach it during the next half cycle, because the ions are much heavier and have a much smaller velocity. That behaviour assumes a high enough operating frequency, which is the case. The net negative charge is prevented from discharging by the blocking capacitor. The resultant negative potential of the electrode and the potential difference across the shield attract the reactive ions, which impinge on the wafer and desorb the exposed layer by this physical reaction. RIE produces a near-anisotropic etch (Fig. 2.20(c)) rather than an ideal anisotropic etch (Fig. 2.20(d)), because the chemical reaction is still present in some measure. The slightly angled shape of Fig. 2.20(c) has an advantage. It allows better coverage by metallization, if required, with fewer failures due to sidewall voids, cracks etc., than the perfectly vertical etch of Fig. 2.20(d).

Summing up, RIE, the established etching process for submicron VLSI, is supplemented by plasma and wet etching in the roles indicated. An alternative to RIE is reactive ion beam etching (RIBE), in which the reaction chamber where the plasma is generated is kept separate from the accelerating section, which drives the reactive ions into the wafer surface. RIBE is still at an early stage of development and more investigation is needed to establish its suitability. Lithography has so far tended to overshadow other IC processes, because these have, with much effort and ingenuity, accommodated the ever decreasing structures. The stage has now been reached where etching rather than lithography could prove to the bottleneck in the advance towards geometries of 0.25 μm and less. Gottscho (1993) persuasively points to the highly empirical approaches to plasma etching, and stresses the need for research to gain a better understanding of reactive plasmas in order to match the lithographical advances which are likely to take place in the decade ahead.

Table 2.10 Levels of interconnection

Level	Interconnections
1	Connections inside component case
2	Connections from package to PCB or wire
3	Connections from PCB to wire, or to another PCB on chassis (usually internal)
4	Connections from internal chassis to another internal chassis
5	Connections from one piece of equipment to another (usually external)

Table 2.11 Levels of packaging

Level	Component
1	Package
2	PCB
3	Backplane

2.6 Packaging and mounting

2.6.1 Introduction

The die of an integrated circuit is encapsulated and the package is mounted on a PCB or another substrate. Packaging is of paramount importance in electronic equipment and is a major engineering undertaking in its own right. It constitutes the first level in an interconnection hierarchy defined over 20 years ago. The five levels of interconnections for an electronic system are described in Table 2.10 (Amey 1989). The definitions in Table 2.10 lead to the levels of packaging given in Table 2.11 (Pinnel and Knausenberger 1987). This section covers level 1 and extends to level 2 in respect of package mounting and the PCB wires, which influence package design.

An outstanding characteristic of ICs is their tremendous reliability compared with the other constituents of an electronic system. The inherent reliability of the die has to be matched as closely as possible by the reliability of the package, and this has to be achieved at an acceptable cost. Packaging cost is typically 20 to 40 per cent of the total production cost. Package development has been governed by the increasing function densities per chip, and the consequent demand for higher pin (I/O terminal) counts, accompanied by the demand for faster performance. System speed is largely governed by the impedance of the wires, especially wire capacitance. The reduction in IC feature size has led to shorter wire within the package, and hence reduced interconnect capacitance and inductance.

A rough initial estimate of the number of pins per package is obtainable using Rent's rule, according to which

$$P = \alpha G^\beta \tag{2.24}$$

where P is the pin count for a module (in our case a package) containing

G blocks, (in the example which follows, equivalent logic gates) (Landman and Russo 1971). Rent's rule, developed empirically by IBM in the mid-1960s, has been applied successfully to yield an approximate estimate of pin count for VLSI random logic chips. It has also been applied to other techniques like partitioning a system into PCBs, and the internal partitioning of a chip into subsystems (Schmidt 1982). Hollis (1987) gives representative values of α (3) and β (0.57 to 0.75) for a typical ASIC gate array fulfilling random logic. This, applying eqn (2.24), leads to pin counts in the range of (228–897) and (848–5045) for chips having 2000 and 20 000 equivalent gates (G blocks) respectively. An equivalent gate is equated to four transistors in this estimate. These numbers although excessive, stress the demand for high pin counts in VLSI. Rent's rule is also used to estimate pin counts in ASIC design. It cannot be applied, for example, to semiconductor memories, which are the driving force behind the increase in chip function density, but require only a very modest pin count, determined mainly by word length rather than the memory capacity. The subsections which follow examine the packages available, package technology, package selection, surface mounting, and future trends.

2.6.2 Packages available

The photograph in Fig. 2.22 illustrates the range of packages available. The photograph, in which packages are shown very close to their actual size, also contains pin counts and dimensions (area and depth in mm) for several of the packages. The dual-in-line package (DIP) is at one end of the spectrum, the pin grid array (PGA), which currently reaches the highest pin count, at the other. Most of the abbreviations are explained in Table 2.13. The M in MQUAD signifies an encapsulation with enhanced power capability, explained in Section 2.6.4. Similarly the Power QUAD, Power PGA, and Power Leaded CC have raised power capability. The chief features to be looked for in assessing a package are the maximum pin count, the dimensions, the pitch (the spacing between adjacent centres of pins or leads), the encapsulating material (plastic or ceramic), the mode of mounting (plated through-hole, TH, or surface mounting, SM), the maximum power dissipation, and the thermal resistance. The DIP (and some of the variations based on it) and the PGA are the only standard packages for TH mounting. All the other packages are for SM. Surface mounting is relatively new and, by virtue of the entrenched TH style, is not yet the major mounting technology, but is the unquestioned recognized standard for new designs. The leads of DIP and PGA packages have a standard pitch of 100 mil (2540 μm). Some more recent DIPs have pitches of 70 mil and even 50 mil, but the 100 mil standard will be maintained. This detail is mentioned, because the DIP and PGA packages have been designed to match the standard 100 mil pitch of PCB through-holes.

The development of packages for the VLSI era has not been confined to raising the pin count and adopting the more efficient square format. Packages with low pin counts have been developed which have a smaller footprint and improved electrical characteristics relative to their forerunners. These include the small outline (SO, also called small outline IC, SOIC) package and an adaptation, the shrunk SOP package (SSOP);

Fig. 2.22 Photograph of IC packages (Courtesy of GEC Plessey Semicondutors)

both these are for SM style. Another variation of the DIP intended for SSI, MSI and semiconductor memories (which need only a modest pin count) consists of single-in-line (SIP), zig-zag-in-line (ZIP), and quad-in-line (QUIP) packages. The SIP and ZIP shown in Fig. 2.23(a) and (b), have a greatly reduced footprint because they are mounted with the thin dimension facing the PCB. The leads of the SIP emanate from that side of the package only. In ZIP, leads are taken from two sides doubling the pin count over an equally-sized SIP. The QUIP reverts to DIP format, but has two rows of staggered pins on each side (Fig. 2.23(c)). With a pitch of 100 mil for each row, the effective pitch is 50 mil, doubling the pin count over an equally-sized DIP.

A summary of package types and their pin counts is contained in Table 2.12, and their acronyms are listed in Table 2.13. The listing embraces established and comparatively recent packages. The pin counts are based on a tabulation given by Striny (1991) and on information obtained from various manufacturers' literature. The encapsulation, where not explicitly defined in the designation, is either plastic or ceramic; the features and relative merits of these technologies are discussed in the next section. The nomenclature is in some cases confusing. A glaring example is the case of the PLCC, which is leaded, and the LCC, which

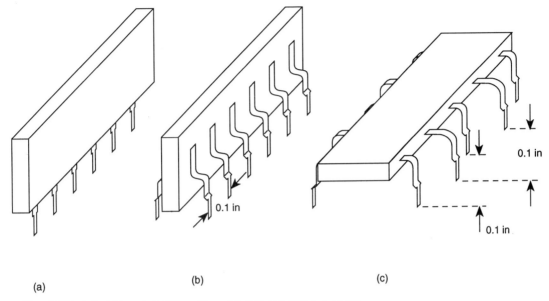

Fig. 2.23 SIP, ZIP and QUIP outlines (a) SIP (b) ZIP (c) QUIP

Table 2.12 Package types

Package	Typical pin count range	Mounting style
Dual-in-line	8–64	TH
Single-in-line	5–40	TH
Zig-zag-in-line	14–28	TH
Quad-in-line	14–64	TH
Small outline	8–32	SM
Chip carrier	16–200	SM
Flat pack	10–300	SM
Pin grid array	68–500+	SM

is leadless. The alternative for LCC, LLCC, is preferred for that reason.

Some general deductions can be made from Fig. 2.22. Flat packs have a smaller footprint than chip carriers for a given pin count. The TQFP has an exceptionally low profile and a very small weight. It is also evident that the lead pitch has been reduced considerably from the 100 mil DIP standard to achieve the high pin counts of the flat pack and the chip carrier. The vendor determines the package style for standard components, although there is occasionally a choice. Some vendors offer alternative packages for SSI/MSI, typically several or all of the following: DIP, SOIC, FP and LCC.

2.6.3 Packaging and mounting technologies

Package fabrication begins where the processes which lead to the multiple die formation on the wafer end. The dies are probe tested to see if they function correctly. The tests are usually confined to dc behaviour,

Table 2.13 Package nomenclature

Type	Abbreviation
Dual-in-line	DIP
Small outline	SO, SOIC
Shrunk small outline package	SSOP
Pressed ceramic glass sealed dual-in-line	CERDIP
Single-in-line	SIP
Zigzag-in-line	ZIP
Quad-in-line	QUIP
Plastic leaded chip carrier	PLCC
Leadless ceramic chip carrier	LCCC
Leaded chip carrier	LDCC
Leadless chip carrier	LLCC or LCC
Flat pack	FP
Quad flat pack	QFP
Ceramic quad flat pack	CQFP
Plastic quad flat pack	PQFP
Pressed ceramic, glass sealed quad flat pack	CERQUAD
Pin grid array	PGA
Plastic pin grid array	PPGA
Ceramic pin grid array	CPGA
Fine pitch quad flat pack	FQFP
Thin quad flat pack	TQFP

and faulty dies are marked with ink to exclude them from further use. The dies are separated either by scribing with a diamond-tipped stylus, followed by breaking the wafer into dies, or by sawing with a diamond-tipped blade with cuts right through the wafer. A third technique, laser scribing, vaporizes the silicon in the kerf (the unused area between the dies). The finely focused laser beam makes a sharper cut than diamond sawing and permits closer spacing of the dies, leading to a larger die count per wafer. Diamond sawing is the currently preferred practice, but laser scribing, which needs more expensive equipment, may gain first place in due course.

The mounting style of the die within the package is either single layer or multilayer. In the single-layer techniques the die is bonded mechanically to a lead-frame paddle or to the base of an embedded lead frame assembly. Fine wires are then connected from the bond pads on the die to the lead fingers. The package is subsequently constructed around the lead frame assembly, and the lead fingers are trimmed and shaped for either through hole (TH) or surface mount (SM) attachment.

The multilayer technique is based on a prefabricated package complete with the connections from its substrate on which the die is mounted, to the package I/O points. The die is mechanically bounded to the package substrate and connections are made from the die pads to the sub-

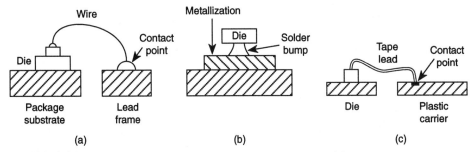

Fig. 2.24 Die interconnections (a) wire bond (b) flip-chip (c) TAB

strate contacts. The multilayer technique applies to flat packs and chip carriers, which have similar structures.

The three most widely used methods of internal package interconnections, sketched in Fig. 2.24, are wire bonding, flip chip (FC) solder bonding, also called controlled-collapse bonding (CCB), and tape-automated bonding (TAB). Wire bonding (Fig. 2.24(a)) is carried out after the die has been mechanically bonded to the appropriate part of the package. Gold or aluminium wires, about 1 mil in diameter, are bonded to the chip pads and the internal package contacts. For FC bonding solder bumps on the die are connected to the carrier or substrate contacts by a controlled collapse reflow soldering process, which was pioneered by IBM. FC bonding virtually eliminates chip-to-package interconnect lead inductance, and this is an important advantage in r.f. applications or operation near the maximum switching speed. Further advantages are that the entire die area, not just the periphery, can be used for die-to-substrate contacts, which are made simultaneously and not serially as in wire bonding. The process is very demanding on the flatness of the mounting surface. There is also the stress on the solder contacts caused by the differences in the coefficients of thermal expansion (CTE) between the silicon die and the package substrate. FC bonding is a demanding technology and has so far not been used on a large scale, but is likely to become more widespread in future, particularly for packages with high pin counts.

TAB was originated in the early days of SSI and MSI for low-cost mass production. The die (Fig. 2.24(c)) has solder bumps which are distributed on its periphery like the die contacts in Fig. 2.24(a). The conductors connecting the substrate to the package pins are formed by an etched copper pattern on a dielectric tape by means of a lithographical etching process similar to that used for the formation of IC metallization layers and PCB wires. Kapton (polymide) is the standard material for the dielectric. Automated production places the die in the central space left for it on the tape, and connects the die solder points to the inner points on the tape. This technique is called inner lead bonding. TAB is set to become the preferred interconnect technique. Apart from its suitability for economic high-volume production, it can operate with a smaller bond pad pitch than wire bonding. It already achieves a pitch of 4 mil compared with 6 mil for wire bonding, and a 2 mil capability is projected for the future (Hoffman 1988). For a 6-mil pitch and die substrate connections 0.1 in long, TAB has about one quarter the lead-to-lead capacitance and one tenth the resistance of wire bond (Burggraaf 1988).

Package encapsulation is generally either ceramic or molded plastic. Ceramic packaging is more expensive, but has the advantage of guaranteed hermetic sealing. Two types of ceramic encapsulation are in use, the pressed refractory ceramic package with a glass seal, used for CERDIP and CERQUAD packages, and the cofired laminated ceramic package, which constitutes the most reliable technology available at present. The former type is cheaper and has adequate reliability for most applications. Molded plastic encapsulation started off with a reputation of not being very reliable, because it allowed an easy ingression of moisture in the initial stage of its appearance. The moisture impaired chip performance by causing corrosion of the aluminium metallization. Several improvements have brought plastic encapsulation to a point where their reliability is much better, although they remain non-hermetic, allowing some penetration of moisture at a very much reduced rate. Hermeticity, or a condition very close to it, can be achieved by passivating the die with glass or oxide (usually some form of silicon nitride), and a plastic package has in such cases acceptable reliability for many applications (Sinnadurai 1985).

Ceramic and plastic materials, the former more than the latter, contain uranium or thorium, which radiate α particles. These cause *soft errors*, which were first reported in 1978 (May and Woods 1979). A soft error arises when an α particle emanating from the package sets a logic circuit of the chip into the wrong state. DRAMs, by virtue of their temporary data storage, are particularly susceptible to soft errors (see Chapter 7). The α radiation is greatly reduced by covering the die with an α-absorbing coating like silicon rubber or polyamide, and such a practice is often adopted, especially for DRAMs and SRAMs.

TH is increasingly giving way to SM attachment. SM packages are either leaded with very short leads about 2 mm long, or are supplied leadless with solder bump contacts. The leads have one of the three shapes shown in Fig. 2.25 with the J-shape being the most common. J leads result in a larger overall footprint than gull-wing leads, but a J-leaded package is easier to test. The butt-lead (also called I-lead) structure is confined to SO dual-in-line packages.

SM has many advantages over TH attachment. The footprint is greatly reduced, allowing an increase of between 50 and 70 per cent in PCB packaging density. That increase comes about not only because the TM package is smaller, but also because the plated through holes with their annular isolation spaces are eliminated. For equal dimensions the SM package achieves the same pin count with a higher lead pitch; alternatively it gives a much higher pin count for the same lead pitch. Furthermore SM packages can be mounted on both sides of the PCB.

SM is the technology of choice, but it has features which call for specialist expertise to realize its advantages. One key problem arises from the nature of the mounting: the solder joints provide both the electrical and mechanical attachment to the PCB. All leads on the package must be coplanar to better than 0.1 mm to ensure reliable and high-yield attachment of the package to the PCB. The mechanical strength of the solder joints is crucial to the reliability. Some problems arise in preserv-

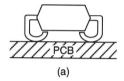
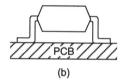
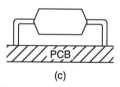

Fig. 2.25 SM package lead geometries (a) J lead (b) Gull-wing (c) Butt lead

ing the integrity of the solder joints under all conditions. TH mounting is, in this respect, superior because the mechanical strength of solder joints is often an order of magnitude higher than the minimum required, whereas the corresponding margin for SM attachment can be as low as 20 per cent.

Thermal stress puts a bigger strain on the solder joints of leadless packages, because the coefficients of thermal expansion (CTEs) for standard epoxy-glass PCB materials and ceramics differ greatly; typical values are (11–16) ppm/°C and (5–7) ppm/°C respectively. The problem can be contained by changing over to special PCB materials. The CTE of a PCB is determined by that of the core, and compounds like copper-invar-copper, copper-clad molybdenum, and polyamide Kevlar have CTEs in the range (5–7) ppm/°C. Leaded ceramic packages have sufficient lead compliance to take up the stress caused by the disparate CTEs of the package and a standard PCB. The choice of whether to use leadless or leaded SM packages is largely governed by economic considerations. Leadless packages are inherently cheaper, easier to handle, and do not require lead coplanarity, but have to be mounted on more expensive special PCBs.

The PCB constitutes a heat sink for SM packages. These, taking typical flat packs and chip carriers of the type shown in Fig. 2.22, have a maximum power dissipation in free mounting of 1–3 W. Two types of construction are adopted for chip carriers and flat packs, the cavity-up and the cavity-down structures. The heat dissipation within the die takes place mainly at its back. The cavity-up package of Fig. 2.26(a) is used where the heat sink capability of the PCB is adequate. The cavity-down package shown in Fig. 2.26(b) has the rear of the die close to the upper surface of the package. The maximum power dissipation can now be extended by forced air-flow and/or a special heat sink mounted on top of the package, from about 2 to 10 W or more.

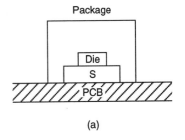

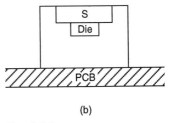

Fig. 2.26 Cavity-up and cavity-down packages (a) cavity up (b) cavity down

2.6.4 Package selection

Package choice is largely, but not entirely determined by economic factors, which extend to PCB production and package mounting. The semiconductor vendor decides on the package style for VLSI chips, offering occasionally alternatives. For example the Intel 486SX microprocessor is available in either a 196-lead QFP or a 168-pin PGA, allowing the user the alternatives of TH or SM attachment. The choice between plastic and the more expensive ceramic encapsulation is governed by the degree of reliability required for the specified operating conditions. The stringent demands for most military and aerospace applications call for ceramic packages. The great majority of commercial applications are satisfied with plastic packages, which also suffice for a significant proportion of military equipments.

Flat packs and chip carriers are both widely used for terminal counts up to about 200, with flat placks extending to higher counts. The PGA continues to lead with the highest I/O, extending at present to beyond 500. It is also widely used at much lower counts, partly because the TH mode is cheap and and well established, and partly because it occupies a smaller footprint than other packages, since the pins are configured in an array and are not confined to the perimeter.

A choice is usually available for SSI, MSI, and semiconductor memory chips (which have a low I/O). The alternatives are to stay with the DIP or to go over to one of the SM packages. On occasions all or most of these SM alternatives are available. A manufacturer of equipment containing little or no VLSI may prefer to stay with DIPs and PGAs as much as possible, in order to maximize on the cheaper and more familiar TH technology. For systems with a substantial mix, SM packages will nearly always be preferred. Their smaller footprint and lead impedance permit faster operation. Flat packs and chip carriers with low pin counts are now generally square-shaped; in their earlier days they were mainly rectangular. The internal lead lengths of rectangular packages are markedly uneven. That disparity is greatly reduced in square-shaped packages, which minimize the skew between internal paths. The latest silicon logic marketed in SSI and MSI, the Motorola ECLinPS family, has a 28-pin square-shaped PLCC package.

ASIC vendors usually offer a generous package choice, which embraces most of the varieties available. The selection is influenced by much the same considerations which apply to standard products, with which they are likely to be combined. The DIP retains its popularity for low pin counts (40 or less) and, together with the PGA, permits the most cost-effective solution for moderate production quantities. SM packages are chosen for medium and large-scale production. Plastic and ceramic flat packs and chip carriers for ASICs are available over a wide range of pin counts, with flat packs reaching a higher maximum.

Last there is the matter of power dissipation. The ASIC vendor often gives the choice of a package in cavity-up or cavity-down orientation, and probably provides a suitable heat sink for the latter. The cavity-up, cavity-down alternative may be made available in a flat pack, a chip carrier, or a pin grid array. The terminal count of a flat pack or a chip carrier is unaffected by the cavity orientation, but with a pin grid array this is not the case. The terminal count for a cavity-down PGA is reduced, because the pins are on the same side as the cavity. Modifications of existing flat packs have emerged to give higher dissipation capability. One of these, the MQUAD flat pack (also called MQUAD, see Fig. 2.22) is similar to the PQFP except for the mould process, which is replaced by laminating an anodized aluminium cap and base to the lead frame with a special epoxy film. A silver-filled epoxy attaches the paddle of the lead frame, which holds the die, to the base of the package, giving it exceptionally good thermal conductivity. Another technique to enhance power capability is the MicroCool QFP, in which a heat slug is attached to an internal miniature PCB which supports a copper lead frame. This QFP is, like many other flat packs with very fine pitch for high density I/O, supported within a moulded carrier ring (MCR) to maintain pin coplanarity (Hamilton *et al.* 1989).

Presently there is a lack of adequately defined standards for packages other than the DIP. Some broad measures of agreement between the major vendors do exist, and standards are being drawn up to aid compatibility within the market. The leading US authority for standards, the former Joint Electron Devices Engineering Council (JEDEC), has now been incorporated within the Electronics Industries Association (EIA). The standards tend to be qualified by the acronym JEDEC rather than by EIA. This is an advantage, because it avoids confusion between EIA and the Electronics Industries Association of Japan (EIAJ). Joint standards are being agreed between EIA and EIAJ, and have already been drawn up for the CQFP, QFP, MQUAD and MicroCool QUAD packages.

2.6.5 Multichip modules

The wire dichotomy between die and PCB stamps the nature of the development being addressed in this section. The pitch of PCB wires has been decreased from 0.1 in, the veteran standard still very much in use, to the finer pitch of SM packages, where needed. The lower limit for reliable operation is 0.01 in for conformally coated, and 0.025 in for uncoated PCBs (Bergman and Ginsberg 1988). Multilevel PCBs, just like multilevel metallization within the die, aid connectivity. Modern multilayer PCBs have from 4 to 10 circuit layers; special, very costly PCBs with up to 60 layers are available. In general the four-layer PCB is the most popular. The wire pitch within a die containing 1 μm structures is about 4 μm. The large pitch ratio for PCB to die wires, ranging from 64 to 640 for the figures just given, is reflected in the large area taken up by the wires on the PCB. The driving force behind packaging development is to minimize that area.

The wires and the spaces between them account for by far the largest part of the PCB surface. The multichip module (MCM) is a means of greatly reducing the surface demanded by packages and their interconnections. It achieves this by containing a number of chips, typically four to six, interconnected within the module by wires which have a far smaller pitch than the PCB wires. The chips within the module are mounted on a substrate which has a multilayer structure like the multilayer PCB. The module, measuring up to about 10 in^2, constitutes a subsystem mounted on the PCB and interconnected via its terminals. It is in effect a miniature PCB within a PCB, which may and often does contain several MCMs. The fine pitch of the wires within the MCM, made possible by thin film and thick film technology, greatly reduces the area taken up by the interconnections. The length of the wires is also greatly reduced and this increases the operating speed.

The definition of what constitutes a MCM, sometimes labelled multichip package, can be inferred from the information just given. It is a connection of two or more bare chips on a single substrate, contained in plastic or ceramic encapsulation. That definition has to be stretched, because in 1992 very few manufacturers were bonding bare chips to the MCM substrate; the chips were nearly all packaged. Bare bonding is, however, becoming established. The composition of some hybrid ICs brings them into the MCM orbit in accordance with the above definiton. However the MCM technology described in this section applies to all-digital and mixed-mode LSI and VLSI chips, assembled in a module in order to achieve the highest possible function density on a PCB.

The effectiveness of a MCM is reflected in the values of two parameters, the packing efficiency P_{eff} and the wire density W_d. P_{eff} is given by

$$P_{eff} = \frac{\text{die area}}{\text{substrate packaging area}} \quad (2.25)$$

66 Fabrication and packaging

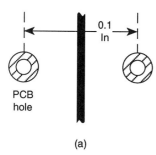

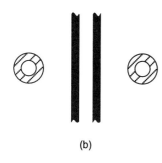

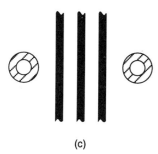

Fig. 2.27 PCB tracks (a) single track (b) two tracks (c) three tracks

For PCB attachment, P_{eff} comes to a few per cent for DIP, and to about 10 per cent for SM packages. Its value within an MCM approaches 50 per cent (Messner 1987). The advantage of the MCM in respect of P_{eff} is evident. W_d is given by

$$W_d = \frac{\text{number of wires per unit length}}{\text{unit area}} \quad (2.26)$$

W_d is expressed either inch/inch^2 or cm/cm^2. Figure 2.27, which illustrates single, dual, and triple tracks running between the plated through holes of a PCB, is an aid for interpreting the formulation of W_D. Assuming uniform conductor and space widths across the PCB W_D, for single, dual, and triple tracks comes to 10, 20, and 30 inch/inch^2 respectively. The smallest acceptable PCB wire pitch, quoted earlier in this section, is 0.01 in, but the smallest value freely used is about 0.025 in, which gives W_d equal to 40 inch/inch^2. For standard PCBs W_d falls in the range 10 to 40 inch/inch^2. Line widths and pitches of wires in MCMs with the smallest interconnect dimensions are about 15 to 25 μm and 50 to 75 μm respectively (Hilbert and Rathmell 1990), leading to W_d in the range 133 to 200 inch/inch^2. The values of W_d just quoted for PCB and MCM wires apply to a single layer; the effective values for multilayer interconnects will be much larger. They indicate the great saving in wire area obtainable with a MCM.

Historically the MCM first appeared in production in 1980 (IBM) and 1984 (NEC and Mitshubi). MCM technology was expensive in those days, but the increase in function density and operating speed justified the cost. The general adoption of the MCM depended on a critical cost analysis. The view was held in those early days that silicon wire was believed to be far cheaper than wire on a laminated PCB or on a ceramic substrate (in hybrid ICs). Knausenberger and Schaper (1984) showed that the normalized interconnection costs, i.e. the cost of wire per unit length, were much the same in all cases. If silicon wire appeared to be cheaper that was only because wire lengths within the die were much shorter. Once the length of the wires interconnecting the chips could be reduced, the cost would decrease proportionally. A further cost reduction would accrue by bare attachment of the chips (chip on board, COB), and this is now becoming established in MCMs.

MCMs are divided into three categories according to the composition of their substrates, ceramic (MCM-C), laminated PCB (MCM-L), and deposited dielectric (MCM-D). The MCM-C type, the oldest of the three, is an adaptation of hybrid IC practice. Dies are attached either bare (COB) or packaged to the substrate. The MCM-L category provides greater chip density and is designed for automated manufacture. Bare die attachment is in the flip-chip or TAB mode. The latest type, the MCM-D has a substrate with a silicon surface for bare die attachment. The first layer of interconnect is formed on the substrate just as it is on the metallization covering the substrate of a chip. An insulator with a low dielectric constant, usually polyimide, isolates the silicon from the next conducting layer. The MCM-D permits a finer pitch of wire than the other two types, and is likely to become the category of choice for larger systems like mainframe components. Various characteristics of the

Table 2.14 Typical MCM characteristics

Characteristic	MCM-L	MCM-C	MCM-D
Technology	Laminated high density PCB	Cofired low dielectric constant ceramic substrate	Silicon die on silicon, thin film, low dielectric constant
W_d (cm/cm^2)	300	800	250–750
Dielectric constant	3.7–4.5	5–5.9	3.5
Pin arrangement (Pitch -mm)	Array (2.54)	Array staggered (1.00–2.54)	Peripheral (0.63)

three types are given in Table 2.14 (Tummala 1992).

Multichip packaging is probably the fastest growing of all packaging technologies. It leads to a reduction by a factor of 5 to 10 in PCB footprint relative to SM attachment, to faster operating speed, and to higher reliability. The overall cost using MCMs (provided bare chip attachment is practiced) is expected to be lower than the cost of implementing a system with single-chip PCB attachment. Confined at first to large computers, MCM technology is now set to spread into consumer and industrial electronics in general.

2.7 Overview

The relentless pursuit of shrinking device dimensions in order to increase chip function density continues. The economic case for doing so, outlined in Chapter 1, is overwhelming. VLSI is (mid-1995) below the micron-submicron boundary, with a standard feature size $\sim$0.8 μm. Some semiconductor memories in full-scale production already have geometries of $\sim$(0.5–0.6) μm, and these are set to become universal for VLSI within the next two to three years. The major thrust is now directed towards feature sizes in the range 0.25 to 0.35 μm.

One cannot but marvel at the resilience of optical lithography. Written off by many years ago for geometries below 2 to 3 μm, optical lithography is not only the mainstay of todays structures, but is confidently expected to achieve feature sizes down to 0.25 μm before long. Much ingenuity and inventiveness are behind that expectation. One innovative idea deserves to be mentioned, the phase shifting mask, which holds out the prospect of achieving a feature size just below 0.2 μm (Lin 1993).

Lithographical capability has to be matched by that of the other processes in IC fabrication i.e. etching, diffusion, implantation etc. These, for geometries down to $\sim$2 μm, could be taken for granted in the sense that they could follow suit. However with geometries of one micron or less, special developments are called for—that applies particularly to etching—so that the advances in lithographical shrinkage can be fully exploited. Every step in the IC fabrication process is a major undertaking, rooted in the mastery of the underlying chemical and physical processes. The enormity of the fabrication effort, which continues to increase in complexity with the penetration into deep submicron VLSI, means that the number of IC manufacturers is going to remain static, or possibly even decrease. When ASICs began to emerge in the early 1980s, silicon foundries supplying them sprang up like mushrooms. They faded away almost as fast as they emerged, and the production of ICs, whether for standard or ASIC chips, is now the prerogative of a comparatively small number of big players, small that is in relation to the huge world market for such products. The processing issues for silicon hold, with minor modifications, for gallium arsenide.

The pre-eminence of VLSI in design and production raises some questions about the role of lower-level ICs. These will always be needed, even if they present a decreasing proportion of the total IC consumption. Silicon foundries will continue to supply them according to demand. The advanced techniques for micron and submicron structures enable the vendors to use the same equipment for fabricating ICs with the larger geometries, which are deployed in SSI/MSI logic and many analogue ICs. The complexity of todays submicron chips, reflected by the number of mask levels, typically between 10 and 15, is used not only for the production of all-digital bipolar or CMOS VLSI, but also for structures which are not confined to VLSI. BiCMOS, a technology used in digital, analogue and mixed-mode chips, is an outstanding example. Another relatively new structure, demanding about a dozen mask levels, is a combination of npn and pnp transistors with high current gain and very good high frequency characteristics. Hitherto the standard bipolar process (see Section 2.2) only yielded pnp transistors with very low current gain and a poor frequency response.

At present IC processsing is ahead of packaging technology. Until VLSI got going, packages with moderate pin counts were adequate. The strain on the DIP began to tell with the arrival of 16-bit microprocessors. The 64-pin DIP, developed for that purpose, stretched the capability of that package to its upper limit. It soon became evident that new approaches to packaging were needed in order to exploit the capabilities of VLSI. SM packages were introduced for the higher I/O counts, and also—to realize their advantages to the full—at low counts as an alternative to the DIP. The result is that an electronic system which previously occupied many PCBs, together with their interconnecting back panels, can now be replaced by a single module. Wafer scale integration (WSI) became a favourite topic some years ago. The intention was to place the entire circuits of a system on a single wafer, which would be in effect a giant IC. Technological difficulties have precluded WSI from becoming a serious contender in VLSI, and WSI is not likely to be in production in the foreseeable future.

Progress in packaging has taken a new turn with the emergence of the MCM, which dramatically decreases space and weight relative to traditional PCB technology, and is a major advance rapidly gaining momentum. Its most striking outlet is where it made its first impact, in mainframe and supercomputers, but it is set to pervade the entire field of electronic equipment. The advances in packaging technology are due to many specific endeavours. The thermal management of single-chip and multichip packages alone is a major technological undertaking.

Various other issues which impact on IC engineering have emerged in this chapter. First and foremost of these is the dominance of wire. The topology of wire is the critical task for chip design. It has been encountered in the multilevel metallization of a chip, in PCB layout and in MCM interconnects. The placement of the wires in all these cases is achieved with computer aided design (CAD). Specially developed software design tools are available for that purpose. The dominant role of wire extends to other sectors of IC engineering, and comes up in many other parts of this text.

Another thread which runs through all facets of IC engineering is simulation. Process simulation is carried out to establish the operating conditions—all computer controlled—for the various fabrication steps like lithography, etching, implantation, etc. Based on models which incorporate the accumulated evidence of extensive experience, the simulations speed up the formulation of the conditions for the various processes and their modifications with a high level of confidence, and reduce the cost of experimental work to optimize IC fabrication.

Virtually all aspects of IC engineering require a background on IC fabrication and packaging, and frequent reference to the material contained in this chapter is made throughout the book.

References

Adams, A.C. and Chang, G.C. (1980). The growth and characterization of very thin silicon oxide films. *Journal of the Electrochemical Society*, **127**, 1787.

Amey, D.I. (1989). In *VLSI Handbook* (ed. J. DiGiacomo), pp.23.3–23.4. McGraw-Hill, USA.

Bean, J.C. (1981). In *Impurity doping processes* (ed. F.Y. Wang), pp.175–215. North-Holland, Amsterdam.

Bergman, D.W. and Ginsberg, G.L. (1988). In *Printed circuits handbook*, (3rd edn, ed. C.F. Coombs), p.5.5. McGraw-Hill, USA.

Brice, J.C. (1986). *Crystal growth processes*. Wiley, New York.

Burggraaf, P. (1988). TAB for high i/o and high speed. *Semiconductor International*, **9**, (6), 72–7.

Deal, B.E. (1980). Standardized terminology for oxide charges associated with thermally oxidized silicon. *IEEE Transactions on Electron Devices*, **27**, 606.

Elliott, D. (1989). *IC fabrication technology*, (2nd edn), p.20. McGraw-Hill, USA.

Flores, G.E. and Kirkpatrick, B. (1991). Optical lithography stalks X-rays. *IEEE Spectrum*, **28**, (10), 24–7.

Glaser, A.B. and Subak-Sharpe, G.E. (1977). *Integrated circuit engineering*, pp.223–4. Addison Wesley, USA.

Gottscho, R. (1993). Plasmas make progress. *Physics World,*. **6**, (3), 39–45.

Hamilton, J., McShane, M., Bigler, C., Casto, J., and Lin, P. (1989). *Molded carrier rings for fine pitch surface mount packages*. Proceedings of IEEE 39th Electronic Component Conference, May 1989, 504.

Hammond, M. (1978) Silicon epitaxy. *Solid State Technology*, **21**, (11), 68–75.

Hashimoto, C., Muramato, S., Shiomo, N., and Nakajiama, O. (1980). A method of forming thin and high reliable gate oxides. *Journal of the Electrochemical Society*, **127**, 129.

Hilbert, C. and Rathmell, C. (1990). *Design and testing of high density interconnection substrates*. Proceedings NEPCON West, pp.567–579.

Hirayama, M., Miyoshi, H., Tsubouchi, N., and Abe, H. (1982) High pressure oxidation for thin gate insulator processes. *IEEE Transactions on Electron Devices*, **29**, 503.

Hoffman, P. (1988). TAB implementation and trends. *Solid State Technology*, **31**, (6), 86.

Hohn, F. (1993). Optical lithography forges ahead. *Physics World*, **6**, (3), 33–7.

Hollis, E.E. (1987). *Design of VLSI gate array ICs*, p.215. Prentice-Hall, USA.

Kennedy, D.P. and O'Brien, R.R. (1965). Analysis of the impurity atom distribution near the diffusion mark for a planar p-n junction. *IBM Journal of Research and Development*, **9**, 179–86.

Knausenberger, W.H. and Schaper, L.W. (1984). Interconnection costs of various substrates—the myth of cheap wire. *IEEE Transactions on Components, Hybrids, and Manufacturing Technology*, **7**, (9), 261–7.

Landman, B.S. and Russo, L.R. (1971). On a pin vs. block relationship for partitions of logic graphs. *IEEE Transactions on Computers*, **20**, 1469–79.

Lee, D.H. and Meyer, J.W. (1974). Ion implanted semiconductor devices. *Proceedings of the IEEE*, **62**, 1242–55.

Lin, B.J. (1993). Phase-shifting masks gain an edge. *IEEE Circuits and Devices Magazine*, **9**. (2), 28–35.

May, T.C. and Woods, M.H. (1979). Alpha particle induced soft errors in dynamic memories. *IEEE Transactions on Electron Devices*, **26**, 2-9.

Messner, G. (1987). Cost density analysis of interconnections. *IEEE Transactions on Components, Hybrids, and Manufacturing Technology*, **10**, (6), 143–151.

Nicollian, E.H. and Brews, J.R. (1982). *MOS physics and technology*, pp.371–422. Wiley, USA.

Pinnel, M.R. and Knausenberger, W.H. (1987). Introduction system requirements and modelling. *AT & T Technical Journal*, **66**, (7/8), 48.

Rea, S.N. (1981). Czochralski silicon pull rate limits. *Journal of Crystal Growth*, **54**, 267.

Schmidt, D.C. (1982). Circuit pack parameter estimation using Rent's rule. *IEEE Transactions on Computer-Aided Design of Integrated Circuits and Systems*, **1**, (4), 186–92.

Sinnadurai, F.N. (ed.) (1985). *Handbook of microelectronic packaging and interconnection technologies*, p.4. Electrochemical Publications, Scotland.

Snow, E.H., Grove, A.S., Deal, B.E., and Sah, C.T. (1965). Ion transport phenomena in insulating films. *Journal of Applied Physics*, **36**, 1664.

Striny, K.M. (1991). In *Electronic packaging and interconnection handbook* (ed. C.A. Harper), p.6.43. McGraw-Hill, USA.

Tummala, R.R. (1992). Multichip packaging—a tutorial. *Proceedings of the IEEE*, **80**, 1924–41.

Uyemara, J.P. (1988). *Fundamentals of MOS digital integrated circuits*, pp.267–9. Addison-Wesley, USA.

3
Component formation

3.1 Introduction

This chapter complements Chapter 2 by describing the structures of transistors and passive components without going into the processing steps, except very briefly on occasions. Transistors far outweigh passive components i.e. resistors and capacitors. CMOS and nMOS logic circuits contain no passive components; these are confined to bipolar logic and to analogue circuits.

The structures described in this chapter are typical for current VLSI with feature sizes down to about 0.8 μm. Such dimensions are not exclusive to VLSI. They are also found increasingly at the lower levels of SSI and MSI on account of their speed and power advantages over transistors with larger geometries. The importance of VLSI feature size and its continuing shrinkage must not be allowed to obscure the existence of transistors produced with larger geometries when called for. The buffer output stages which drive PCB wires with their relatively large capacitances are a case in point. The descriptions are qualitative rather than quantitative with the object of presenting an overview, followed by more details, when called for, in later chapters.

3.2 CMOS

The profile of an n-well CMOS inverter is shown in Fig. 3.1. The inverter is the foundation for CMOS logic. Figure 3.1 has a p substrate and an n well in place of the n substrate and p well for the CMOS pair in Fig. 2.6 of Chapter 2. The change has been made to illustrate that both of these structures are found in practice.

CMOS is susceptible to *latch-up*, a parasitic effect in which a virtual short circuit between the supply line and ground causes local destruction of the stage affected or even self destruction of the chip. The phenomenon is due to the two parasitic transistors Q_1 and Q_2 in Fig. 3.1(a). These operate like the silicon controlled rectifier (SCR) in Fig.3.1(b), which includes two bulk resistances. The three conditions for latch-up are that Q_1 and Q_2 are forward biased, that their common-base current gains α_1 and α_2 satisfy the inequality

$$\alpha_1 + \alpha_2 \geq 1 \qquad (3.1)$$

and that the power supply can furnish a current in excess of the holding current (Troutman 1986). When α_1 and α_2 are both well below unity, the inequality of eqn. (3.1) can be expressed in the form

$$h_{fe1}.h_{fe2} \geq 1 \qquad (3.2)$$

74 Component formation

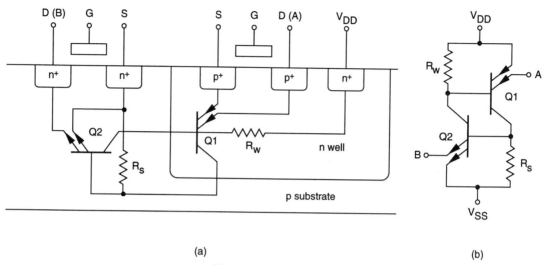

Fig. 3.1 CMOS latch-up (a) profile (b) equivalent circuit

where h_{fe1} and h_{fe2} are the common-emitter current gains. A detailed representation of all parasitic components is more complex, but Fig. 3.1 contains the essential components for explaining latch-up.

In normal operation Q_1 and Q_2 are off with zero base-emitter voltage. Latch-up can be initiated by a current sufficient to cause a voltage build-up of about 0.7 V across R_w or R_s, the bulk resistances of the n-well and substrate sections indicated in Fig. 3.1. Alternatively Q_1 and Q_2 may be put into conduction if the output goes more than about 0.7 V above V_{DD} or below V_{SS} (usually ground). Overshoots and undershoots, electrostatic discharges and noise spikes can all cause such excursions. A similar situation may arise when a signal is applied before the supply voltage has built up to its full value during switch on. Input/output (I/O) structures are particularly prone to latch-up because of the abnormal circuit voltages which may be encountered.

The minimization, better still the complete elimination of latch-up susceptibility is of paramount importance in the design of CMOS structures. Measures to prevent latch-up are taken at the layout design level, the circuit level, and the process level. The lateral spacing between the n^+ and p^+ active (source and drain) regions of the two CMOS transistors is critical. Reduced separation increases transistor density, but also increases the danger of latch-up, largely because of an increase in current gain of the lateral transistor Q_1. Reduction of R_s and R_w is of great help. If one or both of these resistors are small enough to prevent a build up of about 0.7 V across them, latch-up cannot occur. R_s is effectively lowered by using an epitaxial layer on a highly doped substrate, and R_w is decreased by placing a heavily doped n^+ guard ring around the n-well in Fig. 3.1(a). A similar p^+ guard ring is often used to surround source and drain of the pMOSFET. Between them the guard rings of the CMOS process shown in Fig. 3.2 reduce the emitter injection efficiency and base transport factor of Q_1 and Q_2 in Fig. 3.1, thereby decreasing their current gain. Appropriate bias conditions are obtained by taking the p^+ and n^+ guardrings to V_{SS} (usually ground) and V_{DD} respectively.

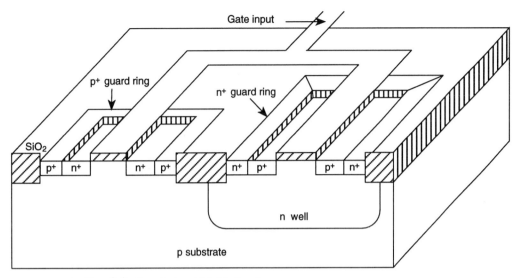

Fig. 3.2 Guard ring CMOS structure

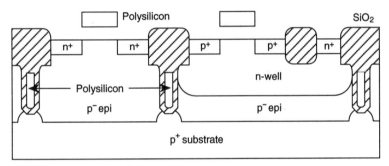

Fig. 3.3 Deep trench isolation CMOS

Two other structures greatly reduce susceptibility to latch-up. The first of these is deep trench isolation, shown in Fig. 3.3 where the trench extends into the substrate (Brown et al. 1986). The n$^+$ contact of the n-well ensures a good ohmic connection to the V_{DD} supply and presents a barrier to lateral carrier flow. The preferred CMOS process is the twin-tub (also called twin-well) structure shown in Fig. 3.4, with the earlier oxide-isolated version of Fig. 3.4(a) giving way to the trench isolation illustrated in Fig. 3.4(b) (Weste and Eshragian 1985; Foster 1987). The CMOS process in Fig. 3.4(b) includes polycide gates and silicide wires from these gates. That practice, described in Section 2.4.6, reduces the resistance of the gate wires by between one and two orders of magnitude relative to polysilicon gates with polysilicon wires. Another reduction in resistance, this time of the active n$^+$ and p$^+$ layers, is obtained by silicide metallization formed beneath a tungsten interface to the first interconnect level with the substrate. The twin-tub process permits independent optimization of the electrical characteristics for the p- and n-type transistors, and avoids the compromises inherent in the n-well or p-well structures. Reinforcing the twin-tub structure with deep trench isolation gives the ultimate protection against latch-up. The claim, sometimes advanced in the literature, namely that latch-up is completely

76 Component formation

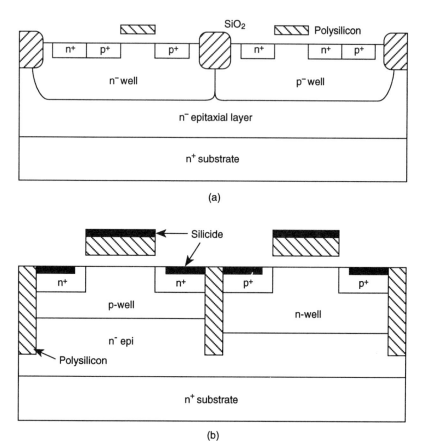

Fig. 3.4 Twin tub CMOS (a) oxide isolated (b) trench isolated

eliminated in the process shown in Fig. 3.4(b), is overambitious, but the probability of a latch-up has certainly been tremendously reduced. Ideally the single-channel MOSFET and the complementary MOSFET pair in CMOS are self-isolating, but oxide or trench isolation is always incorporated in practice to prevent latch-up and to inhibit other effects due to unwanted lateral action. One advantage of twin-tub CMOS is that the substrate foundation and the epitaxial layer grown on it can be either n^+, n as in Fig. 3.4 or p^+, p. This flexibility gives a choice of substrate according to the use of the circuit. A p-type substrate is used when CMOS is combined with nMOS, for example in DRAMs (see Chapter 7).

CMOS and single-channel MOS (for all practical purposes nMOS) circuits have input protection against electrostatic discharge (ESD) built into the device structure. ESD potentials can reach several kV, and the greatest danger they present is to the gate oxide. The critical field for silicon dioxide is $\sim 7 \times 10^6$ V/cm, giving a maximum gate voltage of 14 V for a gate oxide thickness of 200 Å. The function of the input pro-

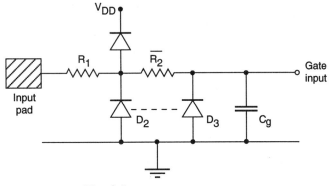

Fig. 3.5 CMOS input protection

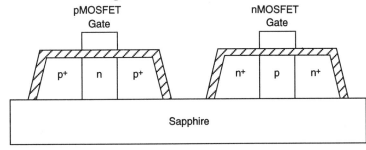

Fig. 3.6 SOS CMOS

tection is to reduce an ESD to a safe level by the time it reaches the gate. It also has to protect against the second destructive mechanism of ESD, the burnout of p-n junctions or polysilicon links. The standard input protection circuits for CMOS is shown in Fig. 3.5; such protection is provided for each input pad on the chip. Diodes D_2 and D_3, and $\overline{R_2}$ are, depending on the technology, intrinsic components represented by a distributed network. Alternatively D_2 is a lumped component, and the junction of D_1 and D_2 is taken to the gate i.e. D_3 and $\overline{R_2}$ are omitted. R_1 is preferably a polysilicon resistor in a p-well process, where a diffused resistor is likely to result in heavier charge injection in the substrate, leading to increased latch-up susceptibility. A diffused resistor is acceptable in an n-well structure.

All ICs described so far have a silicon substrate. An alternative, silicon-on-sapphire (SOS) CMOS, is illustrated in Fig. 3.6. It is an example, in fact the only significant commercial example, of silicon-on-insulator (SOI) CMOS. SOS CMOS has several outstanding advantages. Latch-up susceptibility has been completely removed by virtue of the insulating substrate. The absence of the wells in bulk CMOS permits denser structures, and the reduced capacitances increase circuit speed. Last but not least SOS CMOS has greatly increased resistance to radiation. On the debit side, SOS MOS technology is very expensive, and input protection is more difficult because the substrate is not available for diode formation.

The structure of an nMOSFET is shown in Fig. 3.7. The device is far less complex than the twin-transistor combination of CMOS. The

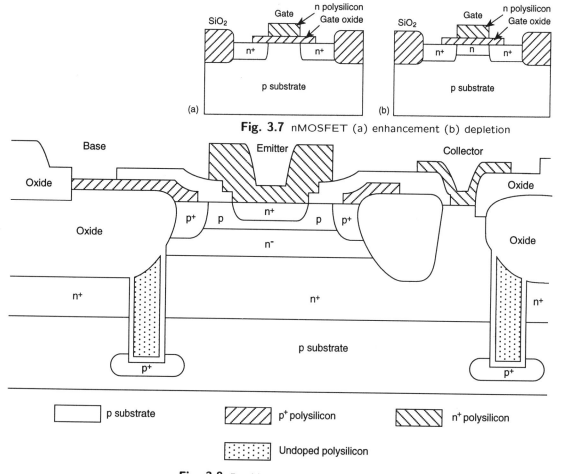

Fig. 3.7 nMOSFET (a) enhancement (b) depletion

Fig. 3.8 Double polysilicon trench isolated bipolar transistor (Courtesy of GEC Plessey Semiconductors)

nMOSFET in Fig. 3.7(a) operates, like all CMOS transistors, in the enhancement mode. It is in the off state for zero gate-source bias and is turned into conduction by a positive gate-source voltage. The depletion mode nMOSFET in Fig. 3.7(b) has an n-implanted channel, giving conduction for zero gate-source voltage and requiring a negative gate-source bias to be turned off.

3.3 Silicon bipolar transistors

The profile of a modern bipolar transistor for geometries in the range (0.5–1.0) μm is shown in Fig. 3.8 (Hunt and Saul 1988). It bears some resemblance to Fig. 1.4(a), but includes numerous refinements. The structure is a self-aligned process which has undergone some variations during the last decade, but is based on the original technique described by Lohstroh (1981). Separation between emitter and base contacts is greatly reduced. Trench isolation demands far less silicon estate than oxide isolation. The polysilicon emitter has become an established feature of modern bipolar IC transistors. Its advantage, a great increase in

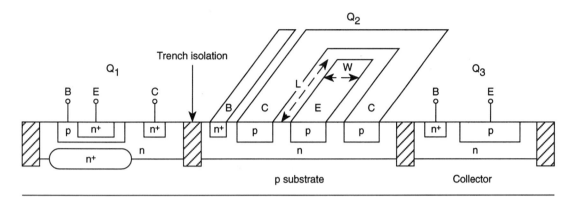

Fig. 3.9 pnp transistor formation

current gain, is discussed more fully in Chapter 12. The double polysilicon structure with p⁺ polysilicon connections to the active base contacts reduces the collector-base area and hence the collector-base capacitance. Figure 3.8 highlights the complex nature of current technology. For equal geometries, an npn is faster than a pnp transistor, and bipolar logic circuits are implemented with npn transistors (and diodes). That position cannot be maintained for analogue circuits where, for example, dc coupling of a multi-stage amplifier is only practicable by combining npn and pnp transistors. The structure of npn transistors is readily adapted for producing pnp transistors simultaneously. The formation of such pnp transistors is shown in Fig. 3.9. The process illustrated applies equally to oxide or reverse-biased junction isolation. The vital feature is that two pnp transistors Q_2 and Q_3 have been formed *without any additional processing steps*. Take the lateral pnp transistor Q_2 first, illustrated with an extended profile to help in understanding its operation. The central emitter stripe and the collector ring surrounding it are formed simultaneously with the p layers for Q_1 base and Q_3 emitter. Likewise there is a simultaneous formation of the four n⁺ layers (Q_1 collector contact, Q_1 emitter, Q_2 base contact, and Q_3 base contact).

The limitations in performance of the pnp transistors in Fig. 3.9 arise from the fact that the processing is directed at optimizing the characteristics of the npn transistors. The designer has a choice over the length and width of the p electrodes of Q_2 and Q_3, but their depth and composition are determined by the requirements for Q_1. Q_1 has a very low current gain: h_{fe} is typically in the range 2 to 10. The high frequency response is also very poor with an f_T of a few MHz compared with an f_T of several GHz for Q_1. This disparate response is caused by the difference in base width. The base width of Q_1 is 0.5 μm or less, that of Q_2 is the feature size plus a margin to allow for production tolerances, lateral spreads etc. For a 5 μm geometry the base width of Q_2 will be about 10 μm. Likewise Q_3, a vertical pnp transistor, has a large base width, in this case the difference between the n epitaxial and p layer depths. Q_3 is constrained to the emitter follower configuration, because the substrate, which is connected to the most negative potential of the

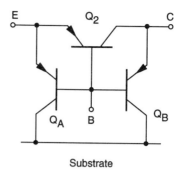

Fig. 3.10 Q_2 (Fig 3.9) and parasitic transistors

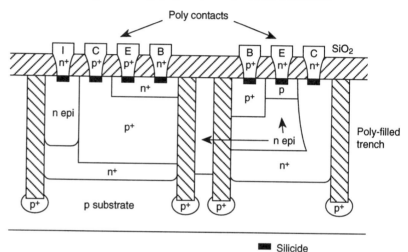

Fig. 3.11 npn-pnp vertical transistor combination (Courtesy of GEC Plessey Semiconductors)

circuit, is grounded for signals. Q_3 achieves a higher current gain than Q_2, with h_{fe} reaching values up to 50. Its main application is in emitter follower output stages of amplifiers.

The equivalent circuit representation of Q_2 in Fig. 3.10 shows that it is accompanied by two vertical pnp transistors, Q_A and Q_B. Q_A is the more important of these, because it will conduct when Q_2 is on, whereas Q_B will be off. The attendant reduction in the current gain of Q_2 is minimized by surrounding its emitter with a collector ring, and by making W in Fig. 3.9 as small as possible in order to get the largest ratio of side wall to bottom area. In spite of their limitations, lateral and vertical pnp transistors like Q_2 and Q_3 in Fig. 3.9 are freely used in analogue circuits.

The technology for pnp transistors and their characteristics have greatly improved in recent years for two reasons. The reduction in feature size leads to structures with comparable lateral and vertical dimensions of electrode layers. Second the ability to make sophisticated device structures with processes involving between 10 and 15 mask levels, thanks to the overlay accuracy attainable with wafer steppers, (Section 2.5) has made it possible to produce vertical npn and pnp high performance structures, which have now come on the market. Figure 3.11

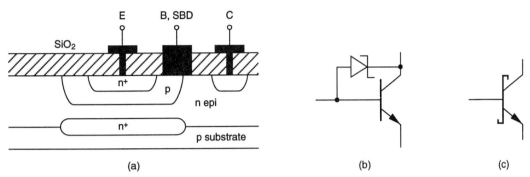

Fig. 3.12 Schottky clamped transistor (a) profile (b) equivalent circuit (c) circuit symbol

gives an example of such a process (GEC Plessey Semiconductors 1993). The f_T and h_{fe} values of the transistors are 2.5 GHz, 50 (pnp), and 10 GHz, 100 (npn) respectively. The dielectric isolation has to be supplemented by a p^+/n (epi) junction isolation, which is reverse biased by connecting I to the most positive potential of the circuit.

Diode formation is straightforward. It is common practice to obtain diodes by joining the base and collector of an npn transistor to form the anode; diodes can be obtained similarly using pnp transistors. One special type of diode, the Schottky diode, is particularly important because of its use in bipolar and GaAs logic circuits. The Schottky diode (alternatively Schottky-barrier diode, SBD) is a metal-semiconductor junction. A Schottky clamped transistor, in which a Schottky diode is in parallel with the base-collector junction, is shown in Fig. 3.12. Assuming the metallization to be aluminium, which is p-type, all metal-p electrode contacts are ohmic, but metal-n electrode contacts are Schottky diodes. The forward voltage of the Schottky diode depends largely on the n impurity concentration. Aluminium-n type junctions are virtually ohmic for heavy (n^+) doping and have been described as such in Section 2.4.6. The Schottky barrier effect becomes marked at low and medium n impurity levels, and in bipolar digital circuits SBDs have a forward voltage about 400 mV less than that of a p-n junction diode. The SBD is a majority carrier device, and the absence of minority carrier charge storage greatly improves the switching speed of circuits which use it.

3.4 BiCMOS

The combination of bipolar and CMOS transistors on a chip, BiCMOS, is a leading technology for both analogue and digital circuits. BiCMOS has, like the high performance vertical npn - pnp combinations in Fig. 3.11, become possible thanks to the capability of processing with a large number of mask levels. The BiCMOS structure in Fig. 3.13 typifies a process adopted for analogue circuits (Gray and Meyer 1993), and is representative of BiCMOS technology in general. The submicron BiCMOS structure in Fig. 3.14 follows the pattern for the trench isolated CMOS and npn transistors in Figs. 3.3 and 3.8 (Yamaguchi and Yuzuriha 1989). An

82 Component formation

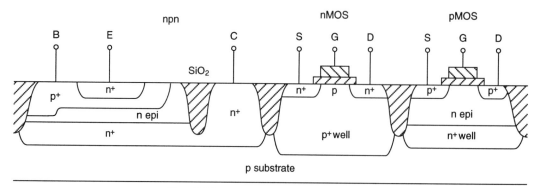

Fig. 3.13 BiCMOS (Reproduced by permission of John Wiley & Sons, Inc.)

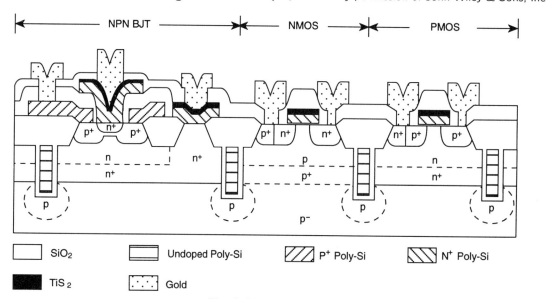

Fig. 3.14 Submicron BiCMOS (Copyright ©1989 IEEE)

important feature is the use of gold for electrode contacts. This practice improves electrical characteristics and is matched by gold for the first metal level wires.

3.5 GaAs transistors

An n-channel metal-semiconductor field effect transistor (MESFET), the mainstay of digital GaAs ICs, is illustrated together with a Schottky diode in Fig. 3.15. The structure resembles a single-channel MOSFET; the essential difference is that the gate is placed directly on the substrate. The substrate is a semi-insulating GaAs layer, which has a high resistivity $\sim 10^6$ to 10^8 Ωcm. That immediately highlights one advantage of GaAs over Si, greatly reduced electrode-substrate capacitances. The gate and the layer beneath it form a Schottky diode, but source and drain contacts are ohmic by virtue of the n^+ doping. The layers, including the isolating oxide, are in current technology all formed by implantation. Separate implants are made for the active channel (n) regions with silicon or selenium ions. These are also acceptable for the

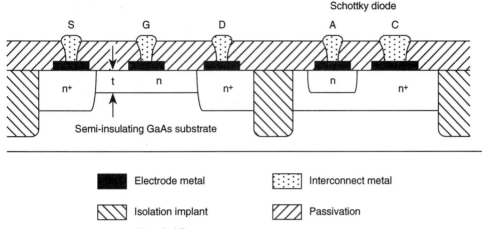

Fig. 3.15 GaAs MESFET

n⁺ source and drain implants, which, however, are often made with sulphur ions. Hydrogen, boron, or oxygen ions are used for the isolation implants. The GaAs MESFET has one feature in common with silicon: the passivation is either silicon dioxide or silicon nitride.

Much development has been undertaken to arrive at suitable metals for the electrode contacts, which are either ohmic (source and drain) or Schottky barrier diodes (gate). The preferred metal for ohmic contacts is an Au/Ge/Ni (gold-germanium-nickel) alloy, or a slightly modified AuGe/Ni/Au combination. Many metals can be used for the Schottky barrier contact. The preferred combination is a Ti/Pt/Au (titanium-platinum-gold) sandwich, with Ti for the Schottky contact, Au for good conductivity on the interconnect side, and Pt in between to inhibit an undesirable reaction between Ti and Au. The avoidance of aluminium, the standard metal for silicon ICs, is due to reliability problems created by an aluminium-gold contact. The first level metallization is consequently gold.

The separate Schottky diode in Fig. 3.15 is a replica of the Schottky barrier gate plus an ohmic contact. Separate Schottky diodes are part of GaAs logic circuits and the combination in Fig. 3.15 is illustrative of the technology for such circuits. It is easy to make a MESFET with multiple Schottky gate inputs, in place of the single gate shown, by placing several gates on top of an elongated channel region between source and drain.

The MESFET in Fig. 3.15 can operate either in the depletion (D-MESFET) or enhancement (E-MESFET) mode, depending on the concentration and thickness of the channel, and the built-in barrier potential of the Schottky gate diode. The mode is set in practice by t, the thickness of the channel, which is typically 1000 to 2000 Å for a D-MESFET, and 500 to 1000 Å for an E-MESFET. The corresponding gate threshold voltages are about -1.5 V to -2.5 V for cutting off the D-MESFET, and 0.3 V for turning the E-MESFET into conduction. D-MESFETs are easier to produce than E-MESFETs, because it is easier to achieve the highly uniform thickness required for low spreads of threshold voltage in production with the larger dimension of the D-MESFET channel.

The speed advantage of GaAs over Si only holds for n-channel MES-

84 Component formation

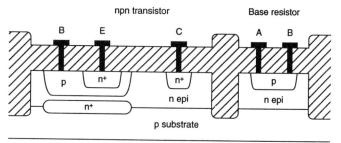

Fig. 3.16 Base resistor

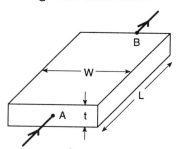

Fig. 3.17 Sheet resistance

FETs; p-channel MESFETs are, if anything, marginally slower than silicon transistors. Hence digital GaAs circuits are confined to nMESFETs and Schottky diodes.

3.6 Resistors and capacitors

Integrated circuit design is directed towards maximising the use of transistors. Digital CMOS, nMOS, and GaAs circuits are built entirely with transistors and diodes. Resistors are required for digital bipolar circuits, and for most analogue circuits. They can be grouped into two categories, one formed within monolithic ICs, and the other composed of film resistors.

Monolithic IC resistors consist of suitably dimensioned layers, which would normally form part of a transistor. The resistivity of such layers is determined by the desired transistor characteristics. The required value of the resistance is obtained by choosing the lateral dimensions of the layer; the vertical dimension is fixed by the process of diffusion, epitaxy, or implantation. The 'base' resistor in Fig. 3.16 is formed by omitting the n^+ emitter layer, and by choosing the length and width for the p base layer to obtain the desired resistance. No additional mask level is needed for that purpose. The selection of the layer depends on the value of resistance required. Usually it is the base layer, with the emitter layer reserved for very low values, and the collector layer—used least of all—for high values.

The resistance is calculated as follows. Let R be the resistance of the layers between A and B in Fig. 3.17. Then

$$R = \frac{\rho L}{tW} \qquad (3.3)$$

when ρ is the resistivity of the layer. For a square surface area, when

$W = L$, R is defined to be the sheet resistance R_s, which, according to eqn (3.3) becomes

$$R_s \equiv \left(\frac{\rho}{t}\right) \qquad (3.4)$$

Re-arrangement of eqn. (3.3) leads to

$$R = R_s \left(\frac{L}{W}\right) \qquad (3.5)$$

The units for R_s are $\Omega/\square$, written Ωsq^{-1} or $\Omega/\square$, because its value holds for the condition of a square ($L = W$) regardless of its size. The reason for parametrizing R in terms of R_s and not ρ is that R_s is set by the desired transistor characteristics. The concept of R_s is also applied to film resistors.

Resistors are formed in CMOS by utilizing an active layer (source or drain), a p- or n-well layer, a polysilicon layer, or an implanted layer. An extra mask level has to be used for an implanted resistor. Polysilicon resistors, which consist of polysilicon strips between the oxide passivating the substrate and another oxide layer, are preferred in CMOS. The larger the value of the resistor, the larger will be the area of silicon required. Very large values, going up to 100 kΩ or even beyond, are obtained with pinch resistors, which demand only a modest amount of silicon in relation to their value. The accuracy of such resistors is poor, but accuracy is not all that important when high values are needed. The pinch resistor, illustrated in Fig. 3.18, consists of a p base layer constricted by an n$^+$ 'emitter' layer, leading to an effective thickness equal to the base thickness of an npn transistor. The decrease in thickness increases the sheet resistance by a factor of about 30 to 50 over that of a p layer on its own (see eqn (3.4)). It is desirable to have all components in square rather than elongated rectangular shape for high component density. Resistors will often have a high (L/W) ratio in order to obtain the desired value, and the pattern in such cases is concertina shaped in order to obtain the square-like form shown in Fig. 3.19.

The major limitations of monolithic IC resistors are poor absolute accuracy, poor matching tolerance, and a large temperature coefficient. Thin film resistors are resorted to when improvements in respect of the above characteristics are essential. An added advantage of thin film resistors is their low voltage coefficient, (5–10) ppm/V. The corresponding

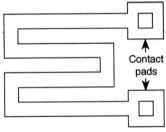

Fig. 3.19 Resistor surface geometry

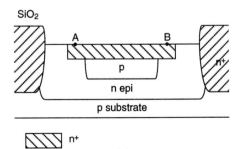

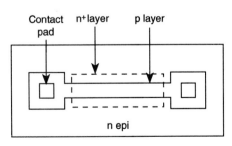

(a)   (b)

Fig. 3.18 Pinch resistor (a) profile (b) top view

Table 3.1 Typical characteristics of resistors

Type	Sheet resistance ($\Omega/\square$)	Absolute accuracy (%)	Matching tolerance (%)	Temperature coefficient (ppm/°C)
Bipolar				
Emitter diffused	3–12	±20	±2	+500
Base diffused	100–250	±20	±2	+1750
Base pinch	(2–10) K	±50	±10	+2500
Ion implanted	(0.1–1) K	±5	±1	±150
CMOS				
Diffused-source/drain	20–100	±30	±2	+1500
Polysilicon	20–200	±30	±2	+1500
Ion implanted	(0.4–1.5) K	±5	±1	±400

coefficients of polysilicon, diffused, and implanted resistors are ~100, 200, and 800 ppm/V respectively. Thin film resistors are combined with one or more monolithic dies into a hybrid IC. Such an IC consists of two or more dies, usually of different composition, which are interconnected with some discrete components, in this case film resistors. That definition can be stretched to hold for an IC consisting of a monolithic die and a bank of film resistors, mounted on an inert substrate. Advanced technologies are available in which film resistors are mounted on the substrate of a monolithic die, part of which is kept free for that purpose. The nature of an IC is often transparent to the user: whether it is monolithic or hybrid cannot always be deduced from the package. Thin film resistors have thicknesses comparable to those of layers in MOS and bipolar ICs, and they tend to follow the feature size of the monolithic dies with which they are associated. The established thin films for resistors are nichrome, tantalum nitride, and chromium-silicon oxide cermets. The values of sheet resistance for nichrome, tantalum nitride, and base layers are rather similar. Nichrome has the edge on other film materials in respect of absolute precision and thermal stability, but is vulnerable to corrosion. Cermet resistors, composed of metal and metal oxides, have a much higher sheet resistance but lack the stability of the other materials. Tables 3.1 and 3.2 give typical characteristics of monolithic and thin film resistors. Polysilicon resistors, listed under CMOS in Table 3.1, are freely used in bipolar and BiCMOS ICs. Garuts et al. (1989) give an example of a self-aligned polysilicon resistor integral with an npn transistor.

Capacitors which take the form of parallel-plate structures, are illustrated in Fig. 3.20. The most widely used type, shown in Fig. 3.20(a), consists of two polysilicon plates separated by silicon dioxide with the lower plate resting on the top of the substrate. The MOS capacitor in Fig. 3.20(b) consists of a diffused or implanted heavily doped layer within the substrate, and a polysilicon or metal plate on top of a thin oxide layer. For MOS capacitors, the gate oxide is usually used for that

Table 3.2 Typical characteristics of thin film resistors (film thickness ~1000 Å)

Material	Sheet resistance ($\Omega/\square$)	Absolute accuracy (%)	Matching tolerance (%)	Temperature coefficient (ppm/°C)	Temperature coefficient tracking (ppm/°C)
Nichrome	20–250	±10	±0.1	±50	±2
Tantalum nitride	20–125	±10	±0.1	±50	±2

Table 3.3 Typical capacitor characteristics (SiO_2 dielectric, thickness 1000 Å)

Type	Capacitance (fF/μm^2)	Absolute accuracy (%)	Matching accuracy (%)	Temperature coefficient (ppm/°C)	Voltage coefficient (ppm/V)
Poly–Poly	~0.35	±20	±0.06	±25	10
MOS	~0.45	±10	±0.06	±25	10

purpose and no extra processing step is needed. Typical characteristics of capacitors are shown in Table 3.3.

3.7 Design rules and device areas

The geometries of active and passive devices were, until the arrival of VLSI, fixed by the semiconductor vendors for each IC pattern. The increasing complexity of VLSI, the drive to minimize component size, and the desire to establish standard layout geometries for a given process have resulted in the establishment of *design rules* (also called *layout rules*) which define device geometries. Process-dependent but pattern-independent these rules, stipulated by the IC vendor for his process technology, permit the production of ICs with good reliability and high yield. The design rules specify all the *minimum* dimensions of devices and wires in terms of line widths and spacing for specified shapes which are either rectangles (including squares) or polygons. Orthogonal patterns are observed throughout. The dimensions are stipulated directly in microns, or—this is now standard practice—in terms of the parameter λ, initiated by Mead and Conway (1980). This normative parameter is linked to the feature size, which is usually taken to be 2λ i.e. λ equals 0.5 μm for a 1 μm geometry. All dimensions are integer, or—rarely—half integer multiples of λ. For example the width of a wire and the spacing between adjacent wires might be 2λ and 1.5λ respectively. The lambda rules are portable and can be readily adapted to any processing line and for scaled geometries.

The design rules, it must be stressed, specify *minimum* dimensions, which are exceeded where appropriate. The channel dimensions of the MOSFET are a case in point. Circuit requirements frequently call for a high width-to-length ratio of the channel: values of 5 or much higher are common. One of the dimensions (L or W) then takes the feature size, the other is much larger (see Fig. 3.2). The design rules lay down

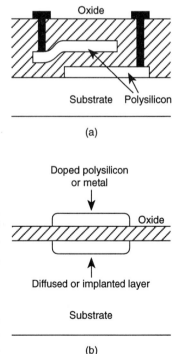

Fig. 3.20 Capacitor structures (a) poly-poly (b) poly/metal

88   Component formation

Table 3.4 Section of lambda based design rules for typical n-well CMOS with 3 μm geometry (all dimensions are minima)

Feature number (Fig. 3.21)	Feature	Mask level	Dimension
1.1	n-well width	n well	4λ
1.2	n-well spacing (wells at same potential)	n well	4λ
1.3	n-well spacing (wells at different potential)	n well	8λ
1.4	Edge distance to internal thinox	n well	3λ
1.5	Edge distance to external thinox	n well	4λ
2.1	Thinox width	Thinox	2λ
2.2	Thinox spacing	Thinox	2λ
2.3	p-thinox to n-thinox spacing	Thinox	8λ
3.1	Source/drain contact area	Contact	$2\lambda \times 2\lambda$
3.2	Contact to contact spacing	Contact	2λ
3.3	Overlap of thinox or poly over contact	Contact	λ
3.4	Contact to gate poly spacing	Contact	2λ

the geometries for all mask levels. They take into account effects on dimensions arising from lateral diffusion, isotropic etching, overlay errors, and other causes. The design rules are drawn up to achieve a balance between an overconservative approach with very generous tolerancing, which is demanding on the silicon estate but gives a very good yield, and an excessively aggressive approach which consumes far less silicon at the cost of unacceptably low yield and reliability. Although the design rules are drawn up for a specific vendor's product, there is a close correspondence between the values of different manufacturers for identical feature size.

The lambda-based design rule concept is illustrated with a selection of dimensions in Table 3.4 and the corresponding layout geometries in Fig. 3.21, which includes the layout of two items outside the lambda-based orbit, the bonding and probe pads. (Probe pads are for tests additional to those made on bonding pads prior to die encapsulation.) These pads do not scale with feature size; the dimensions given reflect established standards. The metal pad includes a protrusion for via contacts to the lower metal level, whilst the opening are for contact cuts through the encapsulating die passivation.

Integrated circuit layout, whether for standard or ASIC chips, is accomplished with the aid of software tools which operate within the constraints imposed by the design rules. These tools include a *design*

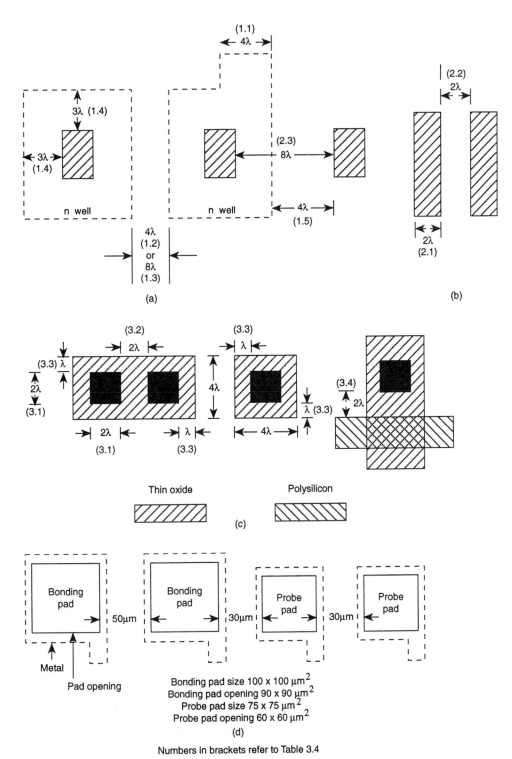

Fig. 3.21 Layout of selected n-well CMOS geometries (Table 3.4) (a) n-well (b) oxide separation (c) source, drain (d) bonding and probe pads

90 Component formation

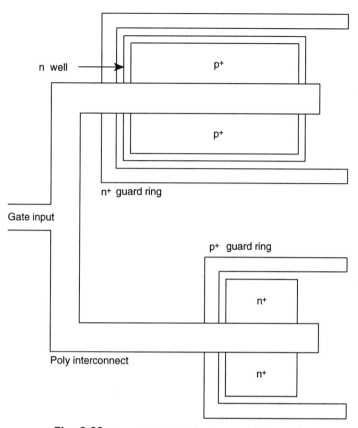

Fig. 3.22 Layout of CMOS structure (Fig. 3.2)

Table 3.6 Estimate of mean wire length $\overline{L}$ in DRAMs

Memory capacity	Feature size (μm)	$\overline{L}$ (μm)
256 Kb	2	23
1 Mb	1	8.8
16 Mb	0.5	3.4

rule checker (DRC), which ensures that the design rules are not violated. An error discovered by the DRC will more likely than not prove to be a basic error in the layout. The degree of layout automation with CAD is such that it is by and large not important for the chip designer to know the design rules in detail, and the matter will not be pursued any further.

In conclusion examples of device surface areas, linked to feature size, will now be given to show orders of magnitude of wire lengths and transistor counts. Figure. 3.22 gives layout for a CMOS structure of the type illustrated in Fig. 3.2. It highlights the fact that the device area is usually at least a magnitude larger than the feature size. Typical transistor (surface) areas have been collected together in Table 3.5, which is representative of established technologies.

A rough estimate of $\overline{L}$, the mean wire length, is given for DRAM chips in Table 3.6. The calculation has been carried out as follows. Taking typical DRAM chip areas (Prince 1991), the area available for wire is the difference between the total area and the area occupied by the cells, calculated by using the values in Table 3.5. This ignores the additional *overhead* circuits for routing, control, etc. Each cell has two wires, not three, because the storage electrode is left open circuit. Lastly the spacing between wires is equated to wire width. $\overline{L}$ in Table 3.6 is of the right order, but very efficient layout design is needed to achieve such values.

Table 3.5 Typical transistor surface areas

Type	Feature size (μm)	Surface area (μm^2)
CMOS	5	2100
CMOS	3	1400
CMOS	1	50
Bipolar	1	20
nMOS DRAM cell	2	84
nMOS DRAM cell	1	30
nMOS DRAM cell	0.5	5

Another hand calculation will now be advanced to estimate the number of transistors on a 1 μm CMOS chip 1 x 1 cm^2. Digital logic cannot attain anything like the efficiency for memory layout, where storage cells allow a high degree of compaction. Making $\overline{L}$ equal to kL_c, where L_c is the chip edge (length) and putting k at 0.01, N, the number of transistors, is calculated by making some further assumptions. The length of wires connecting the transistors to the supply and ground rails and interconnecting the CMOS transistors are neglected. Furthermore there is one wire per CMOS pair shared with another pair. The area occupied by the wires and the spacing between them is $(N/2)$(CMOS pairs) $\times$ 2(wire width) $\times$ 2(spacing) $\times$ 100($\overline{L}$) = 200N μm^2. The CMOS pairs occupy 50N μm^2 (Table 3.5) leading to

$$50N + 200N = 10^8 \qquad (3.6)$$

which gives N equal to 400 000. The biggest conjecture in this calculation has been the estimate of 100 μm for $\overline{L}$. In standard chips with efficient design it may be less, in ASICs it is likely to be much more. The calculation illustrates the critical dependence of transistor density on wire length, and highlights an issue which has already been mentioned and which will crop up again and again, the dominant role of wire in VLSI.

References

Brown, D.M., Ghezzo, M., and Pimbley, J.M. (1986). Trends in advanced process technology - submicrometer CMOS device design and process requirements. *Proceedings of the IEEE*, **74**, 1681.

Foster, D.J. (1987). Advances in CMOS technology. In *Plessey Research and Technology Review* (ed. J.M. Herbert), pp.45–53. Plessey Company, England.

Garuts, V.E., Yu Y-C. S., Traa, E.O., and Yamaguchi, T. (1989). A dual 4-bit 2-Gs full Nyquist analog-to-digital converter using a 70-ps silicon bipolar technology with borosenic–poly process and couplings–base implant. *IEEE Journal of Solid-State Circuits*, **24**, 216–22.

GEC Plessey Semiconductors Short Form Catalogue (1992/93), p.8.

Gray, P.R. and Meyer, R.G. (1993). *Analysis and design of analogue integrated circuits*, (3rd edn), p.174. Wiley, USA.

Hunt, P.C. and Saul, P.H. (1988). Process and circuit innovation in silicon bipolar technology. In *Plessey Research Review* (ed. J.M. Herbert), pp.59–66,. Plessey Company, England.

Lohstroh, J. (1981). Devices and circuits for bipolar (V)LSI. *Proceedings of the IEEE*, **69**, 812–26.

Mead, C. and Conway, L. (1980). *Introduction to VLSI systems*, pp.47–51. Addison-Wesley, USA.

Prince, B. (1991). *Semiconductor memories*, (2nd edn). Wiley, USA.

Troutman, R.R. (1986). *Latchup in CMOS technology*, p.51. Kluwer Academic Publishers, USA.

Weste, N. and Eshragian, K. (1985). *Principles of CMOS VLSI design*, p.89. Addison Wesley, USA.

Yamaguchi, T. and Yuzuriha, T.H. (1989). Process integration and device performance of a submicrometer BiCMOS with 16-GHz f_t double poly-bipolar devices. *IEEE Transactions on Electron Devices*, **36**, 890–96.

4
Device behaviour and modelling

4.1 Introduction

The chief device characteristics which determine IC behaviour are the I-V relationships, the carrier transit time between input and output terminals, the interelectrode capacitances, and the electrode resistances. To these must be added the impedance of the interconnections within the die. Circuit behaviour is expressed in either the time or the frequency domain. The time domain is used for digital, the frequency domain for analogue circuits. The section which follows outlines transient performance and amplifier frequency response in general. These will help the reader to follow the formulations of device behaviour in the subsequent sections, which cover MOSFETs, GaAs MESFETs, and BJTs. The material they contain allows comparison of the various technologies. It is a foundation for the SPICE models which follow, and also for hand calculations. Circuit simulation with SPICE (simulation program with integrated circuit emphasis) is the method of choice for computing circuit performance. Hand analysis does not attain the same accuracy and, depending on the circuit, can be intractable. Nevertheless it leads to an approximate evaluation which gives a good initial indication of performance.

4.2 Transient response and bandwidth

The dominant parameters are transition (rise and fall) times and propagation delay in the time domain, and bandwidth in the frequency domain. Transition and delay times are defined with the aid of Fig. 4.1, which shows the waveforms for an inverter. The transition rise and fall times t_{TLH} and t_{THL} are the intervals in which the amplitude of the signal changes from 10 to 90 per cent of its peak value. The delay times t_{PHL} and t_{PLH} are the intervals between input and output at half amplitude. When transition times are equal i.e. $t_{THL} = t_{TLH}$, or $t_{PHL} = t_{PLH}$, they are designated t_T and t_P.

Expressions will now be obtained by assuming that the response is governed by the single-pole network shown in Fig. 4.2. The transfer function $T(S)$ for Fig. 4.2(a) is given by

$$T(S) \equiv \frac{V_o(s)}{V_i(s)} = \frac{1}{1+s\tau} \qquad (4.1)$$

where $s = j\omega$ and $\tau = RC$. Similarly, for Fig. 4.2(b)

$$T(S) = \frac{g_m R}{1+s\tau} \qquad (4.2)$$

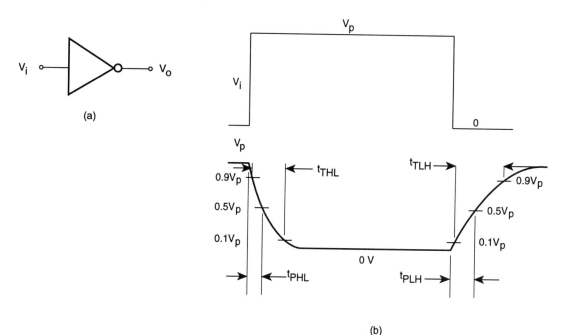

Fig. 4.1 Inverter transition and delay times (a) schematic (b) waveforms

when g_m, the transconductance, is assumed to be independent of frequency, and is defined by

$$g_m \equiv \frac{\partial I_o}{\partial V_i} \qquad (4.3)$$

The solution of eqns (4.1) and (4.2) for a step function input is

$$V_o(t) = V_p \left\{ 1 - \exp\left(\frac{-t}{\tau}\right) \right\} \qquad (4.4)$$

where V_p is the peak amplitude of V_o. It follows from eqn (4.4) that

$$t_T = 2.2\tau \qquad (4.5)$$

and

$$t_P = 0.69\tau \qquad (4.6)$$

Combining eqns (4.5) and (4.6)

$$t_P = 0.31 t_T \qquad (4.7)$$

A near-universal approximation of eqn (4.6) is

$$t_P \simeq \tau \qquad (4.8)$$

Equations (4.5) and (4.8) lead to

$$t_P \simeq 0.45 t_T \simeq 0.5 t_T \qquad (4.9)$$

Equation (4.8) is widely used to obtain an approximate estimate of transient response by hand analysis in terms of the dominant RC time constants, including those formed by the wires.

The transfer function in the frequency domain, $T(\omega)$, is given by

$$T(\omega) = \frac{1}{\sqrt{(1+\omega^2\tau^2)}} \quad (4.10)$$

and

$$T(\omega) = \frac{g_m R}{\sqrt{(1+\omega^2\tau^2)}} \quad (4.11)$$

for Figs 4.2(a) and 4.2(b) respectively.

The 3 dB bandwidth f_{-3dB} is according to eqns (4.10) and (4.11)

$$f_{-3dB} = \frac{1}{2\pi\tau} \quad (4.12)$$

Combining eqns (4.5) and (4.12)

$$t_T = \frac{0.35}{f_{-3dB}} \quad (4.13)$$

The significance of eqn (4.13) is that the transient response is directly related to the bandwidth. The relationship is not so simple for the non-linear, large-signal mode of logic circuits, but eqn (4.13) is nevertheless a useful initial estimate.

Another basic evaluation applies to logic circuits in which the transient performance is largely determined by a load capacitor. Figures 4.2(b) and (c) serve for illustration. The amplifier is switched on or off and the behaviour will be governed by C if RC is very large compared with the time of the amplifier in the 'on' state. The voltage developed across C by a charging current i is given by

$$CdV = idt \quad (4.14)$$

leading to

$$t_{TLH} = \int_{V'_L}^{V'_H} \frac{CdV}{i_i} \quad (4.15)$$

where V'_L and V'_H are the 10 and 90 per cent amplitudes of the output levels V_H and V_L ($V'_L = 0.9V_L + 0.1V_H$, $V'_H = 0.9V_H + 0.1V_L$), and i_i is the current flowing into C.

Similarly for the discharge of C

$$t_{THL} = -\int_{V'_H}^{V'_L} \frac{CdV}{i_o} \quad (4.16)$$

where i_o is the magnitude (positive) of the current flowing out of C. In practice i_i and i_o are generally provided by separate transistors. Assuming quasi-linear operation by putting i_i equal to $g_m v_i$, and taking g_m to be constant

$$t_{TLH} = \frac{C}{g_m v_i}(V'_H - V'_L) \quad (4.17)$$

Furthermore $v_i = V_H - V_L \simeq V'_H - V'_L$ so that

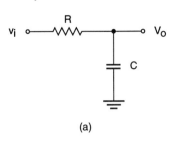

(a)

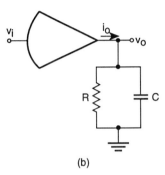

(b)

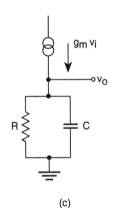

(c)

Fig. 4.2 Single-pole RC network (a) passive network (b) amplifier (c) equivalent circuit of (b)

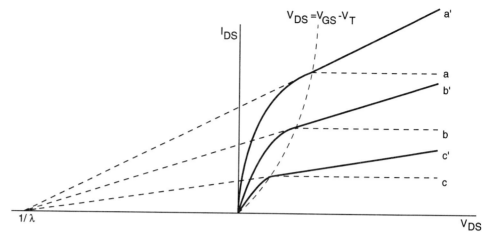

a,b,c: $I_{DS} \propto (V_{GS} - V_T)^2$

a',b',c': $I_{DS} \propto (V_{GS} - V_T)^2(1 + \lambda V_{DS})$

Fig. 4.3 nMOSFET static characteristics

$$t_{TLH} \simeq \frac{C}{g_m} \tag{4.18}$$

Equation (4.18), a very useful approximation for estimating transition time and highlights a parameter of great importance in switching and amplifier circuits, namely C/g_m. Its significance is also evident from the voltage gain A_V for the amplifier in Fig. 4.2(b). Assuming the load to be effectively C ($\omega C R \gg 1$ at all frequencies of interest)

$$|A_v| = \frac{g_m}{\omega C} \tag{4.19}$$

4.3 MOSFET characteristics

4.3.1 DC characteristics

The static characteristics of an nMOSFET are shown in Fig. 4.3, and its channel formation is sketched in Fig. 4.4, which highlights the difference between the *drawn* (lithographic) and *effective* (actual) channel dimensions. The latter have to be used in formulae involving channel length and width. The treatment applies to enhancement MOSFETs, except where stated otherwise.

The I-V relationships for an nMOSFET are

$$I_D = 0 \tag{4.20}$$

$$V_{GS} \leq V_T$$

$$I_D = k'\left(\frac{W}{L}\right)\left\{(V_{GS} - V_T)V_{DS} - \frac{V_{DS}^2}{2}\right\} \tag{4.21}$$

$$V_{GS} \geq V_T, V_{DS} \leq V_{GS} - V_T$$

$$I_D = \frac{k'}{2}\left(\frac{W}{L}\right)(V_{GS} - V_T)^2 \tag{4.22}$$

MOSFET characteristics 97

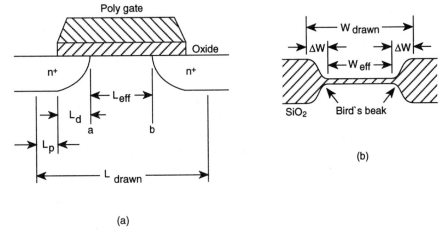

Fig. 4.4 Effective channel dimensions (a) length (b) width

$$V_{GS} \geq V_T, V_{DS} \geq V_{GS} - V_T$$

$$k' = \frac{\mu_n \varepsilon_{ox}}{t_{ox}} \qquad (4.23)$$

where μ_n is the electron mobility, ε_{ox} the permittivity and t_{ox} the thickness of the gate oxide. ε_{ox} is obtained by multiplying the permittivity ε_o of free space (8.854 pF/m, alternatively 8.854×10^{-3} fF/μm), by the *relative* permittivity, the *dielectric constant* of the oxide, which is 3.9, giving $\varepsilon_{ox} = 3.45 \times 10^{-2}$ fF/μm. The static characteristics are demarcated by the locus $V_{DS} = (V_{GS} - V_T)$ into two regions (see Fig. 4.3). The region to the left of the locus is labelled *linear*; the slope $\partial I_D/\partial V_{DS}$ is pretty nearly a straight line for $V_{DS} \ll (V_{GS} - V_T)$. I_{DS} is ideally independent of V_{DS} in the region to the right, the *saturation* region, hence the horizontal dotted lines a,b,c.

The threshold voltage V_T is given by

$$V_{Tn} = V_{FB} + 2|\phi_{Fp}| + \frac{|Q_B|}{C_{ox}} - \frac{Q_{ox}}{C_{ox}} \qquad (4.24)$$

for an nMOSFET, and

$$V_{Tp} = V_{FB} - 2|\phi_{Fn}| - \frac{|Q_B|}{C_{ox}} - \frac{Q_{ox}}{C_{ox}} \qquad (4.25)$$

for a pMOSFET.

V_{FB}, the *flat-band voltage*, is the difference in the work functions of the bulk silicon region and the gate material. ϕ_{Fp} and ϕ_{Fn} are the equilibrium electrostatic potentials at the neutral edge of the depletion region, defined by

$$\phi_{Fp} = \frac{kT}{q} \ln\left(\frac{p_i}{p}\right) \qquad (4.26)$$

for a p-type semiconductor

$$\phi_{Fn} = \frac{kT}{q} \ln\left(\frac{n}{n_i}\right) \qquad (4.27)$$

for an n-type semiconductor; p and n are the equilibrium majority carrier concentrations, and p_i ($= n_i$) is the intrinsic carrier concentration. Q_B is the charge density in the depletion region, and Q_{ox} is the density of the parasitic charge, which exists within the oxide and also at the oxide-silicon interface. C'_{ox}, the gate capacitance per unit area, is given by

$$C'_{ox} = \frac{\varepsilon_{ox}}{t_{ox}} \quad (4.28)$$

Q_{ox} has been greatly reduced by improvements in processing techniques and may be neglected in modern technology (Nicollian and Brews 1982).

Control of threshold voltage is very important in digital MOS. V_{Tp} must be negative and V_{Tn} positive. The former requirement is easily met, because all the terms in eqn (4.25) are negative with the exception of V_{FB}, which is 0.3 V for a polysilicon gate. To obtain a positive V_{Tn} is much more difficult, because $V_{FB} \sim -0.9$ V for an nMOSFET. The second and third terms in eqn (4.24), ignoring the fourth term (~ -50 mV), must be sufficiently large to give the desired positive value. A p layer is often implanted in the channel for that purpose. Typical values for the contributions which make up V_{Tn} in eqn (4.24) are $V_{FB} = -0.85$ V, $2|Q_{Fp}| = 0.58$ V, $|Q_B|/C_{ox} = 1.21$ V, and $Q_{ox}/C_{ox} = 0.05$ V leading to $V_{Tn} = 0.89$ V. These figures show that tight process control is essential to keep the production spread of V_T within acceptable limits.

The treatment so far has assumed zero source-to-substrate bias, V_{SB}. In digital circuits V_{SB} is frequently positive for n and negative for pMOSFETs. V_T, allowing for V_{SB}, is modified to V'_T in accordance with

$$V'_T = V_T + \gamma \left(\sqrt{|V_{SB} - 2\phi_F|} - \sqrt{2|\phi_F|} \right) \quad (4.29)$$

γ, the *body factor* (or *body-effect coefficient*) has values in the range $\sim(0.4 - 1.0)$ in practice. Some further modifications are called for to give a more accurate presentation of MOSFET behaviour; these assume increasing importance with the ongoing device shrinkage into the deep submicron region.

The *effective* channel length and width are smaller than the *drawn* lithographically defined dimensions L and W. The appropriate values to be used in hand calculations and SPICE models are obtained from Fig.4.4.

$$L_{eff} = L_{\text{drawn}} - 2L_p - 2L_d \quad (4.30)$$

$$W_{eff} = W_{\text{drawn}} - 2\Delta W \quad (4.31)$$

The reduction in L arises from processing (L_p) and lateral diffusion (L_d). L_d is about 30 per cent less than the depth of the drain layer (see Section 2.1 and Fig. 2.2). The reduction $2\Delta W$ is due to the formation of the bird's beak (see Section 2.4.1 and Fig. 2.10); ΔW is typically 70–80 per cent of the oxide thickness. The proportional reductions from drawn to effective channel length and width increase with reduced dimensions.

Channel length modulation arises when V_{DS} exceeds $(V_{GS} - V_T)$; see Fig. 4.3). The space charge region at the drain junction is a function of V_{DS}, and the effective channel length is reduced as shown in Fig. 4.5.

The current is no longer constant, but increases with V_{DS}. The plots a', b', c' in Fig. 4.3 distinguish the actual from the ideal square-law behaviour represented by plots, a, b, c. Equations (4.21) and (4.22) modify to

$$I_D = k'\left(\frac{W}{L}\right)\left\{(V_{GS}-V_T)V_{DS} - \frac{V_{DS}^2}{2}\right\}(1+\lambda V_{DS}) \quad (4.32)$$

$$V_{GS} \geq V_T, V_{DS} \leq (V_{GS}-V_T)$$

$$I_D = \frac{k'}{2}\left(\frac{W}{L}\right)(V_{GS}-V_T)^2(1+\lambda V_{DS}) \quad (4.33)$$

$$V_{GS} \geq V_T, V_{DS} \geq (V_{GS}-V_T)$$

The channel-length-modulation parameter λ lies typically in the range 0.01 to 0.1 V^{-1}.

The I-V expressions in eqns (4.21), (4.22), (4.32), and (4.33) are based on the gradual-channel approximation (GCA)(Uyemura 1988, pp.18–22), which postulates that the gate completely controls the channel, and that there will be no conduction if $V_{GS} < V_T$. When conduction takes place with $V_{GS} > V_T$, the channel is said to be in *strong inversion*. In actual fact *sub-threshold* conduction does take place for $V_{GS} < V_T$, and the I-V characteristics for that condition are shown in Fig. 4.6. In *weak inversion*, for which $V_{GS} < V_T$, the I-V characteristic is given by

$$I_D \simeq \left(\frac{W}{L}\right)I_{DO}\exp\left(\frac{qV_{GS}}{nkT}\right) \quad (4.34)$$

where I_{DO} and n are constants which depend on device parameters, with $n \sim (1-2)$. A more accurate analysis shows that the demarcation between strong and weak inversion takes place at a higher gate-source voltage, V_T', given by (Stotts 1989)

$$V_T' \simeq V_T + 2nV_t \quad (4.35)$$

where V_t, the thermal voltage, $\equiv (kT/q) = 26$ mV at $T = 300$ K. For $n = 2$, $V_T' \simeq V_T + 100$ mV. Subthreshold conduction is difficult to model. The best results have been obtained by interposing a region of moderate inversion between weak and strong inversion. The conduction in weak inversion presents a potential hazard in VLSI. If the gate bias of MOSFETs in the off state is not sufficiently below V_T, the sum of all the currents consumed by those transistors could be prohibitively high. The danger is eliminated by ensuring that the gate voltage in the off state is several hundred mV below V_T. This condition, easily met for transistors having $V_T \sim 0.8$ V or higher, can pose problems for low voltage operation, in which $V_{DD} \sim 1$ V, and $V_T \sim 0.4$ V or less.

CMOS analogue circuits have been developed for operating in weak inversion in order to consume very small power. Examples of applications are instrumentation and battery-operated CMOS biomedical equipment, some of which has been designed to operate for up to ten years.

The treatment presented so far has to be modified for *short and/or*

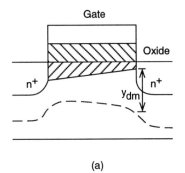

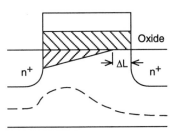

Fig. 4.5 MOSFET channel formation (a) non-saturated (b) saturated

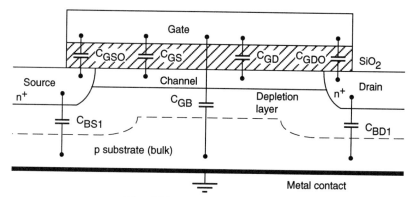

Fig. 4.7 MOSFET capacitances

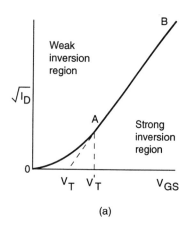

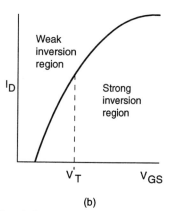

Fig. 4.6 Strong and weak inversion characteristics (a) linear scales (b) logarithmic scale for I_D

narrow channel operation. That terminology applies to a channel length or width comparable with y_{dm}, the maximum depth of the depletion layer in the channel, in Fig. 4.5 (Tsividis 1987, p.168; Uyemura 1988, p.60). Submicron geometries imply generally short and/or narrow channel operation, particularly for *minimum-size* geometries, for which channel length and width are equal. This aspect of MOSFET behaviour is covered in most texts on MOS, including some listed under further reading in the references.

4.3.2 Capacitances

The intrinsic capacitances of an nMOSFET are modelled in Fig. 4.7, in which the shape of the depletion layer applies for $V_S = V_{DS} = 0$ V. The gate-body capacitance C_{GB} is the series combination of the gate-channel capacitance C_{GC} and the depletion layer capacitance C_{dep} between the substrate (bulk) and the channel. C_{BS1} and C_{BD1} are the diffusion capacitances across the source-bulk and drain-bulk pn junctions. The nature of the capacitances and their values are governed by the three modes of operation, the off state, the ohmic and the saturation region. It would be a mistake to think of them purely in terms of parallel-plate or depletion-layer formation. The transistor action leads to far more complex behaviour and modelling.

Taking the off state first, if V_{GS} is biased well below the onset of weak inversion, source and drain are isolated by the p-doped channel; C_{GS} and C_{GD} will be zero if the overlap between the source and drain regions and the gate oxide is ignored. This overlap, represented by C_{GSO} and C_{GDO}, will be allowed for later. C_{GC} will be simply equal to C_{OX}, the gate-oxide capacitance, which is given by

$$C_{ox} = C'_{ox} L_{eff} W_{eff} \tag{4.36}$$

C_{GB} is the series combination of C_{GC} and C_{dep}, i.e.

$$C_{GB} = \frac{C_{GC} C_{dep}}{C_{GC} + C_{dep}} \tag{4.37}$$

V_{GS} in the off state can be positive or negative. For negative V_{GS}, the negative surface potential will attract a large number of holes to the silicon surface. In this state of *accumulation* there is no depletion layer so that an ohmic path exists from the lower surface of the gate oxide to the ground plate at the back of the substrate, leading to $C_{GB} \simeq C_{GS}$. When V_{GS} becomes positive, increasing towards V_T but remaining below it, the depletion layer now formed in the channel increases in depth, causing C_{dep} and C_{GB} to decrease. Increasing V_{GS} further above V_T produces an electron inversion right across the channel for ohmic operation ($V_{DS} < V_{GS} - V_T$), and C_{GC} is divided equally between source and drain, i.e. $C_{GS} = C_{GD} = \frac{1}{2}C_{GC}$. A capacitance C_i is associated with the inversion layer by virtue of the charge induced in it. This capacitance is in parallel with C_{dep}, and is very much larger than either C_{dep} or C_{ox}, so that C_{GB} approaches C_{ox} under these conditions. (Tsivides 1987, pp.66-8). C_{GB} is plotted as a function of V_{GS} in Fig. 4.8.

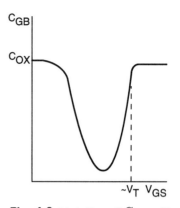

Fig. 4.8 Variation of C_{GB} with V_{GS}

In saturation part of the channel, ΔL in Fig. 4.5(b), is *pinched off*. The conducting part of the channel signifies a finite capacitance C_{GS}, but C_{GD} will be zero. C_{GS} and C_{GB} are closely given by

$$C_{GS} = \frac{2}{3}C'_{ox}WL \qquad (4.38)$$

$$C_{GB} = \frac{\delta_1}{3(1+\delta_1)}C_{ox} \qquad (4.39)$$

$$\delta_i = \frac{dV_T}{dV_{SB}} \qquad (4.40)$$

δ_1 is proportional to the body factor γ, and lies between 0 and 1 (Tsivides 1987, p.125).

The fact that C_{GD} and C_{BD} are both zero follows from the nature of saturation. Ignoring channel modulation, a change in V_{DS} produces no change in the gate and depletion region charges. The value of C_{GS} in eqn (4.38) is arrived at by calculating the change ΔV_{ox} across the oxide for a change ΔV_S. If the inversion layer behaved like the conducting plate of a passive capacitor, ΔV_{ox} would equal ΔV_S. Considerations of the physical action show that $\Delta V_{ox} < \Delta V_S$, leading to the value of C_{GS} in eqn (4.38), a value, which Tsivides (1987, p.321) points out, is independent of ΔL in Fig. 4.5(b).

C_{GB} is approximated to zero in some texts, but consideration of the physical action mentioned above shows that it has a finite value. A change in ΔV_B (substrate voltage) will result in a change of zero at one end, but will increase in the direction of the drain, thereby increasing the gate oxide charge. The saturation value of C_{GB} is given by eqn (4.39).

The voltage-dependent capacitances C_{BS} and C_{BD} are made up of three components, C_{BS1}, C_{BD1}, and C_{BC1}, which represent the depletion layer capacitances between source, drain, and channel respectively. These capacitances are expressed in terms of capacitances per unit area C_j multiplied by the appropriate electrode area in accordance with

$$C_j = \frac{C_{jo}}{\left\{1 - \left(\frac{V_r}{\phi_o}\right)\right\}^{\frac{1}{2}}} \qquad (4.41)$$

where C_{jo} applies to zero bias, V_r is the reverse bias across the junction,

102 Device behaviour and modelling

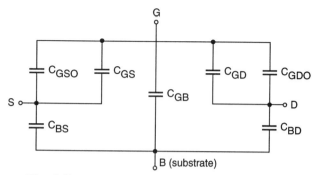

Fig. 4.9 Lumped model of MOSFET capacitances

Table 4.1 Values of MOSFET capacitances

Capacitance	Value		
	Off	Ohmic	Saturation
C_{GD}	$C'_{ox}WL_d$	$C'_{ox}W\left(L_d + \frac{1}{2}L\right)$	0
C_{GS}	$C'_{ox}WL_d$	$C'_{ox}W\left(L_d + \frac{1}{2}L\right)$	$\frac{2}{3}C_{ox}$
C_{GB}	C_{ox}	0	$\frac{\delta_1}{3(1+\delta_1)}C_{ox}$
C_{BS}	C_{BS1}	$C_{BS1} + \frac{C_{BC1}}{2}$	$C_{BS1} + \frac{2}{3}C_{BC1}$
C_{BD}	C_{BDT}	$C_{BD1} + \frac{C_{BC1}}{2}$	0

and ϕ_o is the built-in potential. C_{BC1} will, in the absence of an inversion layer, be zero. The overlap capacitances C_{GSO} and C_{GDO} are given by

$$C_{GDO} = C_{GSO} = W_{eff}.L_d.C'_{ox} \qquad (4.42)$$

assuming a symmetrical geometry (Fig. 4.4).

A lumped model of MOSFET capacitances is shown in Fig. 4.9, and the values of the capacitances are summarized in Table 4.1.

The assessment of capacitance values has been presented in some detail. Even so, the coverage is only an outline which barely scratches the surface of the physical processes underlying MOSFET action. The purpose of giving it a degree of emphasis is to stress the complex issues underlying MOSFET modelling, and the major task required to arrive at an acceptable large-signal equivalent circuit. The treatment adopted in this section applies to small-signal evaluations based on $C \equiv \partial Q/\partial V$, and this is not equal to the large-signal capacitance $C \equiv Q/V$, because all MOSFET capacitances are functions of voltage. It is far easier to achieve high accuracy for small-signal than for large-signal operation, in which, for example in digital circuits, the signal is likely to traverse the off, ohmic, and saturation regions. Errors can become very serious in the simulation of DRAMs, switched-capacitor filters, and charge-scaling data converters, all of which operate on the principle of charge conservation. The attainment of charge conservation is a paramount objective in MOSFET modelling.

4.3.3 Channel transit time

The switching speed of digital circuits is almost entirely governed by the charging and discharging of MOSFET and wire capacitances. The transit time of carriers in the channel is negligible, witness the following calculations:

The carrier channel transit time T_{tr} for an nMOSFET is given by (Muller and Kamins 1986, p.440)

$$T_{tr} \simeq \frac{4}{3} \frac{L^2}{\mu_n (V_{GS} - V_T)} \qquad (4.43)$$

Equation (4.43) holds for operation in strong inversion and saturation; it applies equally for a pMOSFET, with μ_p in place of μ_n. An underlying assumption behind its derivation is that the carrier velocity is proportional to the electric field across the channel. Carrier velocity saturates at electric fields greater than $\sim 10^4$ V/cm, reaching v_{sat} of $\sim 10^7$ cm/s. The transit time under these conditions is given by

$$T_{tr(sat)} \simeq \frac{L}{v_{sat}} \qquad (4.44)$$

Consider an nMOSFET with the following characteristics: $L = 1\ \mu$m, $\mu_n = 650$ cm^2/V-s, $V_{GS} = 5$ V, $V_T = 1.5$ V, $V_{sat} = 10^7$ cm/s. The electric field is 3.5×10^6 V/cm, and velocity saturation evidently applies. The values of T_{tr} and $T_{tr(sat)}$ come to 5.3 ps and 10 ps respectively, and are from one to two orders of magnitude below the switching times of nMOS or CMOS inverters constructed with such devices.

4.3.4 Hot carriers

MOSFET reliability and characteristics are greatly influenced by *hot carriers*, which are generated by strong electric fields; the largest of these is at the silicon-oxide interface in the region of the drain junction. Hot carriers, which may be either *hot electrons* or *hot holes*, are given that name because they attain energies far above the thermal energy of carriers in equilibrium with the ambient temperature. The effect applies to n- and pMOSFETs, but is more pronounced in the former on account of the high electron mobility. *Hot carriers* in an nMOSFET generate electron-hole pairs by impact ionization, and lead to a hole current which becomes part of the substrate current. That current may become sufficiently large to activate Q_1 and Q_2 in Fig. 3.1 and cause latch up (Section 3.2). Some of the hot electrons at the upper range of energies enter the gate oxide where they remain trapped. The oxide charge gives rise to a gate current and also alters the threshold voltage. Hot-carrier effects may thus lead to unacceptable characteristics and latch-up destruction. The effect increases in severity with device shrinkage and special steps in processing are required to contain it in transistors with submicron geometries.

4.4 GaAs MESFET characteristics

GaAs metal-semiconductor field-effect transistor (MESFET) characteristics are derived from those for the junction field-effect transistor (JFET) in SPICE, suitably adapted to allow for MESFET behaviour. One modification due to Curtice (1980) achieves an improved match between calculated and measured I-V characteristics by introducing a hyperbolic tangent function, which leads to

$$I_D = \beta(V_{GS} - V_T)^2(1 + \lambda V_{DS})\tanh(\alpha V_{DS}) \quad (4.45)$$

λ has been defined in Section 4.3.1 (eqns (4.32) and (4.33)), and β is the transconductance parameter for the JFET SPICE model (Quarles et al. 1993). The parameter α impacts on the slope of both the ohmic and saturation regions. The *Curtice* (or hyperbolic tangent) model includes transit time effects and a more accurate representation of C_{GS} and C_{GD}. A further improvement of the I-V relationships and the calculation of the intrinsic capacitances has resulted in the Statz model, alternatively called the *Raytheon* model, because it originated with *that* company (Statz et al. 1987). The drain current is given by

$$I_D = \frac{\beta(V_{GS} - V_T)^2}{1 + b(V_{GS} - V_T)}\left\{1 - \left(1 - \frac{\alpha V_{DS}}{3}\right)^3\right\}(1 + \lambda V_{DS}) \quad (4.46)$$

$$0 < V_{DS} < 3/\alpha$$

$$I_D = \frac{\beta(V_{GS} - V_T)^2}{1 + b(V_{GS} - V_T)}(1 + \lambda V_{ds}) \quad (4.47)$$

$$V_{DS} > 3/\alpha$$

Equations (4.46) and (4.47) give a good fit with measured values; α determines the saturation voltage and λ the output conductance. The Statz model allows for the Schottky gate diode and for velocity saturation, which peaks at an electric field $\sim 3 \times 10^3$ V/cm, a value well below that for silicon. Drain current saturation is due to velocity saturation in GaAs, and the channel pinch off in silicon.

An equivalent circuit of the GaAs MESFET is shown in Fig. 4.10. Statz (1987) pays special attention to the formulation of the depletion layer capacitances C_{GS} and C_{GD}, whose voltage dependence is much more complex than the relationship of eqn (4.41). Like the MOSFET, the MESFET displays subthreshold conduction with a formula similar to eqn (4.34). For $VGS < V_T$

$$I_D \propto \exp(aqV_{GS})/kT) \quad (4.48)$$

a being a constant.

The transient performance is governed almost entirely by the charging and discharging of the nodal capacitances; the channel transit time is negligible in comparison. For a gate length of 1 μm and a saturation velocity of 10^7 cm s^{-1}, the transit time comes to 10 ps in accordance with eqn (4.44), which applies to MOSFETs and MESFETs alike.

A high frequency parameter, which also holds for the BJT, is f_T, the frequency at which the magnitude of the current gain of an FET common-source (or a BJT common-emitter) amplifier with zero drain (or zero collector) load falls to unity. f_T is related to the saturation velocity v_{sat} and L by (Long and Butner 1990)

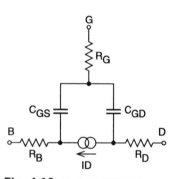

Fig. 4.10 GaAs MESFET equivalent circuit

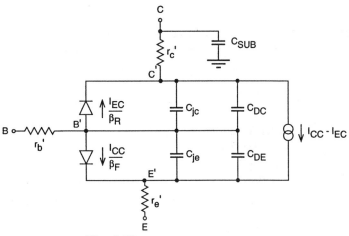

Fig. 4.11 Basic Gummel-Poon model

$$f_T = \frac{V_{sat}}{2\pi L} \tag{4.49}$$

Combining eqns (4.44) and (4.49)

$$f_T = \frac{1}{2\pi T_{tr(sat)}} \tag{4.50}$$

An alternative expression for f_T is

$$f_T = \frac{g_m}{2\pi(C_{GS} + C_{GD})} \tag{4.51}$$

where g_m is the transconductance $\partial I_d/V_{GS}$. In the saturation region $C_{GS} \gg C_{GD}$, and eqn (4.51) reduces to

$$f_T \simeq \frac{g_m}{2\pi C_{GS}} \tag{4.52}$$

Note the re-appearance of the key parameter contained in eqns (4.18) and (4.19), g_m/C.

4.5 BJT characteristics

The BJT is characterized in SPICE by a slightly modified version of the integral charge-control model developed by Gummel and Poon (1970), and described in detail by Getreu (1976, pp.96–123). The Gummel-Poon model, which is shown in Fig. 4.11, covers all four regions of operation, the off state, the active and inverse modes, and saturation. The I-V characteristics are expressed by

$$I_{CC} = I_s \{\exp(V_{B'E'}/V_t) - 1\} \tag{4.53}$$

$$I_{EC} = I_s \{\exp(V_{B'C'}/V_t) - 1\} \tag{4.54}$$

where V_t, the *thermal voltage*, equals kT/q. The junction capacitances C_{je} and C_{jc} vary with voltage in accordance with

Table 4.2 MOSFET SPICE parameters (levels 1, 2, 3, and 6)

Name	Parameter	Units	Default	Example
LEVEL	Model index	-	1	
VTO	Zero-bias threshold voltage (V_{TO})	V	0.0	1.0
KP	Transconductance parameter	A/V^2	2.0e-5	3.1e-5
GAMMA	Bulk threshold parameter (γ)	$V^{1/2}$	0.0	0.37
PHI	Surface potential (ϕ)	V	0.6	0.65
LAMBDA	Channel-length modulation (MOS1 and MOS2 only) (λ)	1/V	0.0	0.02
RD	Drain ohmic resistance	Ω	0.0	1.0
RS	Source ohmic resistance	Ω	0.0	1.0
CBD	Zero-bias B-D junction capacitance	F	0.0	20fF
CBS	Zero-bias B-S junction capacitance	F	0.0	20fF
IS	Bulk junction saturation current (I_s)	A	1.0e-14	1.0e-15
PB	Bulk junction potential	V	0.8	0.87
CGSO	Gate-source overlap capacitance per meter channel width	F/m	0.0	4.0e-11
CGDO	Gate-drain overlap capacitance per meter channel width	F/m	0.0	4.0e-11
CGBO	Gate-bulk overlap capacitance per meter channel length	F/m	0.0	2.0e-10
RSH	Drain and source diffusion sheet resistance	$\Omega/\square$	0.00	10.0
CJ	Zero-bias bulk junction bottom cap. per sq-meter of junction area	F/m^2	0.0	2.0e-4
MJ	Bulk junction bottom grading coefficient	-	0.5	0.5
CJSW	Zero-bias bulk junction sidewall cap. per meter of junction perimeter	F/m	0.0	1.0e-9
MJSW	Bulk junction sidewall grading coefficient (level 1) (level 2, 3)	-	0.50 0.33	
JS	Bulk junction saturation current per sq-meter of junction area	A/m^2		1.0e-8
TOX	Oxide thickness	meter	1.0e-7	1.0e-7
NSUB	Substrate doping	$1/cm^3$	0.0	4.0e-15
NSS	Surface state density	$1/cm^2$	0.0	1.0e-10
NFS	Fast surface state density	$1/cm^2$	0.0	1.0e-10
TPG	Type of gate material +1 opposite to substrate -1 same as substrate 0 Al gate	-	1.0	
XJ	Metallurgical junction depth	meter	0.00	1μ
LD	Lateral diffusion	meter	0.00	0.8μ
UO	Surface mobility	cm^2/Vs	600	700
UCRIT	Critical field for mobility degradation (MOS2 only)	V/cm	1.0e4	1.0e4
UEXP	Critical field exponent in mobility degradation (MOS2 only)	-	0.0	0.1
UTRA	Transverse field coefficient (mobility) (deleted for MOS2)	-	0.0	0.3
VMAX	Maximum drift velocity of carriers	m/s	0.0	5.0e4
NEFF	Total channel-charge (fixed and mobile) coefficient (MOS2 only)	-	1.0	5.0
KF	Flicker noise coefficient	-	0.0	1.0e-26
AF	Flicker noise exponent	-	1.0	1.2
FC	Coefficient for forward-bias depletion capacitance formula	-	0.5	
DELTA	Width effect on threshold voltage (MOS2 and MOS3)	-	0.0	1.0
THETA	Mobility modulation (MOS3 only)	1/V	0.0	0.1
ETA	Static feedback (MOS3 only)	-	0.0	1.0
KAPPA	Saturation field factor (MOS3 only)	-	0.2	0.5
TNOM	Parameter measurement temperature	°C	27	50

© Regents of the University of California

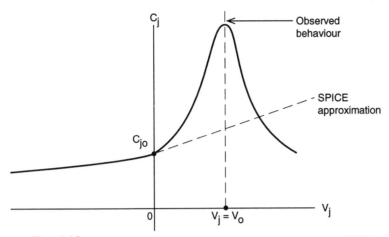

Fig. 4.12 Variation of junction capacitance with voltage—SPICE

$$C_j = C_o \left(1 - \frac{V_j}{\phi}\right)^{-m} \quad (4.55)$$

where C_o is the value of C_j when V_j, the junction voltage, equals zero, and ϕ is the barrier potential; m lies typically between $1/2$ and $1/3$ according to the nature of the junction. Applying eqn (4.55)

$$C_{jc} = C_{jco}\left(1 - \frac{V_{B'C'}}{\phi_c}\right)^{-m_e} \quad (4.56)$$

$$C_{je} = C_{jeo}\left(1 - \frac{V_{B'E'}}{\phi_c}\right)^{-m_e} \quad (4.57)$$

C_{SUB}, the collector-substrate capacitance, is likewise voltage-dependent, but is usually taken to be a constant, a reasonable simplification. Equations (4.55) to (4.57) have to be modified for simulation, because there is a singularity at $V_j = \phi$. Furthermore C_j becomes inaccurate for $V_j > \phi/2$.

The modification adopted in SPICE is to retain eqn (4.54) for negative, but to replace it for positive V_j by (Getreu 1976, pp.32–33)

$$C_j(V_j \geq 0) = C_{jo}\left(1 + \frac{mV_j}{\phi}\right) \quad (4.58)$$

The variation of C_j with V_j in accordance with eqns (4.55) and (4.58) is shown in Fig. 4.12.

The diffusion capacitances C_{DC} and C_{DE} model the mobile minority carrier charges associated with I_{CC} and I_{CE} respectively. These charges are defined by (Getreu 1976, pp.34–36)

$$Q_{DE} \equiv \tau_F . I_{CC} \quad (4.59)$$

$$Q_{DC} \equiv \tau_R . I_{EC} \quad (4.60)$$

τ_F, the total forward transit time, is the sum of the emitter delay and base transit time. τ_R, the reverse transit time, is the sum of the collector delay and the reverse base transit time. The resultant expressions for C_{DE} and C_{DC} are

$$C_{DE} \equiv \frac{Q_{DE}}{V_{B'E'}} = \frac{\tau_F I_{CC}}{V_{B'E'}} \quad (4.61)$$

$$C_{DC} \equiv \frac{Q_{DC}}{V_{B'C'}} = \frac{\tau_R I_{EC}}{V_{B'E'}} \quad (4.62)$$

Table 4.3 MESFET SPICE parameters

Name	Parameter	Units	Default	Example	Area
VTO	Pinch-off voltage	V	-2.0	-2.0	
BETA	Transconductance parameter	A/V^2	1.0e-4	1.0e-3	*
B	Doping tail extending parameter	1/V	0.3	0.3	*
ALPHA	Saturation voltage parameter	1/V	2	2	*
LAMBDA	Channel-length modulation parameter	1/V	0	1.0e-4	
RD	Drain ohmic resistance	Ω	0	100	*
RS	Source ohmic resistance	Ω	0	100	*
CGS	Zero-bias G-S junction capacitance	F	0	5 pF	*
CGD	Zero-bias G-D junction capacitance	F	0	1 pF	*
PB	Gate junction potential	V	1	0.6	
KF	Flicker noise coefficient	-	0		
AF	Flicker noise exponent	-	1		
FC	Coefficient for forward-bias depletion capacitance formula	-	0.5		

© Regents of the University of California

Summing up, C_{jc}, C_{je}, C_{DE}, and C_{DC} represent the stored base charges in the depletion regions and the injected minority carrier charge storage.

τ_F is related to f_T (defined in Section 4.4) by (Muller and Camins 1986, p.357)

$$\tau_F = \frac{1}{2\pi f_T} - \frac{(C_{je} + C_{jc})}{I_c} \quad (4.63)$$

The second term in eqn (4.63) is relatively small, leading to

$$\tau_F \simeq \frac{1}{2\pi f_T} \quad (4.64)$$

an expression very similar to eqn (4.50). τ_F is linked with f_T, because small signal measurements of f_T yield a value for τ_F. The SPICE model contains some refinements which allow for various effects. Both f_T and h_{fe}, the common emitter current gain, vary with I_C, falling off similarly at high and low currents (Fig. 4.13). The fall-off at low currents (region 1) arises from extra components in base current which are ignored in regions 2 and 3. The largest of these is a recombination of carriers in the emitter-base space charge region. The other contributions arise from a

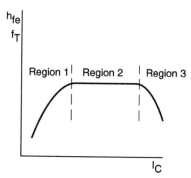

Fig. 4.13 Variation of h_{fe} and f_T with I_C

surface recombination of carriers and the formation of emitter-base surface channels. The decrease in f_T is due to the second term in eqn (4.63), which becomes significant at the low current levels in region 1, but may be ignored in the other two regions. High level injection in the base and the attendant base conductivity modulation, the *Webster* effect, reduce h_{fe} at high currents and also effect transit time, thereby causing the reduction of h_{fe} and f_T in region 3, where a customary approximation in SPICE makes both these parameters inversely proportional to I_C.

The ohmic representation of emitter and collector bulk resistances (r'_e and r'_c in Fig. 4.11) is acceptable, but a modified model is needed for r'_b, which is divided into extrinsic and intrinsic components r_{be} and r_{bi}: the base spreading resistance $r'_b = r_{be} + r_{bi}$. C_{jc} is split between B and B' in accordance with Fig. 4.14, in which X is a parameter dependent on device geometry.

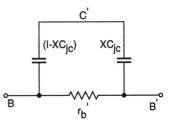

Fig. 4.14 C_{jc} distribution in SPICE

4.6 The role of SPICE

Circuit performance is computed at the design stage, and subsequently at pilot and full production stages, to verify that the stipulated design criteria are met. Small-signal amplification is amenable to analysis by means of linear equations which yield analytical expressions for circuit behaviour. Such an approach cannot be extended to large-signal non-linear operation, where simulation is based on device parameters many of which are functions of voltage and/or current. The alternative is to draw up a computer-aided analysis program, based on device parameters many of which are non-linear. SPICE has been the established standard CAD programme for that purpose since its release by the University of California, Berkeley in 1972 (Nagel 1975). SPICE uses a form of nodal analysis based on a set of non-linear first order differential equations, which are solved by the Newton Raphson method. It caters for large- or small-signal inputs, which can be time-variant or dc. For small signal ac analysis, SPICE automatically selects linearized small-signal models; non-linear effects are excluded. SPICE is a vital design tool for analogue and digital ICs. Initial estimates of circuit behaviour, based on hand-analysis approximations, are followed by SPICE simulations.

SPICE developments since its inception have taken place on a number of fronts. The algorithms have been greatly improved to give better convergence, and the C language has largely replaced Fortran for better efficiency and faster execution. Device models continue to be developed and modified for improved validity. The University of California remains the spearhead and catalyst for much of this work. SPICE was written by engineers and a felicitous co-operation between hardware engineers and computer scientists within the Department of Electrical Engineering and Computer Science at the University of California has resulted in a highly user-friendly programme for circuit designers.

The free availability of SPICE granted by the University of California has been followed by developments, which are still ongoing, at Berkeley, and by a widespread production of proprietary SPICE versions. Vendors like Mentor Graphics, Valid and Cadence, to mention a few suppliers of CAD-CAE tools, incorporate in-house versions of SPICE in their products. Likewise many semiconductor houses include SPICE in their provision for ASIC design. An advanced sophisticated package, HSPICE

Device behaviour and modelling

Table 4.4 BJT SPICE parameters

Name	Parameter	Units	Default	Example	Area
IS	Transport saturation current	A	1.0e-16	1.0e-15	*
BF	Ideal maximum forward beta	-	100	100	
NF	Forward current emission coefficient	-	1.0	1.0	
VAF	Forward Early voltage	V	infinite	200	
IKF	Corner for forward beta high current roll-off	A	infinite	0.01	*
ISE	B-E leakage saturation current	A	0	1.0e-13	*
NE	B-E leakage emission coefficient	-	1.5	2	
BR	Ideal maximum reverse beta	-	1	0.1	
NR	Reverse current emission coefficient	-	1	1	
VAR	Reverse Early voltage	V	infinite	200	
IKR	Corner for reverse beta high current roll-off	A	infinite	0.01	*
ISC	B-C leakage saturation current	A	0	1.0e-13	*
NC	B-C leakage emission coefficient	-	2	1.5	
RB	Zero bias base resistance	Ω	0	100	*
IRB	Current where base resistance falls halfway to its min value	A	infinite	0.1	*
RBM	Minimum base resistance at high currents	Ω	RB	10	*
RE	Emitter resistance	Ω	0	1	*
RC	Collector resistance	Ω	0	10	*
CJE	B-E zero bias depletion capacitance	F	0	2 pF	*
VJE	B-E built in potential	V	0.75	0.6	
MJE	B-E junction exponential factor	-	0.33	0.33	
TF	Ideal forward transit time	s	0	0.1 ns	
XTF	Coefficient for bias dependence of TF	-	0		
VTF	Voltage describing VBC dependence of TF	V	infinite		
ITF	High current parameter for effect on TF	A	0		*
PTF	Excess phase at freq=1.0/(TF*2PI) HZ	deg			
CJC	B-C zero- bias depletion capacitance	F	0	2 pF	*
VJC	B-C built-in potential	V	0.75	0.5	
MJC	B-C junction exponential factor	-	0.33	0.5	
XCJC	Fraction of B-C depletion capacitance connected to internal base node	-	1		
TR	Ideal reverse transit time	s	0	10 ns	
CJS	Zero-bias collector-substrate capacitance	F	0	2 pF	
VJS	Substrate junction built-in potential	V	0.75		
MJS	Substrate junction exponential factor	-	0	0.5	
XTB	Forward and reverse beta temperature exponent	-	0		
EG	Energy gap for temperature effect on IS	eV	1.11		
XTI	Temperature exponent for effect on IS	-	3		
KF	Flicker-noise coefficient	-	0		
AF	Flicker-noise exponent	-	1		
FC	Coefficient for forward-bias depletion capacitance formula	-	0.5		
TNOM	Parameter measurement temperature	°C	27	50	

© Regents of the University of California

from Meta Software, contains an exceptionally large variety of models. PC versions of SPICE include PSPICE of Microsim and IsSPICE of Intusoft. Most proprietary SPICE packages have fast-processor graphical capability, permitting graphical display and analysis of the results. A comparatively recent addition offered by several SPICE vendors is mixed-mode simulation, a combination of analogue (SPICE) and digital simulation with simultaneous graphical display on a common time axis. The syntax for all these SPICE versions is virtually unchanged from the original SPICE syntax, so that users will experience no difficulty when switching from one package to another. It is assumed that readers are familiar with the use of SPICE. It is a skill which can be readily acquired with a moderate amount of initial guidance.

Parameters for the device models of SPICE are usually supplied by the IC vendor. The user is rarely, if at all, in a position to undertake parameter extraction, less still to obtain the statistical distribution of their values in order to compute performance spreads in production.

Substantial developments by semiconductor houses continue in order to obtain the best possible SPICE models for their ICs. These efforts are paralleled by developments to improve the reliability of the numerical methods and to speed up the process of simulation (IEE 1993).

4.7 SPICE models

4.7.1 Introduction

The models presented are based on the SPICE 3 Version 3f3 User's Manual, except where stated otherwise (Quarles et al. 1993). Earlier SPICE Guides from the University of California and proprietary SPICE versions like HSPICE, PSPICE etc. contain models which are similar (Vladimirescu et al. 1981, 1992). The reproduction of information from the SPICE 3 User's Manual, freely granted by the Regents of the University of California, is gratefully acknowledged.

4.7.2 MOSFET models

SPICE offers four models, identified by levels, which are (Vladimirescu and Liu 1980; Sheu et al. 1985):

    LEVEL=1   Shichman-Hodges Model
    LEVEL=2   MOS2
    LEVEL=3   MOS3
    LEVEL=4   Berkeley Short-Channel IGFET Model (BSIM)

Level 1 is a simple FET model, whereas the level 2 model incorporates some second order effects. Level 3 is a semi-empirical model, whose parameters are largely determined by curve fitting. Another two comparatively recent models are available, namely:

    LEVEL=5   New BSIM (BSIM2) (Jeng 1990)
    LEVEL=6   MOS6 (Sakurai and Newton 1990)

MOSFET SPICE parameters are listed in Table 4.2. Explanatory comments for employing these models are confined to Levels 1 to 3, because these are in widest use. The references cited for the BSIM and the MOS6 models, combined with the SPICE 3 Guide, contain adequate information for their use. The BSIM models are for short channel lengths, extending to deep submicron geometries. BSIM, also called BSIM1 by Jeng (1990), is modelled for an effective channel length and gate oxide thickness down to ~ 1 μm and 15 nm respectively. The corresponding dimensions for BSIM2 are ~ 0.2 μm and 3.6 nm.

Accuracy adequate for an inital assessment can be obtained by inserting figures based on measurement for a comparatively small number of key parameters, and using default values, automatically inserted by SPICE in the absence of an insertion, for the remainder.

Additional parameters are specified in the device statement. These are L and W (effective values should be used), AD (area of drain layer),

Table 4.5 Key SPICE parameters—submicron npn transistor

Parameter	Value
BF	100
CJE (fF)	11
CJC (fF)	10
CJS (fF)	25
RB (Ω)	110
RC (Ω)	30
TF (ps)	9

and AS (area of source layer). Default values are 1 m (!) for L and W, and zero for AD and AS. The astronomic default values for L and W are admissible, because the equations for drain current involve the factor (W/L), but not W or L on their own (eqns (4.21) and (4.22) refer). Other values are stipulated in some proprietary SPICE programmes. For example L and W default to 100 μm in PSPICE, and to $1 \times$ SCALE in HSPICE, where SCALE, user-set, is a factor of 10^{-n}, n being an integer. For $n = 6$, L and W default to 1 μm.

The dc characteristics of levels 1 to 3 are largely defined by VTO, KP, LAMBDA, PHI, and GAMMA. These parameters, and the nature of the capacitances, have been explained in Section 4.3, where k' (Section 4.3.1) equals KP in Table 4.2. CBD and CBS in Table 4.2 correspond to C_{BD1} abd C_{BS1} in Section 4.3.2. Some SPICE parameters allow for alternatives like expressing the saturation current in absolute form, IS(A), or as a current density JS(A/m^2). Another alternative exists for CBD and CBS, which may be either quantified directly, or which are computed by SPICE from specified values of CJ, AD and AS. If CJ is omitted in the model list, it is computed from NSUB if provided. When values of CBD, CBS and CJ are provided, they take precedence over computation.

The area factor, available for the junction diode, BJT, JFET and MESFET devices, is a scaling factor which computes the actual magnitude of a parameter by multiplying the value entered (which applies to an area factor of 1.0) by the area factor. Alternatively values can be entered with the area factor at 1.0 (the assumed default setting). It applies to parameters with a starred (*) entry in the area column.

4.7.3 GaAs MESFET model

The Statz model is the SPICE standard, and its parameters are listed in Table 4.3 (Statz *et al.* 1987). An outline of this model has been given in Section 4.4. The Statz model is also found in other SPICE packages, for example PSPICE. HSPICE contains an adaptation of the Curtice model and the Statz model.

Equations (4.46) and (4.47) apply. The dc characteristics are defined by the parameters VTO, B, BETA, ALPHA, and LAMBA.

4.7.4 BJT model

The BJT model is a modification of the integral charge control model of Gummel and Poon, described in Section 4.5. The model parameters are listed in Table 4.4.

The following qualify for an inevitably subjective list of key parameters: IS, BF, RB, RC, CJE, CJC, CJS, and TF. SPICE represents r'_{bb} in Fig. 4.11 by a complex function of RB, RBM, IRB, and XCJC (see Fig. 4.14)

The example values in Table 4.4 apply to a vintage switching transistor of the 1965–1970 era. Table 4.5 gives typical values for key parameters of a contemporary submicron transistor.

4.7.5 Junction diode

The junction diode model parameters are contained in Table 4.6. The junction diode is not used all that much in ICs, but one variant of great interest is the Schottky (barrier) diode (Sbd), whose parameters are es-

Table 4.6 Junction diode SPICE parameters

Name	Parameter	Units	Default	Example	Area
IS	Saturation current	A	1.0e-14	1.0e-14	*
RS	Ohmic resistance	Ω	0	10	*
N	Emission coefficient	-	1	1.0	
TT	Transit time	sec	0	0.1 ns	
CJO	Zero-bias junction capacitance	F	0	2 pF	*
VJ	Junction potential	V	1	0.6	
M	Grading coefficient	-	0.5	0.5	
EG	Activation energy	eV	1.11	1.11 Si	
				0.69 Sbd	
				0.67 Ge	
XTI	Saturation-current temperature exp.	-	3.0	3.0 jn	
				2.0 Sbd	
KF	Flicker noise coefficient	-	0		
AF	Flicker noise exponent	-	1		
FC	Coefficient for forward-bias depletion capacitance formula	-	0.5		
BV	Reverse breakdown voltage	V	infinite	40	
IBV	Current at breakdown voltage	A	1.0e-3		
TNOM	Parameter measurement temperature	°C	27	50	

© Regents of the University of California

sentially covered in Table 4.6. The main changes relative to the junction diode are elimination of charge storage (TT ~ zero), a higher ohmic resistance (~1 kΩ) and a very small junction capacitance. CJO depends on the junction area, but is typically in the range (1–10) fF.

References

Curtice, W.R. (1980). A MESFET model for use in the design of GaAs integrated circuits. *IEEE Transactions of Microwave Theory and Techniques*, **28**, 448-56.

Getreu, I. (1976). *Modeling the bipolar transistor*. Tektronix, USA.

Gummel, H.K. and Poon, H.C. (1970). An integral charge control model of bipolar transistors. *Bell System Technical Journal*, **49**, 827-52.

IEE (1993). *SPICE: Surviving problems in circuit evaluation*. IEE Colloquium, London 1993. Digest No:1993/154.

Jeng, M-C (1990). *Design and modeling of deep-submicrometer MOSFETs.*. Memorandum UCB/ERL M90/90, Electronics Research Laboratory, University of California, Berkeley, USA.

Long, S.I. and Butner, S.E. (1990). *Gallium arsenide digital integrated circuit design*, p.42. McGraw-Hill, USA.

Muller, R.S. and Kamins, T.I. (1986). *Device electronics for integrated circuits*, (2nd edn), p.357. Wiley, USA.

Nagel, L.W. (1975). *SPICE2: A computer program to simulate semiconductor circuits*. Memorandum No. ERL-M520, Electronics Research Laboratory, College of Engineering, University of California, Berkeley, USA.

Nicollian, E.H. and Brews, J.R. (1982). *MOS physics and technology*. Wiley, USA.

Quarles, T., Newton, A.R., Pederson, D.O., and Sangiovanni-Vincentelli, A. (1993). *SPICE3 version 3f3 user's manual*. Department of Electrical Engineering and Computer Science, University of California, Berkeley, USA.

Sakurai, T. and Newton, A.R. (1990). *A simple MOSFET model for circuit analysis and its application to CMOS gate delay analysis and series-connected MOSFET structure*. Memorandum UCB/ERL M90/19, Electronics Research Laboratory, College of Engineering, University of California, Berkeley, USA.

Sheu, B.J., Scharfetter, D.L., and Ko, P.K. (1985). *SPICE2 implementation of BSIM*. Memorandum ERL M85/42, Electronics Research Laboratory, University of California, Berkeley, USA.

Statz, H., Newpan, P., Smith, I.W., Pucel, R.A., and Haus, H.A. (1987). GaAs FET device and circuit simulation in SPICE. *IEEE Transactions on Electron Devices*, **34**, 160–9.

Stotts, S. (1989). Introduction to implantable biomedical IC design. *IEEE Circuits and Devices Magazine*, **5** (1), 12-18.

Tsivides, Y.P. (1987). *Operation and modeling of the MOS transistor*. McGraw-Hill, USA.

Uyemura, J.P. (1988). *Fundamentals of MOS digital integrated circuits*, pp.18-22. Addison Wesley, USA.

Vladimirescu, A. and Liu, S. (1980). *The simulation of MOS integrated circuits using SPICE2*. Memorandum ERL M80/7, Electronics Research Laboratory, University of California, Berkeley, USA.

Vladimirescu, A., Zhang, K., Newton, A.R., Pederson, D.O., and Sangiovanni-Vincentelli, A. (1981). *SPICE Version 2G.5 User's guide*. Department of Electrical Engineering and Computer Science, University of California, Berkeley, USA.

Vladimirescu, A., Zhang, K., Newton, A.R., Pederson, D.O., and Sangiovanni-Vincentelli, A. (1992). *User's guide, SPICE 2G.6*. Department of Electrical Engineering and Computer Science, University of California, Berkeley, USA.

Further reading

Ashburn, P. (1988). *Design and realization of bipolar transistors*. Wiley, UK.

Banzhaf, W. (1989). *Computer-aided circuit analysis using SPICE*. Prentice-Hall, USA.

Conant, R. (1993). *Engineering circuits analysis with PSPICE and probe*. McGraw-Hill, USA.

Keown, T. (1993). *PSPICE and circuit analysis*, (2nd edn). Merill/McMillan, USA.

Long, S.I. and Butner, S.E. (1990). *Gallium arsenide digital integrated circuit design*. McGraw-Hill, USA.

Massobrio, G. and Antognetti P. (1993). *Semiconductor device modeling with SPICE*, (2nd edn). McGraw-Hill, USA.

Muller, R.S. and Kamins, T.I. (1986). *Device electronics for integrated circuits*, (2nd edn). Wiley, USA.

Quarles, T., Newton, A.R., Pederson, D.O., and Sangiovanni-Vincentelli, A. (1993). *SPICE3 version 3f3 user's manual*. Department of Electrical Engineering and Computer Sciences, University of California, Berkeley, USA.

Tsivides, Y.P. (1987). *Operation and modeling of the MOS transistor*. McGraw-Hill, USA.

Vladimirescu, A. and Liu, S. (1980). *The simulation of MOS integrated circuits using SPICE 2*. Memorandum ERL M80/7. Electronics Research Laboratory, College of Engineering, University of California, Berkeley, USA.

5
Digital circuits — techniques and performance

5.1 Introduction

Logic ICs constitute the core of digital electronics and are the main theme of this chapter. Numerous logic families have emerged since the beginning of the IC era around 1960. The popularity of the various families has changed with time. The arrival of LSI, followed by VLSI and the ongoing shrinkage of device geometries, has led to a distinct preference for two silicon families, CMOS and ECL; these are treated in more detail than the other families. Dynamic MOS logic circuits, which rely on temporary storage of data across intrinsic transistor capacitors, were a mainstay of microprocessors in the 1970s, but are now very much on the decline and are outlined only briefly for that reason. The final section contains an extensive survey of IC logic, ranging from SSI to VLSI, and reflects on the capability of deep submicron structures.

5.2 Logic Circuits

5.2.1 Design criteria

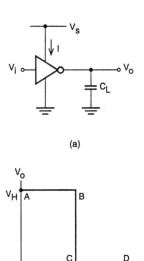

Fig. 5.1 Logic inverter (a) schematic (b) transfer characteristic

Logic circuits are based on the inverter, which is the basic circuit brick of digital ICs. The inverter, logic gates and bistables form the mainstay of SSI. Some prime design criteria will now be explained by considering the logic inverter in Fig. 5.1(a); the explanation applies equally to a logic gate. The ideal transfer characteristic in Fig. 5.1(b) has a vertical section BC at $V_i = 1/2(V_H + V_L)$ halfway between V_H and V_L, the high and low logic levels. That, and the horizontal sections AB and CD ensure equal noise immunity in both logic states. Noise immunity is maximized by making V_H and V_L as closely as possible equal to V_S and ground respectively. Switching speed, a key parameter, is expressed in terms of propagation delay t_{PLH} and t_{PHL} and/or transition times t_{TLH} and t_{THL}. These parameters have been defined in Section 4.2.

The power consumed by the inverter is the sum of two components, P_s, the static input power in the quiescent state, and P_d, the dynamic power demanded during logic transitions, the inverter being exercized at a clock rate f. Figure 5.2 illustrates the nature of the current flow. I_s, the mean current in the *steady state*, is the weighted average of I_{S1} and I_{S2}, and is independent of f. On the other hand I_d, the mean of the dynamic currents which flow during *transitions*, is proportional to f, so that

$$P_d \propto f \qquad (5.1)$$

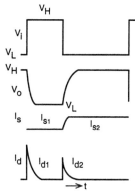

Fig. 5.2 Inverter (Fig. 5.1) current flow

118 Digital circuits — techniques and performance

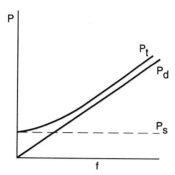

Fig. 5.3 Gate input power vs. frequency

The dynamic currents arise largely from the charging and discharging of C_L (Fig. 5.1), which is the sum of the inverter output capacitance, the capacitance(s) of the interconnect(s), and the input capacitance presented by the following stage(s). Static, dynamic, and total inverter input powers are plotted against frequency in Fig. 5.3. Figures 5.2 and 5.3 are purely qualitative representations. The quantitative evaluation of P_d is dealt with in Section 5.2.4. Switching times of circuits with structures of ~1 μm or less are governed largely by C_L. The higher the charge and discharge currents, the faster will be the switching actions, and circuit design involves a trade-off between speed and power. Speed, although vitally important, is not the only consideration. Gate input power imposes a limit on the function density of a VLSI die, and a compromise is arrived at to achieve high function density with adequate speed.

5.2.2 Overview of logic familes

Logic circuits are associated with *families*, each of which is characterized by its particular technology and basic circuit techniques. Bipolar technology reigned until about 1971 when MOS appeared in earnest with the production of the first microprocessors.

Transistor-transistor logic (TTL) is the leading bipolar family, which has experienced the widest use since its introduction in 1965. The other prominent bipolar family, emitter-coupled logic (ECL), was introduced in 1962. ECL, for a given geometry, achieves a higher speed than TTL at the cost of more power consumption, and is used where its superior speed is required. TTL, ECL, and other logic familes are produced in various ranges with different speeds, with the fastest and most power-hungry ranges used where the highest speeds are called for.

MOS logic arrived with the introduction of complementary MOS (CMOS) in 1963. The tremendous advantage of CMOS logic, virtually no power consumption in the quiescent state, was immediately apparent, but MOS technology at that time was inadequate to allow large-scale exploitation of this new technology. Instead efforts to master MOS processing were at first concentrated on producing single-channel (pMOS or nMOS) logic which, because it consumes far less power than TTL in the quiescent state, was evidently attractive for large-scale integration. CMOS remained confined for a long time to one SSI/MSI family with only very moderate speed, whilst nMOS, preceded by a brief spell—only anecdotally significant—of pMOS, was developed reaching LSI levels in the late 1960s and extending to VLSI soon afterwards. nMOS, it should be noted, was never produced in SSI/MSI format. It now became evident that MOS would sooner or later replace bipolar logic except at speeds beyond its reach. CMOS was set to replace nMOS once its technology had matured, and that stage has now been reached; it has taken tremendous efforts to get there. In consequence CMOS, the preferred technology for VLSI, has also been developed extensively in several ranges of SSI/MSI, where it is replacing (but not exclusively so) TTL.

An interlude occurred with the invention and subsequent development of a new bipolar circuit technique, integrated injection logic (I^2L) in 1972. I^2L was intended to be a bipolar challenge to MOS for low power consumption and experienced moderate success in VLSI (mainly microprocessors), but soon faded out in the light of MOS advances.

Last but not least comes GaAs, which, by the mid-1970s had carved out for itself a commanding niche in microwave amplification. Like CMOS, its realization for digital ICs had to await advances in technology which, as for CMOS, took years of very great efforts to yield the desired results. In the digital realm, GaAs is confined to VLSI.

The mainstay of VLSI is CMOS, with ECL reserved for fast applications beyond the reach of CMOS. TTL is becoming obsolescent for new designs, but is still being consumed in large volume. GaAs logic is the fastest. The extent to which it will establish itself vis-à-vis bipolar logic remains to be seen. The latest newcomer to logic ICs, and one which is gaining tremendous momentum, is BiCMOS, which advances the speed of CMOS but demands less power for doing so than equally fast bipolar logic. BiCMOS is set to become a major contender for high speed VLSI, and has spawned SSI and MSI families. The basic circuits of nMOS, CMOS, ECL, TTL, I²L, BiCMOS and GaAs are described in the sections which follow.

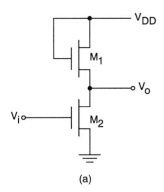

(a)

5.2.3 nMOS

The nMOS inverter shown in Fig. 5.4 illustrates a general characteristic of nMOS and CMOS logic, the all-transistor circuit in which transistors, representing dynamic loads, replace resistors. This results in a great saving of silicon estate and in reduced power consumption.

Analyzing Fig. 5.4, let V_H and V_L be the high and low logic levels; V_{DD} is positive. When $V_i = V_L$ (assumed to be less than V_T) M_2 is off and the only current flowing through M_1 is the very small leakage current of a few nA of M_2, making V_{GS} very nearly equal to V_T, so that

$$V_H \simeq V_{DD} - V_T \quad (5.2)$$

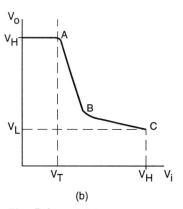

(b)

Fig. 5.4 nMOS inverter (a) schematic (b) transfer characteristic

The traverse of AB in Fig. 5.4(b) takes place with M_1 and M_2 in saturation, commencing at A when $V_i \simeq V_T$, and ending at B when M_2 is on the edge of saturation. The stipulation for operation on the linear/saturation borderline, $V_{DS} = V_{GS} - V_T$ (Fig. 4.3), is now

$$V_o = V_i - V_T \quad (5.3)$$

M_2 is in the linear region from B to C, reaching the terminal voltage V_L when $V_i = V_H$. The slope AB and the values of V_o and V_i at B, can be obtained from eqn (4.22), which becomes

$$(\beta_R)^{\frac{1}{2}}(V_i - V_T) = (V_{DD} - V_O - V_T) \quad (5.4)$$

where the beta ratio β_R is defined by

$$\beta_R \equiv \frac{(W/L)_{\text{driver}}}{(W/L)_{\text{load}}} = \frac{(W/L)_{M2}}{(W/L)_{M1}} \quad (5.5)$$

It follows from eqn (5.4) that AB is a straight line of slope

$$(dV_o/dV_i) = -(\beta_R)^{\frac{1}{2}} \quad (5.6)$$

At B, V_{DS} is $V_o(B)$ where

$$V_o(B) = V_i(B) - V_T \qquad (5.7)$$

Substituting eqn (5.7) in (5.4)

$$V_o(B) = \frac{V_{DD} - V_T}{1 + (\beta_R)^{\frac{1}{2}}} \qquad (5.8)$$

and

$$V_i(B) = V_T + \frac{V_{DD} - V_T}{1 + (\beta_R)^{\frac{1}{2}}} \qquad (5.9)$$

The higher β_R, the steeper will be the slope AB and the smaller will be $V_o(B)$. These are desirable effects for obtaining a very good transfer characteristic (steep slope) and high noise immunity (low $V_o(B)$). Calculation of V_L is more difficult; it demands a method of trial and error in hand analysis (Uyemura 1988, pp.104–5; Hodges and Jackson, 1988, pp.64–66). However $V_o(B)$ is close to V_L in practice for $\beta_R > 10$, and is therefore a good indicator of V_L. Typically $V_{DD} = 5$ V, $V_T = 1.5$ V, leading to the variation of $V_o(L)$ with β_R shown in Table 5.1. nMOS logic is called *ratio-type*, or *ratioed* logic, because V_L depends on β_R. It is demanding on silicon estate, requiring a large area in order to obtain an acceptably low V_L. V_H can be increased by replacing M_1 in Fig. 5.4 with a depletion MOSFET (Fig. 5.5). With M_2 off, M_1 is in good conduction and offers a very low effective resistance, so that $V_H \simeq V_{DD}$. nMOS NAND and NOR gates are shown in Fig. 5.6, from which it can be deduced that an n-input gate demands (n+1) transistors. The logic functions in Figs 5.6 and 5.7 hold for positive logic, which signifies

Table 5.1 V_L versus β_R (Fig. 5.4)

V_L (V)	β_R
1.17	4
0.84	9
0.58	25

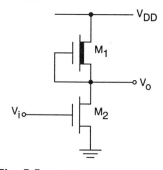

Fig. 5.5 nMOS inverter-depletion load

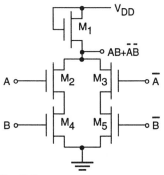

Fig. 5.7 nMOS exclusive NOR gate

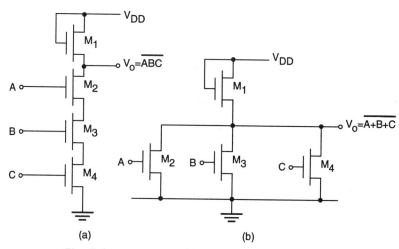

Fig. 5.6 nMOS gates (a) NAND gate (b) NOR gate

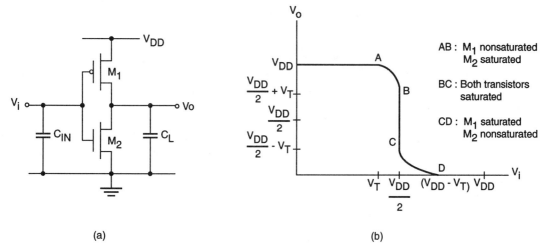

Fig. 5.9 CMOS inverter (a) circuit (b) transfer characteristic

that logic 1 is the more positive of the two levels. This convention is adopted throughout the book, unless stated otherwise. The ability to stack transistors vertically is used to generate functions with AND–OR combinations like the exclusive NOR gate in Fig. 5.7.

nMOS logic makes extensive use of a transistor switch, the *pass transistor* shown in Fig. 5.8. The transistor is a bi-directional switch (the positions of V_i and V_o in Fig. 5.8 can be interchanged), on for $V_G > V_T$ and off for $V_G < V_T$. In practice V_G is V_{DD} (supply voltage) for 'on', and ground for 'off'. The switch operates for an input voltage range from zero to $(V_{DD} - V_T)$, and the output is extremely close to the input voltage over that range.

5.2.4 CMOS

The basic CMOS logic inverter is shown in Fig. 5.9, in which C_{IN} and C_L are the intrinsic input and output capacitances. For $V_i \simeq V_{DD}$, M_2 is on and M_1 is off. M_1 presents a very high load resistance drawing only a very small leakage current. M_2 on the other hand offers a small resistance, ~(100–200) Ω, so that $V_o \simeq$ zero volts. Likewise, when $V_i \simeq 0$ V, M_1 offers a small resistance and M_2 a very high resistance, giving $V_o \simeq V_{DD}$. The transfer characteristic for transistors with identical dc parameters is shown in Fig. 5.9(b). V_H and V_L, the output levels, are within a few mV of V_{DD} and ground respectively. The outstanding feature which has made CMOS the technology of choice, particularly for VLSI, is that there is no current consumption (except for the leakage current of the off transistors) in the steady state: the quiescent input power is virtually zero. In a digital system only a small proportion of the circuits is exercized at any one time. The remainder draw a quiescent current, and the fact that this current is near-zero in CMOS gives it a decisive advantage over other logic for VLSI. Yet another advantage of CMOS, this time in comparison with nMOS, is that the logic levels are independent of transistor dimensions. Unlike nMOS, CMOS is a *ratioless* (or *ratio-independent*) logic.

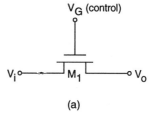

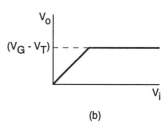

Fig. 5.8 nMOS pass transistor (a) schematic (b) transfer characteristic

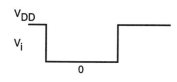
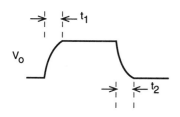
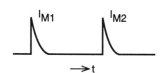

Fig. 5.10 Charge and discharge of C_L in Fig. 5.9

Example 5.1

Calculate V_H and V_L for the inverter in Fig. 5.9 for leakage currents of 2 μA and 40 μA of the 'off" transistor, given that the input levels are V_{DD} and zero volts. $V_{DD} = 5$ V, $|V_T| = 1$ V, and $k'(W/L) = 100$ μA/V² (eqns (4.21) and (4.22)). From the symmetry of the circuit, it is only necessary to calculate V_L or V_H, the other value can be deduced immediately.

We shall accordingly calculate V_L. M_2 will be in the linear region, drawing the leakage current of M_1 which is off. Equation (4.21) applies and, making numerical substitutions, gives

$$V_L^2 - 8V_L + (I_D/50) = 0 \tag{5.10}$$

where I_D is in μA. The solutions are $V_L = 5$ mV and 101 mV for I_D equal to 2 μA and 40 μA respectively. These values of I_D are the maximum specified for the 2-i/p NAND gate of the 54/74HC high speed CMOS family at 25 °C and 125 °C respectively. The corresponding solutions for V_H are 4.995 V and 4.899 V.

Switching circuits demand power for charging the nodal capacitances, and an expression for that dynamic power, P_d, will now be derived for the inverter in Fig. 5.9. C_{IN} is not being considered, because it is charged by the circuit driving the inverter. From considerations of basic electrostatics, the energy required to charge C_L during t_1 in Fig. 5.10 is $C_L V_H^2 \simeq C_L V_{DD}^2$, since $V_H \simeq V_{DD}$. Half of this is stored by C_L, the other half is dissipated in M_1. C_L discharges via M_2 to $V_L \simeq 0$ V during t_2. If the repetition frequency is f, P_d is given by

$$P_d = C_L V_{DD}^2 f \tag{5.11}$$

which is a specific case of the general expression

$$P_d = C_L \delta V^2 f \tag{5.12}$$

where

$$\delta V \equiv V_H - V_L \tag{5.13}$$

$(P_d/2)$, equal to $(1/2)CV_{DD}^2 f$, is dissipated in M_1 during t_1, and in M_2 during t_2. The above expressions are derived with more rigour in Example 5.2.

Example 5.2

Prove that the power required to charge C_L in Fig. 5.9 is $CV_{DD}^2 f$, and explain the power distribution during a complete clock cycle.

The relevant waveforms are shown in Fig. 5.10.

(i) Charging during t_1

$$P_d = \frac{1}{t}\int_0^t V_{DD} i_1 dt \tag{5.14}$$

where $t = 1/f$ and i_1, the charging current flowing *into* C_L, is given by

$$i_1 = C_L \frac{dV_o}{dt} \tag{5.15}$$

Combining (5.14) and (5.15)

$$P_d = \frac{1}{t}\int_0^{V_{DD}} V_{DD} C_L dV_o = C_L V_{DD}^2 f \tag{5.16}$$

The power dissipated in M_1 is

$$\begin{aligned} P_{M_1} &= \frac{1}{t}\int_0^t (V_{DD} - V_o) i_1 dt \\ &= \frac{1}{t}\int_0^{V_{DD}} C_L (V_{DD} - V_o) dV_o \\ &= \frac{1}{2} C_L V_{DD}^2 f \end{aligned} \tag{5.17}$$

(ii) Discharge during t_2
Current is now flowing *out of* C_L, so that

$$i_2 = -C_L \frac{dV_o}{dt} \tag{5.18}$$

The power dissipated in M_2 is

$$P_{M_2} = \frac{1}{t}\int_0^t V_o i_2 dt = -\frac{1}{t}\int_{V_{DD}}^0 V_o dV_o = \frac{1}{2} C_L V_{DD}^2 f \tag{5.19}$$

The results hold equally for the general expression of eqn (5.12) with δV taking the place of V_{DD}.

Additional dynamic power is demanded by the inverter, because M_1 and M_2 conduct simultaneously for a short time during the transitions between logic states. This additional power P_d' is also proportional to f, leading to

$$P_d' = Af \tag{5.20}$$

where A is a circuit parameter, and gives a total dynamic power P_{dt} of

$$P_{dt} = (CV_{DD}^2 + A)f \tag{5.21}$$

P_d' is relatively small in CMOS for the nanosecond and subnanosecond rise times obtained with geometries of a few μm or less, but becomes appreciable for signals with long transition times (Veendrick 1984). It is neglected in this text, because only fast logic is being considered.

Analytical expressions for rise and fall times, based on the charge and discharge of C_L, are (Weste and Eshraghian 1985, pp.137–140)

$$t_{TLH} = \tau_p \left\{ \frac{(2|V_{Tp}| - 0.2 V_{DD})}{(V_{DD} - |V_{Tp}|)} + \ln\left[\frac{1.9 V_{DD} - 2|V_{Tp}|}{0.1 V_{DD}}\right] \right\} \tag{5.22}$$

$$t_{THL} = \tau_n \left\{ \frac{(2V_{Tn} - 0.2V_{DD})}{(V_{DD} - V_{Tn})} \right.$$

$$\left. + \ln\left[\frac{1.9V_{DD} - 2V_{Tn}}{0.1V_{DD}}\right]\right\} \quad (5.23)$$

$$\tau_p = \frac{C_L}{\beta_p(V_{DD} - |V_{Tp}|)} \quad (5.24)$$

$$\tau_n = \frac{C_L}{\beta_n(V_{DD} - V_{Tn})} \quad (5.25)$$

$$\beta_p = \frac{\mu_p \epsilon_{ox}}{t_{ox}}\left(\frac{W}{L}\right)_p, \quad \beta_n = \frac{\mu_n \epsilon_{ox}}{t_{ox}}\left(\frac{W}{L}\right)_n \quad (5.26)$$

For identical dimensions and with V_{Tn} equal to $|V_{Tp}|$, $t_{THL} \sim 2t_{TLH}$ because $\mu_n \sim 2\mu_p$. It is good practice to make t_{THL} and t_{TLH} equal, and this is achieved making $(W/L)_p/(W/L)_n$ equal to (μ_n/μ_p).

Example 5.3

Calculate t_{THL} for the inverter in Fig. 5.9, given the following data: $V_{Tn} = |V_{Tp}| = 0.9$ V, $V_{DD} = 3.3$ V, $t_{ox} = 25$ nm, permittivity of $SiO_2 = 3.45 \times 10^{-2}$ fF/μm, $\mu_n = 450$ cm^2/Vs, $\mu_p = 150$ cm^2/Vs, $(W/L)_n = 1$, $(W/L)_p = 3$, $C_L = 20$ fF.

Note that the (W/L) ratios have been chosen to give equal t_{TLH} and t_{THL}. Substituting in eqns (5.23) and (5.25) $\tau_n = 1.34 \times 10^{-10}$ s, and the two terms in the bracket of eqn (5.23) come to 0.613 and 2.606 respectively, giving $t_{THL} = 0.43$ ns.

C_L has been put at 20 fF, a realistic minimum value for the sum of inverter output capacitance and the input capacitance of a similar stage. This assumes a very short interconnection; wire capacitance has not been allowed for.

Propagation delays can be deduced from transition times by using eqn (4.7) or the more approximate expression of eqn (4.9). These equations presuppose that the output voltage is governed by the single-pole relationship of eqn (4.4). In practice the output waveform tends to be more complex and that, combined with significant turn-on and turn-off details for some logic circuits, frequently leads to propagation delays higher than those deduced from eqns (4.7) or (4.9).

CMOS NAND and NOR gates are illustrated in Fig. 5.11. It can be seen that the transistor count, 2n for an n-input gate, is higher than the corresponding count (n + 1) for nMOS logic. That drawback of CMOS is outweighted by its near-zero static power consumption. CMOS, like nMOS, allows for series stacking and for AND–OR compounding to generate functions like the exclusive-NOR gate in nMOS (Fig. 5.7).

The CMOS counterpart of the nMOS pass transistor (Fig. 5.8) is the CMOS transmission gate shown in Fig. 5.12. The gate, consisting of M_1 and M_2, is controlled by C and is, like the nMOS pass transistor, bidirectional. With C at V_{DD}, V_B equals V_A for $0 \leq V_A \leq V_{DD}$. The

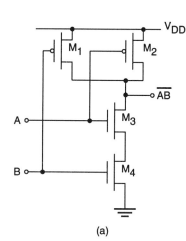

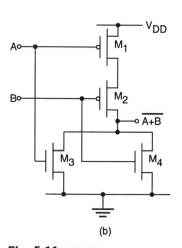

Fig. 5.11 CMOS gates (a) NAND (b) NOR

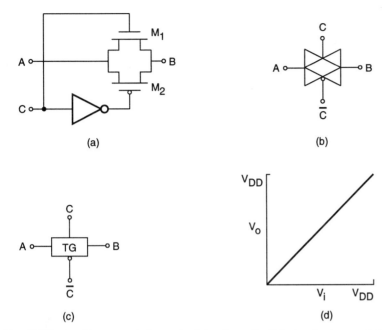

Fig. 5.12 CMOS transmission gate (a) schematic (b) standard representation (c) alternative representation (d) transfer characteristic

transmission gate is an important logic element; in many applications *minimum* transistors ($W = L$) are acceptable for M_1 and M_2, giving economy in size.

An alternative logic arrangement is the *three-state* (or *tri-state*, a widely used proprietary designation by National Semiconductor) output, in which the output of the logic element is either in the normal mode or in the *third state*, an open-circuit condition which is completely independent of the inputs and presents a very high output impedance. Third-state outputs are especially useful when large numbers of logic elements are connected to a data or control bus. All elements, except those required for processing at a particular time, are set into the *third state*. Most logic families have some of their SSI and MSI elements like bistables, shift registers, buffers/line drivers, etc. with three-state outputs.

Two ways of implementing three-state logic in CMOS are shown in Fig. 5.13.

One technique is to interpose a transmission gate between the traditional logic and the output terminal. The operation of the arrangement, shown in Fig. 5.13(b), is self-evident. The alternative in Fig. 5.13(c) consists of two transistors M_2 and M_3 interposed between what would normally be the single output point of a standard logic inverter. The output in Fig. 5.13(c) will be either in the third state (Control = V_{DD}) or $\overline{A}$ (Control = 0).

Logic families with micron and submicron structures have very small gate input and output capacitances, as low as a few fF for geometries of $\sim(0.5$–$0.8)$ μm. The wire capacitances within the chip are, save those of short 'next-neighbour' connections, larger, indeed very much larger on

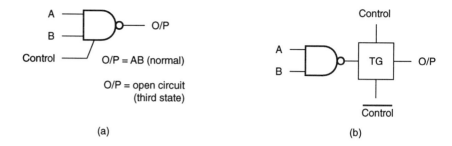

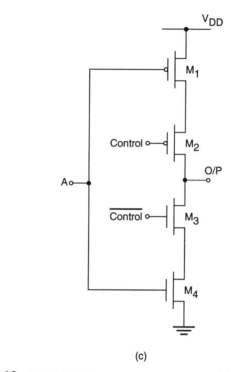

Fig. 5.13 CMOS NAND gate three-state logic (a) logic symbol (b) transmission gate implementation (c) series feed implementation

aggregate, and dominate overall transient response. Even so, they are still one or more orders of magnitude below the capacitances of PCB interconnections. The capacitive load presented by a PCB wire to an IC output driving a signal off chip is likely to be in the range (2–20) pF, the internal capacitance presented to a logic element within the chip might typically be between 20 to 500 fF, depending on wire length and fan-out loading. The ratio of PCB to internal wire capacitance is of the order 1000, and in the absence of special output buffers will lead to a prohibitively large propagation delay. This deduction follows from the basic consideration of propagation delay for the cascaded inverter chain in Fig. 5.14. Assuming the propagation delay to be proportional to the ratio of output to input node capacitances, we can write

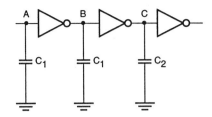

Fig. 5.14 Cascaded inverters

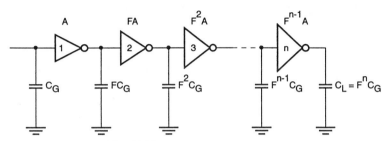

Fig. 5.15 Cascaded output buffer

$$\frac{t_{PBC}}{t_{PAB}} = \frac{t_{PBC}}{t_g} = \frac{C_2}{C_1} \qquad (5.27)$$

where t_g, equated here to t_{PAB}, is the representative delay for a gate with equal capacitance loading at input and output. The customary technique for minimizing propagation delay when driving off-chip for nMOS and CMOS is the cascaded output buffer of Fig. 5.15, where the channel area of the transistors is scaled up by a factor F for adjacent inverters (Mohsen and Mead 1979). The increase in area is achieved by scaling L and/or W, and the underlying assumption is that the node capacitance is effectively the gate capacitance with the exception of C_L, which is the external load, in this case largely contributed by PCB wire. The close proximity of the cascaded inverters justifies, for once, neglecting the internal wire capacitance. The total propagation delay is an arithmetical progression, the increase in area between adjacent inverters a geometrical progression. Each stage in Fig. 5.15 has an identical propagation delay $t_d = Ft_g$ in accordance with eqn (5.27), giving a total delay t_t

$$t_t = nt_d = nFt_g \qquad (5.28)$$

The result is that a small number of inverters with a moderate value for F can accommodate a load capacitance C_L several orders of magnitude larger than C_G with a very much smaller propagation delay than would be the case without a special output buffer.

Optimum values will now be obtained for n and F to give the minimum overall delay for a specified ratio (C_L/C_G). We define a parameter Y by

$$Y \equiv \frac{C_L}{C_G} = F^n \qquad (5.29)$$

Taking logarithms

$$lnY = nlnF \qquad (5.30)$$

and substituting (5.30) in (5.28)

$$t_t = \frac{FT_g \ln Y}{\ln F} \quad (5.31)$$

Differentiating (5.31) with respect to F to obtain the minimum value of t_t leads to

$$F = e \quad (5.32)$$

where e is 2.718, the base of natural logarithms.

Substituting e for F in eqn (5.30) gives $n = \ln Y$, and

$$t_{t(\min)} = e t_g \ln Y \quad (5.33)$$

For $Y = 1000$, to take a typical example, $\ln Y = 6.91$, which is rounded to 7, resulting in $t_t(\min) = 18.8 t_g$ with F equal to 2.72. In minimizing the delay, the number of stages, the silicon estate they occupy, and the power they consume also have to be considered. Fortunately t_t, as a function of F_1 has a broad minimum at $F = e$, and F can be traded for n with only a slight increase in overall delay. If, in the example just given, n is made 4 instead of 7, F comes to 5.62 and t_t to $22.4 t_g$. The number of drivers has been nearly halved at the relatively small cost of increasing t_t by 19 per cent.

The total capacitance C_t of the output buffer (this excludes C_L) equals

$$C_t = C_G \frac{(F^n - 1)}{(F - 1)} \quad (5.34)$$

The optimization for $Y = 1000$ ($n = 7, F = e$) leads to $C_t = 638 C_G$. C_t reduces to $216 C_G$ for $n = 4$, $F = 5.62$.

The total dynamic power P_{dt} demanded by the output buffer is related to P_{dg}, the dynamic power for charging C_G, by

$$P_{dt} = (C_t + C_L - C_G) P_{dg} \simeq (C_t + C_L) P_{dg} \quad (5.35)$$

The term $-C_G$ in eqn (5.35) arises because G in Fig. 5.15 is charged by the stage driving the first inverter. In accordance with eqn (5.35) P_{dt} equals $1637 P_{dg}$ for the optimum $t_t(\min)$, and $1215 P_{dg}$ for n equal to 4.

Example 5.4

Compare the performance of the output buffer in Fig. 5.15, optimized for minimum delay for $Y = 1000$ ($n = 7$, $t_t = 18.8 t_g$) with that of a single buffer dimensioned to have a delay t_g.

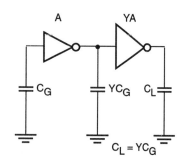

Fig. 5.16 Single buffer driver

The single buffer/driver, shown in Fig. 5.16, has an area YA and a gate capacitance YC_G, with $Y = 1000$. The propagation delay of the 'standard' inverter driving it is $Y t_g$, giving $t_t = (Y + 1) t_g \simeq 1000 t_g$. A standard inverter has been included in Fig. 5.16. It would be completely wrong to equate t_t with t_g, the delay through the buffer. The large propagation delay is incurred by the circuit, in this case the standard gate, feeding the output buffer. C_t equals $1000 C_g$ and P_{dt} comes to $2000 P_{dg}$, compared with figures of $638 C_g$ and $1637 P_{dg}$ for the optimized buffer chain. The cascaded buffer gives an appreciable saving in power.

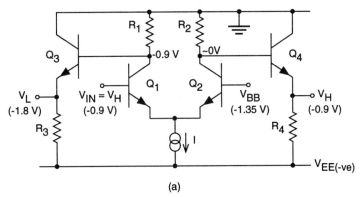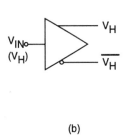

Fig. 5.17 ECL inverter/buffer (a) circuit (b) logic equivalent

What is incomparably more significant is the reduction in delay time achieved with the cascaded output buffer chain. Its total delay is $18.8t_g$, which is 1.9 per cent of the delay for the buffer in Fig. 5.16.

5.2.5 ECL/CML

ECL, a non-saturated circuit technique, is the fastest bipolar logic. ECL was originated by Motorola, and is profusely second-sourced. The basic ECL circuit, the inverter/buffer in Fig. 5.17, consists of a current-mode switch (CMS) Q_1, Q_2, whose outputs feed the emitter followers Q_3 and Q_4. $V_{BE}(\text{on}) \sim 0.9$ V, and the voltage drops across R_1 and R_2 are either ~ 0 or ~ -0.9 V. V_H and V_L are -0.9 V and -1.8 V respectively. V_{BB} is generated on the chip and made equal to $1/2(V_H + V_L) = -1.35$ V. For $V_{IN} = V_H$, Q_1 is on and Q_2 off. However Q_2 has a forward bias: $V_{BE}(Q_2) = \{-1.35-(-0.9-0.9)\}$ V $= 0.45$ V. That bias gives absolutely negligible conduction. It leads to a small differential voltage, 0.45 V, for $\{V_{BE}(\text{on}) - V_{BE}(\text{off})\}$—this applies to Q_1 and Q_2 - and thereby leads to fast switching. Q_3 and Q_4 fulfil the dual role of level-shifting and driving high capacitive loads. The level shifts give the required outputs of $-V_{BE}(V_H)$ and $-2V_{BE}(V_L)$ ignoring, as has already been implied in the stipulation of the voltage drops across R_1 and R_2, the base currents of the emitter followers, whose drive capability maintains the fast transient response for high face-out and interconnect capacitance.

CML consists of a current mode switch like Q_1, Q_2 in Fig. 5.17 without emitter follower outputs. The CML inverter/buffer in Fig. 5.18(a) has typical logic levels of 0 V (V_H) and -0.4 V (V_L); $R_1 = R_2$ and $IR_1 = IR_2 = 0.4$ V. The differential logic swing of 0.4 V is about the maximum permissible amplitude. This constraint is due to the forward collector-base bias of 0.4 V for Q_1 when $V_{IN} = V_H$. The corresponding V_{CB} for Q_2, -0.2 V, is less significant. The forward collector-base bias of 0.4 V turns the transistor into *soft saturation*, where charge storage effects, which delay turn-off, are still small. At higher bias levels, classical saturation seriously deteriorates switching speed.

Example 5.5

Estimate the emitter currents of Q_1 and Q_2 (Fig. 5.18(b)) in the off state and comment.

130 Digital circuits — techniques and performance

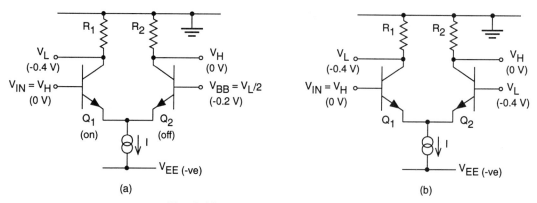

Fig. 5.18 CML inverter/buffer (a) single-ended input (b) differential input

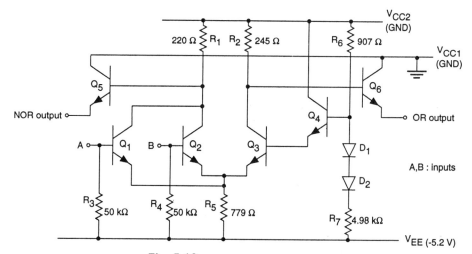

Fig. 5.19 MECL 10K 2-input OR/NOR gate

Let us assume $V_{BE}(\text{on}) = 0.8$ V, $V_{BE}(\text{off}) = 0.4$ V for both Q_1 and Q_2. $\delta V_{BE} \equiv V_{BE}(\text{on}) - V_{BE}(\text{off}) = -0.4$ V. Using the diode equation $I_E(\text{off})/I_E(\text{on}) = \exp(-39\delta V_{BE}) = 1.68 \times 10^{-7}$ ($T = 300$ K).

This shows that the currents of the transistors in the OFF state are negligible.

The differential CML circuit in Fig. 5.18(b), the inputs for which are readily available from a CML single-ended or differential driver stage, is an alternative mode of operation. The advantages of differential compared with single-ended CML are improved noise rejection and the elimination of a bias generator for V_{BB}. CML is attractive for unity fan-out and short interconnections. Under these conditions it is faster than ECL which, however, scores at moderate and high capacitive loads.

A standard MECL two-input OR/NOR gate in the Motorola ECL (MECL) 10K series is shown in Fig. 5.19 (Blood et al. 1980). The currents flowing through the left arm (Q_1 and/or Q_2) and the right arm (Q_3) differ, and this accounts for the inequality in the values of R_1 and R_2, witness the following example.

Example 5.6

Verify that the voltages at the collectors of Q_2 and Q_3 in Fig. 5.19 are equal when these transistors are in conduction. The logic levels are -0.92 V and -1.75 V, the bias voltage at the base of Q_3 is -1.29 V, and $V_{BE}(\text{on})$ is 0.80 V. The base currents of Q_5 and Q_6 may be ignored.

$$V_c(Q_2) = -I(Q_2).R_1 = -\frac{\{5.2 - (0.80 + 0.92)\} \times 220}{779} = -0.98 \text{ V}$$

$$V_c(Q_3) = -I(Q_3).R_2 = \frac{-\{5.2 - (0.80 + 1.29)\} \times 245}{779} = -0.98 \text{ V}$$

Separate ground returns are used, V_{CC1} for the output drivers, and V_{CC2} for the remainder of the circuit. The separation of the V_{CC} returns reduces crosstalk between circuits in a package. More important, it largely eliminates the voltage spike which would occur at Q_4 base when Q_5 and Q_6 draw large currents. The open-circuit emitter follower outputs allow for two types of termination. They are either taken to the V_{EE} line via 50 kΩ pull down resistors by connecting them to the inputs of a similar gate, or to a low-impedance transmission line terminated by a matching resistor which is returned to a -2 V supply (Fig. 5.20).

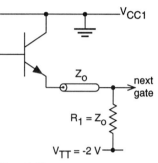

Fig. 5.20 ECL transmission line termination

A separate supply voltage is necessary in order to limit the load current to a safe value. The transmission line can be a coaxial cable, but other types like a microstrip line, or a twisted pair line, are also used. Characteristic impedances are generally ~(50–120) Ω. The transfer characteristic for the ECL gates in Figs 5.17 and 5.19 is sketched in Fig. 5.21(a). The lack of symmetry between the two outputs is due to the change in tail current in the region AB. The symmetry in Fig. 5.21(b) is achieved by replacing R_5 in Fig. 5.19 with a constant current generator, an example of which is given in Fig. 5.22. An alternative current generator is the current mirror, described with the aid of Figs 6.14 and 6.15 in Section 6.2.2.

ECL/CML combinations of stacked logic, in which the all levels are CML with the exception of the top level, which can be either CML or ECL, are commonplace. SSI and MSI chips like bistables and converters have two levels, the lower in CML, the upper in ECL. CML can operate

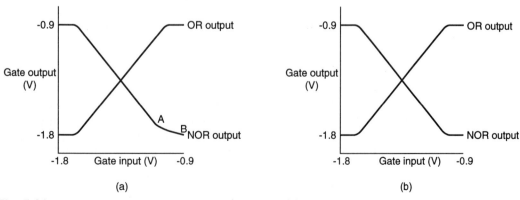

Fig. 5.21 ECL transfer characteristics (a) MECL 10K (b) MECL 10KH

132   Digital circuits — techniques and performance

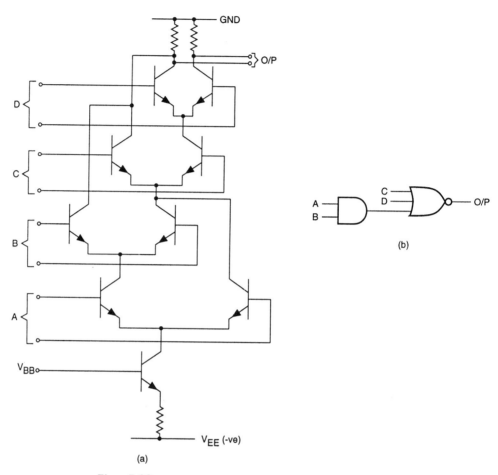

Fig. 5.23 Four-stack logic—CML (a) schematic (b) logic equivalent (©1992 Prentice-Hall)

Fig. 5.22 ECL/CML constant current generator

with supply voltages down to 1 V, and up to four levels are possible with the standard supply voltage of -5.2 V. An example of four-stack logic with CML is shown in Fig. 5.23 (Saul 1992).

The capability of vertical stacking in ECL/CML is an outstanding advantage for VLSI. It represents a great power saving over single-level logic, because only one tail current is required to feed all levels.

5.2.6 TTL

TTL was, at the time of its inception in 1965, the first logic family specifically designed for integrated circuits. It was derived from modified diode transistor logic (MDTL), which was well established at that time. The vital difference between TTL and MDTL is the multiple-emitter input structure in place of input diodes. A two-input NAND gate in the 54/74 series is shown in Fig. 5.24. V_H and V_L are $(V_{CC} - V_{BE} - V_{D1})$ and $V_{CE(SAT)}$, equal to $\sim$3.4 V and $\sim$0.2 V respectively.

Coming to the explanation of the circuit, consider the condition $A = B = V_H$. There will be a conducting path from V_{CC} to ground via R_1,

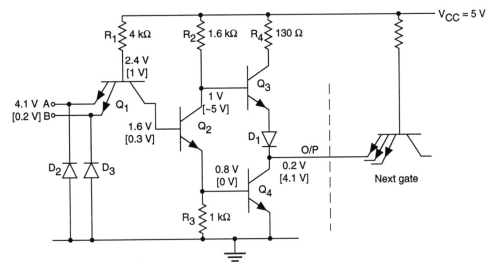

Fig. 5.24 TTL two-input NAND gate (54/74)

$V_{BC}(Q_1)$, $V_{BE}(Q_2)$, and $V_{BE}(Q_4)$, leading to the unbracketed voltages shown in Fig. 5.24. Q_2 and Q_4 are in saturation, and $V_{BE(SAT)} \simeq 0.8$ V, $V_{CE(SAT)} \simeq 0.2$ V. The base-emitter junctions of Q_1 are reverse biased, and this transistor is in the reverse active mode, with the roles of emitter and collector interchanged. D_1 is a hold-off diode. The voltage drop between the base of Q_3 and the output terminal, $(1.0-0.2)V = 0.8$ V, is insufficient to bring Q_3 and D_1 in conduction. The output voltage is determined by Q_4, influenced by the current it sinks from the gate(s) fed by the output. When one or both inputs are at V_L (0.2 V), Q_1 operates in the normal mode and is in saturation. The collector current of Q_1 is the leakage base current of Q_2 and the condition for saturation, namely $I_c < h_{fe}I_B$, is easily satisfied, bearing in mind that I_B is equal to $(5-1)/4 = 1$ mA. $V_{CE(sat)}$ in that condition is ~100 mV, giving $V_C(Q_1) = V_B(Q_2) \simeq 300$ mV. Q_2 and Q_4 are off, because the voltage at the base of Q_2 is insuffient to conduct.

The input protection diodes D_2 and D_3 are a standard feature in TTL, clamping a negative overswing which might have arisen in the signal path. Q_3 and D_1 are drawing the leakage current of the next stage. The typical output voltage quoted in the TTL data is 3.4 V. That figure, which has already been used near the beginning of this section, is specified for a leakage current of 400 μA, the maximum value for a ten-fold fan out at 125 °C. The output for unity fan-out at 25 °C is ~3.9 V. The output stage is designated *totem pole*, a term applied to a series combination like Q_3 and Q_4, which are active pull up and pull down transistors. The low output impedance, either of the emitter follower Q_3 or of the saturated inverter Q_4, allows the gate to drive a high capacitive load with good speed. The transfer characteristic for Fig. 5.24 is given in Fig. 5.25. Region AB spans the range of input voltage for which Q_2 conducts, but Q_1 does not.

The operation of Q_1 is the outstanding feature of TTL. The base current is much the same in both the normal and the reverse mode and hence the base charge remains substantially constant. The result is that

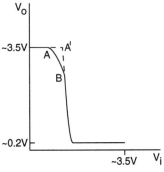

Slope AB = R_2/R_3 (Fig. 5.24)
AA'B: Schottky TTL (Fig. 5.27)

Fig. 5.25 TTL transfer characteristic

134 Digital circuits — techniques and performance

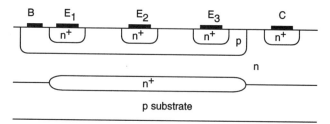

Fig. 5.26 Structure of TTL multi-emitter transistor

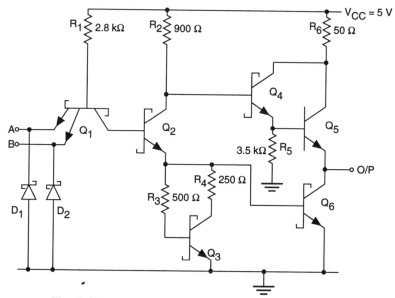

Fig. 5.27 Schottky TTL 2-input NAND gate (54S/74S)

the relatively large time delay associated with the removal of excess base charge in classical saturation is almost entirely eliminated. Moreover the fast transition of Q_1 from the inverse to the normal mode speeds up the extraction of base charge from Q_2 when it is turned off. Technologically the processing of a multi-emitter transistor is quite straightforward, see Fig. 5.26.

A turning point in TTL came with the introduction of the Schottky-clamped transistor (described in Section 3.3) in which classical saturation and the excess base charge associated with it are eliminated. The introduction of Schottky TTL in 1968 was originated with the 54S/74S series, represented by the two-input NAND gate in Fig. 5.27.

All the transistors which undergo saturation and the input protecting diodes are Schottky clamped. Reverse collector-base bias for Q_5, when in conduction, is ensured by $V_{CE(SAT)}$ of Q_4, which fulfils the hold-off role of D_1 in Fig. 5.24. It also provides extra base drive for the output emitter follower, thereby decreasing the turn-on time. Q_3 'squares up' the transfer characteristic by giving a simultaneous onset for the conduction of Q_2 and Q_6. The region AB for standard TTL is replaced by AA'B in Fig. 5.25.

Yet another leap forward came with the emergence of low-power Schottky (LS) TTL in the form of 54LS/74LS series introduced in 1974.

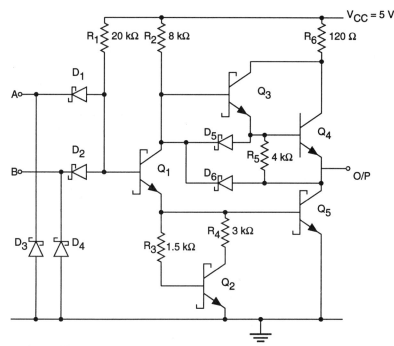

Fig. 5.28 Low power Schottky TTL 2-input NAND gate (54LS/74LS)

The striking innovation was the abandonment of the multi-emitter transistor. Instead the circuit, a schematic of which is shown in Fig. 5.28, has reverted essentially to MDTL, with Schottky diodes replacing the multi-emitter input transistor. All TTL families brought out since the introduction of 54LS/74LS TTL series have followed the input circuit of Fig. 5.28, either directly or, as in the 54/74ALS Advanced Low-Power Schottky (ALS) family, indirectly. There the Schottky input diodes are replaced by pnp emitter followers, which impose only light loading on the drivers at the inputs by virtue of their small base currents. The generic name TTL has however been retained for all the Schottky families which, with the exception of the 54S/74S series, follow MDTL practice.

5.2.7 I²L

Integrated injection logic (I²L), alternatively labelled merged transistor logic (MTL), is a bipolar technique which was announced in 1972 and emerged in (V)LSI on a moderate scale for a few years. It posed a potential challenge to nMOS logic for a brief spell, but was soon completely overtaken by MOS logic in respect of speed and the speed-power product. More recently some I²L developments have been reported with GaAs heterojunction bipolar transistors. GaAs heterojunction integrated injection logic (HI²L) circuit practice is very similar to I²L, which will now be described (Roberts et al. 1989).

The I²L 'foundation' circuit, the inverter in Fig. 5.29, consists of two transistors, the current generating *injector* Q_1 and the multiple-collector inverter Q_2. Thus the basic I²L circuit has one input, but multiple (collector) outputs; the number of collectors provided is determined by the fan-out requirement. This contrasts with all other logic circuits,

136 Digital circuits — techniques and performance

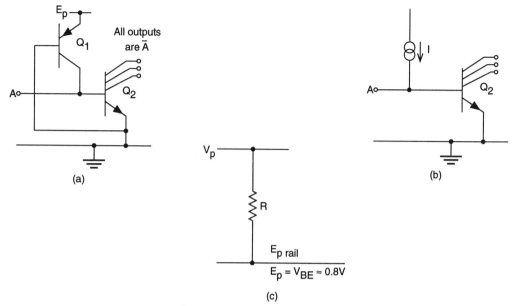

Fig. 5.29 I²L inverter (a) schematic (b) equivalent circuit (c) E_p voltage supply rail

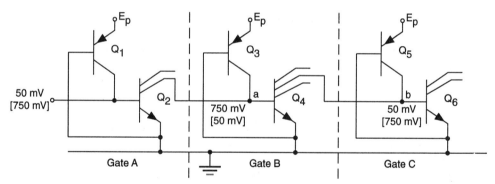

Fig. 5.30 Cascaded I²L inverters

which have multiple inputs, but only one output. V_H and V_L are equal to $V_{BE(SAT)}$ (~750 mV) and $V_{CE(SAT)}$ (~50 mV) respectively. Q_1 is in saturation when driving the inverter Q_2 on, or in the active mode, when it is the load for the inverter transistor connected to A. Voltage levels for three cascaded gates are given in Fig. 5.30. The E_p rail is V_{BE} or $V_{BE(SAT)}$, ~(700–800) mV. I²L is powered by supply voltage taken to the E_p rail via an external resistor, whose value controls the current of the injector(s) (Fig. 5.29(c)). The choice of current is a trade-off between speed and power. The simple structure of I²L, sketched in Fig. 5.31, is one of its striking features. The multi-collector resembles the multi-emitter of a TTL input transistor in Fig. 5.26, but the I²L transistor operates in the inverse mode. Q_1 in Fig. 5.29 is a lateral pnp transistor. Its base is returned to ground via the emitter region of Q_2, and the need of surface (mask) metallization for ground connections is

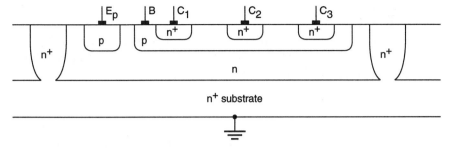

Fig. 5.31 Profile of I²L structure

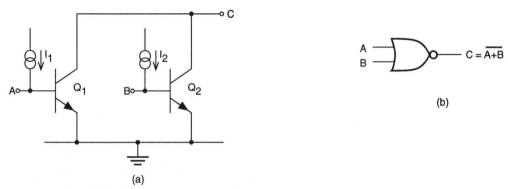

Fig. 5.32 I²L NOR gate (a) circuit (b) logic equivalent

thereby eliminated. Q_1 can, and often does serve as an injector for many gates; Q_1, Q_3, and Q_5 in Fig. 5.30 could be one transistor. The circuit of a logic gate includes a wired AND/OR output, because the basic circuit of Fig. 5.29 can only accept one input. The two-input NOR gate in Fig. 5.32 illustrates this feature.

The speed of I²L can be increased by reducing the differential logic swing. An elegant way of doing this is Schottky I²L, illustrated in Fig. 5.33. The levels are $V_H = V_{BE(SAT)}$—as before—and $V_L = V_{CE(SAT)} + V_{SD} \simeq (50 + 400) = 450$ mV, V_{SD} being the voltage across a Schottky diode. The differential swing is ~300 mV in place of ~700 mV for I²L.

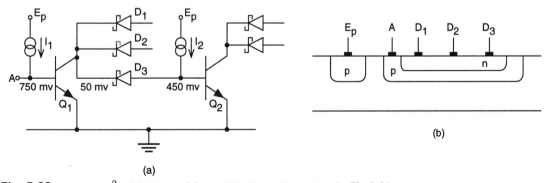

Fig. 5.33 Schottky I²L (a) circuit (b) modification of structure in Fig 5.31

5.2.8 BiCMOS

Bipolar circuits are capable of faster switching than CMOS or nMOS, largely because the BJT has a higher transconductance than the MOSFET. Equation (4.17) leads to a general approximation, namely

$$t_T \propto \frac{C_L}{g_m} \quad (5.36)$$

C_L, for a given size, identical fan-out, and identical wire lengths, will have similar values for bipolar and MOS circuits, so that g_m will be the deciding factor regarding speed. The respective values of g_m for bipolar and MOS transistors are

$$g_m(\text{BJT}) = \frac{qI_c}{kt} = 39I_c \ (T = 300 \,\text{K}) \quad (5.37)$$

and

$$g_m(\text{nMOSFET}) = \frac{\mu_n \epsilon_{ox}}{t_{ox}} \left(\frac{W}{L}\right)(V_G - V_T) \quad (5.38)$$

Equations (5.37) and (5.38) are derived from eqns (4.53) and (4.22) respectively. g_m (BJT) equals 39 mA/V ($I_C = 1$ mA) and 3.9 mA/V ($I_C = 100 \,\mu\text{A}$). The transductance g_m (nMOSFET) equals 149 μA/V for a transistor with the parameters given in Example 5.3. The difference between the BJT and MOSFET transconductance values is enormous. Even if it is narrowed down by MOSFETs with different parameters, or by lower BJT currents, the comparison shows beyond any doubt the superior speed-potential of the BJT.

The object of a BJT–CMOS combination is to improve on the speed of CMOS, retaining its extremely high input resistance and following a CMOS input with a BJT output stage which, like CMOS, consumes no current in the steady state. The totem pole output stage of the TTL gate in Fig. 5.24 operates that way, but the remainder of the circuit demands quiescent input power. The 'trick' of obtaining the speed of BJTs without incurring steady state power is to use a Class B type output stage driven by a modified CMOS inverter. The basic BiCMOS inverter shown in Fig. 5.34 was in use during the initial introduction of BiCMOS. When $V_i = V_L(V_L < V_T)$ M_1 is on and M_2 off, turning Q_1 into conduction and maintaining Q_2 off. V_H is, neglecting the very small voltage across M_1, given by

$$V_H \simeq V_{DD} - V_{BE} \quad (5.39)$$

V_{BE} in this case will be small, $\sim$500 mV, because the current flowing through Q_1 is the leakage current of Q_2, which is off. When $V_i = V_H(V_H > V_T)$, M_1 and Q_1 will be off and M_2 turns on Q_2, giving

$$V_L \simeq V_{BE} \quad (5.40)$$

Once again, a small voltage drop, in this case across M_2, has been neglected and $V_{BE} \simeq 500$ mV. Equations (5.39) and (5.40) show that the logic swing is less than V_{DD} by an amount $2V_{BE}$. The basic inverter is

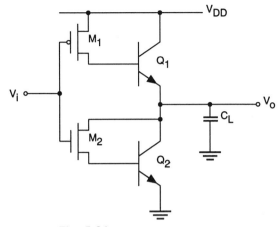

Fig. 5.34 Basic BiCMOS inverter

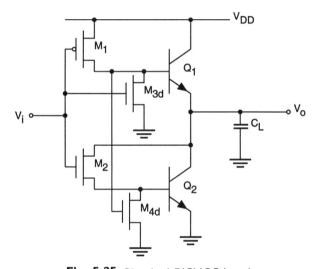

Fig. 5.35 Standard BiCMOS inverter

modified by adding two nMOS transistors, M_{3d} and M_{4d}, which lead to a fast turn off for Q_1 and Q_2 (Fig. 5.35).

The transistors M_{3d} (d for discharge) and M_{4d} assist with the removal of the base charges from Q_1 and Q_2 respectively. The logic levels given by eqns (5.39) and (5.40) still hold. M_1 and M_2 are dimensioned in accordance with the established criteria for CMOS logic. Minimum size transistors ($W = L$) are generally acceptable for M_{3d} and M_{4d}. A BiCMOS NAND gate incorporating that practice is shown in Fig. 5.36.

The reduced logic levels of BiCMOS relative to the full-rail swing of CMOS are a serious disadvantage at low supply voltages. Difficulties then arise, because the noise margins are reduced. MOSFETs for operation at V_{DD} of (2–3.3) V have $V_T \simeq 1$ V or even less, and $V_{BE}(\text{on}) \simeq V_T$. The noise margin may become prohibitively small, especially when BiCMOS drives CMOS. Several techniques are available for realizing full-swing (alternatively rail-to-rail) operation. A standard circuit for that purpose is shown in Fig. 5.37, which should be compared with Fig. 5.35

140 Digital circuits — techniques and performance

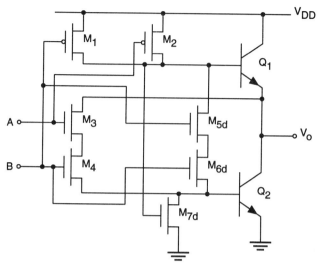

Fig. 5.36 BiCMOS 2-input NAND gate

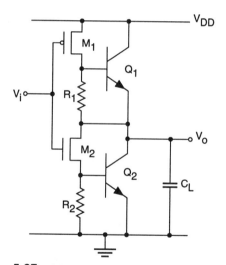

Fig. 5.37 BiCMOS inverter with full-rail output

(Uyemura 1992). Consider the transition of the output from V_L to V_H. The currents through R_1 and Q_1 charge C_L until V_o is within V_{BE} (on) of V_{DD}. The transistor Q_1 is soon turned off by a further rise in V_o, and R_1 now charges C_L on its own until $V_o \simeq V_{DD}$; the small voltage drop across M_1 is negligible. The mechanism for the discharge of C_L is similar, with R_2 taking over completely, shortly after V_o has fallen to $\sim V_{BE}$ (on). A compromise has to be reached in choosing the values of the resistors. They have to be small enough for adequately fast action when Q_1 or Q_2 is off, and yet large enough to avoide undue reduction in the drive currents of these transistors. The full-rail output becomes particularly important for a supply voltage around 2 V, when the $2V_{BE}$ loss in the logic swing of standard BiCMOS looms large. Full-swing operation can also be obtained with an all-transistor adaptation of the circuit in Fig. 5.37 (Embabi *et al.* 1993).

The propagation delays and transition times of CMOS and BiCMOS are proportional to C_L, but the slope of t_p versus. C_L is considerably steeper for CMOS (Fig. 5.38). For very small values of C_L, a relevant situation for unity fan-out and a short local interconnect, CMOS is faster than BiCMOS, because the latter incurs the added delay due to the bipolar output stage. BiCMOS comes into its own with increasing C_L, overtaking CMOS in speed at a technology-dependent cross-over point which occurs at $C_L \sim (100\text{–}500)$ fF.

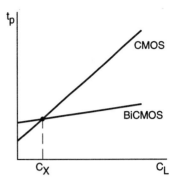

Fig. 5.38 CMOS and BiCMOS gate delays

5.2.9 GaAs

The three most widely used GaAs logic circuits are buffered FET logic (BFL), Schottky diode FET logic (SDFL), and direct coupled FET logic (DCFL).The superior speed of GaAs over Si derives largely from its much higher mobility, which however only applies to n-channel GaAs MESFETS. Since p-type GaAs has a smaller mobility than p-type Si GaAs logic is confined entirely to n-channel MESFETs. Likewise the comparison between GaAs and Si coming up in Section 5.5.3 applies to n-type GaAs. BFL and SDFL are composed of D-MESFETs, whereas DCFL is a combination of D- and E- MESFETs. The values for various parameters and the characteristics of the circuit contained in this section apply to MESFETs with 1 μm feature size. The circuit of a BFL NOR gate is shown in Fig. 5.39. V_H and V_L are ~ 0.6 V and -1.2 V, and V_T is in the range -0.6 V to -1.2 V. Two supply voltages are required, and these are typically ± 2.5 V. Input and output levels are made equal by level-shifting Schottky diodes D_1 and D_2 in conjunction with the current sink, Z_5. The source follower Z_4 has good drive capability and gives BFL superior speed over SDFL and DCFL for high load capacitance. A typical transfer characteristic is shown in Fig. 5.40.

The SDFL gate in Fig. 5.41 also has Schottky level-shifting diodes, this time at the input. They perform a role like the input diodes of the Low Power Schottky TTL gate in Fig. 5.28. $V_H \simeq 2.3$ V and $V_L \simeq 0.3$ V for $V_T \simeq -1$ V. The transfer characteristic is shown in Fig. 5.42.

A NOR gate for DCFL, the first and simplest GaAs logic, is sketched in Fig. 5.43. The schematic includes a two-fold fan-out to gates represented by Z_4 and Z_5. The fan-out is shown in order to explain the nature of the transfer characteristic (Long and Butner 1990, pp.211–214). The

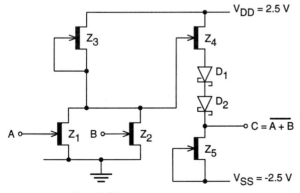

Fig. 5.39 BFL 2-input NOR gate

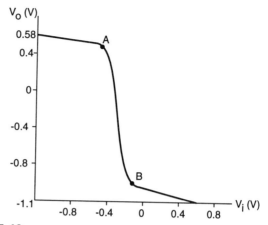

Fig. 5.40 Transfer characteristic for BFL NOR gate (Fig. 5.39)

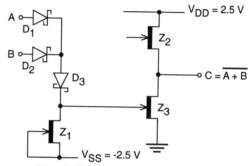

Fig. 5.41 SDFL 2-input NOR gate

circuit is powered by a single supply voltage and resembles an nMOS inverter with a depletion load (Fig. 5.5). Here V_H is ~0.7 V and V_L is ~0.15 V with $V_T \sim 0.25$ V for Z_1 and Z_2, and ~ -0.6 V for Z_3. The transfer characteristic, Fig. 5.44 will now be explained with the aid of Fig. 5.45 (Long and Buttner 1990, pp.211–214). The dynamic load line (a) of Z_3 is superimposed on the I-V characteristics of Z_1 (or Z_2). The load line of the following input stages is the straight-line approximation

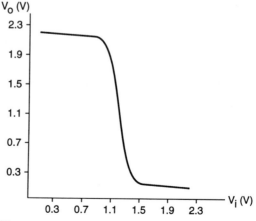

Fig. 5.42 Transfer characteristic for SDFL NOR gate (Fig. 5.41)

Logic Circuits 143

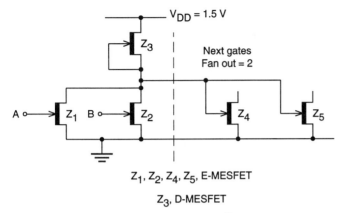

Fig. 5.43 DCFL two-input NOR gate (©1990 McGraw-Hill)

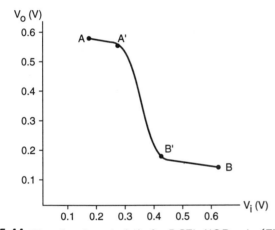

Fig. 5.44 Transfer characteristic for DCFL NOR gate (Fig. 5.43)

The static characteristics are for Z_1 and Z_2. The dynamic load line of Z_3 is given by (a), and that of the input presented by the next stage by (b).

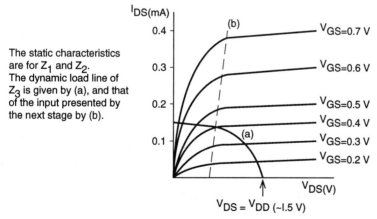

Fig. 5.45 Static characteristics and load line for DCFL NOR gate (Fig. 5.43) (©1990 McGraw-Hill)

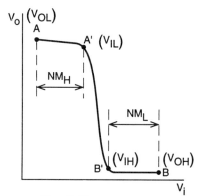

Fig. 5.46 Noise margins

(b). The Schottky gate input characteristic has been linearized, because the current will be largely determined by R_S, the bulk source resistance of Z_3, leading to a slope of FO/R_s, where FO is the fan-out, for (b). The intersection of (a) and (b) gives V_H (A in Fig. 5.44), and the intersection of (a) with the I-V characteristic for $V_{GS} = V_H$ gives V_B (B in Fig. 5.44). The plot AA'B'B in Fig. 5.44 is obtained from the intersections of (a) with the I-V characteristics; V_{GS} and V_{DS} represent V_i and V_o respectively. The gate threshold voltage V_{TH} (B') is ~0.45 V, although V_T for Z_1 and Z_2 in Fig. 5.43 is only ~0.25 V. The nature of the dynamic load presented to the driver transistors Z_1 and Z_2 in Fig. 5.39 and to Z_3 in Fig. 5.41 is similar, as is the method of obtaining the transfer characteristics for these circuits.

The noise immunity of a logic gate is an important indicator of its quality, and the defining parameters are indicated in Fig. 5.46, which is a transfer characteristic not tied to any particular logic circuit. A' and B' are the changeover points at which dV_o/dV_i equals -1, and the corresponding inputs are V_{IL} and V_{IH}. The value V_{IL} is the absolute maximum for a logic high, and V_{IH} the absolute minimum for a logic low output. The noise margins NM_H and NM_L are defined by

$$NM_H = V_{IL} - V_{OL} \tag{5.41}$$

$$NM_L = V_{OH} - V_{IH} \tag{5.42}$$

Typical characteristics for the logic families described in this section are given in Table 5.2

The choice of logic is largely governed by speed, power consumption, and gate area. BFL is the fastest logic with significant fan-out (three or more), but not surprisingly consumes most power. BFL and SDFL have similar noise margins; in this respect they are superior to DCFL. The advantages of DCFL are a single supply voltage and low power consumption, permitting the highest function density and making this logic a favourite for VLSI. The logic families are compared in Table 5.3.

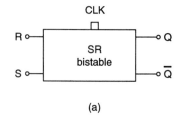

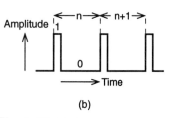

Fig. 5.47 RS bistable (a) schematic (b) clock waveform

Table 5.2 Typical characteristics for GaAs logic

Parameter (V)	BFL	SDFL	DCFL
V_T	-1.00	-1.00	0.25(E), -0.7(D)
V_H	0.60	2.20	0.60
V_L	-1.20	0.30	0.18
NM_H	0.65	0.75	0.12
NM_L	0.85	0.80	0.20
V_{DD}	2.50	2.50	1.50
V_{SS}	-2.50	-2.00	0.00

Table 5.3 Comparison of GaAs logic

BFL	SDFL	DCFL
Highest speed with fan-out	Medium speed with low fan-out	Medium speed with low fan-out
Highest input power per gate	Medium input power per gate	Lowest input power per gate
Two supply voltages	Two supply voltages	One supply voltage
D-MESFETs	D-MESFETs	E-MESFETs and D-MESFETs. Processing more complex
Good noise immunity (∼0.7 V)	Good noise immunity (∼0.7 V)	Lowest noise immunity (∼0.2 V)
Gate delay $(FO = 3)$ ∼80 ps	Gate delay $(FO = 3)$ ∼200 ps	Gate delay $(FO = 3)$ ∼100 ps

5.3 Bistables

A bistable is a sequential memory element composed of logic gates. The core elements are the reset-set (RS) and toggle (T) flip-flops, and RS latches. The behavioural schematics of the RS and T flip-flops are shown in Figs 5.47 and 5.48. The definitive equation for the RS flip-flop is

$$Q_{n+1} = S_n \quad (5.43)$$

$$R_n \neq S_n$$

Equation (5.43) needs to be interpreted in accordance with Fig. 5.47(b). It defines the Q output in the interval $(n + 1)$ following the arrival of a clock pulse (CLK) in terms of the S, R inputs during the preceding interval n. Q and $\overline{Q}$ are logic opposites. Their designation implies this, but cases occur—as will be pointed out—where they are equal. There are two further S, R combinations other than the two ($R_n = 0, S_n = 1$ and $R_n = 1, S_n = 0$) covered in eqn (5.43). The equations for these are

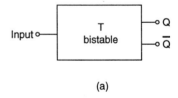

(a)

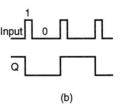

(b)

Fig. 5.48 T bistable (a) schematic (b) waveforms

146 Digital circuits — techniques and performance

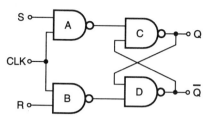

Fig. 5.49 Clocked NAND gate bistable

$$Q_{n+1} = Q_n \qquad (5.44)$$

$$R_n = S_n = 0$$

and

$$Q_{n+1} = X \text{ (don't care)} \qquad (5.45)$$

$$R_n = S_n = 1$$

Equation (5.44) signifies that $R_n = S_n = 0$ is a no-change condition; equation (5.45) expresses a situation which is of no use. The T bistable toggles, i.e. the outputs change state each time there is an input pulse, see Fig. 5.48(b). The definitive equation is

$$Q_{n+1} = \overline{Q}_n \qquad (5.46)$$

A clocked NAND bistable is sketched in Fig. 5.49 and its truth table is given in Table 5.4. The bistable consists of the control gates A and B, and the latch formed by the cross-coupled gates C and D. The conditions (a) to (c) in the truth table confirm equations (5.43) and (5.44), whilst condition (d) obeys eqn (5.45) with Q equal to $\overline{Q}$. The action depends on the width of the clock pulse, which must be less than the propagation delay through one of the NAND gates, so that the control gates are disabled by the time the outputs have taken up their (n + 1) states; otherwise the outputs will be indeterminate.

IC implementations of the RS and T bistable are the JK and D flip-flops. The JK flip-flop, chronologically the first, is sketched in Fig. 5.50. It is a NAND bistable modified by adding logic feedback from Q and $\overline{Q}$ to the R and S control gates, which are renamed K and J respectively. The truth table in Table 5.5 shows that the RS action, defined by eqns (5.43) and (5.44) is obeyed with J and K taking the place of S and R. The effect of the feedback is to modify the operation for $J = K = 1$. In that state the flip-flop operates in the T mode. Consequently the JK flip-flop can function in the RS mode ($R \neq S$) or the T mode ($J = K = 1$; input to CLK terminal).

Table 5.4 Truth table for NAND bistable (Fig. 5.49)

	R_n	S_n	Q_{n+1}	$\overline{Q}_{n+1}$
(a)	0	1	1	0
(b)	1	0	0	1
(c)	0	0	Q_n	$\overline{Q}_n$
(d)	1	1	1	1

Example 5.7

Prove the T action of the JK flip-flop in Fig. 5.50 with the aid of Boolean algebra.

When the clock goes from 0 to 1

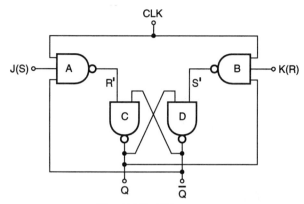

Fig. 5.50 JK flip-flop

Table 5.5 Truth table for JK flip-flop (Fig. 5.50)

J_n	K_n	Q_n	R'	S'	Q_{n+1}
0	1	0	1	1	0
0	1	1	1	0	0
1	0	0	0	1	1
1	0	1	1	1	1
0	0	0	1	1	0
0	0	1	1	1	1
1	1	0	0	1	1
1	1	1	1	0	0

$$S' = \overline{KQ_n} \text{ and } R' = \overline{J\overline{Q}_n}$$

Putting $J = K = 1$, $S' = \overline{Q}_n$ and $R' = Q_n$. Applying eqn (5.43), $Q_{n+1} = \overline{Q}_n$, demonstrating the T mode.

The D flip-flop (D for delay), shown in Fig. 5.51, is a neat modification of the RS flip-flop, and automatically gives logic opposites at the inputs of the RS latch. Only a single input is required. Applying eqn (5.43)

$$Q_{n+1} = D_n \tag{5.47}$$

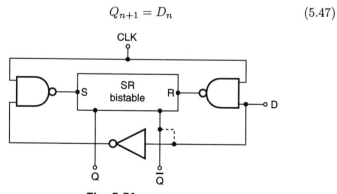

Fig. 5.51 D flip-flop

148 Digital circuits — techniques and performance

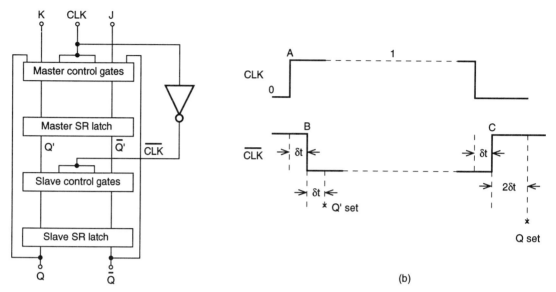

Fig. 5.52 Master-slave flip-flop (a) schematic (b) waveforms

It follows immediately from eqn (5.47) that the T mode is obtained by connecting $\overline{Q}_n$ to the D terminal (the dotted connection in Fig. 5.51) and applying the input to the CLK terminal. T mode operation of the JK and D flip-flops of Figs 5.50 and 5.51 depends, like the operation of the bistable in Fig. 5.49, on the signal pulse width being less than the delay through a gate. That condition is violated in practice, because the duration of the input by far exceeds the gate delay in very many applications. The JK and D flip-flops cannot be used as they stand, and have to be substantially modified. One established approach is the master-slave bistable shown in Fig. 5.52; the logic can be applied equally well to a D bistable. The circuit operates as follows. The slave latch is disabled at B, shortly after the arrival of the clock pulse at A. The interval δt equals the delay through a gate (or the inverter). The J, K inputs set the master latch outputs Q', $\overline{Q'}$ in accordance with Table 5.5. That setting is passed to the RS inputs of the slave control gates at C, by which time the master control gates are disabled. The result is that when the new (n + 1) states of Q and $\overline{Q}$ are fed back to the master control gates, the slave latch will be disabled. The lock-out logic ensures that the outputs can be set once only (toggling once only in the T mode with $J = K = 1$) for each clock pulse, independent of its transition times and duration. The master-slave JK flip-flop in Fig. 5.52 is negative-edge level triggered: the outputs change shortly after the transition from 1 to 0 of the clock. Figure 5.52(a) can easily be modified for positive-edge triggering. All that needs to be changed is to relocate the clock input and the inverter, applying the clock to the slave control, and the inverted clock to the master control gates. Usually standard TTL and CMOS master-slave flip-flops are negative- and, ECL master-slave flip-flops are positive-edge level-triggered. The operation of some sequential logic circuits based on

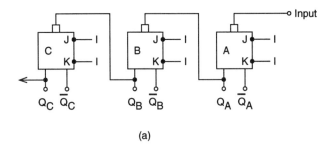

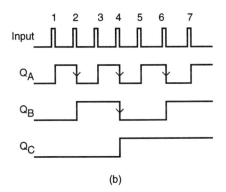

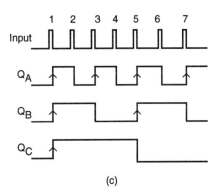

Fig. 5.53 Asynchronous counter (negative-edge triggered bistables) (a) schematic (b) up-counting (c) down-counting

JK and D flip-flops is governed by whether these are positive-edge or negative-edge triggered. The asynchronous up-counter in Fig. 5.53 can be changed into a down-counter by connecting the $\overline{Q}$ instead of the Q output to the input of the next flip-flop.

An alternative lock-out logic is adopted in edge-triggered flip-flops. Before explaining it, a comment about the distinction between level- and edge-triggered flip-flops. The operation of a level-triggered flip-flop is governed by both transitions of the clock pulse, being initiated by one and completed by the other. An edge-triggered flip-flop, on the other hand, requires only one (defined i.e. positive or negative edge) transition of the clock to initiate the action,, which is then completed without further reference to the clock. The schematic for a positive-edge-triggered D flip-flop, shown in Fig. 5.54 is a simplified version of the logic for the TTL 54/7474 D flip-flop. That logic has been adopted for many flip-flops in TTL and CMOS SSI. Latches A and B consist of cross-coupled NAND gates like the C, D pair in Fig. 5.49. The action is initiated on the positive edge of the clock, with latch A taking up the input D after the gate delay time δt. Latch B will be set to $\overline{D}$ after a further gate delay at a time $2\delta t$. The outputs will take up the states in accordance with eqn (5.47) at a time $3\delta t$. The feedback from the output of latch B to the input of latch A, shown bold, reduces the effective clock pulse width to $2\delta t$, and maintains it at that value regardless of any subsequent change in the input from D to $\overline{D}$, ensuring lock-out and preventing multiple toggling in the T mode, when $\overline{Q}$ is connected to D. The assumption underlying the waveforms in Fig. 5.54(b) is that the clock input is a step function. With increasing rise time, the shape of the effective internal clock changes un-

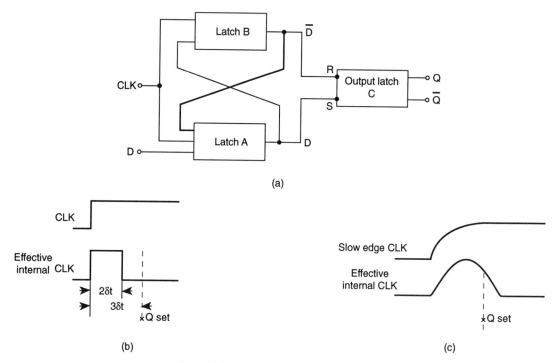

Fig. 5.54 Edge-triggered D flip-flop (a) schematic (b) clock waveforms—fast edge (c) clock waveforms—slow edge

til it ultimately reaches a point, illustrated in Fig. 5.54(c), where Q is set with the effective internal clock still in the logic 1 state, rendering the lock-out ineffective. For that reason, the edge-triggered flip-flop imposes a maximum permissible transition time on the clock pulse. Similar considerations apply to the D input signal. A typical value is 90 ns for the TTL7474 D bistable which has a maximum toggle rate of 25 MHz. It might be thought that this matter is of purely academic interest, since transition times of clock and data inputs in data processing are way below the permitted maxima. In instrumentation, however, cases do occur where rise times of inputs are excessive. A simple method of speed-up is to pass such a signal through a cascaded chain of buffers/inverters. A signal with a very slow rise time will emerge, after transmission through three or four buffers, with the characteristic rise time of the logic family for a step function input.

5.4 Dynamic CMOS

Dynamic MOS circuits—pMOS first, then nMOS—were the mainstay of computer and memory logic when MOS technology ushered in the era of microprocessors and (V)LSI. The designation (V)LSI applies to chips which may fall into one or other of these categories. Dynamic circuits rely on temporary storage of data across node capacitances, in practice the electrode capacitances of MOSFETs. Dynamic logic is synchronous, and the clock pulses enable the dynamic circuit for only part of the cycle, thereby saving a substantial amount of power and raising the function density. The nodal capacitances hold the data for that part of the clock

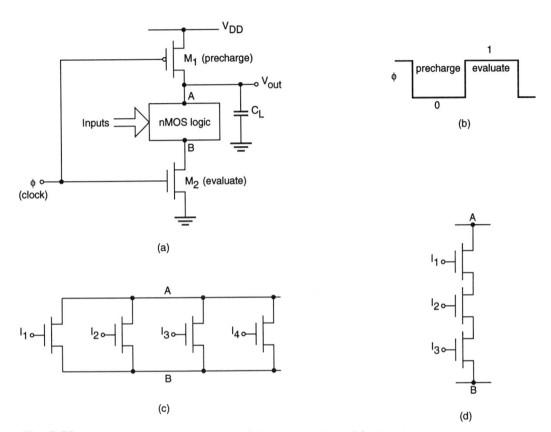

Fig. 5.55 Dynamic CMOS (a) schematic (b) clock waveform (c) 4-input NOR gate (d) 3-input NAND gate

cycle, 0 or 1, which disables their section of the circuit. Clearly the clock rate must be sufficiently high to prevent the inevitable leakage of charge from turning a 1 into a 0 during quiescence. The near open-circuit input resistance of a MOSFET greatly helps in charge conservation. The high speeds of data processing mean that the constraint on clock rate presents no problem. It is possible to construct GaAs dynamic circuits, but this avenue has so far not been explored in depth. Reports of GaAs dynamic logic since their potential was outlined over a decade ago (Rocchi and Gabillard 1983) are few and far between. CMOS has removed the need for saving power by dynamic operation, but dynamic CMOS has the advantage of a reduced transistor count compared with static CMOS. It is used occasionally in full custom ASICs, hence its inclusion here. Another vitally important category of dynamic circuits, dynamic storage in DRAMs, is covered in Chapter 7. Dynamic CMOS logic is illustrated in Fig. 5.55 (Uyemura 1988, pp.541–576; Weste and Eshraghian 1985, pp.163–174). The schematic contains only one pMOSFET, enough to justify the designation CMOS. With the clock at 0 (0 V), M_1 conducts, M_2 is off, and C_L which, as in Fig. 5.1 and other diagrams in this chapter, includes the capacitances of wire and inputs of the following stages, charges quickly and unconditionally during this *precharge* phase to 1

(V_{DD}). When the clock goes to 1, M_2 is turned into conduction and constitutes a virtual short-circuit between B and ground, whilst M_1 is turned off. During this *evaluation* phase the output quickly takes up the logic state determined by the nMOS block. Figs 5.55(c) and 5.55(d) give examples of logic blocks which turn Fig. 5.55(a) into a 4-input NOR gate and a 3-input NAND gate respectively. Alternatively the nMOS logic might consist of an AND–OR combination like transistors M_2 to M_5 in Fig. 5.7, forming a dynamic exclusive–NOR gate. Power saving, it has already been pointed out, is irrelevant in CMOS. The advantage of dynamic CMOS is a reduction of the transistor count from, for example, $2n$ to $(n + 2)$ for an n-input gate.

Operation with a single-phase clock presents a problem in cascaded circuits. Consider the operation of Fig. 5.56, and let us postulate that the n-block consists of a NOR combination like Fig. 5.55(c). It is assumed that all inputs are 0, and are in that state at the commencement of the evaluation phase with the exception of I_n, which has suffered a delay and is still at is precharge value of 1, changing to 0 after a time t_d. The result will be an output with a glitch, see Fig. 5.56(b). More complex malfunctioning is possible with a spread in the timing delays of the inputs. A number of solutions is available. One of these is *domino* CMOS, shown in Fig. 5.57 in which dynamic CMOS is followed by a static CMOS inverter. The precharge output is now zero, and the delay of I_n in Fig. 5.56 would simply lengthen the '0' state of the output by t_d without producing a glitch. Another alternative is modified domino logic, which reverts to Fig. 5.55. The modification consists of cascading alternate stages containing nMOS and pMOS logic, with ϕ clocking for the n-stages and $\overline{\phi}$ clocking for the p-stages (Weste and Eshraghian 1985, pp.171–2).

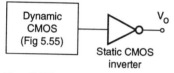

Fig. 5.57 Domino CMOS

5.5 Survey of logic circuits

5.5.1 Overview

The evolution of logic circuits was outlined in Section 5.2.2. The overview presented here explains the position of the various silicon logic families

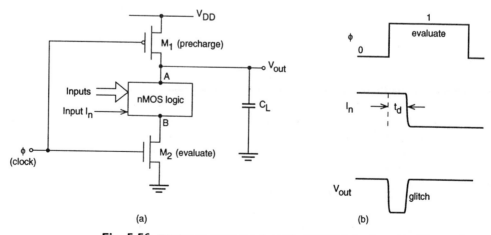

Fig. 5.56 Timing in cascaded dynamic CMOS (a) schematic (b) waveforms

Table 5.6 Count rates of silicon logic families (5 V supply voltage)

Family	Bistable count rate (MHz)	Test load
TTL LS	45	15 pF
TTL S	125	15 pF
TTL FAST	125	50 pF
TTL ALS	50	50 pF
TTL AS	200	50 pF
CMOS HC/HCT	50	15 pF
CMOS AC/ACT	160	50 pF
BiCMOS ABT	200	50 pF
MECL III	<u>500</u>	50 Ω (Fig. 5.20)
ECL 10K	160	50 Ω (Fig. 5.20)
ECL 10KH	<u>250</u>	50 Ω (Fig. 5.20)
ECLinPS	1400	50 Ω (Fig. 5.20)

(Courtesy of Motorola, Philips Semiconductors, and Texas Instruments)

The TTL, CMOS, and BiCMOS families are second-sourced. The four ECL series were originated by Motorola. ECL 10K and ECL 10KH are second-sourced. The other two families are, at the time of writing, probably produced only by Motorola. Where minimum but no typical values are supplied, these are shown underlined

with specific reference to their speed. It is cast in terms of SSI and MSI, which are good initial indicators of performance. The three mature technologies in SSI/MSI are TTL, ECL, and CMOS. Moreover, inspite of the maximization of VLSI, electronic equipments will continue to contain substantial SSI and MSI chips, which are known by the name of *glue*. (An earlier jargon, *jelly beans*, appears to have fallen into disuse.) The families differ in respect of speed and (input) power. The faster the range for a given technology and geometry the greater is the power it consumes. Table 5.6 contains the count rates of the fastest bistables in the preferred logic families, which have geometries of 3 μm or less. These count rates give a relative indication of speed. All tabular data in this and subsequent sections applies to typical values at 25 °C, unless stated otherwise. System speed is determined by fan-out and the interconnect impedance, in practice the capacitive load presented by the following stages and the PCB wires. The TTL and CMOS familes are designed for substantial capacitive loading and are specified with C_L equal to 15 or 50 pF. Very fast transition times can sometimes cause ringing and malfunction due to impedance of the wires. The maximum speed is only obtained by careful engineering of the layout. The entries in Table 5.6 show the progression of TTL, where the mature LS family, the preferred choice for over a decade after its introduction in 1974 (Section 5.2.6), is accompanied by four other Schottky families. The smaller geometries of the FAST, ALS, and AS series lead to either similar count rates with a high C_L, 50 in place of 15 pF, or a higher count rate with the same load of 50 pF. The two CMOS families, the HC/HCT and AC/ACT series, were specifically designed to be equivalent to the TTL LS and AS families. The elements of an earlier family are usually adopted for other ranges, and CMOS logic has taken over very many of the TTL SSI/MSI functions. The functional identity has been retained in the nomenclature. For example, the quad 2-input NAND gate in the original TTL family, the SN54/7400, becomes the SN54/74L500, SN54/74ALS00, etc. in the other TTL families. The CMOS equivalents are the CD54/74AC00, CD54/74ACT00 (Harris Semiconductor), MC74AC00, MC74ACT00 (Motorola 74AC/ACT), and HC/HCT00 (Philips 74HC/HCT), etc.

Vendors make additions to a logic family from time to time. The existing ranges are by and large adequate with regard to speed and functional variety. CMOS is evidently set to replace TTL in the fullness of time. The huge volume of existing equipment constructed with TTL, and the vast engineering experience gained with it, have assured that TTL will continue to be used for some time to come. Nevertheless it has been rendered obsolescent by the capability of CMOS. The majority of TTL and CMOS SSI/MSI is offered in DIP. There has been a move away from this originally exclusive package, and the families in Table 5.6 are available in a variety of encapsulations, with surface mounting beginning to gain pre-eminence.

The logic famiIes in Table 5.6 all operate with a 5 V supply, positive except for ECL, where it is negative and slightly larger (-5.2 V). The alternative of a smaller supply voltage for submicron ICs has been a pressing need for a long time, and a standard level of 3.3 V was established by JEDEC several years ago. The use of a smaller supply voltage is highly desirable because of the great saving in dynamic power, which is proportional to the square of the supply voltage (see eqn (5.11)). Another issue is device breakdown voltage, which makes operation below 5 V highly desirable for 0.8 μm, and essential for smaller CMOS structures.

Vendors stipulate a 3.3 V supply for VLSI, either in total or for part of the die (in which case it is usually generated on-chip from the 5 V supply), where necessary, and have come up with a number of low voltage (LV) CMOS and BiCMOS SSI/MSI families. It is of course possible to operate 5 V CMOS at lower voltages. The CMOS families in Table5.6 have their characteristics specified at several supply voltages from 2 V and 5.5 V (the maximum permitted); a specification of AC characteristics at 3.3 V is quite common. The drawback of operating CMOS below its design voltage is the loss of speed. The count rate of a bistable operated at 3.3 V is about 70 per cent of its value at 5 V. The LV CMOS and BiCMOS families are listed in Table 5.7. They either achieve or surpass the speed of the TTL and CMOS logic in Table 5.6 by virtue of their structures which go down to 0.6 μm. The striking feature is the 350 MHz count rate of the bistable in the HLL and ALVC families. This is far faster than anything in 5 V TTL or CMOS, and is a convincing example of the advances possible with submicron CMOS.

Some ECL chips have considerably higher speeds than the standard

Survey of logic circuits 155

Table 5.7 Low voltage (LV) logic (3.3 V supply voltage)

Family	Feature size (μm)	Supply voltage range (V)	Bistable count rate (MHz)	Comments
LV-CMOS	2.0	1.0–3.6	77	LV version of CMOS/HCT
LVC	0.8	1.2–3.6	125	LV CMOS compatible with TTL FAST (AS)
HLL	0.6	1.2–3.6	350	Fastest LV CMOS
ALVC	0.6	1.2–3.6	350	CMOS Multibyte version of HLL
LVT	0.8	2.7–3.6	150 (min.)	LV BiCMOS version of ABT

(Courtesy of Philips Semiconductors and Texas Instruments)

bistables. These are usually counters, special bistables, and frequency dividers, where the lower rate of the output is amenable to transmission via a PCB microstrip line. An example of this is the Motorola MC 12090 bistable, intended for high speed prescaling in phase locked loops (PLLs). It has a guaranteed count rate of 750 MHz (min.). Another outstanding example is the range of fixed modulus frequency dividers by GEC Plessey Semiconductors (GPS), extending up to 3.5 Ghz. TTL and CMOS contain a veritable plethora of chips which covers a comprehensive spectrum of functions. Table 5.8 lists the broad core of the elements available. MECL families have a similar but more restricted and smaller range of elements.

The BiCMOS ABT series in Table 5.6—ABT stands for Advanced BiCMOS Technology—is specifically aimed at the bus interface. Bus width has increased over the years in order to raise the system bus bandwidth (SBW), expressed by

$$SBW = \frac{\text{number of bytes transferred}}{\text{cycle time x (clock cycles/transfer)}} \quad (5.48)$$

Wide bus interfaces are covered by the *widebus* (TI trademark) alternatively *multibyte* (Signetics/Philips Semiconductors trademark) chips which cater for 2-byte or higher logic. A selection of ABT elements is shown in Table 5.9. The flip-flops are directed at bus-oriented applications. The 8 (or 16) bistables within a package are independent of one another except for common clock and enable (3-state) inputs. The Universal Bus Transceivers (UBTs; proprietary TI trademark) permit three modes of operation: transparent, latched, or clock. Taking data flow from A to B, the two ports (or from B to A), the outputs follow the A inputs after the propagation delay in the transparent mode. In the latched mode, the A data is latched by the latch control, which overrides the clock. Normal clock control is achieved by the appropriate settings of the output enable, latch enable, and clock signals. Bidirectional op-

Table 5.8 TTL/CMOS elements SSI and MSI

Category	Typical features, Modes
Gates	NAND, NOR, AND, OR, exclusive-OR/NOR, inverters
Buffers, drivers, bus transceivers	3-state outputs often available. Inverting, noninverting
Flip-flops	JK, D, edge triggered, level triggered, single, dual, quad, hex, octal
Latches	Quad, 8-bit, 9-bit, 10-bit, 3-state outputs, true and complementary
Arithmetic elements	4-bit adders, 4-bit ALU/function generators, look ahead carry generators
Counters	Asynchronous, synchronous, binary, decade, up-down, with registers, frequency dividers, rate multipliers
Shift registers	Parallel-in parallel-out, serial-in parallel out, parallel-in serial out, bidirectional, input latches, output latches
Multivibrators	Monostable, Schmitt-trigger inputs, retriggerable
Decoders, encoders, data selectors, multiplexers	16-to-1, 8-to-1, 4-to-1, 2-to-1, decoders/demultiplexers, 4-to-16, 4-to-10 (BCD to decimal), 3-to-8
Comparators	4-bit, 8-bit, parity generators, checkers odd/even parity
Display decoders/drivers	BCD to decimal, BDC to seven-segment

Table 5.9 Selection of ABT devices (All devices have 3-state outputs)

Device	Function
5N54/74ABT374	Octal edge-triggered D-type flip-flops
5N54/74ABT16379	16-bit edge-triggered D-type flip-flops
5N54/74ABT245	Octal bus transceivers
5N54/74ABT16240	16-bit bus drivers
5N54/74ABT16500A	16-bit universal bus transceivers
5N54/74ABT16657	16-bit transceivers with parity generators/checkers
5N54/74ABT16833	Dual 8-bit to 9-bit parity bus transceivers
5N54/74ABT32373	32-bit transceivers with D-type latches
5N54/74ABT32500	36-bit universal bus transceivers

(Courtesy of Texas Instruments)

eration permits transmission from A to B, or B to A. The dual 8-bit to 9-bit parity bus transceivers are designed for communication between data buses.

Coming to the 3.3 V logic families in Table 5.7, the initial ranges of the functions are modest, but are increasing rapidly. There is some emphasis on chips for bus interfacing. Before long, 3.3 V may well contain a variety of elements comparable with 5 V logic. Within the broad spectrum of digital ICs, 5 V is being complemented by 3.3 V logic at all levels, VLSI very much included. On account of reliability 3.3 V logic is essential for structures of 0.6 μm or less. Another vital advantage of LV logic, its low power consumption, makes it a preferred choice for the rapidly growing market of portable battery-operated electronic data processing (EDP) equipment. Indeed there is a strong demand for 1 V microsystems powered by a single-cell battery or solar cells (Machi and Chatterjee 1994). A 1 V (or 1.2 V) level is obtainable, but only just (because it is right at the end of the lower limit) by all the low voltage CMOS families in Table 5.7. We may expect new low voltage logic for that purpose. The most probable development is the production of 2.2 V CMOS, which could take 1 V operation in its stride.

5.5.2 Characterization and performance

The ambient standard temperature ranges for electronic equipment are:

(i) commercial (C): 0 °C to 70 °C;

(ii) industrial (I): -40 °C to +85 °C;

(iii) military (M): -55 °C to +125 °C.

Table 5.10 contains the temperature and supply voltage ranges of standard silicon SSI/MSI available in industrial, commercial, and military temperature ranges. The supply voltage ranges express tolerance limits for all families except the CMOS, AC, HC, and the LV series. There the supply voltage can be chosen to suit requirements. It is quite acceptable, for example, to operate CMOS AC and HC at the 3.3 V level of the LV families, or even lower, provided the reduced speed of operation is acceptable. Care should be taken not to exceed the the 5.5 V limit in TTL; chips with internal feedback like latches and flip-flops are prone to breakdown at slightly higher voltages. Tables 5.11 and 5.12 contain the main dynamic characteristics of standard SSI/MSI. Families are enhanced by the addition of new chips which either fulfil new functions or improve the performance of existing elements. One example is the addition of bistables or counters which advance previous speeds.

The most popular ECL families are ECL 10K and ECL 10KH. The latter, ECL 10KH, is about twice as fast as, and consumes half the power of ECL 10K. It has largely ousted MECL III except for the fastest applications, which call on the superior speed of MECL III bistables. The latest addition to ECL, the ECLinPS family, has a nearly threefold improvement in speed, consuming only about half the input power of ECL 10KH. It is available in two versions, 10E (compatible with 10KH) and 100E (compatible with ECL 100K, which is not listed in Table 5.6). ECL 100K, originated by Fairchild, is similar to the other ECL families but incorporates greatly improved compensation against changes in supply voltage and temperature.

The CMOS HC/HCT and AC/ACT series compare in speed with the TTL ALS and AS families. The qualitative comparison of CMOS and TTL gate input power in Fig. 5.58 shows that the dynamic component in TTL power dominates at the higher clock rates, where there is little difference in the input powers to TTL and CMOS. The overriding ad-

Table 5.10 Temperature and supply voltage ranges of SSI/MSI logic

Family	Temperature range (C,I,M or °C)	Supply voltage (V)	Supply voltage range (V)
TTL (all series)			
54	M	5.0	4.5 to 5.5
74	I	5.0	4.5 to 5.5
CMOS (all series)			
54	M	5.0	2.0 to 6.0 (AC, HC)
74	C	5.0	4.5 to 5.5 (ACT, HCT)
BiCMOS ABT			
54	M	5.0	4.5 to 5.5
74	C	5.0	4.5 to 5.5
ECL 10KH	0 to 75	-5.2	-4.94 to -5.46
ECL 10K	-30 to +85	-5.2	-4.68 to -5.72
MECL III	-30 to +85	-5.2	-4.68 to -5.72
ECLinPS			
10E	0 to 75	-5.2	-4.94 to -5.46
100E	0 to 85	-4.5	-4.20 to -5.46
LV-CMOS	C	3.3	1.0 to 3.6
	M		
CMOS LVC	C	3.3	1.2 to 3.6
CMOS ALVC	C	3.3	1.2 to 3.6
BiCMOS LVT			
54	C	3.3	1.2 to 3.6
74	M	3.3	2.7 to 3.6

(Courtesy of Motorola, Philips Semiconductors, and Texas Instruments)

Table 5.11 TTL and ECL gate propagation delay and input power, and bistable count rate

Series	T_P (ns)	P_{IN} (mW)	Speed-power product (pJ)	Bistable count rate (MHz)
TTL LS	9.5	2	19	45
TTL S	3.0	19	57	125
TTL ALS	4.0	1	4	50
TTL AS	1.5	20	30	200
ECL 10K	2.0	25	50	<u>125</u>
ECL 10KH	1.0	25	25	<u>250</u>
MECL III	1.0	60	60	<u>500</u>
ECLinPS	0.35	39	14	<u>1400</u>

Underlined values are minima
(Courtesy of Motorola and Texas Instruments)

Table 5.12 CMOS gate propagation delay, C_i, C_{PD}, and bistable count rate

Family	t_p (ns)	C_i (pF)	C_{PD} (pF)	Bistable count rate (MHz)
HC/HCT	7.5	$\overline{10}$	25	50
AC/ACT	4.8	4.5	30	160
LV-CMOS	9.0	3.5	22	77
LVC	4.0	3.5	20	125
HLL,ALVC	2.1	3.0	35	350

(Courtesy of Philips Semiconductors, and Texas Instruments)

C_i for HC/HCT is the maximum value
C_i, C_{PD} for HLL, AVLC apply to octal buffer/line driver with 3-state output

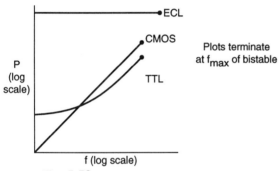

Fig. 5.58 Gate input power vs. frequency

vantage of CMOS remains: it consumes virtually no power in the steady state. This advantage, vital in VLSI, is also important at SSI/MSI/LSI levels. The distinction between HC/AC and HCT/ACT CMOS is that HC/AC CMOS can accept CMOS but not TTL inputs. The outputs of all these families are TTL/CMOS compatible. The necessary interface conditions, which apply to all logic families, are (Fig. 5.46)

$$V_{OH}(min) > V_{IH}(max) \tag{5.49}$$

$$V_{OL}(max) < V_{IL}(min) \tag{5.50}$$

For TTL driving AC/HC CMOS, eqn (5.49) is readily satisfied, because $V_{OL}(max)$ of TTL, 0.5 V, is far below $V_{IL}(min)$ of CMOS, 1.5 V (a 5 V supply voltage assumed). Equation (5.49) points to the problem: $V_{OH}(min)$, 3.0 V for TTL, is less than $V_{IH}(max)$, 3.5 V for standard CMOS. The condition of eqn (5.49) is satisfied by modifying CMOS to give V_{IH} of 1.5 V (typ), 2.0 V (max) for the ACT/HCT series at the cost of a minimal, insignificant reduction in speed.

Preference is given to specifying propagation delay rather than transition time, which is omitted altogether in many data books. The ideal relation and the approximations relating t_P and t_T in eqns (4.7) ($t_P \simeq$

0.31t_T) and (4.9) ($t_P \simeq 0.5t_T$) are rarely satisfied in practice, nor is it important they they should be. They are based on assuming the simple waveform, given by eqn (4.4), for the output and they ignore the influence of overshoots, undershoots, turn-on and turn-off delays. However one aspect relevant to the topic under consideration is the deliberate slowing down of t_T to reduce crosstalk at high speeds, when the very short transition times, which give the signal a rich harmonic content, lead to strong coupling between adjacent circuits. In ECL t_T can be controlled by internal time constants in a manner which only slightly affects t_P, and ECL is expressly designed for $t_P \simeq t_T$. Experience has shown that this equality gives optimum performance in high-speed systems.

The input power to a gate (or to a separate section of it, for example one of several independent gates), is the sum of the static and dynamic input powers, P_s and P_d, given by

$$P_s = V_{CC}.I_{CC} \tag{5.51}$$

and, using eqn (5.11) for P_d, the total power P_t is

$$P_t = P_s + P_d = V_{CC}I_{CC} + C_L V_{CC}^2 f \tag{5.52}$$

V_{CC} is the supply voltage and I_{CC} the mean dc current at a low clock rate, at which P_d is negligible. V_{CC} and I_{CC} have been retained, although V_{DD} and I_{DD} would be more appropriate for CMOS, because data books adopt this nomenclature for bipolar, BiCMOS, and CMOS families alike.

Equation (5.52) implies that the differential logic swing equals V_{CC}, and for CMOS, which is now being considered, this is the case. For all practical purposes, the input power for CMOS equals P_d.

Dynamic power in TTL becomes significant well within the operating range (Fig. 5.58), but the frequency at the point of intersection is largely of academic interest, because the preference for CMOS over TTL is governed by other factors, which have already been mentioned. Propagation delay, bistable count rate, and the speed–power product, are shown for TTL and ECL in Table 5.11. The improvement in both speed and the speed-power product for the newer families, ALS and AS in TTL, 10KH and ECLinPS in ECL, stands out. The power quoted from the data books for TTL is P_s. ECL power is virtually independent of the exercising frequency right up to f_{max} (bistable count rate), because P_s far outweighs P_d. CMOS characteristics are contained in Tables 5.12 and 5.13. The capacitances C_i and C_{PD} are the gate input and *power dissipation* capacitances respectively. The dynamic gate input power for a chip is given by

$$P_d = C_{PD} V_{cc}^2 f_i + \sum (C_L V_{cc}^2 f_o) \tag{5.53}$$

where f_i is the input, and f_o the output frequency. For a gate, f_i and f_o are equal; f_o equals $f_i/2$ for a flip-flop in the toggle mode. P_d in Table 5.13 (p.161) has been calculated for two conditions. First C_L is equated to C_i (unity fan-out), so that P_{da} equals $(C_{PD} + C_i)V_{CC}^2 f$. Second the value of C_L is equated to the test load value shown in Table 5.6

Table 5.13 CMOS gate input power at 40 MHz (2-input NAND gate)

Family	P_{da} (mW)	P_{db} (mW)
HC/HCT (5 V)	35	40
AC/ACT (5 V)	35	80
LV-HCMOS (3.3 V)	11	31
LVC (3.3 V)	10	30
HLL, ALVC (3.3 V)	17	37

$P_{da} : C_L = C_i$
$P_{db} : C_L =$ Test load (Table 5.5) = 15 pF
(HC/HCT), 50 pF (all other families)

and $P_{db} = (C_{PD} + C_L)V_{CC}^2 f$. Table 5.13, in which C_L equals 15 pF for the HC/HCT and 50 pF for all other families, bears out the striking reduction in P_d for 3.3 V CMOS. That difference accounts for the relatively low P_{db} of the HC/HCT series. A similar comparison is made in Table 5.14 for the octal D flip-flop 534, with C_L equal to the test value in Table 5.6. Equation (5.53) for this condition becomes

$$P_d = V_{cc}^2 f \left(C_{PD} + \frac{C_L}{2} \right) \quad (5.54)$$

in which C_L equals 15 pf for the HC/HCT and 50 pf for all the other families. Once again, the reduction in P_d for 3.3 V CMOS stands out.

A different approach has to be taken for comparing the input powers of BiCMOS with CMOS and TTL. Equation (5.52) still holds, but the circuits of the BiCMOS ABT family are more sophisticated than the basic BiCMOS arrangements sketched in Figs 5.34 to 5.37, which draw no static power. Furthermore BiCMOS ABT caters for bus interfacing and does not contain the basic gates. A comparison of power consumption versus frequency of BiCMOS ABT, CMOS AC, and BiCMOS LVT (the 3.3 V version of ABT) is shown in Fig. 5.59. It applies to the octal transceiver 245, a highly popular benchmark. The plots indicate the greatly reduced power consumption of the ABT family at frequencies above the break point, ~20 MHz, due to the superior drive capability of the bipolar BiCMOS output stage. The 3.3 V LVT family performs better still with near-zero static input power and a power consumption of ~210 mW compared with ~340 mW for ABT at 100 MHz. The ABT and LVT familes achieve much the same speed; the D flip-flops in both familes have a typical upper count rate of ~200 MHz. The validity of the comparison in Fig. 5.59, which is taken from the Phillips Semiconductor ABT Data Handbook (1991), depends on adopting an identical test procedure for all families. In this case the input power applies to a test in which each output has a 50 pF load, and for a binary count from 00000000 to 11111111 at an input frequency of 100 MHz.

Table 5.14 Dynamic input power at 40 MHz for octal D-bistable 534 (T mode)

Family	P_d (mW)
HC/HCT (5 V)	42
AC/ACT (5 V)	65
LV-HCMOS (3.3 V)	22
LVC (3.3 V)	24
HLL, ALVC (3.3 V)	23

$P_d = C_{PD} V_{cc}^2 f_i + C_L V_{cc}^2 (f_i/2)$
(C_L as in Table 5.5)

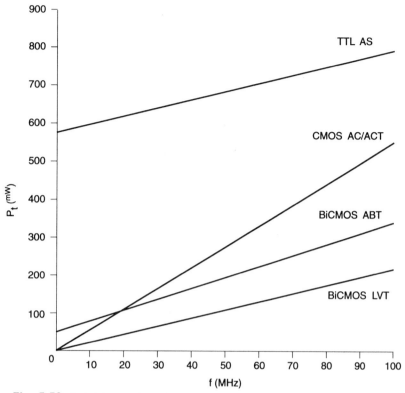

Fig. 5.59 Octal transceiver 245-dynamic input power vs. frequency (Courtesy of Philips Semiconductors)

Example 5.8

Calculate the input power for the CMOS CD54/74AC245 octal transceiver at 100 MHz and compare the result with Fig. 5.59.

The capacitance C_{PD} for the chip equals 57 pF (Harris Semiconductor AC/ACT Data Book 1988), and each output pin is loaded with 50 pF, giving a total capacitance of 107 pF for calculating dynamic power. The eight output pins change state in accordance with the binary up-count. If f is the input frequency, the LSB changes at a rate f, the next bit at $(f/2)$, the bit next to that at $(f/4)$ and so on. The power dissipation A of the least significant bit at 100 MHz equals $107 \times 10^{-12} \times 25 \times 10^8 = 267.5$ mW.

The total power, in accordance with an adoption of eqn (5.53) is a geometric progression which comes to $A(1 - r^n)/(1 - r)$ where $r = 0.5$ and $n = 7$. Making the numerical substitutions, the total power equals 531 mW, which is in fair agreement with P_t in Fig. 5.59.

The BiCMOS ABT family exemplifies the evolution of digital, linear, and mixed-mode standard ICs. Chips are emerging with increasing processing power and function density. It is best to think in terms of chip function without being too concerned about placing the chip into a specific category like MSI, LSI, or VLSI. Categorization has, and will continue to have its place. However the chips now available and documented in data books of a specific logic, e.g. ABT, often fulfil complex functions

Table 5.16 Maximum CMOS, BiCMOS and TTL output currents

Family	I_{OH} (mA)	I_{OL} (mA)
CMOS HC/HCT	-8	4
CMOS AC/ACT	-24	24
BiCMOS ABT	-32	64
BiCMOS LVT	-32	64
LV-HCMOS	-6	6
LVC, HLL, ALVC	-24	24
TTL LS	-0.4	8
TTL ALS	-0.4	8
TTL AS	-2	20
TTL S	-1	20

(Courtesy of Philips Semiconductors, and Texas Instruments)

and are close to the LSI/VLSI border, sometimes possibly even within VLSI. The pin counts can also be larger. The SN54/74ABT32245 36-bit bus transceiver is contained in a 100-pin flatpack. The SN54/74FB2032 9-bit TTL/BTL competition transceiver with a pin count of 52 translates signals between TTL and backplane transceiver logic (BTL), and performs arbitration when there is competition for bus access. The incorporation of intelligence is another feature, no longer confined to VLSI level, in standard components. *Intelligent* (US: *smart*) chips now exist not only in VLSI, but also in LSI. We are witnessing a continuation of standard ICs which straddles across MSI and LSI, possibly even into VLSI without concern about boundaries, and categorization in terms of transistor count is not all that significant.

The TTL, CMOS, and BiCMOS families are all specified for 50 pF loading of the output (with the exception of TTL LS and CMOS HC/HCT, whose dynamic performance is specified for a load of 15 pF). The output current capability has steadily increased with the production of new families. The increase is partly aimed at interfacing CMOS and BiCMOS with TTL, but is also directed at heavily loaded buses and at interconnections like twisted pair wires or transmission lines, which are terminated by resistors close to their characteristic impedance, ~(200–300) Ω. ECL is in a category of its own, with the facility of driving a 50 Ω matched transmission line (Fig. 5.20). The advance of CMOS and BiCMOS output current capability is highlighted in Tables 5.15 and 5.16, in which the standard sign convention is adhered to. Currents are positive when they flow into a transistor, and negative when they flow out of a transistor terminal. Consider now a CMOS LVC gate fanning out to TTL S. In the high state, TTL S draws a small input current, 50 μA. and fan-out presents no problem. When the output is low, CMOS LVC can sink 24 mA, which is enough to support a twelve-fold fan-out. The 24 mA level for I_{OH} and I_{OL} level has become the norm, with an increase to 32 mA (I_{OH}) and 64 mA (I_{OL}) available for the 5 V and 3.3 V

Table 5.15 Maximum TTL input currents

TTL family	I_{IH} (μA)	I_{IL} (mA)
LS	20	-0.4
S	50	-2
ALS	20	-0.1
AS	20	-0.5

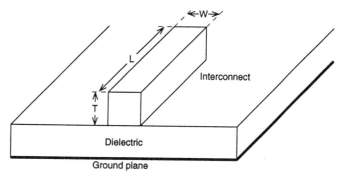

Fig. 5.60 Interconnect structure

BiCMOS families.

Two aspects of VLSI performance with CMOS and ECL will now be looked at in brief, the performance of CMOS and ECL with the existing geometries (∼0.8 to 1.0 μm), and the projected performance with reduced geometries down to ∼0.5 μm. The essential factor to be considered when assessing SSI/MSI versus VLSI is the nature of the wires. SSI/MSI chips have PCB interconnections, whose capacitances far exceed the wire capacitances within the chip. VLSI designs call for transistors with areas reduced from their size in SSI/MSI, for which they need to be large in order to cater for the high capacitive loads of PCB wires. The technique for driving a large capacitive load has been explained in Section 5.2.4. Regardless of whether the driver is the cascaded chain of Fig. 5.15, or an alternative arrangement like a superbuffer (Glasser and Dobberpuhl 1985, pp.26-7, Uyemura 1988, pp.221–2), large-size transistors with large electrode capacitances are needed in order to avoid a prohibitive loss in speed. Heavy capacitive loading is not confined to driving off-chip. Within a chip high fan-out and long interconnections can present large capacitances, and internal buffer drivers feature within a chip to handle such conditions. Transition times and propagation delays of structures on the micron/submicron boundary or smaller are largely determined by the wire capacitance, which is proportional to wire length. Expressions will now be quoted for interconnect capacitance in terms of wire geometry.

Chip and PCB wires can be represented by the microstrip line sketched in Fig. 5.60. Sakurai and Tamaru (1983) give the following expression, which is adopted in various texts (Annaratone 1986) for the wire capacitance C_W,

$$C_W = \varepsilon_o \varepsilon_r \left\{ 1.15 \left(\frac{LW}{H} \right) + 1.4(2W + 2L) \left(\frac{T}{H} \right)^{0.222} \right.$$

$$\left. + 4.12H \left(\frac{T}{H} \right)^{0.728} \right\} \tag{5.55}$$

where ε_o is the (*absolute*) permittivity of free space, and ε_r is the *relative* permittivity (dielectric constant) of the insulating layer (dielectric). Dielectric constants and permittivities for dielectrics of interest are given in Table 5.17. Equation (5.55) contains three modifications of the for-

Table 5.17 Permittivity of dielectrics

Dielectric	Dielectric constant ε_r	Dielectric permittivity $\varepsilon_o \varepsilon_r$ (fF/μm)
SiO$_2$	3.9	3.45×10^{-2}
Si	11.9	1.05×10^{-1}
Si$_3$N$_4$	7.5	6.64×10^{-2}
GaAs	13.1	1.16×10^{-1}
Fibre glass (PCB)	4.5	3.98×10^{-2}

$\varepsilon_o = 8.854 \times 10^{-3}$ fF/μm (free space permittivity)

mula for a parallel-plate capacitor with plates of equal size. The factor 1.15 allows for the lower plate being a near-infinite plane. The second term is a perimeter contribution, and the third a corner contribution; both these arise from the finite thickness of the wire. Practical interest is confined to the condition $L \gg W$. Then the third term may be neglected, and C'_W, the capacitance per unit length, becomes

$$C'_W \simeq \varepsilon_o \varepsilon_r \left\{ 1.15 \left(\frac{W}{H}\right) + 2.8 \left(\frac{T}{H}\right)^{0.222} \right\} \quad (5.56)$$

Equation (5.56) is accurate to within a few per cent for $0.3 < (W/H) < 30$. An alternative expression by Yuan et al. (1982) gives

$$C'_W = \varepsilon_o \varepsilon_r \left\{ \frac{W}{H} + 2.42 - 0.44\frac{H}{W} + \left(1 - \frac{H}{W}\right)^6 \right\} \quad (5.57)$$

for $W \geq H$

$$C'_W = \frac{2\pi \varepsilon_o \varepsilon_{eff}}{\ln\left(\frac{8H}{W} + \frac{W}{4H}\right)} \quad (5.58)$$

for $W \leq H$

where

$$\varepsilon_{eff} = \frac{\varepsilon_r + 1}{2} + \frac{\varepsilon_r - 1}{2}\left(1 + \frac{10H}{W}\right)^{-\frac{1}{2}} \quad (5.59)$$

Three cases will be considered with the wire on:
(i) a silicon substrate;
(ii) a GaAs substrate;
(iii) a PCB dielectric.

Figure 5.61 represents the structure for silicon and GaAs substrates. Fig. 5.61(b) is the equivalent model for Fig. 5.61(a), because the silicon substrate offers a very low resistance, typically ~ 10 Ω, between its surface and ground, so that the silicon dioxide–silicon boundary behaves like a ground plane. The picture is very different for GaAs, where the

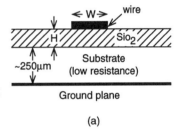

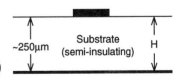

Fig. 5.61 Wires on substrate (a) Si substrate (b) equivalent representation of (a) (c) GaAs substrate

Table 5.18 C'_W (Fig. 5.60)

Dielectric	Dimensions (μm)	C'_W (fF/μm)
SiO$_2$	$W = H = T = 1$	0.136
SiO$_2$	$W = 2, H = T = 1$	0.176
GaAS	$W = T = 1$ $H = 250$	0.052
GaAS	$W = 2, T = 1$ $H = 250$	0.058
Fibre glass (PCB)	$H = 500, T = 37$ $W = 343$	0.094
Fibre glass (PCB)	$H = 500, T = 37$ $W = 30$	0.065

Table 5.19 C'_W silicon dioxide

Formula	C'_W (fF/μm)	
	$W = T = H = 1$ μm	$W = 2$ μm, $T = H = 1$ μm
Sakurai and Tamaru (1983)	0.136	0.176
Yuan et al. (1982)	0.103	0.146
Bakoglu (1990)	0.140	0.174

substrate has a very high impedance and constitutes a dielectric. The substrate is typically $\sim$250 μm thick and with $W \ll H$ ($W \simeq$(1–2) μm, $H \simeq$ 250 μm) C'_W is given by eqns (5.58) and (5.59). The PCB wire forms a capacitance in accordance with the expressions which hold for wire on silicon, with H, the thickness of the oxide below the wire, being replaced by the thickness of the fibre glass. The computations of C'_W, based on eqns (5.56), (5.58), and (5.59) are given in Table 5.18. Equation (5.56) has been used in preference to eqn (5.57) for C'_W (SiO$_2$ and fibre glass), because it allows for the thickness of wire; eqn (5.57) does not. The values obtained for C'_W using eqns (5.56) and (5.57) are similar (Table 5.19), especially for the case $W = 2$ μm, $H = T = 1$ μm, which reflects general practice for 1 μm geometries, although it is possible to make W equal to 1 μm in that case, using aggressive design rules. The comparison in Table 5.19 includes a formula quoted by Bakoglu (1990) and based on a paper by Yuan and Trick (1982).

The great suprise in Table 5.18 is the value of C'_W for GaAs. Taking a given geometry, GaAs is the fastest logic. That much was said in Sections 1.3.6 and 5.2.2, and is the key theme taken up in the next and final section of this chapter, in which silicon is compared with gallium arsenide. It has also been pointed out that wire capacitance dominates the speed of VLSI logic. On the face of it, wires on GaAs should have a far smaller capacitance than wires on silicon. Thinking purely in terms of a parallel plate capacitor, the dielectric thickness for GaAs ($\sim$250 μm) compared with that for Si (~1 μm) should reduce C'_W by a factor of 80 or thereabouts, allowing for the values of the dielectric constants (Table 5.17). That would give GaAs VLSI a phenomenal advantage. The euphoria of such an understandable but mistaken initial approach is cut short by examining the wire structure. The narrow wire (W ~2 μm) is separated by a dielectric, whose depth (~250 μm) far exceeds W, from the 'infinite-plane' lower plate. In the event C'_W for GaAs is lower than C'_W for Si by a factor of about 3 (Table 5.18), which represents a considerable improvement but is nowhere near the reduction one might expect at first sight.

Gate behaviour will now be assessed with the aid of SPICE simulations, augmented by some hand analysis. The estimates of wire capacitance, based on the calculations summarized in Table 5.18, are allowed for in several of the simulations. The transistor structures considered are:

(i) large geometries for SSI/MSI, in order to drive PCB wires;
(ii) small geometries for VLSI.

The reasons for making transistors large in order to drive the relatively large capacitances of PCB wires have already been given. It is worth recapitulating on the cascaded buffer described in Section 5.2.4 with the aid of Figs 5.14 to 5.16. Although this treatment is confined to CMOS, it holds in a general way for bipolar logic. The intrinsic speed of a cascaded inverter with equal or not too dissimilar input and output loads, like for instance the inverter between A and B in Fig. 5.14, can be quite high. The load capacitance for unity fan-out is, ignoring the influence of wire, the gate capacitance of the next stage. The high capacitive load of a wire can still give a fast response for a single inverter whose transistors have been made deliberately large for that purpose. The typical input capacitance of a CMOS SSI gate is (3.5–4.5)pF, which is comparable to the load offered by a short PCB wire, whose capacitance is ~1 pF/cm (Table 5.18). Transition and delay times are similar, and eqns (5.22) to (5.26) can be applied to both of them for estimating the influence of various parameters. This leads to

$$t_P, t_T \propto \frac{C_L}{\mu \left(\frac{\varepsilon_{ox}}{t_{ox}}\right) \left(\frac{W}{L}\right)} F(V_{DD}, V_{TH}) \quad (5.60)$$

where $F(V_{DD}, V_{TH})$ is defined in eqns (5.22) and (5.23). For large transistors, whose input capacitance outweighs the wire capacitance, a good approximation is to equate the input capacitance to C_{ox} (Table 4.1), which is given by

$$C_{ox} = \left(\frac{\varepsilon_{ox}}{t_{ox}}\right) WL \quad (5.61)$$

an expression which follows from eqn (4.28). Combining eqns (5.60) and (5.61)

$$t_P, t_T \propto L^2 \quad (5.62)$$

which implies that for cascaded transmission with identical transistors t_P and t_T are approximately proportional to L^2 and independent of W. The transistor area can be quite large by making W large but keeping

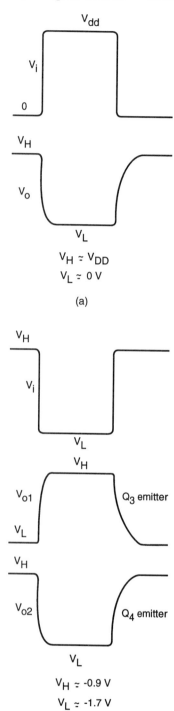

Fig. 5.62 Waveforms—SPICE simulations (Tables 5.20 and 5.21) (a) CMOS inverter (Fig. 5.9) (b) ECL inverter (Fig 5.17)

L as small as possible to obtain a fast transient response. That happens in practice, where very high ratios of (W/L), up to 100 or more, are used in SSI/MSI. Reverting to the situation in VLSI, where C_L is composed of significant contributions from both wire and the following stage(s), eqn (5.60) shows that (W/L) should be high and t_{ox} small for fast switching. A compromise is called for, because transistor size increases and function density decreases with (W/L).

The results of some CMOS and ECL inverter SPICE simulations are contained in Tables 5.20 and 5.21. Input and output waveforms for the purpose of identifying t_P and t_T are sketched in Fig. 5.62. The CMOS simulations span geometries from 0.8 to 2 μm; they cover both large and small transistors. The two sets of SPICE computations of ECL are likewise for relatively large SSI and small VLSI structures. All simulations hold for unit fan-out (with one exception, when the loading was confined to a 50 pF capacitance). Mean values, the average of two for the CMOS, and four for the ECL simulations, were used to compute t_P and t_T, which are expressed to the nearst 5 ps if less than 1 ns, and to the nearest 10 ps if larger. Wire capacitances of 5.2 fF and 260 fF have been included in some of the simulations. They correspond, in accordance with Table 5.18, to a fairly short local interconnection 20 μm long, and to a 1 mm interconnection, which is a typical average for a VLSI die 1×1 cm^2. The parameters used for the transistors are listed in an appendix at the end of this chapter. (Remember that SPICE inserts default values where necessary.) The CMOS simulations were made at level 3, which is reckoned to give better results than levels 1 and 2 for MOSFETs with structures of 2 μm or less.

The dimensions of the *SIM1* transistors hold for a 2 μm geometry; *SIM1A*, *SIM1B*, and *SIM1C* compare the performance at level 3 (*SIM1A*), level 2 (*SIM1B*) and level 1 (*SIM1C*). Levels 2 and 3 give similar results, level 1 (*SIM1C*) is distinctly slower. The values of W, CBS, and CBD have been greatly increased for *SIM1D*, moving the transistors to SSI dimensions. There is no change in t_p and t_T relative to *SIM1A*, a result which confirms the validity of eqn (5.62), because $C_L \sim C_{ox}$. Another simulation was made in the *SM1* group. The load and parameters were as for *SIM1D* except that CBD and CBS, instead of being entered directly, were computed automatically by SPICE in accordance with its algorithm from the alternative input parameters CJ, AD, and AS, chosen to give precisely the same value of CBD and CBS as for *SM1D*. The results agreed exactly with those for SIM1D, and this check simulation is not listed in Table 5.20. If both sets of data (CBD, CBS, and CJ, AD, AS) are supplied, the direct entries (CBD and CBS) override (Vladimirescu and Liu 1980; Quarles et al. 1993).

The large dimensions of W for the *SIM2* simulations were chosen to give C_{ox} similar to the gate input capacitance of CMOS AC/ACT ($\sim$4.5 pF). CBD and CBS have been raised to 2.5 pF and TOX is 25 nm in place of the default value (100 nm) used in *SIM1*. The calculated values of C_{ox} are 3.45 pF (pMOS) and 8.63 pF (nMOS). The transient response is still very good with $t_P \sim 0.5$ ns for FO = 1 (*SIM2A*). *SIM2A* demonstrates like *SIM1D* the capability of transistors with large dimensions to switch very fast. These dimensions, however, rule out such transistors for VLSI. A load of 50 pF (*SIM2B*) causes an increase in t_P to

Table 5.20 SPICE simulations for CMOS inverter (Fig. 5.9) (FO = fan-out). (All simulations are at level 3, unless stated otherwise)

Simulation	Load	Parameter outline	t_P (ps)	t_T (ps)
SIM1A	FO = 1	$L = 2$ μm, $W = 5$ μm (pMOS) $L = 2$ μm, $W = 2$ μm (nMOS) $CBD = CBS = 25$ fF	550	956
SIM1B	FO = 1	Level 2, otherwise as for SIM1A	590	1060
SIM1C	FO = 1	Level 1, otherwise as for SIM1A	385	795
SIM1D	FO = 1	W increased to 200 μm (pMOS) 500 μm (nMOS) $CBD = CBS = 2.5$ pF, otherwise as for SIM1A	550	965
SIM1E	FO = 4	Same as for SIM1D	730	1330
SIM2A	FO = 1	$L = 2$ μm $W = 2500$ μm (pMOS) $W = 1000$ μm (nMOS) $CBD = CBS = 2.5$ pF $t_{ox} = 25$ nm	490	880
SIM2B	No FO $C_L = 50$ pF	Same as for SIM2A	1710	3380
SIM3A	FO = 1	Parameterized for typical 2 μm VLSI. $L = 2$ μm, $W = 10$ μm (pMOS) $W = 5$ μm (nMOS) $t_{ox} = 22.5$ nm	250	385
SIM3B	FO = 4	Same as for SIM3A	600	1210
SIM3C	FO = 1 $C_L = 5.2$ fF	Same as for SIM3A	265	400
SIM3D	FO = 1 $C_L = 260$ fF	Same as for SIM3A	890	1690
SIM3E	FO = 1	$V_{DD} = 3.3$ V, otherwise as for SIM3A	360	685
SIM3F	FO = 1	Level 1, otherwise as for SIM3A	230	375
SIM3G	FO = 1	Level 2, otherwise as for SIM3A	260	370
SIM4A	FO = 1	Parameterized for typical 0.8 μm VLSI. $L = 0.8$ μm, $W = 10$ μm (pMOS) $W = 2$ μm (nMOS)	100	155
SIM4B	FO = 4	Same as for SIM4A	225	480
SIM4C	FO = 1 $C_L = 5.2$ fF	Same as for SIM4A	120	190
SIM4D	FO = 1 $C_L = 260$ fF	Same as for SIM4A	910	2100
SIM4E	FO = 1	$V_{DD} = 3.3$ V, otherwise as for SIM4A	145	220
SIM4F	FO = 1	Level 1, otherwise as for SIM4A	100	155
SIM4G	FO = 1	Level 2, otherwise as for SIM4A	95	145

Table 5.21 CMOS inverter transient response

Simulation	Load	t_T (ps)	
(Table 5.19)		SPICE	Hand analysis
SIM3A	FO = 1 C_L = zero	0.39	0.19
SIM3D	FO = 1 C_L = 260 fF	1.69	1.15
SIM4A	FO = 1 C_L = zero	0.15	0.045
SIM4D	FO = 1 C_L = 260 fF	2.10	1.26

1.71 ns, approaching the CMOS AC/ACT gate speed, which also holds for a 50 pF load.

SIM3 computations are based on a 2 μm VLSI geometry, the parameters of which are based on extracts from various sources, combined with guestimates. These have been made here and elsewhere to put forward a set of parameters which reflects standard practice. The superiority of *SIM3A* over *SIM1A* is largely due to the reduced t_{ox} for the former. The effect of loading the output is barely noticeable for C_L = 5.2 fF, but becomes significant for C_L = 260 fF. The simulation for a 3.3 V supply voltage, *SIM3E*, indicates an increase of about 50 per cent for t_P. Lastly simulations at levels 1 and 2 (*SIM3F* and *SIM3G*) show only a negligible change relative to *SIM3A*.

The final CMOS simulations, *SIM4*, are for a feature size of 0.8 μm. The propagation delay for FO = 1 is only 100 ps, increasing strongly with fan-out (*SIM4B*) or capacitive loading (SIM4C and SIM4D), as is to be expected. It is highly significant that t_P and t_T (FO = 1), are very similar for the 2 μm and 0.8 μm geometries of *SIM3* and *SIM4* for C_L = 260 fF, the value for a 1 mm wire (*SIM3D* and *SIM4D*). It highlights the dominance of wire in VLSI. At the same time it would by rash to conclude that there is little point, if any, in going to smaller structures in order to speed up switching times. Critical parts of a chip, with respect to speed (for example counters and multiplexers) have local and not global interconnections, and a speed improvement at local interconnect level is worth having, even if global wires result in reduced system speed. The loss of speed when going from 5 V to 3.3 V is brought out in *SIM4E*. It amounts to an increase of about 50 per cent in t_P, as is the case in *SIM3E*. *SIM4F* and *SIM4G* support the findings of *SIM3F* and *SIM3G*: there is very little difference between the results obtained with levels 1, 2, and 3.

Table 5.21 compares some of the simulations in Table 5.20 with hand analysis based on eqns (5.22) to (5.26). The results show that the divergence between SPICE and hand calculated transition times is considerably reduced where C_L dominates transient performance. Even then the difference is large and the value of hand analysis in this case is confined to obtaining an initial estimate.

Table 5.22 SPICE simulations of ECL inverter (Fig. 5.17). (FO = fan out)

Simulation	Load	Parameter outline	t_P (ps)	t_T (ps)
SIM5A	FO = 1	Parameters broadly correspond to ECL 10KH. Tail current = 4 mA, emitter follower current = 2.4 mA, emitter area $\simeq 2 \times 15$ μm^2	390	485
SIM5B	FO = 1, $C_L = 260$ fF	Same as for SIM5A	425	460
SIM5C	FO = 1, Transmission line termination (Fig. 5.20) $Z_o = R_1 = 50$ Ω	Same as for SIM5A	980	1220
SIM6A	FO = 1	Parameterized for typical 0.6 μm VLSI geometry. Total current = 400 μA, emitter follower current = 240 μA, emitter area $\simeq 0.6 \times 2.4$ μm^2	55	80
SIM6B	FO = 1, $C_L = 5.2$ fF	Same as for SIM6A	65	95
SIM6C	FO = 1, $C_L = 260$ fF	Same as for SIM6A	280	470

The first set of the ECL simulations in Table 5.22, SIM5, is for transistors with parameters and a geometry corresponding broadly to the ECL 10KH family. The emitter dimensions are $\sim 2.5 \times 15$ μm^2. Subnanosecond propagation delay is obtained for FO = 1 (SIM5A) and the gate can easily drive a load of 260 fF with only slight loss in speed (SIM5B). There is however a marked slowing down when simulating the circuit for the standard ECL test condition, the 50 Ω transmission line termination with matching resistors (Fig. 5.20). This gives a transient response close to that of the 10KH family (SIM5C).

SIM6 is based on a 0.6 μm geometry (emitter area $\sim 0.6 \times 2.4$ μm^2) near the frontier of technology (Garuts et al. 1989), and the response is in fair agreement with the delay quoted in that reference, bearing in mind that some of the parameters and the dc operating conditions are guestimates. Like the 0.8 μm CMOS family (SIM4), the 0.6 μm ECL of SIM6 is highly sensitive to capacitance loading (SIM6B and SIM6C).

Returning to MOS, Dennard et al. (1974) have formulated MOSFET scaling rules which remain a powerful guide to this day. The one parameter which has escaped scaling until comparatively recently is the supply voltage. Table 5.23 lists the effects of 'ideal' and 'constant voltage' scaling. The effects of the scaling on the various parameters are straightforward, bearing in mind the following:

(i) $I_{DS} \propto (1/t_{ox})(W/L)(V_{DD}, V_{TH})^2$ in accordance with eqns (4.21) and (4.22);
(ii) the gate capacitance is equated to C_{ox};
(iii) $t_P \propto (C_{ox}.V_{DD})/I_{DS}$.

172 Digital circuits — techniques and performance

Table 5.23 MOS scaling rules ($S > 1$)

Parameters	Scaling factor (Constant field)	Scaling factor (Constant voltage)
Dimensions (W, L, t_{ox})	$1/S$	$1/S$
Substrate doping	S	S
Voltages (V_{DD}, V_{TH})	$1/S$	1
Device current (I_{DS})	$1/S$	S
Gate capacitance (C_{ox})	$1/S$	$1/S$
Gate input power (P)	$1/S^2$	S
Gate delay (t_P)	$1/S$	$1/S^2$
Speed-power product ($t_P \times P$)	$1/S^3$	$1/S$
Device area ($A \propto WL$)	$1/S^2$	$1/S^2$
Gate input power density (P/A)	1	S^3

(©1974 IEEE)

MOSFET scaling is considered further in Chapter 12, together with scaling of the interconnections and of bipolar transistors. For the present, note the striking effect of omitting to scale the supply voltage. The 5 V level has remained unchanged in standard ICs until recently. Considering the overall aspects of system design that is understandable, and the 5 V level will continue to remain side by side with the 3.3 V level, which has emerged in earnest.

At a given lithographical capability, bipolar will switch faster than MOS transistors, primarily because of their superior transconductance. The vital expression for comparison is eqn (4.17), which states that $t_T \propto (C_L/g_m)$. Leaving aside speed, MOSFETs have the vital advantage of a higher function density because of their lower power consumption, which is purely dynamic in CMOS. Another comparison, that between Si and GaAs, is the subject of the following section, which concludes this chapter.

5.5.3 Comparison of GaAs with Si

The case for GaAs (V)LSI rests largely on its speed advantage over Si. Speed, however, is not the only consideration. Cost, reliability, function density, and physical properties which have a general bearing on the perfomance of GaAs vis-à-vis Si have to be considered. Switching speed is determined by the transit time of the charge carriers within the transistor and by the currents charging and discharging C_L. Carrier transit time is, as has already been pointed out, becoming small compared with the charge and discharge times for C_L for transistors which have geometries of ~2 μm or smaller, even when C_L is only the input capacitance of the next stage.

The electron mobility of GaAs at room temperature is about five times that of Si (Table 5.24). The notion that GaAs is five times as fast as Si stems from this. It cannot be sustained on close examination, although the higher mobility of GaAs does contribute to its superior speed. The figure of merit governing transient response, g_m/C_L, has been quoted several times previously. Recall that C_L of a loaded circuit includes its

Table 5.24 Properties of GaAS and Si (300 K)

Property	GaAs	Si
Energy gap (eV)	Direct : 1.42	Indirect : 1.12
Electron drift mobility (cm^2/Vs)	8500	1500
Thermal conductivity (W/cm °C)	0.44	0.9
Dielectric constant	13.1	11.9
Substrate resistivity (Ω-cm)	10^8	$\sim$(1–10)

output capacitance, the wire capacitance, and the input capacitance of the next stage(s). The first step is to compare the MESFET, MOSFET, and BJT transconductances.

The Shockley model is the basis for MESFET and MOSFET analysis. The transconductance for the MESFET is given by

$$g_m = \frac{\varepsilon \mu W}{a L_g}(V_{GS} - V_p) \tag{5.63}$$

where ε is the permittivity, μ the mobility, W the width of the channel, a the channel depth below the gate, and L_g the gate length. V_{GS} and V_p are the gate-source and pinch-off voltages respectively. The similarity between MESFET and MOSFET expressions is evident from eqn(5.38), which is quoted below for the purpose of comparison.

$$g_m = \frac{\varepsilon \mu}{t_{ox}}\left(\frac{W}{L}\right)(V_G - V_T) \tag{5.64}$$

The electron drift velocity v is given by

$$v = \frac{\mu}{L_g}(V_{GS} - V_p) \tag{5.65}$$

Combining eqns (5.63) and (5.65)

$$g_m = \frac{\varepsilon v W}{a} \tag{5.66}$$

Figure 5.63 gives the velocity field characteristics for GaAs and Si. At low electric fields, $v(\text{GaAs}) \gg v(\text{Si})$, with $\mu(\text{GaAs}) \sim 5\,\mu(\text{Si})$; see Table 5.24. The velocity of GaAs peaks around 3 keV/cm (0.3 V/μm) and then decreases asymptotically to a saturation value v_s. The field strength for saturation is substantially in excess of 0.3 V/μm and eqn (5.66) becomes

$$g_m = \frac{\varepsilon v_{sat} W}{a} \tag{5.67}$$

The plots in Fig. 5.63 hold for undoped semiconductors. The value of v_{sat} for GaAs and Si depends on the field strength, the doping level of the substrate, and the transistor dimensions. It differs from the predictions based on the steady state velocity field characteristics in Fig. 5.63 for reasons of the physical operation, explained by Welch et al. (1985), Long and Butner (1990, pp.41–42) and Sze (1981). Transistors for logic

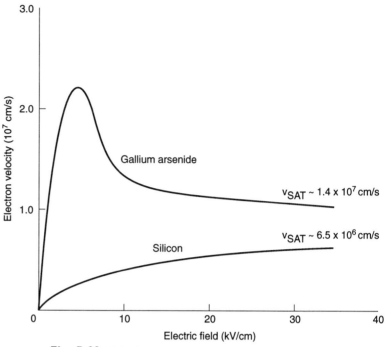

Fig. 5.63 Velocity field characteristics for GaAs and Si

circuits operate at fields with doping levels and geometries which give $v_{sat} \sim 1.4 \times 10^7$ cm/s and $\sim 1.0 \times 10^7$ cm/s for GaAs and Si respectively in the references just cited, leading to a g_m ratio of 1.4 to 1. The steady-state saturation velocities at a field strength of ~ 2 V/μm are, according to Fig. 5.63, 1×10^7 cm/s for GaAs and $\sim 4 \times 10^6$ cm/s for Si. Hence, although GaAs retains a much higher mobility than Si for doped substrates with typical values of 5000 cm^2/Vs (GaAs) and 800 cm^2/Vs (Si) at $N_d \simeq 1 \times 10^{17}$/cm^3 (Long and Butner 1990), the g_m ratio is only $\sim$1.4. The input and output voltages of a logic circuit traverse the low-field and high-field (saturation) regions. Transition (and delay) time is composed of two components, one governed by low-field, and the other by saturated operation. The result is that the actual switching speed turns out to be somewhere between what might be expected by considering either low-field operation with its 'five-to-one' mobility advantage or saturated operation alone, with its 1.4 to 1 g_m ratio. Accurate transient analysis has to be based on a SPICE-type non-linear large-signal model.

The comparison so far has been between a GaAs MESFET and a Si MOSFET. The effective g_m of a BJT is less than the intrinsic value in eqn (5.37), by virtue of the overdrive it furnishes in ECL. The 0.6 μm ECL gate in SIM6 of Section 5.5.2 (Garuts et al. 1989) operates with a tail current of $\sim$400 μA and a logic differential voltage of $\sim$0.9 V, which results in an effective g_m of $\sim$0.45 mA/V. The g_m for a MESFET with $W = 10$ μm, $a = 0.1$ μm, $L_g = 1$ μm, and $v_{sat} = 1 \times 10^7$ cm/s, comes to 1.15 mA/V, using eqn (5.67). These simple hand calculations are in accord with the general state of affairs, backed by detailed analysis, namely that GaAs MESFET and BJT transconductances are similar, and that GaAs has the edge over Si at submicron dimensions.

The other issue to consider in the figure of merit (g_m/C_L) is the magnitude of C_L. Leaving out wire capacitance for the case of a very short local interconnection, $C_L \sim C_i$, the MESFET input capacitance. Unlike the BJT and the MOSFET, the MESFET has no pn junctions around its drain and source regions and its interelectrode capacitances are therefore significantly smaller for a given feature size. Equally important, the interconnect capacitance is greatly reduced. The expectation of a two- to threefold reduction relative to Si, voiced in Section 5.5.2 with reference to Table 5.18, is not fulfilled in practice. It applies to a single wire, whereas the interconnections in VLSI consists of a multitude of wires close together with the spacing between them more or less equal to their width. Under these conditions C'_W is still less for GaAs than for Si, but by a much smaller factor. Another advantage of the GaAs substrate is its elimination of 'slow-wave' propagation, which results in a lossy transmission of signals up to several GHz in a silicon substrate (Long and Butner 1990, p.17).

GaAs has a number of other advantages. The high g_m of the MESFET at low gate voltages gives a low dynamic input power and consequently a high transistor count. Indeed the speed–power trade-off is sometimes used to produce (V)LSI GaAs with ECL speed—possibly even a little slower—but with less power consumption. Inductors deposited on the substrates of GaAs analogue microwave ICs can maintain a high Q, something which is not possible with the low-resistance silicon substrate (Saul 1989). GaAs has superior radiation hardness and can stand ten times the dose of CMOS and bipolar ICs.

Not all advantages are with GaAs. The much higher thermal conductivity of Si (Table 5.24) makes for easier and cheaper packaging and allows greater power dissipation. Si technology is far more mature, and is inherently more reliable.

Digital GaAs ICs are likely to occupy only a very small share of the market in the immediate future, and will be reserved largely for special applications like ultrafast prescalers, fast arithmetic units, etc. GaAs heterojunction bipolar transistor (HBT) logic is potentially capable of making a bigger inroad into ultrafast silicon in the long term.

One outstanding characteristic of GaAs is its capacity to emit light by virtue of being a direct-gap semiconductor. The direct gap signifies that the minimum energy separation between the conduction and valence bands occurs at the same momentum. In Si these two minima are separated in momentum. Electrons excited into the conduction band decay with the emission of infra-red light in the case of pure GaAs, and visible light in the case of ternary compounds. The property of light emission is well suited to the full, compact integration of electronic and optic components. Moves to do so are still at an early stage, but have great potential.

Appendix—SPICE parameters

The parameters are mainly contained in the .MODEL statement, but a few like L and W are entered on the device line.

SIM1A
Device line pMOSFET : $L = 2U$, $W = 5U$.
Device line nMOSFET : $L = 20U$, $W = 2U$.

.MODEL pMOSFET : LEVEL=3 VTO=-1 GAMMA=0.4 KP=1.0E-5 CBD=25F CBS=25F.

.MODEL nMOSFET : LEVEL=3 VTO=1 GAMMA=0.37 KP=2.5E-5 CBD=25F CBS=25F.

SIM1B
Level=2 in .MODEL statement, otherwise as for SIM1A.

SIM1C
Level=1 in .MODEL statement otherwise as for SIM1A.

SIM1D
Device line pMOSFET : $L = 2U$, $W = 500U$.
Device line nMOSFET : $L = 2U$, $W = 200U$.

.MODEL pMOSFET : LEVEL=3 VTO=-1 GAMMA=0.4 KP=1.0E-5 CBD=2.5P CBS=2.5P.

.MODEL nMOSFET : LEVEL=3 VTO=1 GAMMA=0.37 KP=2.5E-5 CBD=2.5P CBS=2.5P.

SIM1E
As for SIM1D.

SIM2A
Device line pMOSFET : $L = 2U$, $W = 2500U$.
Device line nMOSFET : $L = 2U$, $W = 1000U$.

.MODEL pMOSFET : LEVEL=3 VTO=-1 GAMMA=0.4 KP=1.0E-5 CBD=2.5P CBS=2.5P TOX=250E-10.

.MODEL nMOSFET : LEVEL=3 VTO=1 GAMMA=0.37 KP=2.5E-5 CBD=2.5P CBS=2.5P TOX=250E-10.

SIM2B
As for SIM2A.

SIM3A
Device line pMOSFET : $L = 2U$, $W = 10U$, $AD = 40P$, $AS = 40P$.
Device line nMOSFET : $L = 2U$, $W = 5U$, $AD = 20P$, $AS = 20P$.

.MODEL pMOSFET : LEVEL=3 VTO=-0.75 KP=2.5E-5 PHI=0.7
+GAMMA=0.7 TOX=225E-10 CJ=2.5E-4 CGDO=2.5E-10 CGBO=2.5E-10
+CGSO=2.5E-10 CJSW=1.5E-9 UO=230 XJ=5.2E-7.

.MODEL nMOSFET : LEVEL=3 VTO=0.75 KP=5E-5 PHI=0.7
+GAMMA=0.7 TOX=225E-10 CJ=1.5E-4 CGDO=2.5E-10 CGBO=2.5E-10
+CGSO=2.5E-10 CJSW=5.8E-10 UO=300 XJ=3.5E-7 LD=2.5E-7.

SIM3B, SIM3C, SIM3D, SIM3E
As for SIM3A.

SIM3F
Level=1 in .MODEL statement, otherwise as for SIM3A.

SIM3G
Level=2 in .MODEL statement, otherwise as for SIM3A.

SIM4A
Device line pMOSFET : $L = 0.8U$, $W = 5U$, $AD = 8P$, $AS = 8P$.
Device line nMOSFET : $L = 0.8U$, $W = 2U$, $AD = 3.5P$, $AS = 3.5P$.

.MODEL pMOSFET : LEVEL=3 VTO=-0.85 TOX=200E-10 NSUB=5.7E16
+UO=155 CJ=410E-6 CJSW=4E-10 CGDO=1.15E-10 CGSO=1.15E-10
+CGBO=1.2E-9 RD=900 RS=900.

.MODEL nMOSFET : LEVEL=3 VTO=0.85 TOX=200E-10 NSUB=6.7E16
+UO=435 CJ=490E-6 CJSW=5E-10 CGDO=2E-10 CGSO=2E-10
+CGBO=1.2E-9 RD=500 RS=500.

SIM4B, SIM4C, SIM4D, SIM4E
As for SIM4A.

SIM4F
Level=1 in .MODEL statement otherwise as for SIM4A.

SIM4G
Level=2 in .MODEL statement, otherwise as for SIM4A.

SIM5A
.MODEL statement : IS=1.30E-16 BF=100 VAF=75 IKF=0.05
+ISE=0.06E-13 NE=2.00 NR=1.50 RB=300 RE=5 RC=45
+CJE=2E-13 VJE=0.80 MJE=0.5 TF=3.0E-11 XTF=1.0 VTF=10
+ITF=6E-2 CJC=1.5E-13 VJC=1.5 MJC=0.5 CJS=1.1E-13
+XTB=1

SIM5B, SIM5C
As for SIM5A.

SIM6A
.MODEL statement : IS=5E-18 BF=100 CJE=6.7F CJC=7.5F CJS=9.0F
+TF=1E-11 RE=10 RB=200 RC=50 VAF=50

SIM6B, SIM6C
As for SIM6A.

References

Annaratone, M. (1986). *Digital CMOS circuit design*, pp.137–43. Kluwer Academic Publishers, USA.

Bakoglu, H.B. (1990). *Circuits, interconnections, and packaging for VLSI*, pp.137–8. Addison-Wesley, USA.

Blood, W.R. et al. (1980). *MECL system design handbook*, p.1. Motorola Semiconductor Products, USA.

Dennard, R.H., Gaensslen, F.H., Yu, H.N., Rideout, V.L., Bassous, E., and LeBlanc, A.R. (1974). Design of ion implanted MOSFETs with very small physical dimensions. *IEEE Journal of Solid-State Circuits*, **9**, pp.256–68.

Embabi, S.H.K., Bellaouar, A., and Elmasry, M.I. (1993). *Digital BiCMOS integrated circuit design*, pp.295–8. Kluwer Academic Publishers, USA.

Garuts, V.E., Yu Y-C.S., Traa, E.O., and Yamaguchi, T. (1989). A dual 4-bit Gs/s full Nyquist analog-to-digital converter using a 70-ps silicon bipolar technology with borosenic-poly process and coupling-base implant. *IEEE Journal of Solid-State Circuits*, **24**, pp.216-22.

Glasser, L.A. and Dobberpuhl, D.W. (1985). *The design and analysis of VLSI circuits*. Addison-Wesley, USA.

Harris Semiconductor Databook (1988). SSD-283A, p.210. Harris Semiconductor (formerly RCA), USA.

Hodges, D.A. and Jackson, H.G. (1988). *Analysis and design of digital integrated circuits*, (2nd edn). McGraw-Hill, USA.

Long, G.S.L. and Butner, S.E. (1990). *Gallium arsenide integrated circuit design*. McGraw-Hill, USA.

Machi, S. and Chatterjee, P. (1994). 1-V microsystems. *IEEE Circuits and Devices*, **10** (2), pp.13–17.

Mohsen, A.M. and Mead, C.A. (1979) Delay time optimization for driving and sensing of signals on high-capacitance paths for VLSI systems. *IEEE Journal of Solid-State Circuits*, **14**, pp.462–70.

Philips Semiconductor ABT Data Handbook (1991). *ABT multibyte advanced BiCMOS bus interface logic*, **23**, p.13. Philips Semiconductors, Holland.

Quarles, T., Newton, A.R., Anderson, D.O., and Sangiovanni-Vincentelli, A, (1993). *SPICE 3 version 3f3 user's manual*, p.38. Department of Electrical Engineering and Computer Science, University of California, Berkley, USA.

Roberts, D.A., Schmitz, N.A., and Watkins, J.D. (1989). Bipolar gallium arsenide heterojunction integrated injection logic. In *VLSI handbook* (ed. J. DiGiacomo), pp.15.1 to 15.4. McGraw-Hill, USA.

Rocchi, M. and Gabillard, B. (1983). GaAs digital dynamic IC's for applications up to 10 GHz. *IEEE Journal of Solid-State Circuits* **18**, pp.369–76.

Sakurai, T. and Tamaru, K. (1983). Simple formulas for two- and three-dimensional capacitances. *IEEE Transactions on Electron Devices*,

30, pp.183–5.

Saul, P.H. (1989). Technology comparison: gallium arsenide vs. silicon. In *GaAs technology and its impact on circuits and systems* (eds D. Haigh and J. Everard), pp.165–83. Peter Peregrinus, IEE, UK.

Saul, P.H. (1992). Bipolar circuit techniques. In *High speed digital electronics*, (ed. L.J. Herbst), p.132. Prentice-Hall, England.

Sze, M.S. (1981). *Physics of semiconductor devices*, (2nd edn), pp.45–6. Wiley, USA.

Uyemura, J.P. (1988). *Fundamentals of MOS digital integrated circuits*. Addison-Wesley, USA.

Uyemura, J.P. (1992). *Circuit design for CMOS VLSI*, pp.426–33. Kluwer Academic Publishers, USA.

Veendrick, H.J.M. (1984). Short circuit dissipation of static CMOS circuitry and its impact on the design of buffer circuits. *IEEE Journal of Solid-State Circuits*, **19**, pp.468–73.

Vladimirescu, A. and Liu, S. (1980). *The simulation of MOS integrated circuits using SPICE 2*, p.7. Department of Electrical Engineering and Computer Science, University of California, Berkley, USA.

Welch, B.M., Eden, R.C., and Lee, F.S. (1985). In *Gallium arsenide, materials, devices, and circuits*, (eds M.J. Howes and D.V. Morgan), pp.518–24. Wiley, UK.

Weste, N.W. and Eshraghian, K. (1985). *Principles of CMOS VLSI design*. Addison-Wesley, USA.

Yuan, C.P. and Trick, T.N. (1982). A simple formula for the estimation of the capacitance of two-dimensional interconnects in VLSI circuits. *IEEE Electron Device Letters*, **3**, pp.391–3.

Yuan, H-T., Lin, Y-T., and Chiang, S-Y. (1982). Properties of interconnection on silicon, sapphire, and semi-insulation gallium arsenide substrates. *IEEE Transactions on Electron Devices*, **29**, pp.639–44.

Further reading

Alvarez, A.R. (1993). *BiCMOS technology and applications*, (2nd edn). Kluwer Academic Publishers, USA.

Bakoglu, H.B. (1990). *Circuits, interconnections, and packaging for VLSI*. Addison-Wesley, USA.

Embabi, S.H.K., Bellaouar, A., and Elmasry, M.I. (1993). *Digital BiCMOS integrated circuit design*. Kluwer Academic Publishers, USA.

Glasser, L.A. and Dobberpuhl, D.W. (1985). *The design and analysis of VLSI circuits*. Addison-Wesley, USA.

Hodges, D.A. and Jackson, H.G. (1988). *Analysis and design of digital integrated circuits*, (2nd edn). McGraw-Hill, USA.

Long, G.S.L. and Butner, S.E. (1990). *Gallium arsenide integrated circuit design*. McGraw-Hill, USA.

Uyemura, J.P. (1988). *Fundamentals of MOS digital integrated circuits.* Addison-Wesley, USA.

Weste, N.W. and Eshraghian, K. (1985). *Principles of CMOS VLSI design.* Addison-Wesley, USA.

6

Analogue circuits—techniques and performance

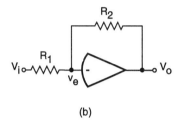

6.1 Introduction

The universal building block in analogue electronics is the amplifier. It is hard to think of electronic equipment without one. A standard configuration with negative feedback, the operational amplifier (op amp) is the established IC chip for that purpose, and is given the special attention it deserves. Originally in electronics equipment which came into service towards the end of the Second World War, it fulfils a number of basic operations other than amplification, namely summation, integration, and differentiation, hence its name.

Next in importance is data conversion. We live in an analogue world, and we process data more and more digitally. Analogue-to-digital (ADC) and digital-to-analogue (DAC) converters are the means of entering and leaving the digital realm. The countless uses they have in all kinds of instrumentation and communications have been augmented by digital signal processing (DSP) with which, largely thanks to the tremendous function capacity of VLSI, signals are processed far more effectively and economically than they were previously in an all-analogue domain. That places a burden on data converters to achieve high resolution, and special attention is paid to this in their description. Data converters form one part, and filters another in a section on analogue signal processing.

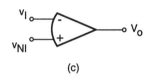

The two main sections of this chapter cover operational amplifiers and analogue signal processing, in accordance with the outline just given. Further sections deal with voltage comparators, and with voltage references and voltage regulators.

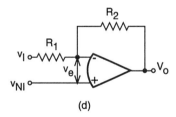

Fig. 6.1 Op amp configurations
(a) inverting amplifier
(b) operational amplifier
(c) differential amplifier
(d) operational amplifier differential-input

6.2 Operational amplifiers

6.2.1 Basic concepts

The nature of the operational amplifier is stamped by the feedback shown in Fig. 6.1. The prerequisite is a phase-inverting amplifier with a high voltage gain. The behaviour of the inverting amplifier in Fig. 6.1(a) is defined by

$$V_o = -AV_i \qquad (6.1)$$

where A is the *open loop* gain, to be distinguished from the *closed loop* gain with negative feedback. The form of feedback which turns Fig. 6.1(a) into an operational amplifier is shown in Fig. 6.1(b). Assuming an infinite input resistance, so that no signal current flows into the input terminal

$$\frac{V_i - V_e}{R_1} = \frac{V_e - V_o}{R_2} \tag{6.2}$$

and

$$V_o = -A V_e \tag{6.3}$$

in accordance with eqn (6.1). Combining eqns (6.2) and (6.3)

$$\frac{V_o}{V_i} = -\frac{R_2}{R_1 + \frac{(R_1 + R_2)}{A}} \tag{6.4}$$

(V_o/V_i) in eqn (6.4) is the closed loop gain. If $(R_1 + R_2)/A \ll R_1$, which amounts to $A \gg (R_2/R_1)$

$$\frac{V_o}{V_i} \simeq -\frac{R_2}{R_1} \tag{6.5}$$

The gain given by eqn (6.5) is the closed loop gain (A_{CL}). The stipulation above eqn (6.5) applies in general, because $A \sim (10^4 - 10^5)$ and R_2/R_1 is in most cases two to three orders of magniture smaller. Equation (6.5) shows that the closed loop gain is to all intents and purposes independent of A.

Example 6.1

Calculate the change in closed loop gain for the amplifier in Fig. 6.1(b) when A increases from 10 000 to 50 000. (R_2/R_1) equals 100.
Since $R_2 \gg R_1$, eqn (6.4) modifies to

$$\frac{V_o}{V_i} \simeq -\frac{A}{1 + A\left(\frac{R_1}{R_2}\right)} \tag{6.6}$$

Carrying out the numerical substitutions the ratio of the closed loop gains for $A = 10^4$ and $A = 10^5$ comes to 0.991, demonstrating that a 500 per cent change in A leads to a change of only 0.9 per cent in the closed loop gain.

Equations (6.3) and (6.5) lead to

$$V_e \simeq \frac{V_o \left(\frac{R2}{R1}\right)}{A} \tag{6.7}$$

and for $(R_2/R_1) \ll A$, $V_e \ll V_i$. There is virtually no movement (change) in voltage at the amplifier input terminal, which is called a *virtual earth*, signifying that $V_e \simeq$ zero. This approximation is used profusely in calculations of op amp performance.

Further advantages of the op amp are its low input and output resistances due to the nature of the feedback. In Fig. 6.2(a) δV gives rise to a current $i = (\delta V - V_o)/R_2$ which, combined with eqn (6.3), leads to

$$r_i = \frac{\delta V}{i} = \frac{R_2}{(1 + A)} \simeq \left(\frac{R_2}{A}\right) \tag{6.8}$$

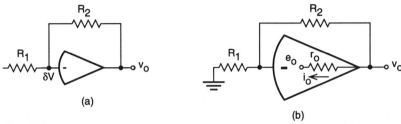

Fig. 6.2 Op amp input and output resistance (a) input resistance (b) output resistance

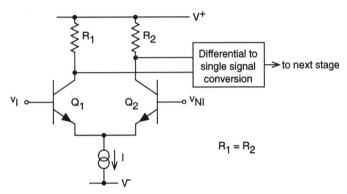

Fig. 6.3 Differential input stage—op amp

r_i being the input resistance. The effective output resistance with feedback, r'_o, is given by (Fig. 6.2(b))

$$r'_o = \frac{v_o}{i_o} \tag{6.9}$$

But

$$i_o = \frac{v_o - e_o}{r_o} = \frac{v_o\left(1 + \frac{AR_1}{R_1+R_2}\right)}{r_o} \simeq \frac{v_o A\left(\frac{R_1}{R_2}\right)}{r_o} \tag{6.10}$$

leading to

$$r'_o \sim \frac{r_o}{A\left(\frac{R_1}{R_2}\right)} \tag{6.11}$$

signifying that the output resistance is greatly reduced with feedback. IC op amps are dc coupled (although external ac coupling is sometimes used) and a change in dc bias conditions leads to by a change in output voltage. Transistor bias is sensitive to changes in supply voltage and—more so—to variations in temperature. The only effective means of assuring adequate bias stability of the input stage, which has the greatest effect on the amplifier output, is the differential input amplifier sketched in Fig. 6.3. Provided that the two transistors are identical and that their outputs can be combined into a single signal for onward transmissions, drifts in the dc bias of the two transistors will cancel out. The functional equivalent of Fig. 6.3 is shown in Fig. 6.1(c), for which eqn (6.3) is modified to

$$V_o = -A(V_I - V_{NI}) \tag{6.12}$$

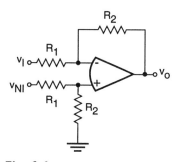

Fig. 6.4 Op amp-differential amplification

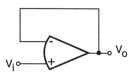

Fig. 6.5 Voltage follower

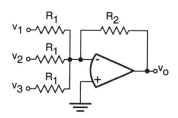

Fig. 6.6 Summing amplifier

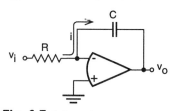

Fig. 6.7 Integrator

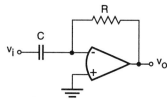

Fig. 6.8 Differentiator

The op amp feedback is shown in Fig. 6.1(d). Combining eqns (6.2) and (6.12), and making the approximations which led to eqn (6.5)

$$V_o \simeq -V_I \left(\frac{R_2}{R_1}\right) + V_{NI} \left(\frac{R_1 + R_2}{R_1}\right) \quad (6.13)$$

The virtual earth at the input in Fig. 6.1(b) gives way to a virtual shortcircuit (signal-wise), between the inverting (−) and non-inverting (+) inputs, and this approximation is used freely in evaluating amplifier performance.

The differential op amp, universal for all ICs, achieves either inverted or non-inverted amplification by grounding the (−) or (+) terminal respectively. Another alternative is to amplify a differential voltage with the circuit in Fig. 6.4. The amplification, using eqn (6.13) and assuming $(R_2/R_1) \gg 1$, is given by

$$V_o = -\left(\frac{R_2}{R_1}\right)(V_I - V_{NI}) \quad (6.14)$$

The op amp becomes a voltage follower with the 100 per cent feedback in Fig. 6.5, for which the gain is extremely close to unity in accordance with

$$\frac{V_o}{V_i} = \frac{A}{1+A} \simeq 1 \quad (6.15)$$

an expression derived from eqn (6.12) by putting $V_I = V_o$. The voltage follower has a very high input resistance, because the signal is applied to the non-inverting terminal, which follows the output, and the output is virtually identical to the input. The output resistance is extremely small because of the heavy feedback. These characteristics make the voltage follower an ideal buffer in many applications.

Further op amp functions which will now be briefly described are summing (adding), integration, differentiation, and log and antilog amplification. The circuit for summing is sketched in Fig. 6.6. The output is given by

$$V_o = -\frac{R_2}{R_1}(V_1 + V_2 + V_3) \quad (6.16)$$

Fig. 6.6 also serves for subtraction, which results when inputs are of opposite polarity. For the integrator in Fig. 6.7

$$\frac{V_i}{R} = -C\frac{dV_o}{dt} \quad (6.17)$$

Equation (6.17) leads to

$$V_o = -\frac{1}{CR} \int V_i dt \quad (6.18)$$

Note the use of the assumption $V(-) = V(+)$ to obtain eqn (6.17). Similarly the output voltage for the differentiator in Fig. 6.8 is given by

$$V_o = -CR\frac{dV_i}{dt} \quad (6.19)$$

Logarithmic amplification is obtained with the circuit in Fig. 6.9, whose behaviour is formulated in accordance with

$$I_c(Q_1) = \frac{V_i}{R} = I_s \exp\left\{\left(\frac{V_{BE}}{V_t}\right) - 1\right\} \simeq I_s \exp\left(\frac{V_{BE}}{V_t}\right) \quad (6.20)$$

But

$$V_o = -V_{BE} \quad (6.21)$$

leading to

$$V_o = -V_t \ln\left(\frac{V_i}{I_s R}\right) \quad (6.22)$$

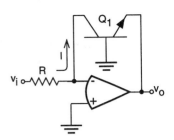

Fig. 6.9 Log amplifier

The amplifier is unipolar, functioning only for positive inputs; a pnp transistor is required for negative inputs. The antilog amplifier in Fig. 6.10 has the transistor in the input path and operates according to

$$V_o = -IR = -RI_s \exp\left(\frac{V_{BE}}{V_t}\right) \quad (6.23)$$

which can be expressed in the form

$$V_o = -RI_s \operatorname{antiln}\left(\frac{V_i}{V_t}\right) \quad (6.24)$$

since $V_i = V_{BE}$. Once again the amplifier is unipolar; a pnp transistor serves for positive, an npn transistor for negative outputs. Diodes could be used instead of transistors in Figs. 6.9 and 6.10, but the latter are preferred because the logarithmic response holds for a wider range of input voltage. (Faulkenberry 1977). Another op amp application, included because of its wide usage, is the current-to-voltage converter sketched in Fig. 6.11, which converts I to a voltage $-IR$; the operation of the circuit is self-evident.

Table 6.1 Ideal op amp characteristics

Characteristic	Value
Open loop gain A	∞
Output resistance	0
Input resistance	∞
Offset current	0
Offset voltage	0

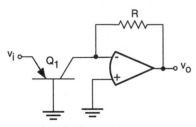

Fig. 6.10 Antilog amplifier

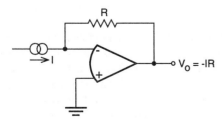

Fig. 6.11 Current-to-voltage converter

186 Analogue circuits—techniques and performance

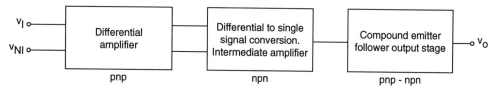

Fig. 6.13 Typical op amp block schematic

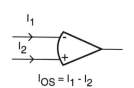

Fig. 6.12 Offset current

The ideal characteristics of an op amp are listed in Table 6.1. Good performance is still obtained with values differing from the ideal. Example 6.1 demonstrates the excellent gain stability for $A \sim 10\,000$, which is on the low side for IC op amps. The postulation of open-circuit input resistance is generally not important for the inverting input terminal, which is virtually at zero when the non-inverting terminal is grounded or at a fixed voltage. The output resistance has a bearing on the maximum output voltage and output power obtainable. The specifications of offset current and voltage are linked to the differential input stage like that shown in Fig. 6.3. The offset current I_{OS} is the difference between the (ideally equal) input currents I_1 and I_2 in Fig. 6.12. It gives rise to a differential input voltage which depends on the values of the external resistors making up the amplifier circuit. The offset voltage V_{OS} is another parameter reflecting the imbalance of the input transistors. It is the differential input voltage which must be applied to obtain zero output in the absence of external inputs. Typical values of I_{OS} and V_{OS} for bipolar op amps are 15 nA and 3 mV respectively.

6.2.2 Circuit techniques

The circuits described in this section are the basic constituents of an op amp, but are also widely used elsewhere, so that their description has general value. A schematic showing the various stages of a typical op amp is given in Fig. 6.13. The input stage is a differential amplifier for reasons given in the previous section, and reinforced in this section by a detailed explanation of its operation. The differential outputs are transformed into a single-ended signal, the input to the intermediate amplifier, which drives the compound emitter–follower output stage. The circuit is designed so that for perfectly matched input transistors and precise design values of transistor I–V characteristics, supply voltages and resistors the output is zero for zero inputs (V_I and V_{NI} grounded). That condition entails a dc level shift within the amplifier, obtained by a pnp input stage feeding the npn intermediate amplifier. The compound emitter follower produces no change in dc level.

The operation of the differential amplifier and the conversion of the differential signals into a single signal are governed by a *current mirror*, an ingenious circuit which generates a constant current, and can also convert differential antiphase signals into a single signal. That step is essentially the transformerless equivalent of changing the push–pull signals of an audio amplifier into a single-ended signal driving the loudspeaker.

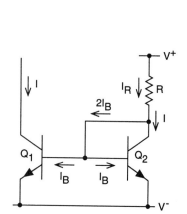

Fig. 6.14 Current mirror

The current mirror derives its name from the relation between the currents in Fig. 6.14. If the transistors are matched, a balance closely achieved in IC processing, their collector currents are equal and they *mirror* one another, because they have a common emitter-base voltage. One can see straight away that $I_R \simeq I$, because the base currents are negligible in comparison. Q_2, a diode-connected transistor, presents a low impedance to I_R. Q_1, on the other hand, is a current source because the collector has a high output resistance, which means that the current is virtually independent of the circuit it supplies. I_R, generated by a resistor connected to V^+, is given by

$$I_R = \frac{V^+ + |V^-| - V_{BE}}{R} \qquad (6.25)$$

The deduction of $2I_B$ from I_R results in

$$\frac{I}{I_R} = 1 + \frac{2}{h_{FE}} \qquad (6.26)$$

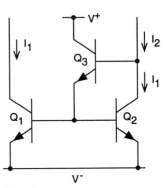

Fig. 6.15 Improved current mirror

IC transistors have $h_{FE} \sim 100$ or more, and I equals I_R within two per cent. A great improvement is obtained with the circuit in Fig. 6.15, in which I_2 corresponds to I_R in Fig. 6.14, whose resistor R has been omitted in order to signify that I_2 may be generated differently. It can be seen from inspection that the effective h_{FE} is $h_{FE(Q3)}$, multiplied by $h_{FE(Q2)}$ and leading to

$$\frac{I_1}{I_2} = 1 + \frac{2}{h_{FE(Q2)}(1 + h_{FE(Q3)})} \simeq 1 + \frac{2}{(h_{FE})^2} \qquad (6.27)$$

where, to establish an order of magnitude, $h_{FE(Q2)}$ has been equated with $h_{FE(Q3)}$. A challenge to this approximation is that Q_3 operates at a much smaller current; its emitter current is the sum of Q_1 and Q_2 base currents. On the other hand h_{FE} holds up fairly well at low currents for the type of IC transistors used. I_1 and I_2 are balanced to within 0.1 per cent for $h_{FE} \simeq 50$.

The way is now clear for explaining the representative op amp input stage in Fig. 6.16. A differential input voltage V_i changes the currents in

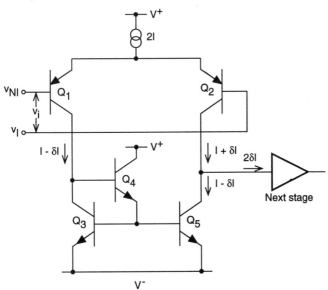

Fig. 6.16 Simplified op amp input stage

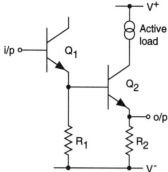

Fig. 6.17 Op amp intermediate amplifier

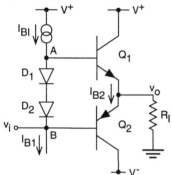

Fig. 6.18 Compound emitter follower output stage

Q_1 and Q_2 from I to $(I-\delta I)$ and $(I+\delta I)$. Two current mirrors are used, one—not shown—to generate the bias current $2I$, and another, consisting of Q_3, Q_4, and Q_5, to convert the antiphase differential amplifier outputs into a single-ended signal, and to present active collector loads for Q_1 and Q_2. These result in a high differential voltage gain, ~1000 or more, for the input stage.

The standard circuit for the intermediate amplifier is an emitter–follower feeding a common-emitter stage with an active load, see Fig. 6.17. The emitter follower minimizes the loading of the input amplifier (Q_2 collector in Fig. 6.16), and the feedback resistor R_2 stabilizes the gain of Q_2. The output stage (Fig. 6.18) is a compound emitter follower consisting of Q_1 and Q_2 and operating in Class AB with a slight forward bias emitter current I_{B2} of ~(30–50) μA. I_{B2} is set by the bias network consisting of D_1, D_2, and I_{B1}. At first sight it may seem surprising that an amplifier intended for highly linear operation has a Class AB output stage. This near-universal practice is adopted in order to conserve input power, which is very small in quiescence and increases with signal level. Such power economy is vital in ICs. The inherent distortion is small, because the stage operates in push–pull and is linearized by the heavy negative feedback.

The circuit of the 741 op amp is shown in Fig. 6.19. This amplifier has been chosen because, inspite of its incredible longevity—it was introduced in 1966—it is still produced by many vendors, enjoys great popularity, and reflects the approach to bipolar op amp circuitry. There are variations, introduced by some manufacturers, of the circuit shown in Fig. 6.19, but these are minor and do not alter the essential design philosophy. The input stage is a cascode combination consisting of the npn emitter followers Q_1 and Q_2, and the pnp common-base amplifiers Q_3 and Q_4. The poor frequency response of pnp lateral transistors (Section 3.3) is mitigated by the inherent wide bandwidth of a common-base stage. The current mirror Q_8, Q_9 generates a bias current of ~20 μA for the input pair Q_1, Q_2. The small current leads to a high input resistance. Q_8 and Q_9 have identical collector currents equal to the collector current of Q_{10}. Q_{10}, Q_{12}, and R_4 form a *Widlar* current source, which generates a low current with a moderate resistance, unlike the current mirror in Fig. 6.14, for which R would have to be 1.5 MΩ for ± 15 V supply voltages in order to generate 20 μA (Widlar 1965, 1969)! The operation of the Widlar circuit will now be explained with the aid of Fig. 6.20 (extracted from Fig. 6.19), in which $I \gg I_g$, thereby creating a base-emitter differential voltage ΔV_{BE}. Equating emitter and collector currents of Q_{10}.

$$I_g = \frac{\Delta V_{BE}}{R_4} \tag{6.28}$$

and using the standard I–V expression for a BJT (eqn (6.20))

$$\Delta V_{BE} = V_t \ln\left(\frac{I}{I_g}\right) \tag{6.29}$$

leading to

$$I_g = \frac{V_t}{R_4} \ln\left(\frac{I}{I_g}\right) \tag{6.30}$$

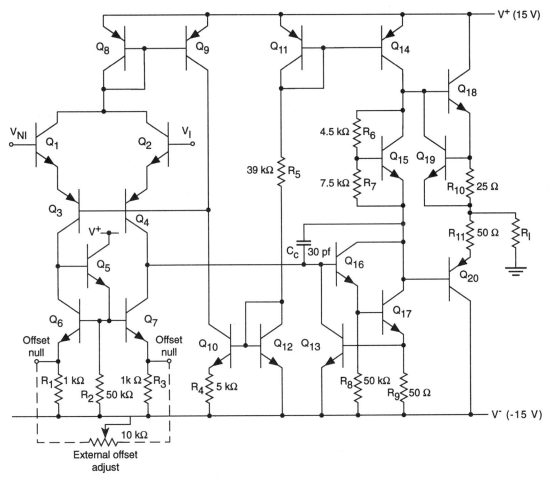

Fig. 6.19 Schematic of 741 op amp

The transcendental equation (6.30) is solved by trial and error for R_4. I comes to 733 μA in accordance with the values of R_5 and the supply voltages in Fig. 6.19, leading to $I_g = 19$ μA for $R_4 = 5$ kΩ.

The third current mirror in Fig. 6.19, Q_{11}, Q_{14}, injects the current through R_5, 733 μA, into the bias network for the output stage, in which Q_{15} is an alternative to the diodes D_1 and D_2 in Fig. 6.18. Its collector-emitter voltage constitutes the sum of the V_{BE} drops for Q_{18} and Q_{20}, ignoring the negligible voltages across R_{10} and R_{11} in quiescence. The op amp output stage is biased for Class AB operation with a very small quiescent current.

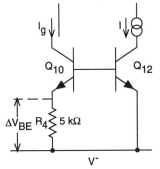

Fig. 6.20 Widlar current source (from Fig. 6.19)

Example 6.2

Explain the bias generation by Q_{15} in Fig. 6.19 and estimate the quiescent emitter current of Q_{18} and Q_{20}.

The bias circuit, extracted from Fig. 6.19, is shown in Fig. 6.21. A current of 733 μA flows into Q_{15} and the resistive network. Assume $V_{BE} = 0.6$ V, and neglect all base currents. The circuit is a V_{BE} multiplier, because $V_{R6} = (R_6/R_7)V_{BE}$, making V_{CE} equal to

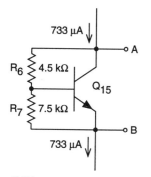

Fig. 6.21 Bias generator for 741 output stage (from Fig. 6.19)

$\{1 + (R_6/R_7)\}V_{BE} = 1.6V_{BE}$. $I_{R6,R7} = (1.6 \times 0.6)/12 = 80 \ \mu A$, giving $I_{Q15} = 733 - 80 = 653 \ \mu A$. Assume that Q_{15}, Q_{18}, and Q_{20} are identical. V_{BE} for Q_{18} and Q_{20} equals $(1/2) \times 1.6V_{BE}(Q_{15}) = 0.8V_{BE}(Q_{15})$. Q_{15} takes a current of 650 μA, and applying the standard I–V equation, I_{18} ($= I_{19}$) works out to be 6.4 μA. In practice Q_{18} and Q_{20} are made larger than the other transistors, because they have to supply the output power and I_{18} is correspondingly higher ($\sim$30 to 50 μA).

The output has to be protected against an accidental short circuit, which might demand excessive current and could easily lead to the destruction of Q_{18} and Q_{20} (Fig. 6.19). R_{10} and Q_{19} afford protection for a positive-going output. A current of $\sim$20 mA brings Q_{19} to the edge of conduction, and the transistor soon saturates when the current increases further. The protection for a negative output swing is indirect, because a pnp transistor connected across Q_{20} would deteriorate its inherently poor frequency response. The overload protection for Q_{20} operates by virtue of the excessive drive current demanded from Q_{17} and sensed by R_9. The voltage across R_9 causes Q_{13} to saturate and clamps the input drive for Q_{16}. Some direct protection is however provided by R_{11}.

The description of the 741 op amp has been given in some detail, because the techniques it incorporates find wide applications in many other analogue circuits. Furthermore the schematic arrangement of Fig. 6.13 is the norm.

The bias current of 20 μA for Q_1 and Q_2 is acceptable for general-purpose usage, but is far above the level desirable for applications like integrators, sample-and-hold circuits, and many others. The reduction of input current is achieved in a number of ways. One of these is an input stage made up of *superbeta* BJTs, which have an h_{FE} in the range 2000 to 5000 at collector currents of a few μA. The very high current gain is largely due to a very narrow basewidth, $\sim$0.1 μm compared with $\sim$(0.5–1) μm for the conventional BJTs in general-purpose op amps. That structure results in a very small collector-base breakdown voltage, and the superbeta transistors are preferably biased to zero V_{CB}. Figure 6.22 is a schematic of an op amp input stage in which the superbeta transistors Q_1 and Q_2 are connected in cascode to the common-base stages Q_5 and Q_6, whose outputs are taken to an active-load signal-conversion current mirror like Q_5 to Q_7 in Fig. 6.19 and represented here for the sake of simplicity by Q_5, Q_6, R_1, and R_2. Zero quiescent collector-base bias is maintained for Q_1 and Q_2 by the two V_{BE} drops of the common-mode feedback transistors Q_3 and Q_4. Superbeta input transistors operate with I_1 in the range $\sim$(3–6) μA. They achieve an offset current of $\sim$(50–200) pA, decreasing to $\sim$0.5 pA for a Darlington amplifier.

Input and offset currents can be reduced further by a JFET or MOSFET input stage. JFET matured before MOSFET technology, and JFET op amps were first to appear on the scene. The profile of a p-channel JFET is sketched in Fig. 6.23, and a schematic of a JFET op amp is given in Fig. 6.24. The input amplifier is followed by bipolar stages which conform to the practice in Fig. 6.19. A MOSFET input stage similar to Fig. 6.24 has the lowest input current and I_{OS}, with typical values of 0.04 pA and 0.01 pA respectively. There is much variety

Operational amplifiers 191

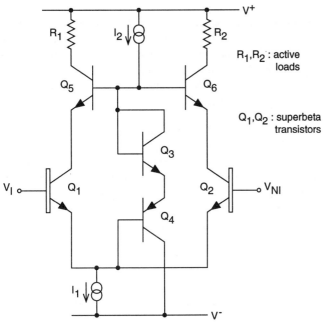

Fig. 6.22 Superbeta transistor input stage

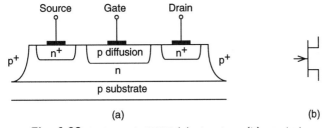

Fig. 6.23 n-channel JFET (a) structure (b) symbol

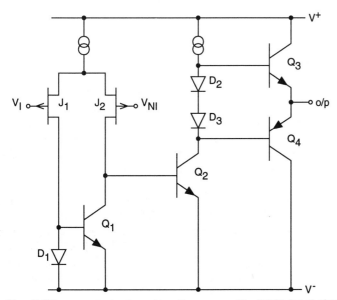

Fig. 6.24 Simplified schematic of op amp with JFET input stage

in MOS-bipolar combinations for op amps. It is possible to have an op amp constructed entirely with MOSFETs, but the great majority of op amps with MOSFET input and output stages contain BJT intermediate amplifiers. MOSFET current sources are sketched in Fig. 6.25. The transistors operate in the saturation region in accordance with eqn (4.22), and the condition to do so is,

$$V_{DS} \geq V_{GS} - V_T \qquad (6.31)$$

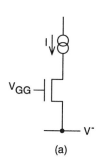

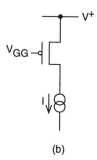

Fig. 6.25 MOS current generator (a) current sink (b) current source

for the current sink in Fig. 6.25(a), with a similar expression for the current source in Fig. 6.25(b). The preferred CMOS output stage is identical to the logic inverter in Fig. 5.9. MOSFET input stages consist of a differential pMOSFET or nMOSFET pair. MOSFET current mirrors are widely used, but the preferred intermediate amplifier is generally bipolar. The op amps described have a compound emitter–follower output stage, whose output resistance approaches closely the ideal of zero, stipulated in Table 6.1.

Another type of op amp, the operational transconductance amplifier (OTA) shown in Fig. 6.26, differs from the traditional op amp in two respects. It has a very high (ideally infinite) output resistance and its characteristic is defined by

$$I_o = I_8 - I_9 = g_m V_i \qquad (6.32)$$

where

$$V_i \equiv V_I - V_{NI} \qquad (6.33)$$

Evidently

$$V_o = I_o R_l = g_m V_i R_l \qquad (6.34)$$

R_l being the external load. The distinguishing feature of the OTA is that its g_m is directly proportional to an input bias current I_{ABC} i.e.

$$g_m = A I_{ABC} \qquad (6.35)$$

A being a constant. Combining eqns (6.34) and (6.35)

$$V_o = A I_{ABC} R_l V_i \qquad (6.36)$$

The significance of eqn (6.36) is that the OTA voltage gain is directly proportional to I_{ABC}. Examples of OTA applications are variable gain and current controlled amplifiers, multiplexers, and multipliers. The validity of eqn (6.35) and the value of A will now be established with reference to the simplified circuit of a typical OTA given in Fig. 6.26.

The core of the circuit is the transconductance amplifier, formed by Q_1 and Q_2 in association with D_1 and Q_3. Assume that collector and emitter currents are equal, and that base currents may be neglected. The current mirror Q_4, D_2 gives $I_5 = I_1$, and the injection of I_5 into the current mirror Q_7, D_4 makes I_9 equal to I_5, and hence to I_1. The third current mirror Q_6, D_3 makes I_6 and I_8 equal to I_2. This results in

$$I_8 - I_9 = I_2 - I_1 \qquad (6.37)$$

Operational amplifiers

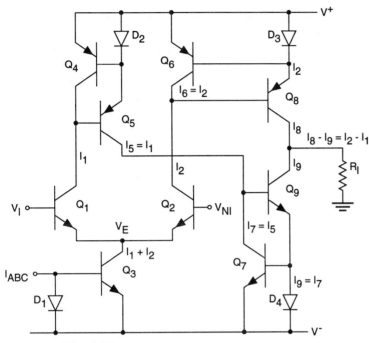

Fig. 6.26 Simplified schematic of OTA amplifier

Using eqn (6.20)

$$I_1 = I_s \exp\left(\frac{V_I - V_E}{V_t}\right) \tag{6.38}$$

$$I_2 = I_s \exp\left(\frac{V_{NI} - V_E}{V_t}\right) \tag{6.39}$$

But

$$I_3 = I_1 + I_2 = I_{ABC} \tag{6.40}$$

because D_1 and Q_3 are in effect a current mirror. Combining eqns (6.38) and (6.39)

$$I_1 - I_2 = I_{ABC} \tanh\left(\frac{V_i}{2V_t}\right) \simeq I_{ABC}\left(\frac{V_i}{2V_t}\right) \tag{6.41}$$

for $(V_i/2V_t) \ll 1$. The validity of the approximation $\tanh(x) \simeq x$, implied in eqn (6.41), can be assessed from

$$\tanh x \simeq x - \frac{x^3}{3} \simeq x \tag{6.42}$$

$$x \ll 1$$

It holds within one per cent for $x \leq 0.17$, or $V_i \leq 8.8$ mV, when V_t equals to 26 mV. Such a range is acceptable for many applications. Equations (6.32), (6.37) and (6.41) make A in eqn (6.35) equal to $(1/2V_t)$. If I_{ABC} is generated by an external voltage V_G in series with a resistor R_1

Table 6.2 Characteristics of selected op amps

Parameter	LM741[1] General purpose	HA-5127[2] Ultra-low noise	CA3160[2] General purpose
Technology	Bipolar	Bipolar	BiMOS MOSFET input CMOS output
V_{OS} (V)	1 mV	30 µV	2 mV
I_{OS}	20 nA	12 nA	0.5 pA
A_{OL} (V/mV)	200	1500	320
GBP (MHz)	1.2	8.5	4
Slew rate (V/µs)	0.5	10	10
v_n (white) (nV/$\sqrt{Hz}$)	20	3.2	30
CMRR (dB)	90	120	90
i_n (white) (pA/$\sqrt{Hz}$)	0.5	0.4	~Zero
Input resistance	2 MΩ	4 MΩ	1.5 TΩ
Supply voltage (V)	±15	±15	±7.5
Max output (V) $R_l \geq 2$ kΩ	±13	±13.5	±6.7
$\partial V_{OS}/\partial T$ (µV/°C)	15	0.4	8
Input current	80 nA	15 nA	5 pA

[1] National Semiconductor, [2] Harris Semiconductor
GBP: gain bandwidth product
(Courtesy of National Semiconductor and Harris Semiconductor)

$$I_{ABC} = \frac{V_G - V_{BE}}{R_1} \simeq \frac{V_G}{R_1} \qquad (6.43)$$

$$V_G \gg V_{BE}$$

making g_m proportional to V_G and turning the circuit into a multiplier, i.e. $V_o \propto V_i V_G$. The circuit can function as an amplifier with a linear gain control V_G. The input stage of Fig. 6.26 is in fact a two-quadrant multiplier: V_i can be positive or negative, I_{ABC} and hence V_G must be positive. It is the foundation for a four-quadrant extension, the Gilbert cell, which accepts positive and negative voltages for both V_i and V_G (Gilbert 1968).

6.2.3 Specification and performance

The main characteristics of three op amps are listed in Table 6.2. Many op amps are second-sourced by numerous suppliers. The specifications for a given type may differ a little here and there between manufacturers, but such variations are not significant. The 741 is a popular op amp which probably holds the record for longevity of any standard analogue IC. The HA-5127 presents an advance on the 741 in several respects. It has a much higher bandwidth, is less noisy and has a much lower drift of offset voltage with temperature. The CA3160 is a general purpose BiMOS op amp whose outstanding feature is a virtually open-circuit input resistance (because the input bias current is extremely small).

I_{OS} and V_{OS} depend on the balance of the input stage. It is easier to achieve low V_{OS} for BJTs than for JFETs or MOSFETs. Not only is V_{OS} less for a BJT, but the spread between average and maximum values is much smaller. FETs are superior regarding I_{OS}, which reduces from nA levels for BJTs to pA levels for JFETs and MOSFETs, which have a very high input resistance (~ 1 TΩ). The output resistance, ~ 50 Ω for a bipolar and ~ 300 Ω for a CMOS stage, is not of critical importance. The open loop gain A (alternatively A_{VOL}), defined in eqn (6.12), is called *large signal gain* in most data sheets, because the customary test condition specifies a large output voltage, as a rule ± 10 V for supply voltages of ± 15 V. A_{OL} is expressed directly or in dB. The quotation for A_{OL} of the 741 op amp in Table 6.2 can take one of the following forms: 200 V/mV, 2×10^5, or 106 dB. The magnitude of A_{OL} is generally more than adequate. The advantage there might be in having a gain in excess of about 500 000 is nullified by an increase in noise level.

The explanation of the differential amplifier in Section 6.2.1 postulated perfectly balanced input transistors; that is the assumption underlying eqn (6.12). In fact the outputs of a differential amplifier contain another component arising from the *common mode gain*, which is the amplification when identical voltages are applied to the amplifier input terminals. Simple analysis shows that the differential gain is very much higher—by several orders of magnitude—than the common mode gain A_{CM}, defined by

$$A_{CM} \equiv \frac{V_o}{\frac{1}{2}(V_I + V_{NI})} \tag{6.44}$$

giving

$$V_o = \frac{1}{2} A_{CM}(V_I + V_{NI}) \tag{6.45}$$

The actual op amp output voltage is the sum of the V_o values given by eqns (6.12) and (6.45). Equation (6.45) shows that the common mode V_o is zero for an antiphase (push–pull) differential input of $V_I = -V_{NI}$, but finite otherwise.

For perfectly balanced input transistors, the common mode contribution will be eliminated by the action of Fig. 6.16. A common mode input voltage for $V_I = V_{NI}$ which gives zero differential voltage, will cause identical increases δI_{CM} in the collector currents of Q_1 and Q_2. The current flowing into the next stage is now the difference between $(I + \delta I_{CM} + \delta I)$ and $(I + \delta I_{CM} - \delta I)$, which is $2\delta I$ as before. The common mode contribution would thus cancel out, i.e. $A_{CM} = 0$, were it not for the inevitable imbalance between the transistors. The common mode rejection ratio (CMRR) is defined by

$$\text{CMRR} \equiv \frac{A_{OL}}{A_{CM}} \tag{6.46}$$

Tighter process control over the years has led to higher values of CMRR, which is rarely below 80 dB and more likely close to 100 dB, or even higher. BJTs are superior here, because their characteristics can be matched better in production than those of either JFETs or MOSFETs. The CMRR is unimportant in the many applications where the signal is applied to one of the inputs with the other input at ground. It becomes

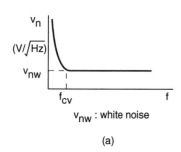

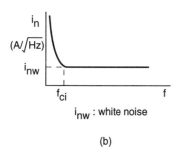

Fig. 6.27 Spectral noise distribution (a) voltage (b) current

significant when both V_I and V_{NI} take up large and nearly identical values as for example in a voltage follower. Another parameter, the power supply rejection ratio (PSRR), expresses the change in output voltage for supply voltage variations in accordance with

$$\text{PSRR} = \frac{\Delta V_o}{\Delta V_S} \qquad (6.47)$$

where ΔV_S signifies symmetrical changes in the positive and negative supply voltages. The magnitudes of the CMRR and the PSRR are similar, because they both apply to common mode signals.

Noise is specified in terms of noise voltage and noise current spectral and power densities, $v_n(f)$, $i_n(f)$, $v_n^2(f)$, and $i_n^2(f)$. The spectral densities are rms values of the power densities. The total noise voltage and current over a bandwidth $\Delta f \equiv (f_H - f_L)$ are given by

$$V_n = \left\{ \int_{f_L}^{f_H} v_n^2(f) df \right\}^{\frac{1}{2}} \qquad (6.48)$$

and

$$I_n = \left\{ \int_{f_L}^{f_H} i_n^2(f) df \right\}^{\frac{1}{2}} \qquad (6.49)$$

The two common forms of noise are *1/f noise*, for which the power density varies as $1/f$, and *white noise*, which is independent of frequency (Fig. 6.27). White noise as a rule dominates in semiconductors, because 1/f noise is confined to very low frequencies. The white noise values V_{nw} and I_{nw} are, in accordance with eqns (6.48) and (6.49)

$$V_{nw} = v_{nw}\sqrt{\Delta f} \qquad (6.50)$$

$$I_{nw} = i_{nw}\sqrt{\Delta f} \qquad (6.51)$$

The total noise arising from several contributions is determined by adding the noise powers. Bipolar are less noisy than JFET and MOSFET op amps, because v_n, which contributes far more noise than i_n in most cases, is considerably smaller. FETs and MOSFETs have a much lower i_n than BJTs, with MOSFETs possessing the lowest noise currents. V_n is $\sim$(10–20) nV/$\sqrt{Hz}$ but can be as low as $\sim$(2–3) nV/$\sqrt{Hz}$ for bipolar input stages. Typical white noise values of i_n are 0.4 pA/$\sqrt{Hz}$, 0.01 pA/$\sqrt{Hz}$, and 0.002 pA/$\sqrt{Hz}$ for bipolar, JFET, and MOSFET input stage respectively.

An equivalent circuit for op amp noise is shown in Fig. 6.28. The injection of v_n in the non-inverting path is arbitrary; it could have been placed equally well in the inverting path. The noise contribution can be allowed for by modifying eqn (6.12) to

$$V_o = -A\left\{(V_I - V_{NI}) - v_{ni}\right\} \qquad (6.52)$$

remembering that v_{ni} is the rms value of the various contributions to noise. R_g is the voltage generator resistance. R_3 is included in order

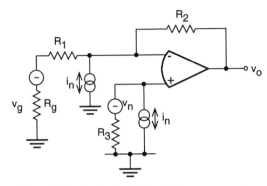

Fig. 6.28 Equivalent circuit for op amp noise

to make the dc resistances in the input paths equal for the purpose of balancing input currents. This step, not shown in Fig. 6.1(d), is good engineering practice; R_3 is made equal to $R_1 R_2/(R_1 + R_2)$. The total noise spectral density is made up of three components:

(i) v_n;
(ii) voltages developed across resistors in the input paths by i_n;
(iii) thermal noise voltages across resistors, given by $v_{th}^2 = 4\text{kTR}$ for a resistor R.

The effective noise resistance R_I at the (-) input terminal, is expressed by

$$R_I = \frac{(R_1 + R_g)R_2}{R_1 + R_g + R_2} \qquad (6.53)$$

The total noise spectral density v_{ni} is given by

$$v_{ni} = \{v_n^2 + (R_I^2 + R_3^2)i_n^2 + 4kT(R_I + R_3)\}^{\frac{1}{2}} \qquad (6.54)$$

Example 6.3

Calculate v_{ni} for the op amp in Fig. 6.28. $R_1 = 5$ kΩ, $R_2 = 100$ kΩ, $R_g = 5$ kΩ, $R_3 = 9.1$ kΩ, $v_n = 20$ nV/$\sqrt{Hz}$, $i_n = 0.4$ pA/$\sqrt{Hz}$, $T = 300$ K. Assume the noise is white.

Substituting $v_n^2 = 4 \times 10^{-16}$, $(R_I^2 + R_3^2)i_n^2 = 2.65 \times 10^{-17}$, $4kT(R_I + R_3) = 3.0 \times 10^{-16}$—all in V^2/Hz. Hence, adding these three componants of eqn (6.54)

$$v_{ni} = 27\text{nV}/\sqrt{Hz}$$

The result in Example 6.3 is representative of the input noise level in a general purpose op amp. Noise and signal outputs are functions of gain and bandwidth. Op amp negative feedback affects signal and noise similarly, although there may be slight differences, and the signal-to-noise ratio(SNR) is largely independent of feedback. Amplifier bandwidth and the magnitudes of the resistors are minimized in applications where a high SNR is important.

A realistic estimate of the SNR is possible by evaluating the total noise allowing for the amplifier bandwidth and by considering white noise

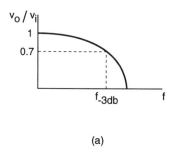

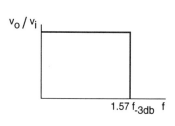

Fig. 6.29 Signal and noise response of single pole network (a) signal (b) noise

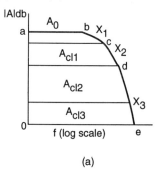

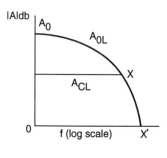

Fig. 6.30 Op amp frequency response (a) piece-wise linear approximation (b) actual response

only. Assuming that the amplifier response is governed by a single-pole network (Fig. 4.2) and applying eqns (4.10) and (4.12) to eqn (6.48), the output noise voltage V_{no} comes to

$$V_{no} = V_{ni} \left[\int_0^\infty \frac{df}{\left\{1 + \left(\frac{f}{f_{-3dB}}\right)^2\right\}} \right]^{\frac{1}{2}}$$

$$= V_{ni} \sqrt{\frac{\pi}{2} \cdot f_{-3dB}}$$

$$= V_{ni} \sqrt{1.57 f_{-3dB}} \qquad (6.55)$$

where f_{-3dB} is the op amp bandwidth. Equation (6.55) shows that the effective noise bandwidth of a single-pole network is given by $1.57 f_{-3dB}$; this is the value for Δf in eqns (6.50) and (6.51). Equivalent noise and signal bandwidth are compared in Fig. 6.29. The result of eqn (6.55) will be used after the coverage of op amp frequency response, which now follows.

Amplifier gain $A(s)$ can be expressed in a general form by

$$A(s) = \frac{A_o}{\left(1 + \frac{s}{p_1}\right)\left(1 + \frac{s}{p_2}\right) \cdots \left(1 + \frac{s}{p_n}\right)} \qquad (6.56)$$

which supposes that the amplifier consists of n cascaded single-pole networks. That is a fair assumption for op amps with a schematic like Fig. 6.13. Figure 6.30(a) contains a Bode plot in the form of a piecewise linear approximation for a three-pole transfer function of such an amplifier. The regions bc, cd, and de give phase shifts of −90 °C, −180 °C, and −270 °C respectively relative to the intended phase reversal (−180 °C) in the low frequency region ab. The Nyquist criterion stipulates that an amplifier is unconditionally stable if the phase shift is less than 180 °C at the point X where the closed loop intersects the open loop gain. A_{CL1} will accordingly be stable, A_{CL2}, which is on the borderline of stable operation, and A_{CL3} will be unstable. Taking the general case in Fig. 6.30(b), the phase shift at X, the intersection of the intended A_{CL} with A_{OL}, must be less than 180 °C, guaranteed by a substantial phase margin which can be maintained in production. This condition becomes exceptionally stringent for a voltage follower, whose unity gain (0 dB) means that A_{CL} is the x axis with the intercept at X' at the point of maximum phase shift. Frequency compensation is achieved in a number of ways (Franco 1988, pp.252–290: Irvine 1981).

Compensation may be either internal or external, with pins provided on the package should external compensation be available. Internal compensation generally ensures that the amplifier is unconditionally stable at unity gain. The standard technique adopted for that purpose is dominant pole compensation, in which an additional pole is created enforcing a slope of −6 dB/octave (−90° phase shift) over the entire high frequency region. In this condition $A_{OL}(jw)$ is related to the low frequency gain

A_o by

$$A_{OL}(jw) = \frac{A_o}{1 + \frac{jf}{f_b}} \simeq \frac{A_o f_b}{jf} \quad (6.57)$$

Equation (6.57) leads to

$$|A_{CL}| f \simeq A_o f_b \quad (6.58)$$

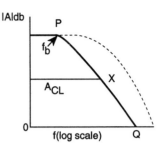

Fig. 6.31 Dominant pole compensation

in the high frequency region PQ of Fig. 6.31, over which $|A_{CL}| f$, the gain bandwidth product (GBP) is constant and equal to $A_o f_b$. For the 741 op amp, $A_o \sim 200\,000$ and GBP, documented in terms of bandwidth at unity gain, is ~ 1 MHz, giving $f_b \sim 5$ Hz. That figure is not as alarming as it sounds, because A_{CL} is likely to be several orders of magnitude below A_o. If, for example, A_{CL} equals 100, the bandwidth comes to 10 kHz and this is acceptable for many applications. Nevertheless dominant pole compensation severely curtails amplifier bandwidth. A simple calculation shows how the low value of f_b comes about. Dominant pole compensation is effected by C_c in Fig. 6.19. The Miller effect makes C_c equivalent to an input capacitance C_{in}, equal to $C_c(1 + A_v)$, at the base of Q_{16}, A_v being the gain of the Darlington stage Q_{16}, Q_{17}. With $C_c \sim 30$ pF and $A_v \sim 500$, $C_{in} \sim 15\,000$ pF. The resistance across C_c consisting of the parallel combination of the output resistances of Q_4 and Q_7, and the input resistance of Q_{16}, amounts to ~ 2 MΩ. Hence f_{-3dB} of the dominant pole $\simeq 1/(2\pi \times 15 \times 10^{-9} \times 2 \times 10^6)$ Hz $= 5$ Hz. Alternative methods of compensation achieve a higher bandwidth with stable operation over a restricted range of A_{CL}.

Example 6.4

Find the SNR of the op amp in Example 6.3 for a sinusoidal input of V_g of 3 mV$_{p-p}$ at 80 kHz, given that the GBP is 8 MHz and that the amplifier has dominant pole compensation.

The calculation of noise bandwidth must allow for R_g, and A_{CL} for this purpose equals $R_2/(R_1 + R_g) = 10$, giving $\Delta f = 800$ kHz. Utilising the result of Example 6.3, eqn (6.50), and applying the effective bandwidth formula of eqn (6.55), $V_{in} = 27\sqrt{1.57 \times 8 \times 10^5}$ nV $= 30.3\,\mu$V. The effective value of R_1 in Fig. 6.1(d) is $(R_g + R_1)$ in Fig. 6.28, and $V_g = 3$ mV$_{p-p} = 1.06$ mV$_{rms}$. This leads to a SNR of 1.06 mV/30.3 μV$= 35$ (31 dB).

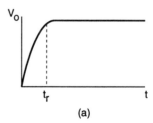

(a)

Op amp small signal frequency response is specified in terms of the GBP or t_T for unity gain (voltage follower) operation. Manufacturers' information contains GBP or t_T (exceptionally both), and the expression GBP$= 0.35/t_T$, a repeat of eqn (4.13), is nearly always quoted for obtaining one parameter from the other. Large signal transient behaviour differs because of the *slew rate*, a phenomenon due to the nature of the negative feedback and the frequency compensation. The slew rate is usually specified for unity gain, i.e. the voltage follower configuration. In large signal operation with low A_{CL}, the situation under consideration, the output is a linear ramp with a rise time t'_r much larger than t_r for small signal amplification (Fig. 6.32). The slew rate SR for Fig. 6.32(b)

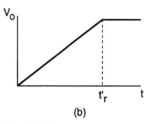

(b)

Fig. 6.32 Voltage follower response for step function input (a) small signal (b) large signal

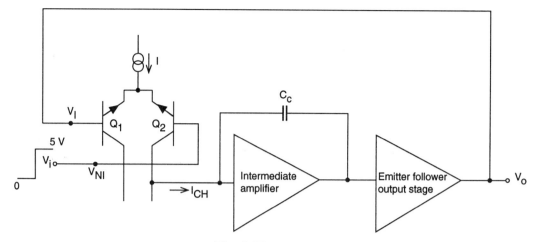

Fig. 6.33 Schematic for evaluating slew rate

is given by
$$SR = \frac{dV_o}{dt} \tag{6.59}$$

For the 741 op amp $t'_r \simeq 20 t_r$. The interpretation of what constitutes 'small-signal' or 'large-signal' operation is given at the end of this analysis.

Consider a voltage follower with a large-signal (5 V) step function input. The output, initially at zero, cannot respond instantaneously, and the 5 V differential at the input by far exceeds the amount for diverting the tail current of the input stage into one transistor, cutting off the other (Fig. 6.33). Linear operation cannot take place, and is only restored when the output signal reaches an amplitude close to the input. The time for it to do so is governed by the charging of the compensating capacitor C_c in Figs. 6.19 and 6.33 (or a similar capacitor, which features in most compensated op amp circuits). The input has diverted the entire tail current into Q_2, I_{CH} equals I, and C_c charges at a constant rate resulting in

$$SR = \frac{dV_o}{dt} = \frac{I}{C_c} \tag{6.60}$$

which establishes the linear nature of V_o in Fig. 6.32(b). The factors determining the SR and methods for its improvement are described by Franco (1988, pp.224–31) and Gray and Meyer (1993, pp.642–54). Slew rate assumes special importance in voltage followers and other large-signal applications like voltage comparators. Its value for general purpose op amps is $\sim(0.5–1)$ V/μs, but high speed op amps with a much faster SR are available.

The slew rate imposes limitations on both pulse and sinusoidal signals. A sinusoid like
$$V(t) = V_o \sin wt \tag{6.61}$$

has a rate of change
$$\frac{dV}{dt} = \omega V_o \cos wt \tag{6.62}$$

whose maximum is

$$\left(\frac{dV}{dt}\right)_{max} = \omega V_o \qquad (6.63)$$

The slew rate must accommodate that maximum for an undistorted output of the sinusoid, i.e.

$$SR \geq \omega V_o \qquad (6.64)$$

Equation (6.64) identifies the maximum undistorted output at a specified frequency, or alternatively the maximum frequency for a specified V_o.

The other constraint on V_o is the output capability of the amplifier, which is governed mainly by the supply voltages and the load resistance. A plot of V_o against frequency (with the customary stipulation of a load resistance of 2 kΩ or higher, for which the maximum output is close to the supply voltages) is shown in Fig. 6.34. It applies to the following parameters, which hold for op amps like the 741: ±15 V supply voltages, SR= 0.8 V/μs, low-frequency $V_o(\text{max}) = 26V_{p-p}$. The maximum frequency for full undistorted output is, in accordance with eqn (6.64), ~10 kHz. Figure 6.34 quantifies the trade-off between $V_o(\text{max})$ and frequency in eqn (6.64), which also indicates the border region between small-signal (Fig. 6.32(a)) and large-signal (Fig. 6.32(b)) operation. Taking a voltage follower with a GBP of 1.2 MHz and an SR of 0.5 V/μs, V_o, which equals V_i for small-signal operation, comes to 66 mV; the demarcation between the two regions is in the range of 50 to 100 mV.

$V_o(\text{max})$ in the low frequency region of Fig. 6.34 holds for an external load equal to or greater than 2 kΩ. The output is governed by the maximum permissible load current and the bias voltages, linked to the supply voltages. The maximum current for bipolar op amps is ~20 mA; the current drawn by a 2 kΩ load at $V_o(\text{max})$ in Fig. 6.34 is ±6.5 mA. The maximum current is reached with decreasing R_l; protection circuits like those described for the 741 amplifier in Section 6.2.2 sharply clamp the output voltage once the maximum current has been reached (Fig. 6.35).

The capabilities of op amps will now be illustrated with reference to their specification. A veritable plethora of products is available. Catalogues of the twenty or so leading vendors of analogue ICs contain up to 100 to 150 different types of op amps. Granted that there will be some duplication of identical or near-identical types, for example the alternative of a single, a dual, or a quad op amp IC, each containing one or more identical amplifiers, the choice is still vast and can be bewildering. There are good reasons for this state of affairs. The amplifier is the universal constituent of analogue electronics and the op amp is the established IC implementation. The huge consumption of the many popular types makes profuse second-sourcing worthwhile. The National Semiconductor (NS) Operational Amplifiers Data Book (1993) lists twenty IC manufacturers, (American, European, and Japanese) who supply some of their analogue products, and contains a full list of all ICs concerned. Some of these will have been originated, others will be second-sourced by NS. To give an example, Analog Devices supply 90, and Burr-Brown 80 ICs which are identical or very similar to NS products. Types with slightly differing specifications probably come from the same production line,

Table 6.3 Op amp categories

General purpose
Low bias current
Low noise
Low power
Precision
Transconductance
Voltage follower
Wideband

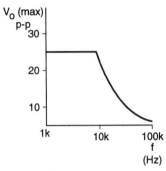

Fig. 6.34 Maximum voltage versus frequency

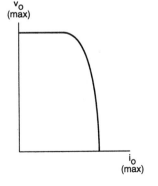

Fig. 6.35 Ouput voltage versus output current

being segregated on testing. An IC manufacturer is therefore likely to build up an inventory with good variety without having to lay down an excessive number of different production lines.

Table 6.3 contains the main categories of op amps. These groupings do not represent rigid, self contained departments. It is quite usual to find a specific product listed under more than one of the categories shown. The remaining tables in this section extend the introductory data on the three op amps in Table 6.2 under some of the categories in Table 6.3, giving a few representative examples of the spectrum available in each case. Many op amps are, like ICs in general, offered in several temperature ranges. The temperature in this context is understood to be the ambient temperature throughout this text, unless stated otherwise. There have been a few alterations in the definitions of temperature ranges over the years. A new range was added a few years ago, and the information at the beginning of Section 5.5.2 needs to be expanded. The standard ranges at present are:

(i) Commercial (C): 0 °C to 70 °C;
(ii) Industrial (I)/Automotive: −40 °C to 85 °C. The designation 'automotive' arises from the growing volume of ICs for the automotive market. The industrial range quoted above emerged in the late 1980s and was first called the *extended* industrial range to distinguish it from the earlier, but long established and still current original industrial range, which has a span from −25 °C to 85 °C. The term 'extended' is being dropped and industrial coverage is now understood to stretch from −40 °C to 85 °C.
(iii) Extended: −40 °C to 125 °C. The prime reason for introducing this range is automotive instrumentation, a dynamic market with tremendous growth. Typical temperature requirements for the automotive market are −40 °C to 85 °C for dashboard and interior locations, and −40 °C to 125 °C for locations in the engine compartment (Prince 1991, p.74). The interpretation of what constitutes the automotive range is ambiguous. Some data books published in 1993 equate it to the industrial, others published in 1994 to the extended range. The equating of the automotive and industrial ranges in (ii) predominates at present. One can look on the extended range as an extended automotive range, which in the fulness of time may well be called the automotive range. However the range of − 40 °C to 85 °C is likely to remain for many automotive ICs, just as the original industrial range of −25 °C to 85 °C is going to exist for years to come.
(iv) Military (M): −55 °C to 125 °C.

The great majority of op amps still have a very moderate bandwidth with a GBP ∼(1–2) MHz. It is worth reflecting on this state of affairs in the light of the continuing thrust for higher speeds, and the successes in producing faster analogue and digital ICs. The truth of the matter is that a very substantial proportion of the market for op amps is satisfied with the moderate GBP of general purpose types like the 741. Another issue in amplifier performance is the trade-off between speed and accuracy. The faster an analogue IC, the less accurate is its performance. Advances in technology have made op amps with a GBP

Table 6.4 High speed operational amplifiers

Type	GBP (MHz)	Slew rate (V/μs)	A_{OL}	Minimum stable A_{CL}	v_n (nV/$\sqrt{Hz}$)	i_n (pA/$\sqrt{Hz}$)
OPA27[1]	8.5	2	1 500 000	1	3	1.5
HA2841[2]	50	240	50 000	1	16	2.0
OPA37[1]	63	12	1 500 000	5	3	1.5
LM6164[3]	175	300	250 000	5	8	1.5
HA2840[2]	600	625	25 000	10	6	6.0
LM6165[3]	725	300	10 500	25	5	1.5

[1] Burr-Brown, [2] Harris Semiconductor, [3] National Semiconductor
(Courtesy of Burr-Brown, Harris Semiconductor and National Semiconductor)

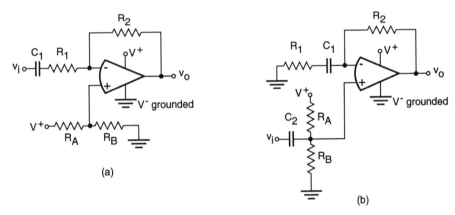

Fig. 6.36 Single supply op amp (a) inverting (b) non-inverting

approaching 1 GHz possible and Table 6.4 contains devices with a GBP up to 750 MHz. The minimum A_{CL} equal to 1 is obtained for GBPs up to ~10 MHz, and gives way to higher closed loop gains at increased GBPs: that is to be expected.

Low noise op amps are shown in Table 6.5. The noise voltage is

Table 6.5 Low noise op amps

Type	v_n (nV/$\sqrt{Hz}$)	i_n (pA/$\sqrt{Hz}$)	GBP (MHz)
Harris 5147	3.0	0.4	120
NS LM 627/637	3.0	0.4	14 (627), 65 (637)
Phillips 5534	4.0	0.6	10

The noise figures apply at $f = 1$ kHz, where the noise is either within or at the edge of the white spectrum.
(Courtesy of Harris Semiconductor, National Semiconductor, and Philips Semiconductors)

Table 6.6 Precision op amps

Device	V_{OS} (mV)	$\partial V_{OS}/\partial T$ (μV/°C)	GBP (MHz)
LM 11[1]	0.100	1.0	~1
LM 627/637[1]	0.015	0.2	14 (627), 65 (637)
HA-5127[2]	0.010	0.4	8.5
HA-2548[2]	0.300	4.0	150

[1] National Semiconductor [2] Harris Semiconductor
(Courtesy of Harris Semiconductor and National Semiconductor)

the more important of the two noise parameters, because it makes the dominant contribution in the majority of applications. Franco (1988), p.617) reports an op amp with exceptional low noise, the LT 1028 of Linear Technology, for which v_n equals 0.9 nV/$\sqrt{Hz}$.

The levels of v_n in Table 6.5 are not all that much lower than those in Table 6.4. It is i_n which is greatly reduced.

The term precision amplifier signifies constant gain under all operating conditions; the key parameter is the drift in offset voltage with temperature. A selection of precision op amps is given in Table 6.6. The LM11 has been included because of its exceptionally low I_{os} and $\partial I_{os}/\partial T$, 0.5 pA and 20 fA/°C respectively. However by and large $\partial V_{os}/\partial T$ is more important.

Op amps can be used as voltage followers provided they are stable with a closed loop gain of unity by virtue of internal or external compensation. Other attributes looked for are a low dc input current, a very high input resistance, and a high slew rate. A modest variety of voltage followers was marketed in the early stages of linear IC development, but this practice has ceased. The availability of op amps with adequate static and dynamic characteristics for voltage follower usage has rendered the provision of special elements for that purpose unnecessary.

Low bias current is readily obtained with JFET or MOSFET input stages, yielding op amps whose I_B lies in the range (1–10) pA. However the bipolar NS LM11 has I_B and I_{os} equal to 25 pA and 0.5 pA.

Op amps are designed to operate with two equal positive and negative power supplies, ±15 V being the norm. However the usual type of op amp circuit does not require an earth connection but only a minimum potential *difference* between the V^+ and V^- inputs. Op amps can generally function satisfactorily with a single supply voltage, and many are documented to that effect. Single supply operation, and the lowering of the supply voltage(s) are topical considerations in the present climate of low-voltage system design. Single supply voltage has the attractions of saving one power supply and being compatible with the digital supply voltage. Development is moving strongly in the direction of a single supply voltage which powers the analogue and digital sections of a system. The established standard levels are 3.3 V and 5 V. Two more, 1.5 V and 2.2 V, already in use, are likely to become standards before long.

The obvious limitation of a single supply operation, for which a pos-

Table 6.7 Micropower op amps

Parameter	CA 3078	CA 3440
Technology	Bipolar	BiMOS
Input stage bias	Adjusable	Adjustable
Supply voltage range	±0.75 V to ±15 V	±5 V to ±15 V
GBP (kHz)	63	63
Standby power	~1 μW	300 nW

(Courtesy of Harris Semiconductor)

itive voltage is applied to the V^+ input with the V^- input grounded, is the clipping of a negative output in excess of a few hundred mV. This limitation can be overcome by ac coupling the signal and by a dc offset of the output. (Jacob 1982). Equation (6.13) gives the clue about how to proceed. The signal is ac coupled to one of the input terminals, and a dc level shift is applied to the other, see Fig. 6.36. The dc offset of the output equals the dc input, $V^+ R_B/(R_A + R_B)$, because the amplifier behaves like a voltage follower for the dc signal.

Regardless of single or dual supply operation, there is a limitation on the maximum output swing obtainable with a compound emitter–follower stage (Figs. 6.18 and 6.19), due to the minimum bias V_{BE} required for each transistor. This limitation leads to a maximum output equal to the differential supply voltage less $2V_{BE}$. For V_{BE} equal to 0.8 V, the maximum output would be 28.4 V_{p-p} for ±15 V supplies, and 3.4 V for a 5 V supply. These figures illustrate the increasing proportion of loss in output with reduction of the supply voltage. Other factors lower the maximum output a little further (see Table 6.2).

A CMOS output stage allows a full rail-to-rail output swing to within a few mV of the supply voltage(s) for a very high output load (infinity in the test specification). The CMOS CA3160 op amp in Table 6.2 gives a maximum output of ±6.7 V for ±7.5 V supply voltages with R_l equal to 2 kΩ, the standard test load. The maximum output swing is very close to the supply voltage(s) if $R_l \geq 100$ kΩ.

Op amps with low quiescent input power are experiencing an upsurge, because there is a rapidly growing demand for micropower systems with low power supplies, low power consumption, and low standby power. Micropower op amps are designed for that purpose and two such amplifiers are listed in Table 6.7

The main parameters, shown with one entry because they happen to be identical, of the Harris CA3080 and the NS LM13600 transconductance amplifiers are contained in Table 6.8. There are some differences in performance. The transconductance is adjustable over six decades for the LM13600, and over three decades for the CA3080. Another feature specific to the LM13600 is the modification of the input stage by the addition of linearizing diodes which extend the input range $(V_I - V_{NI})$ for linear operation in Fig. 6.26, (see also eqn (6.33)) from a few mV to several volts. (Geiger et al. 1990, pp.738–43).

Table 6.8 CA 3080[1] and LM13600[2] transconductance amplifiers

Parameter	Value
g_m (μmho)	9600
Peak output current ($R_i = 0$) μA	500
GBP (MHz)	2
Slew rate (V/ms)	50
Unity gain, compensated Input resistance (kΩ)	26

[1] National Semiconductor [2] Harris Semiconductor
Test conditions: ±15 V supply voltages, $I_{ABC} = 500$ μA
(Courtesy of Harris Semiconductor and National Semiconductor)

6.3 Voltage comparators

The voltage comparator is a differential amplifier without the negative feedback which turns it into an operational amplifier. All IC op amps come into that category and can be used for voltage comparison. The comparator action is illustrated in Fig. 6.37. The variable input V_i is compared with a fixed reference V_{REF}; the connections to the amplifier input terminals are interchangeable. The point of changeover, the threshold, is evidently V_{REF}, neglecting unbalances due to I_{os} and V_{os}. The output levels are given by

$$V_o = V_{oH} \tag{6.65}$$

$$V_i \leq V_{REF}$$

$$V_o = V_{oL} \tag{6.66}$$

$$V_i \geq V_{REF}$$

leaving aside the transition region ΔV, whose magnitude is evidently

$$\Delta V = \frac{V_{oH} - V_{oL}}{A_{OL}} \tag{6.67}$$

Taking a standard op amp with ±15 V supplies, A_{OL} equal to 200 V/mV, and V_{OH}, V_{OL} equal to ±14 V, ΔV—shown on an enlarged scale to highlight its existence—comes to 140 μV.

These figures illustrate the sensitivity of discrimination: a change in V_i of 140 μV is sufficient to swing the output from V_{oL} to V_{oH} or vice versa. The threshold window ΔV is influenced by the input offset voltage and current and their temperature coefficients. For a temperature change ΔT and a specified $\partial V_{os}/\partial T$, ΔV changes by $(\partial V_{os}/\partial T)\, \Delta T$ i.e. the threshold stability changes by 0.25 mV for $\Delta T = 50$ °C and $\partial V_{os}/\partial T = 5$ μV/°C.

The internal frequency compensation of op amps limits the slew rate and in so doing the response time of the comparator, which operates in the open-loop mode and is likely to be stable without frequency com-

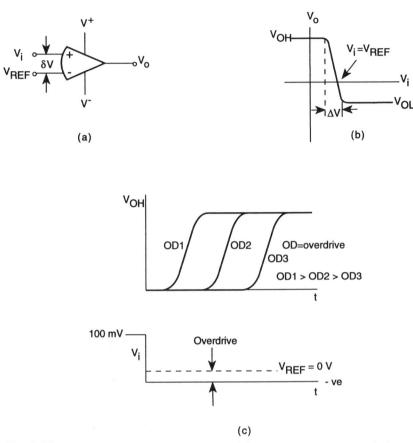

Fig. 6.37 Voltage comparator (a) schematic (b) static transfer characteristic (c) transient response

pensation. This is one of the reasons for producing comparators distinct from op amps. Another is the output voltage swing. The output will, more often than not, interface with digital logic. Op amp output levels are within a volt or so of the supply voltages. They can be modified to TTL, CMOS, or ECL outputs with external circuits, but it is far tidier to design a comparator with the required logic output levels. Summarizing a comparator is designed:

(i) without frequency compensation;
(ii) to give output levels compatible with specified logic (TTL, CMOS, or ECL).

The transient response is expressed by t_P for a specified overdrive in Fig. 6.37(c). Table 6.9 contains a performance summary for a selection of voltage comparators. One very important area of immediate interest is the role of the comparator in data converters, which are covered in Section 6.5. A critical parameter in such applications is the constancy of the threshold under all operating conditions. The fastest comparators in Table 6.9, the LM160 and LM6685, consume most power. At the other end of the spectrum the low power LM139 has the slowest response. The LM139 and LF111 are TTL/CMOS compatible for 5 V operation. The general purpose comparators LM111 and LF111 input stage virtually

Table 6.9 Comparator characteristics

Parameter	LM 139	LM 111	LF 111	LM 160	LM 6685
Supply voltage (V)	+5	+5	+5		+6
	±15	±15	±15	±5	−5.2
Positive supply current (mA)	0.8	5	5	18	15
Negative supply current (mA)	−9	−4.1	−4.1	−9	−7
V_{OS} (mV)	5	0.7	0.7	2	1.9
I_{OS}	5 nA	4 nA	5 pA	0.5 μA	1.0 μA
Input bias current	25 nA	60 nA	20 pA	5 μA	4 μA
V_{oH} (V)	5	5	5	3	−0.96
V_{oL} (V)	0.25	0.23	0.23	0.25	−1.85
t_{PLH} (ns)	1300	200	200	14	2.6

(Courtesy of National Semiconductor)

Notes
(1) The LM 160 and LM6685 have complementary outputs
(2) The LM6685 has ECL outputs, with 50 Ω drive capability as in Fig. 5.20.
(3) Split supply or single supply operation is available for the the LM139, LM111, and the LF111. The output levels and the supply current for these elements apply for single supply (5 V) operation. Split supply operation gives V_{oH}, V_{oL} very close to V^+, V^-.
(4) The propagation delay t_{PLH} applies for the standard test signal in Fig. 6.37(c), a negative-going 100 mV step function with 5 mV overdrive.

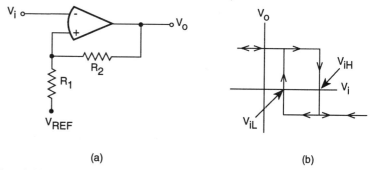

Fig. 6.38 Voltage comparator with hysteresis (a) schematic (b) transfer characteristic

eliminates errors arising from input currents.

An application of op amps which deserves to be singled out is the Schmitt trigger, a voltage comparator with hysteresis. The hysteresis is obtained by positive feedback from the output to the (+) input. A schematic and a transfer curve for a comparator with hysteresis are shown in Fig. 6.38. The amount of hysteresis, δV_i, is defined by

$$\delta V_i = V_{iH} - V_{iL} \tag{6.68}$$

R_1 and R_2 are chosen to give the desired value in accordance with

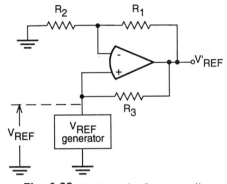

Fig. 6.39 Scaling of reference voltage

$$V_{iH/L} = \left(\frac{R_1}{R_1+R_2}\right)(V_{oL/H} - V_{REF}) + V_{REF} \quad (6.69)$$

and substituting V_{oL}, V_{oH} for the desired V_{iH}, V_{iL} respectively. Hysteresis has two advantages. The positive feedback speeds up the transitions and gives a much faster transient response. It also leads to a more decisive changeover when an input signal hovers near the threshold because of, for example, noise. Against that, the introduction of hysteresis demands external resistors, which lower the accuracy of the threshold and increase its variation with temperature.

6.4 Voltage references and regulators

Accurate voltage supplies are a prerequisite for electronic equipment. Digital meters, test equipment, voltage comparators, and data converters are some examples where the specified performance is closely linked to the quality of the supply voltages. The key characteristics of a voltage reference are its absolute accuracy, long-term stability, and temperature coefficient. Its dependence on load current is of secondary importance, because voltage references are designed for modest currents of a few mA, possibly even less; higher currents are catered for by voltage regulators. These, like voltage references, provide an accurate voltage from a raw supply, but for much higher currents with an upper limit in the range (0.1–10) A. They cannot achieve the accuracy of a voltage reference, although their performance is still very good. Instead the qualities looked for in a voltage regulator are line regulation, the stability of the output with variations in the raw supply voltage, and load regulation, i.e. output stability with varying load current.

6.4.1 Voltage references

The two sources for accurate voltage references are the regulator breakdown diode and the bandgap voltage reference. The principle of obtaining an output voltage from a reference voltage is illustrated in Fig. 6.39, in which the op amps scales up the reference voltage according to

$$V'_{REF} = \left(\frac{R_1+R_2}{R_2}\right) V_{REF} \quad (6.70)$$

The prime task is to produce a reference voltage with a low temperature coefficient (TC). Regulator diodes, loosely called Zener diodes,

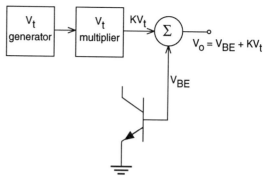

Fig. 6.41 Band gap reference voltage

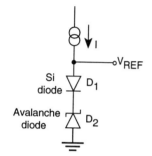

Fig. 6.40 Composite diode reference voltage

break down in accordance with one of two mechanisms. Zener breakdown occurs at up to ∼(5–6) V, avalanche breakdown at higher voltages. Avalanche has a positive and Zener breakdown a negative temperature coefficient. The series combination of a silicon junction diode with its negative TC and an avalanche breakdown diode yields a voltage with a near-zero TC at a specified current (Fig. 6.40). Such a *composite* diode (to be accurate, composite diode combination) gives a reference voltage with a TC of a few ppm/°C over a wide range of temperature. The ultimate in stability is achieved by temperature-stabilization of the die containing the composite diode, or at least that section of the die where it is located. In that way it is possible to achieve a TC of ∼1 ppm/°C over a temperature range of ∼50 °C. (Remember that temperature in this context stands for ambient temperature). For the up-scaling by R_1 and R_2 in Fig. 6.39 the tracking coefficients of the resistor pair with temperature are vitally important. The value of 2 ppm/°C quoted in Table 3.2 can be bettered exceptionally with special processing.

Composite diodes make the best voltage references. Their limitations are the voltage range covered and the inherent high noise level in avalanche diodes. The high noise level of the avalanche diode can be reduced by RC filtering and by special processing. The minimum reference voltage in Fig. 6.40, in excess of 6 V, is too high for the mainstream of digital and a growing proportion of analogue ICs. The feature sizes of today and the deep submicron dimensions of tomorrow call for supply voltages from 5 V downwards. The move towards 1 V systems for personal communications is in full swing. (Malhi and Chatterjee 1994). The limitation imposed by the excessive minimum reference voltage of the composite diode is overcome by means of the bandgap reference, which has an intrinsic value of ∼1.25 V, is capable of being increased by means of the circuit in Fig. 6.39, and at the same time can be used directly for 1 V systems. The principle of the bandgap reference is shown in Fig. 6.41. V_{BE} is combined with a multiple KV_t of the thermal voltage V_t to give V_{REF} in accordance with

$$V_{REF} = V_{BE} + KV_t \qquad (6.71)$$

V_{BE} has a positive, V_t a negative TC, and K is set for $\partial V_{REF}/\partial T$ to be zero at a specified temperature, frequently 300 K. V_t equals (kT/q) by definition. Differentiating V_t

$$\frac{dV_t}{dT} = \frac{k}{q} = 8.625 \times 10^{-2} \text{ mV/°C} \quad (6.72)$$

The transistor I–V relationship is, neglecting the (−1) term in the diode equation

$$I_E = AT^3 \exp\left\{\frac{-(V_{GO} - V_{BE})}{V_t}\right\} \quad (6.73)$$

where A is a constant and V_{GO}, equal to 1.205 eV, is the bandgap voltage of silicon extrapolated to absolute zero. Differentiating eqn (6.73) for constant current

$$\frac{\partial V_{BE}}{\partial T} = -\frac{1}{T}\{3V_t + (V_{GO} - V_{BE})\} \quad (6.74)$$

For zero TC at constant current

$$\frac{\partial V_{BE}}{\partial T} + K\frac{dV_t}{dT} = 0 \quad (6.75)$$

Substituing eqns (6.72) and (6.74) in eqn (6.75)

$$K = 3 + \frac{V_{GO} - V_{BE}}{V_t} \quad (6.76)$$

and combining eqns (6.71) and (6.76)

$$V_{REF} = V_{GO} + 3V_t \quad (6.77)$$

Equation (6.77) holds for zero TC at a specified temperature and current. Taking typical values of $V_{GO} = 1.205$ eV, $V_{BE} = 0.65$ V and $V_t = 26$ mV (25 °C), K and V_{REF} come to 24.3 and 1.283 V respectively.

Two established circuits for generating a bandgap reference voltage are shown in Fig. 6.42, in which Q_1 and Q_2 form a Widlar current source (Fig. 6.20). Taking Fig. 6.42(a), I_2, using eqn (6.29), is given by

$$I_2 = \frac{V_t}{R_3} \ln\left(\frac{I_1}{I_2}\right) \quad (6.78)$$

(I_1/I_2) is set to the desired value by R_1, R_2, and R_3 (which has a second-order effect). Equation (6.78) gives V_{R2} equal to $(R_2/R_3)V_t \ln(I_1/I_2)$. This equals KV_t and R_2 is chosen to give the magnitude of K required by eqn (6.76). V_{REF} in Fig. 6.42(a) equals $V_{BE3} + KV_t$ whereas eqn (6.71) demands it to equal $V_{BE1} + KV_t$. The equality of V_{BE} for Q_1 and Q_3 is ensured by choosing R_4 to give equal collector currents for these transistors. V_{REF} is scaled up to the desired V'_{REF} by R_5 and R_6.

The action of the op amp in the circuit of Fig. 6.42(b), which generates an unscaled bandgap reference voltage, is to force the equality

$$I_1 R_1 = I_2 R_3 \quad (6.79)$$

by making I_1/I_2 equal to R_1/R_3. $I_1 = (V_t/R_2)\ln(R_1/R_3)$ and

$$V_O = V_{BE2} + I_2 R_3 = V_{BE2} + I_1 R_1$$

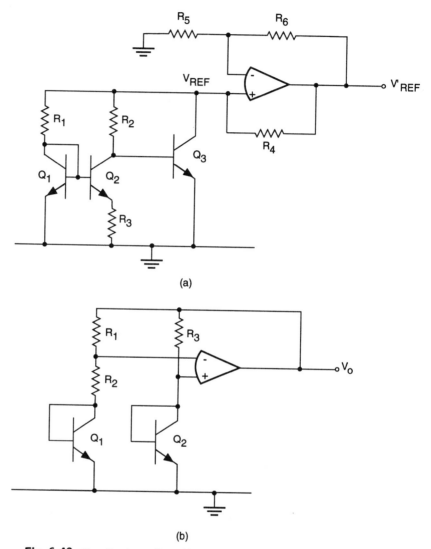

Fig. 6.42 Circuit schematics of band gap reference voltage generator (a) and (b) are alternative schematics

$$= V_{BE2} + \left(\frac{R_1}{R_2}\right) V_t \ln\left(\frac{R_1}{R_3}\right) \qquad (6.80)$$

CMOS bandgap references take two forms. One approach is to obtain bipolar transistors with a p-well structure like Fig. 3.4, with the drain and source layers for the n electrodes. This leads to a conventional bipolar circuit. The other technique combines bipolar and MOS transistors. The former furnish V_{BE} and the latter, operating in weak inversion, where MOSFETs have diode-like I–V characteristics (see eqn (4.34)), provide KV_t (Geiger et al. 1990, pp.369–71; Allen and Holberg 1987, pp.594–9).

Example 6.5

Find the TC at 125 °C for the bandgap reference designed earlier in this section for zero TC at 27 °C, with the following data and results: $V_{GO} = 1.205$ V, $V_{BE} = 0.65$ V, $K = 24.4$, $V_{REF} = 1.283$ V, $V_t = 26$ mV.

Recalculate the reference voltage to improve the performance over the range 27 °C to 125 °C by designing for zero TC within that range.

This example has been chosen because its solution embraces a number of valuable steps in analogue circuit design.

The first step is the derivation of an expression for V_{BE} as a function of current and temperature, and to use it for calculating $(\partial V_{REF}/\partial T)$ at constant current. The required relationship is obtained by applying eqn (6.73), slightly modified with n in place of 3 for the exponent of T, formulating I_E, the emitter current at T_1, and I_{EO}, the emitter current (here not a reverse bias leakage current) at T_O, dividing one by the other and taking logarithms. The result, first publicized by Widlar (1971) is

$$V_{BE} = V_{GO}\left(1 - \frac{T}{T_o}\right) + V_{BE0}\left(\frac{T}{T_o}\right)$$
$$+ \frac{nkT}{q}\ln\left(\frac{T_o}{T}\right) + \frac{kT}{q}\ln\left(\frac{I_E}{I_{EO}}\right) \qquad (6.81)$$

The stipulation of constant current makes I_E and I_{EO} equal and the last term vanishes. Differentiating eqn (6.81)

$$\left(\frac{\partial V_{BE}}{\partial T}\right)_T = \left(\frac{\partial V_{BE}}{\partial T}\right)_{T_o} + \frac{nk}{q}\ln\left(\frac{T_o}{T}\right) \qquad (6.82)$$

a reduction obtained by making use of eqn (6.74).

Differentiating eqn (6.71)

$$\left(\frac{\partial V_{REF}}{\partial T}\right)_T = \left(\frac{\partial V_{BE}}{\partial T}\right)_T + \frac{Kk}{q} \qquad (6.83)$$

$$\left(\frac{\partial V_{REF}}{\partial T}\right)_{T_o} = \left(\frac{\partial V_{BE}}{\partial T}\right)_{T_o} + \frac{Kk}{q} = 0 \qquad (6.84)$$

Combining eqns (6.82), (6.83) and (6.84)

$$\left(\frac{\partial V_{REF}}{\partial T}\right)_T = \frac{nk}{q}\ln\left(\frac{T_o}{T}\right) \qquad (6.85)$$

Substituting $n = 3$, $T_o = 300$ K, $T = 398$ K, $(\partial V_{REF}/\partial T)_{125\,°C} = -0.0731$ mV/°C. This is the maximum reached over the specified range 27 °C to 125 °C. Using eqns (6.71) and (6.81), $V_{REF}(125\,°C) = 1.279$ V.

The coefficient $\{(\partial V_{REF}/V_{REF})/\partial T\}$ is accordingly –57 ppm/°C. The mean coefficient $(\Delta V_{REF}/V_{REF})/\Delta T$ for the interval 27 °C to 125 °C, works out to be –32 ppm/°C ($\Delta V_{REF} = -4$ mV, $V_{REF} = 1.281$ V, $\Delta T = 98$ °C).

Table 6.10 Voltage references

Type	Output voltage (V)	Technology	Temperature stability (ppm/°C)	Operating temperature	Max. load current (mA)
AD589[2]	1.235	B	10–100	C, M	5
LM185[3]	1.25–5.30	B	30–150	C, I, M	20
AD586[2]	5	R	5–25	C, M	10
REF10[1]	10	R	1–6	C, M	10
REF-10[2]	10	B	5–25	I, M	20
AD2710/2712[2]	10/±10	R	1–5	C	5

[1] Burr-Brown Corporation, [2] Analog Devices, [3] National Semiconductor
B = Bandgap, R = Regulator diode
(The information is reproduced by permission from data published by these companies.)

The TC can clearly be improved by choosing an optimum T_o for zero TC within the range T_1 (300 K) and T_2 (398 K) such that the TCs at T_1 and T_2 are numerically equal and opposite. The logarithmic nature of eqn (6.85) gives

$$T_o = \sqrt{T_1 T_2} \qquad (6.86)$$

and T_o comes to 72.5 °C. Carrying out the computations

$$\left(\frac{\partial V_{REF}}{\partial T}\right)_{27\,°C} = 0.0365 \text{ mV}/°C$$

$$\left(\frac{\partial V_{REF}}{\partial T}\right)_{125\,°C} = -0.0365 \text{ mV}/°C$$

and hence V_{REF} equals 1.2954 V (27 °C), 1.2944 V (72.5 °C), and 1.2937 V (125 °C).

Table 6.10 contains information on selected standard voltage references and spans the spectrum from the highest performance figures to more relaxed specifications, which are adequate for many applications. The temperature stability is expressed by a band and not by a typical figure, because most of the types are brought out in various versions which cover different ranges of operating temperature and/or different specifications of temperature stability. Regulator diodes can achieve a lower TC than bandgap references, which are however easier to process and give a performance adequate for many applications. Some references, like the AD589, have a direct bandgap voltage output, which satisfies the demand for a supply voltage around 1 V and can easily be scaled up by adding a circuit like that in Fig. 6.39.

The steps taken to achieve the temperature stabilities quoted in Table 6.10 include special sophisticated circuit techniques which improve the TC of the bandgap reference, special processing to yield regulator diodes with improved characteristics, particularly low noise, and—on occasions—heating of the die to maintain it at a constant temperature

Table 6.11 Typical performance of voltage references

Parameter	Value
Max operating curent (mA)	5–20
Load regulation (μV/mA)	50–100
Line regulation (μV/V)	100–600
Long term stability (ppm/1000 hrs)	15–50
Initial accuracy (%)	0.05–2

independent of the environment. The R category is either a composite diode or a regulator (avalanche breakdown) diode on its own. The established technology for voltage references is a hybrid composition within a standard IC package. The accuracy of the initial output and the TC are adjusted by active laser trimming of the thin film resistors. Table 6.11 gives a broad overview of the performance obtainable with standard ICs. Line regulation is usually more important than load regulations, because voltage references are likely to drive a fixed load.

6.4.2 Voltage regulators

The spectrum of standard IC voltage regulators extends in round figures to load currents from 100 mA to 10 A, and output voltages from 1.25 V to 50 V, exceptionally beyond.

Voltage regulators are either *series* or *switching* regulators. The series regulator operates by controlling a *pass* transistor between the input and the output. The basic operation is shown in Fig. 6.43. The pass transistor is in the active mode and operates like an emitter follower with V_{REF} applied to its base. The output voltage is given by

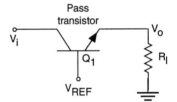

Fig. 6.43 Principle of voltage regulator

$$V_o = V_{REF} - V_{BE} \quad (6.87)$$

making it proportional to V_{REF} and independent of V_i, and explaining the alternative description of *linear* regulator for the series mode. The arrangement is readily adapted for negative voltages. IC regulators are fixed or adjustable; the adjustment in either case is in principle the arrangement in Fig. 6.39, where R_1 and R_2 are internal for fixed, and external for variable output.

The operation of a switching regulator is entirely different. A power transistor is switched at a high frequency, commonly 20 kHz, and the width of the pulses is controlled by the reference voltage. This process of pulse width modulation gives an output which is filtered and rectified to produce a dc voltage whose level is controlled by the type of feedback shown in Fig. 6.39. Voltage regulators do not, by and large, obtain the stability (long term, and with changes of temperature) of voltage references. They fulfil the more onerous function of catering for a wide range of output current with high efficiency, and do not have to satisfy the more fastidious specifications of voltage references with their maximum output capability of 20 mA or less.

Series regulators are based on the schematic shown in Fig. 6.44. The maximum output current is governed by the power capability of Q_1 and

216 Analogue circuits—techniques and performance

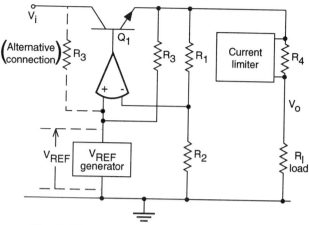

Fig. 6.44 Schematic of series voltage regulator

the output capability of the op amp. An output current of 1 A, supplied by a transistor with an h_{FE} of 50, demands a base drive of 20 mA, just about the maximum available from a standard op amp. Replacing Q_1 with a Darlington pair lowers the base current to ∼0.4 mA, which is well within op amp capability. Darlington pairs are frequently used in series regulators and the aforesaid consideration is one of the reasons. The ability to mount the transistor carrying the bulk of the series current external to the chip on a heat sink, if necessary, is another. Series regulators incorporate some (more likely all) of the following protection circuits:

(i) current overload short circuit protection. The protection mechanism is in principle similar to Q_{19} and R_{10} in Fig. 6.19.
(ii) thermal overload protection (alternatively thermal shutdown protection). The circuit is activated and reduces the base drive to the pass transistor when the temperature of the pass transistor is close to the permitted maximum.
(iii) safe operating area (SOA) protection for Q_1, ensuring that the power rating of the transistor is not exceeded.

In addition a start-up circuit may have to be incorporated. All op amps have a stable state with $V_I = V_{NI} = V_O$. Some regulator circuits move from that undesired state to proper operating conditions when switched on. Others may need an auxiliary circuit to do so.

Table 6.12 contains data on series voltage regulators. Fixed and adjustable types are available for positive and negative voltages. Fixed regulators have either a single or multiple-choice output level. The setting of V_O for an adjustable regulator is made by the choice of R_1 and R_2 in Fig. 6.44. The entries for load regulation temperature stability, and dropout voltage apply for maximum output current. The TL783C is unique. It has by far the highest voltage range and this has come about by combing a standard bipolar circuit with a double-diffused MOS (DMOS) pass transistor (Sze 1981, pp.489–90; Burns and Bond 1987). This MOSFET is capable of withstanding far higher voltages than a BJT and, like other types of MOSFET, is not subject to the BJT phenomena of secondary breakdown and thermal runaway. The stabilization process

Table 6.12 Series voltage regulators

Type	Mode	Output voltage (V)	Max load current (A)	V_{DO} (V)	Line regulation (%/V)
LM117[1]	Adjustable	1.25 to 37	1.5	2.3	0.010
LM138[1]	Adjustable	1.25 to 32	5.0	2.7	0.020
LM123[1]	Fixed	5	3.0	1.8	0.01
TL783C[2]	Adjustable	1.25 to 125	0.7	13.0	0.001

[1] National Semiconductor [2] Texas Instruments
(Courtesy of Texas Instruments and National Semiconductor)

of the series regulator inevitably extends to the ripple superimposed on the input voltage (100 Hz in Europe, 120 Hz in the US), for a full wave rectifier output). The reduction ratio of input-to-output ripple voltage is ∼(60—70) dB for series voltage regulators.

Series regulators can operate over a wide range of input voltage. The minimum $V_{CE}(= V_i - V_o)$ for which the pass transistor Q_1 can function properly, is called the *dropout* voltage (V_{DO}) and is ∼(2–3) V for load currents of (1—5) A. V_{CE} includes the voltage drop across the collector bulk resistance r'_c in Fig. 4.11 and is expressed by

$$V_{CE} = I_c r'_c + V_{CB} + V_{BE} \tag{6.88}$$

If V_{CB} is made zero (an acceptable condition)

$$V_{CE} = I_c r'_c + V_{BE} \tag{6.89}$$

V_{DO} is governed by V_{BE} and the $I_c r'_c$ drop, which is the major contributor at output currents of (1–10) A. Regulators with a low dropout voltage, LDO regulators, were initially stimulated by automotive electronics, and are in demand for all types of battery equipment. A low dropout voltage minimizes the power loss across the $(V_i - V_o)$ voltage differential. It permits operation when the battery output falls below its full level. Take automotive electronics. The drain of the starter motor reduces the output of a 6 V battery to 5.6 V or thereabouts, momentarily and possibly longer, till the engine has recharged it. An LDO regulator can almost certainly maintain a regulated 5 V output, whereas a conventional regulator cannot, under these conditions. Another demand for a battery-operated system is protection against reverse battery connection.

The lowering of the dropout voltage is achieved by substituting a lateral pnp in place of the npn pass transistor, and driving it with a common emitter stage. The schematic for such a regulator is shown in Fig. 6.45 (Menniti and Storti 1984). The lateral pnp is processed together with npn transistors without any extra masking levels (Fig. 3.9). r'_c is less for a lateral pnp than for an npn transistor with a similar geometry, because its collector layer is more heavily doped. The near-symmetrical lateral structure gives high breakdown voltages BV_{CBO} and BV_{CEO} in excess of ∼50 V, allowing the regulator to withstand high positive or negative

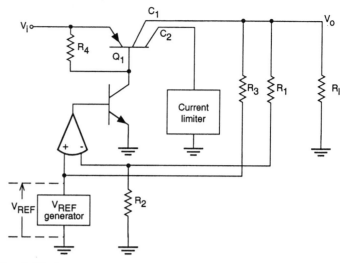

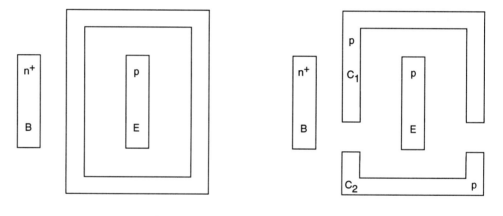

Fig. 6.45 Schematic of series voltage regulator—pnp pass transistor

E, B, C, are emitter, base, and collector contact strips

(a) (b)

Fig. 6.46 Top view of lateral pnp transistor (a) single collector (Fig. 3.9) (b) split collector

overvoltages. V_{DO} is lowered further by operating the pass transistor in soft saturation, a mode already met in CML (Section 5.2.5, Fig. 5.18). This move can be applied to both npn and pnp pass transistors. With soft saturation, and re-writing eqn (6.89) in the form

$$|V_{CE}| = |I_c r'_c| + |V_{BE}| - |V_{CB}| \qquad (6.90)$$

$|V_{CB}|$ and $|V_{BE}|$ have opposite signs, making V_{CE} smaller than the value computed from eqn (6.89). A further advantage obtainable with the lateral pnp transistor is the addition of a second collector C_2, again without an extra masking level, which has a far lower h_{FE} than the main collector C_1 (Figs. 6.45 and 6.46). This facilitates an overload/short circuit output protection which is superior to the current limiter in Fig. 6.44,

Table 6.13 LDO regulators

Type	Mode	Output voltage (V)	Output current (mA)	V_{DO} (mV)
LP2980	Fixed	3.0, 3.3, 5†	50	120
LM2926	Fixed	5	500	350
LP2952	Variable	1.23 to 29	250	470
LM2940	Fixed	5, 8, 12, 15†	1000	500

(Courtesy of National Semiconductor)
† Standard products are available with these voltages.

where the loss in output voltage across R_4 increases with load current.

The lowering of V_{DO} in LDO regulators is impressive, witness Table 6.13, in which the emergence of 3.3 V regulators is in keeping with the coming on stream of low voltage (mainly 3.3 V) logic, op amps, and data converters. The technique of using pnp in place of npn pass transistors is, however, not all plain sailing. The pnp pass transistor in Fig. 6.45 is an inverting amplifier unlike the unity gain emitter–follower in Fig. 6.44, and this combined with its larger input capacitance makes the feedback loop prone to oscillation. Another feature of Fig. 6.45 is a quiescent output current larger than for Fig. 6.44. The LDO regulator demands some care and attention for stable operation, and conventional regulators are recommended for applications where their larger V_{DO} is acceptable.

Fixed voltage regulators have three terminals, input, output, and common (usually ground). Adjustable regulators are set to the desired output voltage by two external resistors R_1 and R_2 (Figs. 6.44 and 6.45). Rather than add a fourth terminal, the majority of regulators are arranged for three-terminal adjustment. The circuit and regulator connection schematics are shown in Fig. 6.47. I_{ADJ} is the current through the voltage reference, and the output is

$$V_o = V_{REF}\left(1 + \frac{R_2}{R_1}\right) + I_{ADJ}R_2 \qquad (6.91)$$

Capacitors C_2 and C_3 in Fig. 6.47(b) are recommended where a low impedance at high frequencies (C_3) and improved ripple injection (C_2) are important—C_1 is only used if the power supply generating V_i is about 20 cm or more away from the regulator.

Example 6.6

Prove the relationship in eqn (6.91).
Let V_A be the voltage at the ADJUST point in Fig. 6.47(a). The virtual equality of the input voltages to the op amp signifies that

$$V_o = V_{REF} + V_A \qquad (6.92)$$

The voltage drops across R_1 and R_2 give

220 Analogue circuits—techniques and performance

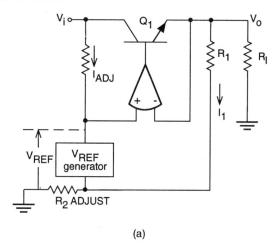

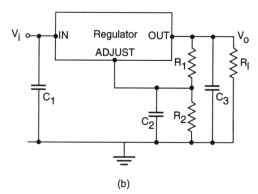

Fig. 6.47 Adjustable series regulator (a) schematic (b) external connections

$$V_o = I_1 R_1 + (I_1 + I_{ADJ})R_2 \tag{6.93}$$

But
$$V_A = R_2(I_1 + I_{ADJ}) \tag{6.94}$$

Eliminating I_1 in eqn (6.94) with the aid of eqn (6.93)

$$V_A = \frac{R_2(I_{ADJ}R_1 + V_o)}{R_1 + R_2} \tag{6.95}$$

Substituting for V_A from eqn (6.95) in eqn (6.92) gives eqn (6.91).

An outstanding advantage of the switching over the series voltage regulator is its higher efficiency. The efficiency is given by

$$\eta = \frac{V_o}{V_i} \tag{6.96}$$

η can be quite small for a series regulator, and the efficiency only becomes high when the difference between input and output voltages approaches the dropout voltage. The operation of a switching regulator is outlined in Fig. 6.48. The control voltage, derived from the reference and output voltages, determines the width of the pulses which drive the power switch Q_1. The output pulses are filtered to give a smooth dc voltage, and the feedback loop relates V_o to V_{REF} by

$$V_o = V_{REF}\left(1 + \frac{R_2}{R_1}\right) \qquad (6.97)$$

The high efficiency, also called conversion efficiency, is due to the nature of switching Q_1, which is either OFF or ON in saturation. $V_{CE(SAT)}$ is far lower than V_{CE} for the pass transistor of a series regulator and that accounts for the higher efficiency, which is comparatively independent of V_i and V_o. A further advantage of the switching regulator is the high frequency of operation, selectable by the user and in practice usually set in the range (20–50) kHz. Inductors and capacitors of the switching filter have much smaller values than those for operation at the standard mains frequency (50 to 60 Hz). However the square wave has to be well filtered in order to reduce the output ripple to an acceptable level, and the series regulator is superior on that score.

Figure 6.48 is the schematic of a step-down switching regulator: V_i must exceed V_o. The modified schematics shown in Fig. 6.49 allow operation in the step-down, step-up, or inverting mode. The diode in Fig. 6.49(a) prevents the build-up of a reverse voltage, which could destroy the transistor during turn-off. Voltage step-up in Fig. 6.49(b) comes about because a voltage of opposite polarity to and larger than V_i is induced in the inductor when Q_1 is turned off and the current flows in the same direction. Regarding inversion, when Q_1 in Fig. 6.49(c) is ON current flows through the inductor and therefore out of the capacitor. Phase reversal is maintained during the off period by the induced voltage across L. Other arrangements include transformers which can give step-up or inverse operation. All these modes are well documented in manufacterers' data.

Switching regulators are not confined to high power applications. They are used extensively in personal computer (PC) power supplies, where their high efficiency makes them particularly attractive. For the same reason they have outlets in micropower equipment like electronic (quartz) clocks and watches, calculators, medical instrumentation and heart pacemakers. The attraction for micropower usage is the high efficiency combined with flyback action, leading to a voltage step-up. The input can decrease substantially due to say, an ageing battery, and yet the output is maintained with good accuracy and high efficiency.

The characteristics of some switching regulators representative of modern practice are contained in Table 6.14. Typical temperature ranges are –40 °C to 125 °C and -55 °C to 150 °C. Current limiting and thermal shutdown protection are nearly always provided. The adjustable voltage range is quoted for step-down (only) regulators i.e. it is confined to the step-down mode. It is clearly more flexible for regulators with step-up and step-down modes and for that reason is not defined in the data for such chips.

Flyback regulation converts the output into dc via a transformer which stores the energy delivered by the power switch transistor during

222 Analogue circuits—techniques and performance

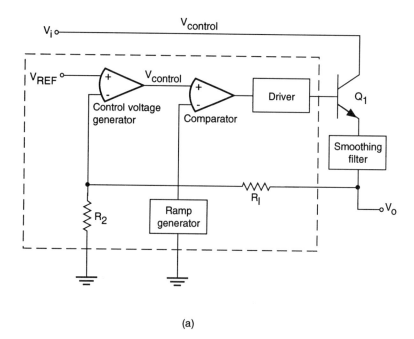

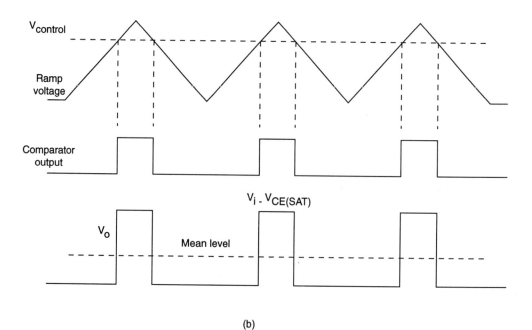

Fig. 6.48 Switching regulator (a) schematic (b) waveforms

Voltage references and regulators

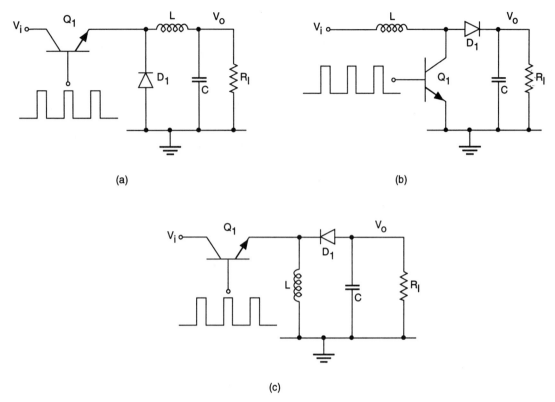

Fig. 6.49 Switching regulator modes (a) step-down (b) step-up (c) inverting

Table 6.14 Switching regulator characteristics

Device	Maximum switched current (A)	Modes	Switch frequency (kHz)	V_i (V)	V_O (V)
LM1524	0.2	Step-up Step-down Flyback Inverting	1–550	5–40	Adjustable
LM 78540	1.5	Step-up Step down Inverting	0.1–100	2.5–50	Adjustable 1.25–40
LM 2576	3.0	Step-down		4–40	Fixed 3.3, 5, 12, 15 Adjustable 3.0–30
LH 1605	5.0	Step-down	6–100	8–35	Adjustable 3.0–30
HS7067	7.0	Step-down Flyback Invert	25–200	10–60	Adjustable

(Courtesy of National Semiconductor)

224 Analogue circuits—techniques and performance

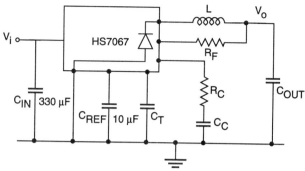

Fig. 6.50 HS 7067 switching regulator—step-down mode (Courtesy of National Semiconductor)

Table 6.16 HS switching regulator—component values (Fig. 6.50)

Parameter	Switch frequency	
	25 kHz	200 kHz
L (μH)	86	21
C_T	0.0039 μf	330 pf
C_C (μF)	0.2	0.068
R_F (kΩ)	4	4
R_C (kΩ)	5.7	5.7
C_{out} (μF)	1500	680

(Courtesy of National Semiconductor)

Table 6.15 Conversion efficiency of LM2576 switched regulator (load current =3.0 A)

V_O	η
3.3	75
5	77
12	88
15	88

(Courtesy of National Semiconductor)

its ON state and transfers it to the capacitance across the load during the OFF state. The flyback mode can implement either step-up or step-down regulation. The conversion efficiency depends slightly on the output voltage and the load current. The extent of the variation is shown for the LM2576 in Table 6.15. In general the conversion efficiency of switched regulators is in the region (70–90) per cent.

Finally Fig. 6.50 shows the external connections for the HS 7067 regulator in Table 6.14, and Table 6.16 gives parameter and recommended component values at two switch frequencies.

6.5 Analogue signal processing

6.5.1 Introduction

We live in an analogue world and analogue signal processing is central to electronic activity. It embraces many fields like speech and music transmission and receception, instrumentation, telecommunications, and radar, to mention only a few. The basic ingredients of signal processing are signal conditioning by means of filters, spectrum analysis, and data conversion. Filtering fulfils the functions of selecting one signal from several contained in the input, and reducing noise by limiting transmission bandwidth. Transmission, reception, and processing involve spectral analysis and spectral transformation.

The role of an ADC in all-analogue processing is confined to storage

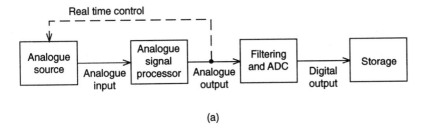

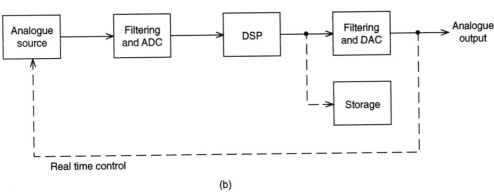

Fig. 6.51 Signal processing (a) analogue (b) digital

and display. Signal processing is becoming increasingly digital, especially in communications. In telephone networks and much of telecommunications the analogue signal is converted into a pulse code, transmitted digitally, and reconverted into analogue form at the receiver. Fig. 6.51(a) contains a schematic for all-analogue signal processing in which the ADC gives a digital output for storage, which includes a display like, for example, the polarity and magnitude of a voltage. The ADC may also feed an off-line computer. Another possible connection is feedback from the output to the input for real time (on-line) control. Digital signal processing is represented by the schematic in Fig. 6.51(b) which, like Fig. 6.51(a), shows paths for storage and on-line control. Digital control has been practiced for some time prior to the arrival of VLSI. The digital signal processor (DSP) is a specially designed microprocessor in which digital signals undergo arithmetic processing in real time. The DSP may also be part of a microcontroller, a microprocessor with specialized hardware and algorithms for on-line control.

The ADCs and DACs shown in Fig. 6.51 include various filters required to fulfil their functions. Filters are integral components for data conversion and have of course many other uses in signal processing. They are covered in the next section, which is followed by three sections on data converters. The first two of these deal with mature, established types. The third covers an up-and-coming category, oversampled data converters, which operate in a distinctly different mode and are treated separately for that reason.

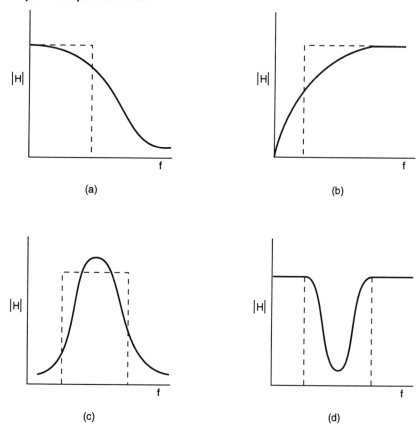

Ideal characteristics are
shown by dotted lines

Fig. 6.52 Filter transfer characteristics (a) low pass (b) high pass (c) band pass (d) notch (band stop)

6.5.2 Analogue filters

The four categories being considered are low pass, high pass, bandpass, and bandstop (notch) filters which have the characteristics shown in Fig. 6.52, in which $|H|$ is the magnitude of the transfer function V_o/V_i. The generalized transfer characteristic for a low pass filter in Fig. 6.53 illustrates a number of features which have to be considered in filter design. These are ripples in the pass and stop bands, and the steepness of the transition region.

Passive filters, i.e. filters consisting of passive components have a number of limitations. The size of the components rules them out for IC implementation in most cases. Inductors are really only practical at microwaves in monolithic microwave integrated circuits (MMICs). Active filters consist of an RC network, part of which is in the feedback path of an operational amplifier. One striking advantage is that active filters achieve the response of passive bandpass and notch filters, but without the use of inductors. The four types of active filters adopted from passive types are the Butterworth, Chebyshev, Bessel, and Cauer filters. The transfer function of all filters can be expressed by

$$H(s) = \frac{a_0 + a_1 s + a_2 s^2 + \cdots a_n s^n}{b_0 + b_1 s + b_2 s^2 + \cdots b_n s^n} \quad (6.98)$$

where $s = jw$ and n is the order of the filter. The most widely used type, the Butterworth filter, has $a_0 = 1$, and a_1, to $a_N = 0$. The higher the order of a filter, the closer will its transfer characteristic be to the ideal shown in Fig.6.52. In practice IC filters are second order or higher. The approximation for an n order Butterworth filter is

$$|H(f)| = \left\{ 1 + \left(\frac{f}{B}\right)^{2n} \right\}^{-\frac{1}{2}} \quad (6.99)$$

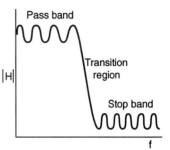

Fig. 6.53 Low pass filter transfer characteristic

where B is the 3 dB bandwidth. The Butterworth filter has a near-flat response over the pass band, and is labelled 'maximally flat' because of that. The Chebyshev achieves a sharper transition than the Butterworth filter for a given order at the expense of ripples in the pass band. The roll-off can be traded for the amount of ripple which is acceptable. A steeper transition is obtained with the Cauer (also *elliptic*) filter, which contains ripples in both pass and stop bands. The quality of a filter is determined not only by its amplitude, but also by its phase response; the importance of the latter depends on the application. Audio systems are an example where phase distortion has to be minimized. All the filters mentioned so far have a poor phase response, which is strongly frequency dependent. If the phase shift can be made to vary linearly with frequency, its effect is confined to delaying the signal without distortion. The Bessel filter achieves such a response and has a much smoother transition band. Examples of low pass, high pass, and band pass active filters are sketched in Fig. 6.54, and can be adapted, with additions and modifications, to give Butterworth, Bessel, Chebyshev, and Cauer filters. Detailed filter design is outside the scope of this text and is covered extensively in the literature on analogue circuits (Allen and Sanchez-Sinencio 1984; Huelsman 1993).

The elimination of inductors, although an outstanding advantage, is not the only advantage of the active filters. The input and output impedances of passive filters present problems, especially at low frequencies. The low input impedance, which varies with frequency, loads the source. At the other end the high output impedance limits the drive capability of the filter. There is no isolation between the load impedance and the filter, making the load a component of filter design. Active filters largely overcome these disadvantages. The low output impedance of the op amp allows a good output amplitude and makes the filter response practically independent of load. The input impedance is far higher for an active than for a passive filter.

A further leap forward in filter design came with the introduction of switched capacitors, which replace resistors and have added a new dimension to active filters. Switched capacitor (SC) techniques are not confined to active filters. They are used for many analogue circuits, including data converters (Allen and Sanchez-Sinencio 1984). SCs are implemented in MOS technology which lends itself to the formation of parallel-plate capacitors. The nature of the action causing a switched capacitor to behave like a resistor will now be described with the aid

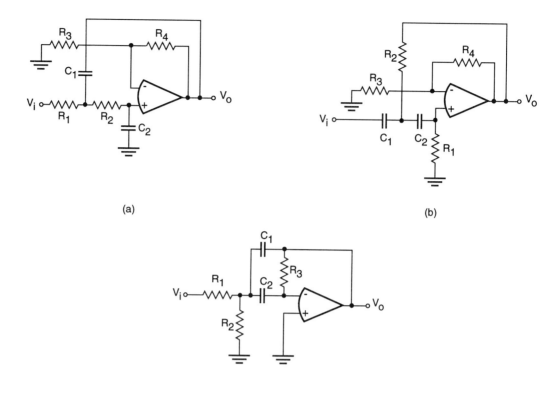

Fig. 6.54 Active filter schematics (a) low pass (b) high pass (c) multiple-feedback band pass

of Fig. 6.55, in which ϕ_1 and ϕ_2 are electronic switches, in practice MOSFET pass transistors (Fig. 5.8). Assume $V_1 > V_2$. C will initially be charged to either V_1 or V_2, depending on wheter ϕ_1 or ϕ_2 is closed first after switch-on. From then on the action repeats each cycle. Suppose C is charged to V_2. The next ϕ_1 closure will charge C to V_1 by a flow of charge into C along AB. Similarly there will be a flow of charge out of C along BC i.e. in the same direction as AB, during the following closure of ϕ_2. The result is a net flow of charge $C(V_1 - V_2)$ in time T_c, and the average current is evidently

$$I_{av} = \frac{C(V_1 - V_2)}{T_c} = f_c C(V_1 - V_2) \qquad (6.100)$$

where f_c is the clock frequency. Hence C is equivalent to a resistor R_{eq} given by

$$R_{eq} = \frac{1}{f_c C} \qquad (6.101)$$

Alternative clocking with true and inverse pulses ϕ and $\overline{\phi}$, shown in Fig.6.55(c), achieves the same action provided the overlap during the changeover can be ignored. The expression in eqn (6.101) can be deduced from the formula for the dynamic input power of a CMOS inverter, P_d

Analogue signal processing   229

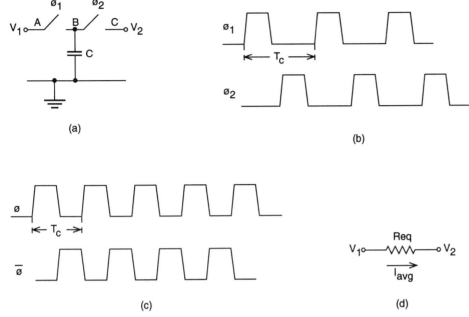

Fig. 6.55 Switched capacitor realization of resistor (a) capacitor switch (b) two-phase clock waveforms (c) single-phase clock waveforms (d) equivalent resistor

in eqns (5.11) and (5.16): V_2 in this case is zero. The dynamic power $C_L V_{dd}^2 f$ leads immediately to an equivalent resistance $1/fC_L$, which is identical to eqn (6.101). Various alternative resistor realizations are shown in Fig. 6.56.

Example 6.7

Verify the expressions for R_{eq} in Figs. 6.56(b) to (d).

The left plate of C in Fig. 6.56(b) is, like its lower plate in Fig. 6.56(a) held at a fixed potential, in this case V_1. The switching action transfers a charge $C(V_1 - V_2)$ from V_1 to V_2 during each cycle, and R_{eq} has the value obtained for Fig. 6.56(a) in eqn (6.101).

C_1 and C_2 in Fig.6.56(c) have one of their plates at a fixed potential, V_1 for C_1, zero for C_2. The junction of the other plates is switched from V_1 to V_2 and the charge transfer per cycle is $(C_1 + C_2)(V_1 - V_2)$; the result follows.

Finally the voltage difference across C in Fig. 6.56(d) is changed from $(V_2 - V_1)$ to $(V_1 - V_2)$ each time ϕ_1 is closed, and from $(V_1 - V_2)$ to $(V_2 - V_1)$ each time ϕ_2 is closed. The total charge transfer per cycle is $2 \times 2(V_1 - V_2)C$, giving $R_{eq} = 1/4f_c C$.

The outstanding advantage of an SC circuit is the tremendous saving in silicon estate, because a very large resistance can be realized with a small capacitance and a modest clock frequency. Monolithic resistors occupy large areas even for moderate values, witness Table 3.1, which also highlights another weakness, their poor temperature coefficient. The SC realization of resistors amounts to a dramatic improvement in both

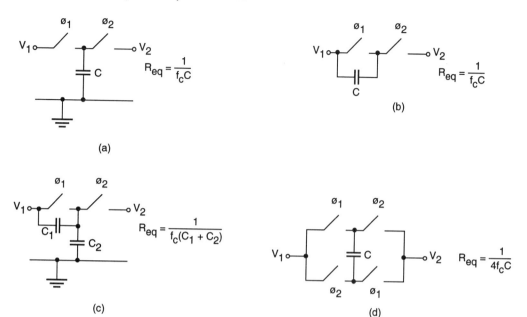

Fig. 6.56 Switched capacitor resistor realizations (a) parallel capacitor (b) series capacitor (c) series-parallel combination (d) bilinear

respects. A simple example will drive home the saving in silicon estate. Take the case of the RC low pass filter in Fig. 6.57(a), with a 3 dB bandwidth of 50 Hz and implemented by the SC filter in Fig. 6.57(b). C_s and C_d are the switched and discrete capacitors, and C_d is made equal to 15 pF, a reasonable size which does not demand an unduly large silicon area. Using eqn (6.101)

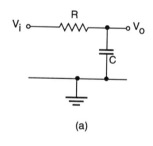

$$RC = \frac{C_d}{f_c C_s} = \frac{1}{2\pi f} = 3.18 \times 10^{-3} \text{ s} \qquad (6.102)$$

and for $C_d = 15$ pF, $f_c = 1$ kHz, C_s comes to 4.7 pF. Were the filter composed of a discrete resistor and C_d (Fig. 6.57(a)), R would come to 212 MΩ!

In SC technology the preferred capacitance range is (1–10) pF but this can be extended to ~ (0.1–100) pF. The SC filter greatly improves the accuracy of a passive filter. The ratio C_d/C_s in eqn (6.102) can be very closely controlled, to an accuracy approaching 0.1 per cent. Differentiating eqn (6.102) and putting τ for RC

$$\frac{d\tau}{\tau} = \frac{dC_d}{C_d} - \frac{dC_s}{C_s} \qquad (6.103)$$

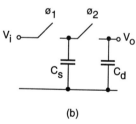

Fig. 6.57 Low pass filter (a) passive (b) switched capacitor

The temperature coefficients of C_d and C_s are virtually identical and the variation of τ with temperature is an order of magnitude or more smaller than for a passive filter.

The silicon estate taken up by capacitors will now be estimated for a 0.8 μm geometry. Putting ε_{ox} and t_{ox} in eqn (5.61) at 3.45×10^{-2} fF/μm (Table 5.17) and 200×10^{-4} μm (Appendix, Chapter 5, SIM4A simulation), the capacitance per unit area works out to be 1.73 fF/μm², demanding 578 μm² and 5780 μm² for capacitances of 1 pF and 10 pF

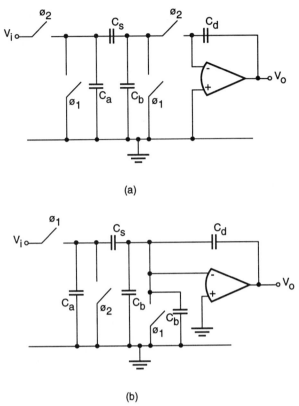

Fig. 6.58 Stray-insensitive switched capacitor integrator (a) inverting (b) non-inverting

respectively. These are economical areas, bearing in mind that the typical area of a VLSI die is 10^8 μm².

The derivation of the resistor realization in eqn (6.101) is based on the assumption of V_1 and V_2 being in the steady state, whereas in the majority of cases they are time-varying functions. The range of validity for eqn (6.101) is assessed by a detailed analysis of the sampling mechanism. The result is that eqn (6.101) holds for frequencies up to $\sim 0.02 f_c$, but becomes increasingly inaccurate at higher frequencies (Geiger et al. 1990, p.702). The upper limit of f_c is governed by the response of the op amps and MOSFET switches. SC circuits, whether used for filters or data converters, are linked to op amps. General purpose op amps have a GBP of about 1 MHz, giving a modest maximum frequency of 20 kHz for f_c. The speed is greatly increased by faster op amps, which achieve GBPs of up to ~ 700 MHz (Table 6.4). The lower limit of f_c is set by the unwanted discharge of switched capacitors, caused by the leakage currents of the MOSFETs and the input currents of op amps. It is in the range (50–100) Hz.

The inevitable parasitic capacitances at each circuit node reduce the accuracy of SC circuits. This effect can be eliminated by stray-insensitive circuits like the SC integrator in Fig. 6.58, where C_a and C_b represent parasitic capacitances. C_a is charged and C_b discharged by ϕ_1, the charge of C_s transferred during that phase is unaffected by C_a or C_b. This

Table 6.17 Active filters

Type	Mode	Function	Maximum order	Frequency range (kHz)
AF151	Dual State-variable	Universal	4th	0.1–10
LM40	SC low pass	Butterworth	4th	0.1–40
LM60	SC low pass	Butterworth	6th	0.1–30
LMF100	SC Dual	Universal	4th	0.1–40
LMF120/121	SC Biquad Mask programmable	Universal	12th	0.1–100
LMF90	SC Elliptic	Notch	4th	0.1–30

(Courtesy of National Semiconductor)
All types, with the exception of the AF151, are SC filters.

section concludes with a brief performance outline of some standard active filters given in Table 6.17.

All types, with the exception of the AF151, are SC filters. The universal filters have an architecture which allows them to realize just about any filter response. The mask-programmable LMF 120/121 is an interesting variation. It has the versatility of the other universal filters in Table 6.17 and is customized with the aid of highly automated design to meet a user's specific requirement.

6.5.3 ADCs

The ideal converter characteristic in Fig. 6.59 applies to ADCs and DACs alike. It is shown for a very modest resolution of 3 bits in order to illustrate the action. IC converters generally have resolutions of 8 bits or more. The accuracy of n-bit conversion is, ignoring noise and non-linearities, $1/2^n$ signifying that a 10-bit converter, to take an example, has a resolution of about 0.1 per cent. The conversion process can be formulated by

$$V = K_v V_{REF}(a_1 2^{-1} + a_2 2^{-2} + \cdots a_n 2^{-n}) \quad (6.104)$$

where V equals V_i, the normalized input voltage for an ADC, or V_o, the normalized output voltage for a DAC. V_{REF} is a reference voltage, K_v a scaling factor (frequently set to unity) and a_1 to a_n are the digital output (ADC) or digital input (DAC) coefficients (either 1 or 0). Each converter has specified values for K_v and V_{REF}, and $K_v V_{REF} \simeq V(max)$, the maximum input (ADC) or output (DAC). Re-arranging eqn (6.104) to represent DAC action

$$(a_1 2^{-1} + a_2 2^{-2} + \cdots a_n 2^{-n}) = \frac{V_o}{K_v V_{REF}} \quad (6.105)$$

which expresses the conversion of the digital input into an analogue output voltage $V_o/(K_v V_{REF})$.

The quantization error voltage v_n is, to a first approximation expressed by

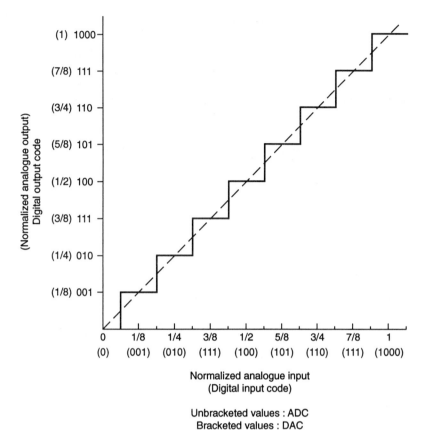

Fig. 6.59 Ideal data converter transfer characteristic (a) characteristic (b) quantization noise

$$v_n \simeq \pm \frac{1}{2}\frac{K_v V_{REF}}{2^n} = \pm \frac{K_v V_{REF}}{2^{n+1}} \quad (6.106)$$

Equation (6.106) is an approximation because it ignores the nature of the noise spectrum by equating v_n to the amplitude of (1/2) LSB. The signal-to-noise ratio (SNR), with noise being confined to quantization noise unless stated otherwise, is in accordance with eqns (6.104) and (6.106)

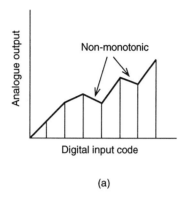

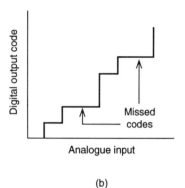

Fig. 6.60 Excessive differential non-linearity
(a) non-monotonicity (b) missing codes

$$\text{SNR} = \left(\frac{2^{n+1}}{K_v V_{REF}}\right) V \qquad (6.107)$$

According to eqn (6.107), the SNR has maximum and minimum vales of 2^{n+1} and 2 respectively. The influence of quantization noise on conversion and its nature are illustrated in Fig. 6.59. The dotted line in Fig. 6.59(a) shows the ideal conversion characteristic, the solid plot allows for the quantization uncertainty. The average of the sawtooth noise signal in Fig. 6.59(b) is zero. If it is taken to be white noise with an equal probability of lying anywhere in the band $\pm(1/2)V_n$, its root mean square value $\bar{v}_n$ becomes (Candy and Temes 1992, p.2)

$$\bar{v}_n = \frac{K_v V_{REF}}{2^{n+1}\sqrt{12}} \qquad (6.108)$$

The full significance of eqn (6.108) emerges later in connection with oversampling (Section 6.5.6). Increasing n decreases quantization noise and improves the accuracy of conversion. The limit to increasing resolution is governed by the constraint of obtaining adequate linearity.

Errors take a number of forms. Differential non-linearity, the error for an incremental input change, must not exceed one LSB. At that level, the incremental change can range from 0 to 2 LSB in place of the ideal 1 LSB. A higher differential non-linearity leads to the malfunctions of non-monotonicity (DAC) or missing codes (ADC), shown in Fig. 6.60. Such behaviour can cause havoc in control applications.

Integral non-linearity, gain error, and offset error, all illustrated in Fig. 6.61, are other conversion errors. Integral non-linearity is expressed in terms of the maximum deviation from the ideal, and amounts to about two LSBs in Fig. 6.61(a). Gain and offset errors are less serious and can easily be corrected.

Analogue-to-digital conversion is either static (non-sampling) or sampling. In a static ADC the entire signal is converted in a single process, and must be either dc or vary slowly with time for the conversion to be meaningful. Such signals arise in the measurement of dc voltages and currents, in biomedical applications etc. A far larger number of applications have signals which change rapidly with time and which can only be converted by sampling. The signal is sampled at a frequency f_s, and t_s (equal to $1/f_s$) is the interval between successive samples in Fig. 6.62. The conversion of the analogue signal into a digital code expressing the magnitude of the samples A to N in Fig. 6.62 is a form of pulse amplitude modulation (PAM). The signal is sampled by means of a sample-and-hold (S/H) amplifier, the principle of which is illustrated in Fig. 6.63. During the sample interval S (ϕ closed) V_o follows V_s. In the hold-interval H (ϕ open), the signal is stored across C_H and converted. The constancy of the signal in the hold interval is very good in practice. The droop in V_c is ~ 1 μV/μs in standard S/H amplifiers. The acquisition time is the time taken by the amplifier to track the signal after being switched from 'hold' to 'sample'. It is a function of the amplifier's dynamic response and is allowed for by making the S interval in Fig. 6.63(b) sufficiently long. (Zero acquisition time has been assumed in Fig. 6.63(b) in order to simplify the presentation). The aperture time t_{ap} in Fig. 6.63(c) is the interval between the receipt of the hold command

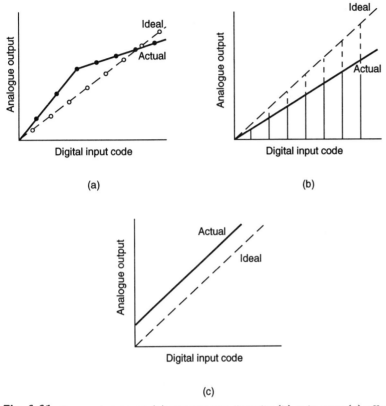

Fig. 6.61 Conversion errors (a) integral non-linearity (b) gain error (c) offset error

and the commencement of the hold action. It is largely due to the delays through the (electronic) switch ϕ and the circuit which drives it, and is in the range of $\sim$(20–60) ns. The consequence of t_{ap} is that the S/H amplifier will hold the signal at X', not at X. That in itself would not matter were t_{ap} constant throughout the conversion process. An error arises because of the uncertainty in t_{ap}, the aperture jitter δt_{ap}, which is typically 0.5 ns.

δt_{ap} imposes a limit on the signal bandwidth. Its maximum must be less than the amplitude of one LSB for correct resolution. Consider a sinusoidal input $V_m \sin w_m t$. $(\partial V_i/\partial t)_{max}$ is, in accordance with eqn (6.63), $2\pi V_m f_m$. The amplitude change $\delta V_i(max)$ caused by δt_{ap} is given by

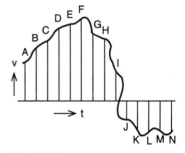

Fig. 6.62 Sampling of analogue signal

$$\delta V_{i(max)} = \delta t_{ap} \left(\frac{\partial V_i}{\partial t}\right)_{max}$$

$$= 2\pi V_m f_m \delta t_{ap} \qquad (6.109)$$

The magnitude of the LSB is $V_m/2^n$, giving the condition

$$2\pi V_m f_m \delta t_{ap} < \frac{V_m}{2^n} \qquad (6.110)$$

which results in

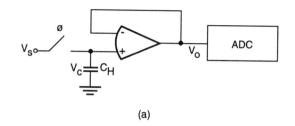

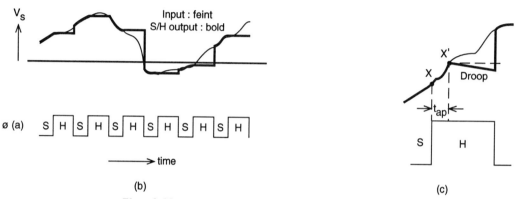

Fig. 6.63 Sample and hold action (a) circuit schematic (b) waveform (c) aperture time

$$f_m < \frac{1}{2^{n+1}\pi \delta t_{ap}} \qquad (6.111)$$

For a 12-bit ADC and a typical aperture jitter of 0.5 ns, f_m comes to 77.7 kHz. Although the approach taken applies to a sinusoid, it is a good guide to the limitation on signal bandwidth in general.

The analogue input can be reconstructed from the digital output containing the PAM information shown in Fig. 6.62 in accordance with Nyquist's sampling theorem, which stipulates that

$$f_s \geq 2f_m \qquad (6.112)$$

A real signal includes a spectrum of frequencies. The filtering of an analogue signal will not give a precise cut off completely eliminating all frequencies above f_m. All the same, if the spectral content above f_m, now interpreted to be the 3 dB bandwidth of the filtered signal, can be made very small, the analogue signal can be reconstructed from the digital data to very nearly its original form. The accuracy of reproduction is evidently increased by *oversampling* with f_s in excess of $2f_m$, which is called the Nyquist frequency. A modest degree of oversampling is common. More drastic oversampling is a feature of sigma–delta conversion, which is described in Section 6.5.6. The sampling process brings about *aliasing*, the production of undesired frequencies within the passband $0 < f < f_m$. If, for example, f_s and f_m are 4 kHz and 3 kHz respectively, an alias of 1 kHz will appear within the passband. Satisfying eqn (6.112) does not entirely remove the effect of aliasing, because the signal spectrum extends, albeit with small amplitude, beyond f_m. Anti-aliasing filters reduce alias levels to negligible proportions.

The last phenomenon considered is a fundamental feature in converting a signal from a frequency to a time domain and vice versa. The conversion leads to an amplitude error in the form

$$H(f) = \frac{\sin(\pi f/f_s)}{\pi f/f_s} \qquad (6.113)$$

$H(f)$ represents the magnitude at f in the frequency spectrum and f_s is the sampling rate, as before.

Equation (6.113) contains the sinc function, which is defined by

$$\mathrm{sinc}\,x \equiv \frac{\sin x}{x} \qquad (6.114)$$

It can be said that 'generally, a rectangular pulse in the time or frequency domain deforms into a sinc function in the other domain' (Lynn and Fuerst 1990). The sinc function occurs again and again in signal processing. One specific example is the *exact* frequency response $H(jf)$ of the non-inverting SC integrator in Fig. 6.58, which can be expressed by (Franco 1988, p.567; Allen and Holberg 1987, p.559)

$$H(jf) \propto \frac{1}{\mathrm{sinc}(\pi f/f_c)} \qquad (6.115)$$

The sinc$x$ or *roll-off*, as it is sometimes called may, depending on the signal and sampling frequencies involved, turn out to be negligibly small. If not, it is corrected either by digital filtering in the DSP or by analogue filtering following the DAC. The latter form of correction is frequently incorporated in a data interface or a companding encoder/decoder (codec) for voice frequency. The sinc$x$ error is given in Table 6.18, which holds for voice-band transmission with the standard values of 8 kHz and 3.4 kHz for f_s and f_m respectively. Oversampling by a factor of 3 reduces the error to 3 per cent. Correction filters reduce it to about 1 per cent over the entire pass band in voice-band codecs.

Table 6.18 Sinc$x$ roll-off $f_m = 3.4$ kHz ($x = \pi f_m/f_s$)

f_s (kHz)	Sinc$x$	Sinc$x$ (dB)
8	0.73	−2.73
16	0.93	−0.63
24	0.97	−0.26

This brief introduction to general aspects of data conversion is now followed by an outline of ADC circuit techniques. Commencing with the S/H amplifier, this element is usually incorporated within the ADC, although separate ICs for that purpose are available. The customary arrangement is a buffer voltage follower driving the sample-and-hold amplifier feeding another buffer follower with a storage capacitor connection to its (+) terminal (Fig. 6.63(a)); C_H is either an internal or an external capacitor. Its value is a compromise between fast acquisition time and large amplifier bandwidth (low C_H), and low droop rate combined with a small sample-hold offset (high C_H). Otherwise the S/H amplifier design follows op amp practice. Bipolar FET technology is used sometimes to have the advantages of low offset voltage and wide bandwidth. A JFET tends to be preferred to a MOSFET for the 'hold' stage, because it is less noisy.

ADC circuits are made with bipolar, BiMOS, or CMOS technologies; the latter include SC techniques. The illustrations of ADC circuit

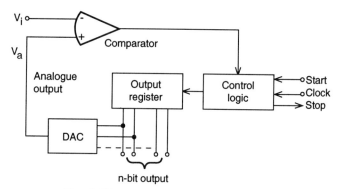

Fig. 6.64 Successive approximation ADC

practice in this, and DAC circuit practice in the next section represent the mainstream of circuit techniques for mature products. The successive approximation ADC in Fig. 6.64 owes its great popularity to the simple fast conversion process. An unexpected feature is the inclusion of a DAC. Conversion commences with the MSB of the output register set to 1, and all other bits set to 0. The DAC converts the register output into an analogue voltage, in this case equal to $\sim V_{max}/2$, where V_{max} is the maximum output with all bits of the register set to 1 (eqn (6.104)). Depending on whether V_i is greater or less than V_a, the MSB setting remains at 1 or will be changed back to 0. The next cycle consists of setting the second MSB to 1 and repeating the comparison. The output register will contain the result of the conversion after n cycles.

An entirely different approach is taken in the dual-slope integrating ADC shown in Fig. 6.65. Action commences with integrating V_i for a fixed time T_1 (S_1 closed), set by the maximum count N_1 of the gated counter supplied with a clock at f_c

$$T_1 = N_1 f_c \qquad (6.116)$$

The overflow of the counter causes the control logic to change the integrator input from V_i to one of two reference voltages $\pm V_{REF}$, chosen to have the opposite polarity of V_i. Integration takes place in the opposite direction until V_o has reached its original level. The time T is related to the count N by

$$T = N f_c \qquad (6.117)$$

$$V_o = \frac{V_i T_t}{RC} = \frac{V_{REF} T}{RC} \qquad (6.118)$$

Combining eqns (6.116) to (6.118)

$$V_i = V_{REF} \frac{N}{N_1} \qquad (6.119)$$

Equation (6.119) shows that V_i is directly proportional to N and that R, C, and f_c have been eliminated. The voltage follower buffer at the input is not absolutely essential but highly desirable, and its presence in standard ADCs can be taken for granted. V_{TH} is a voltage controlled by

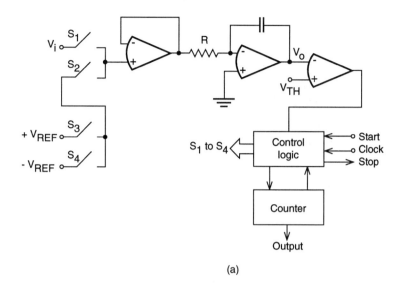

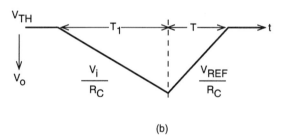

Fig. 6.65 Dual slope ADC (a) schematic (b) integrator output

an auto-zero correction circuit, which ensures zero ADC output for zero input by correcting for comparator and op amp input offset errors. The ADC can be made to reject harmonics of dc voltages derived from an ac supply at f_{ac} by choosing T_1 equal to $1/f_{ac}$. The conversion rate to achieve harmonics rejection is consequently very low; with a maximum time of $2/f_{ac}$ (T being equal to T_1), it comes to (17–20) ms at frequencies of 50 to 60 Hz. Dual slope integrators are capable of very high resolution, but have a low conversion rate.

The *flash* (or parallel) ADC, a 3-bit schematic of which is shown in Fig. 6.66, is at the opposite end of the spectrum as regards speed. The input is applied to all comparators, which have linearly scaled values of V_{REF} applied to the other input terminals. The encoder converts the non-standard output into standard binary code. The action is very fast with conversion rates well in excess of 100 MHz for 6-bit and 8-bit converters. The accuracy of conversion depends strongly on the matching of the resistors in the chain fed by V_{REF}. The upper limit of resolution is around 8-bits. Higher resolution at the cost of reduced speed is achieved by the *half-flash* converter, a combination of two flash converters, a subtractor, and a DAC (Fig. 6.67). The reduced speed is still superior to what can be achieved with a successive-approximation ADC. The output

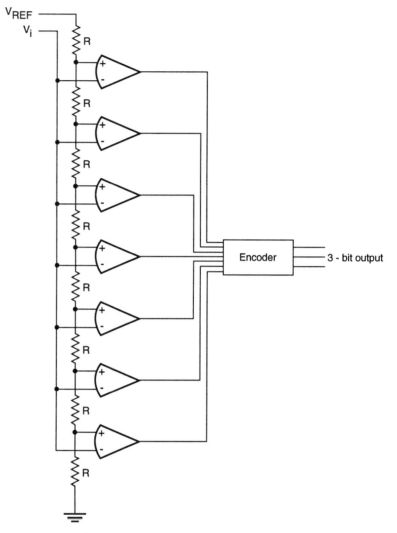

Fig. 6.66 Flash ADC

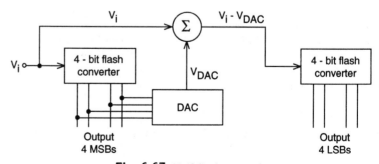

Fig. 6.67 Half flash converter

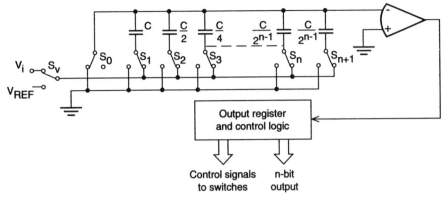

Fig. 6.68 Charge-scaling successive approximation ADC

of the first flash converter, which provides the four MSBs, is changed by the DAC into an analogue voltage $V_{DAC} \simeq V_i$. The subtractor, a circuit similar to the adder in Fig. 6.6 combined with a unity gain inverter, gives the residue $(V_i - V_{DAC})$, resolved into the four LSBs by the second flash converter. The 4-bit DAC must be accurate to at least 8 bits in order to give 8-bit ADC resolution.

Switched-capacitor (SC) converters are, like SC filters, firmly established in MOS technology. The advantages of SC resistor realizations have been described in Section 6.5.2. A charge-scaling (charge-balancing, alternatively charge-distribution) ADC is shown in Fig. 6.68 (Franco 1988, pp.588-9). It is paralleled by a similar DAC to be described in Section 6.5.4. The operation consists of three cycles. During the first cycle all switches occupy the positions shown in the diagram. S_o grounds the top plates of the capacitor bank and S_v applies V_i to all the bottom plates, charging the entire capacitance bank to V_i. During the next cycle, S_o is first opened and switches S_1 to S_{n+1} are switched to ground, causing the common top plates and hence the comparator input to equal $-V_i$. S_o and S_{n+1} remain in these positions during the final cycle, when S_v is connected to V_{REF}, which must have the same polarity as V_i. Switches S_1 and S_n are switched sequentially, starting with S_1, from ground to V_{REF}. If the comparator output changes state, the switch is returned to ground, if not its connection to V_{REF} is retained. Considering the action in terms of $S_m (m \leq n)$ switching, S_m changes the comparator input voltage V_{COMP} by an amount $V_{REF}(C/2^{m-1})/C_t$ where C_t is the total capacitance of the bank. C_t equals $2C$, and the change in V_{COMP} is therefore $V_{REF}/2^m$. The process is one of successive approximation and the final state of the switches gives the n-bit digital output.

Example 6.8

Prove the statement that switching S_m from ground to V_{REF} increases V_{COMP} by $V_{REF}/2^m$.

The increase δV_{COMP}, is evaluated with the aid of the equivalent circuit in Fig. 6.69. The total capacitance of the bank is $2C$ exactly, thanks to the additional capacitance $C/2^{n-1}$, equal to the LSB capacitance. δV_{COMP} is given by

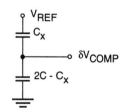

Fig. 6.69 Switching action for ADC (Fig. 6.68)

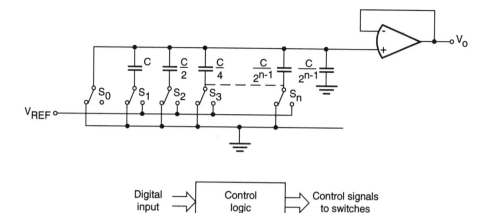

Fig. 6.70 Charge scaling DAC

$$\delta V_{comp} = \frac{C_x}{(2C - C_x) + C_x} V_{REF} = \frac{C_x}{2C} V_{REF} \quad (6.120)$$

C_x, switched by S_m, equals $C/2^{m-1}$, and substitution in eqn (6.120) leads to $\delta V_{COMP} = V_{REF}/2^m$.

Capacitance ratios can be accurate to about 0.1 per cent which makes for a 10-bit resolution. However that accuracy applies for ratios around unity, and becomes less for increasing ratios. Another factor to be considered is the large silicon estate taken up by the capacitor bank. A combination of voltage scaling for the most significant, combined with charge scaling for the least significant bits, is customary practice and greatly increases the resolution possible with this technique (Allen and Holberg 1987, p.562–4).

6.5.4 DACs

The description of DAC circuit techniques begins where the previous section left off. The charge-scaling DAC in Fig. 6.70 and the ADC in Fig. 6.68 have identical capacitor banks and Example 6.8 holds for both of them. All switches in Fig. 6.70 are initially in the positions shown and the capacitor bank is completely discharged. S_o is then opened and switches S_1 to S_n are set in accordance with the digital input to give the required output.

In DACs with resistive techniques (Fig. 6.71) a current I_o is generated by the addition of the binary-weighted currents via switches S_1 to S_n, and converted into V_o by the type of op amp connection shown and illustrated previously in Fig. 6.11. This leads to

$$V_o = -I_o R_f = -\frac{V_{REF} R_f}{R}(a_1 2^{-1} + a_2 2^{-2} + \cdots a_n 2^{-n}) \quad (6.121)$$

an expression in the form of eqn (6.104). The coefficients a_1 to a_n are 1 or 0 according to the setting of the switches. The reference voltage is either produced within the DAC or supplied externally. The output of the DAC is the product of a scaled reference voltage and a digital input code. The generic term *multiplying* DAC is reserved for converters which

Analogue signal processing 243

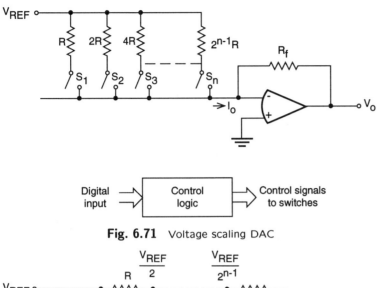

Fig. 6.71 Voltage scaling DAC

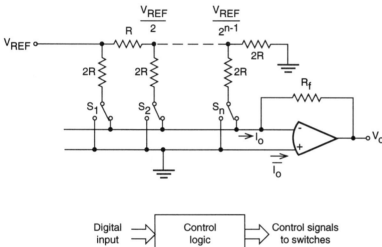

Fig. 6.72 Voltage scaling R–2R ladder DAC

operate with an external reference, giving the user control of both the product terms which make up the analogue output. A multiplying DAC is in effect a digitally adjustable potentiometer and can function as a gain control. The external reference voltage can be either dc or ac. If it is ac, the DAC can, for example, function as a digitally controlled audio gain adjustment.

The resolution obtainable is determined predominantly by the accuracy of the resistors and the resistance of the switches. Resistor characteristics are contained in Section 3.6. The accuracy of monolithic resistors is inadequate except for low resolution. DACs are frequently hybrids which contain a monolithic die and a thin film resistor assembly in one package. The demand for accuracy is satisfied best with the R–2R ladder network of the voltage scaling DAC in Fig. 6.72. The greatest accuracy for a resistance ratio is obtained when the ratio itself is close to unity. The total current $(I_o + \bar{I}_o)$ equals $(V_{REF}/R)(1 - 2^{-n})$ regardless of the switch settings. Hence the currents are complementary (in binary weighting),

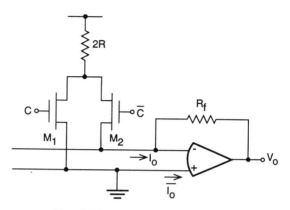

Fig. 6.73 MOS switch (Fig. 6.72)

which accounts for the designations I_o and $\overline{I}_o$. The electronic switches consist of nMOSFETs with scaled geometries (Fig. 6.73). The transistor dimensions are chosen to give on-resistances with binary weights. Typical values are 20 Ω for S_1, 40 Ω for S_2, 80 Ω for S_3 etc. If I_{S1} equals 0.5 mA, I_{S2} will be 0.25 mA, I_{S3} 0.125 mA etc. The IR products of the switches are identical (10 mV) and the desired proportionality of currents in the branches has been preserved with an effective reference voltage ($V_{REF} - 10$ mV) in place of V_{REF}, a reduction which is easily compensated for by adjusting R_f. The customary practice is to incorporate scaled MOSFET switches for the first four to six MSBs; the IR drops across the switches are insignificant at the lower currents.

The circuit techniques for bipolar and MOS DACs are similar. Bipolar technology has the advantage of faster switching, and the BJT is a current source with a high output resistance, which allows immediate translation of the output current into an output voltage by means of a resistive collector termination. The current-to-voltage conversion by means of an op amp is however preferred as a rule.

Two schematics of DACs based on R–2R ladder networks are reproduced in Fig. 6.74. The emitter-base voltages of all transistors must be identical in order to maintain accurate current ratios, and this is achieved by scaling the emitter areas as shown in Fig. 6.74(a). The current *densities* of all the transistors and hence their emitter-base voltages are identical. The scaling of transistors is avoided with the circuit in Fig. 6.74(b). All transistors supply identical currents which are scaled by the R–2R ladder in their collector paths.

6.5.5 Specification and performance

The comments about the plethora of op amps available, made in Section 6.2.4, are equally relevant here. The user is confronted with a bewildering choice, and has to make some effort in order to decide on the best selection for a given application. The two foremost key parameters are speed and resolution. Speed is expressed in terms of conversion time or throughput rate for an ADC, and settling (conversion) time or update rate for a DAC. It can be argued that settling time is simply the reciprocal of throughput rate, but examination of data when both these parameters are supplied shows that the latter is sometimes slightly less than might be expected from the reciprocal relationship. This is largely

Analogue signal processing

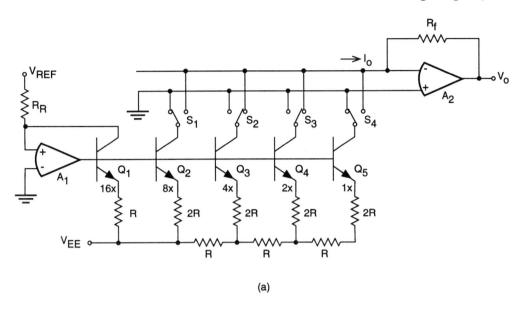

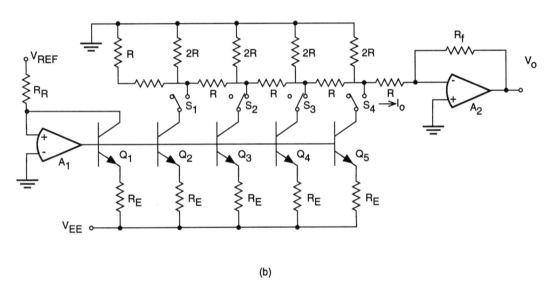

Fig. 6.74 4-bit bipolar DAC (a) scaled emitters (b) identical transistors

because an allowance has to be made for set-up time.

The lowest resolution of commercial converters is usually 8 bits, although there are a few 6-bit converters for ultrafast operation. For very many applications, data converters with moderate speeds are, like amplifiers with moderate bandwidths, acceptable. The characteristics of standard converters bear this out: the number of elements with moderate speeds far outweighs the faster types. The advances in technology have not been directed only to further speed and resolution. There has been more effort to incorporate other features like compatibility with microprocessors, multi-channel inputs and multiplexing for channel se-

Table 6.19 Sampling ADC characteristics

Type	Resolution (bits)	Update rate (KSPS)	S-H amp bandwidth (kHz)	Max. analogue input (V)
AD7821	8	1000	100	(± 2.5)
AD7569/AD7669[†]	8	400	200	(± 2.5)
AD7579	10	50	25	5 or ± 2.5
AD779	14	128	1000	10 or ± 5
AD7884	16	166	83	± 5

(Courtesy of Analog Devices)
Temperature ranges: C = Commercial, I = Industrial, M = Military
All the ADCs have TTL/CMOS outputs
[†] ADC characteristics of AD7569/AD7669 8-bit Analog I/O System
KSPS: Kilosamples per second

lection, self-calibration, and software-related facilities. Data converters are more frequently becoming parts of data interfaces and data acquisition systems which include these facilities.

Tables 6.19 to 6.23 list converters in order of resolution, and include chips at the spearhead of speed. Most types are produced in different versions which cater for various temperature ranges and packages. The standard temperature ranges have been discussed and defined in Section 6.2.3. Sample-and-hold amplifiers are usually included in sampling ADCs and this is the case for all the ADCs in Table 6.19. Two 8-bit ADCs have been included. The second of these is part of a complete 8-bit analogue I/O system on a single monolithic chip with BiCMOS technology, including a single (AD7569) or dual (AD7669) DAC output. Input and output ranges up to 1.25 V and 2.5 V, positive for a 5 V supply, and bipolar for ± 5 V supplies, may be programmed externally. The chip is interfaced via an 8-bit I/O port and standard microprocessor control lines.

Successive approximation is the most popular for lower, flash conversion the most widely used method for higher speeds. The combination of precision bipolar circuits with CMOS logic gives the best compromise of resolution with minimum power consumption. CMOS converters, available for resolution from 8 to 16 bits, are slower but considerably cheaper and more economic in power. The conversion rate decreases with resolution for a given technology and feature size. It is axiomatic that all data converters have a differential linearity of ± 1 LSB or better, signifying that there are no missing codes in an ADC or instances of non-monotonicity in a DAC. It becomes progressively difficult to achieve such linearity at high resolutions and over a wide temperature range. Self-calibrating techniques make it possible to obtain 16-bit and even higher bit ADCs with confidence. One well established technique for successive approximation ADCs is based on a calibration DAC which has a higher

Table 6.20 High-speed flash ADCs (All the converters have ECL outputs and are available for the C and M temperature ranges.)

Type	Resolution (bits)	Update rate (MSPS)	Supply voltages (V)	V_{REF} E= External I = Internal (V)	Max. analogue input (V)
AD9016	6	550	5, −5.2	±1, E	±1.000
AD9038	8	300	−5.2	−2, E	−2.000
AD9060	10	75	5, −5.2	±1.75, E	±1.750
AD9032	12	25	5, −5.2	±1.024, I	±1.024

MSPS: megasamples per second. (Courtesy of Analog devices)

Table 6.21 Static ADC characteristics

Type	Resolution (bits)	Conversion time (μs)	Temperature ranges	Mode S-A: successive approx. I: integration
AD7574	8	15	C, I, M	S-A
ADC-910	10	6	C, I, M	S-A
AD774B	12	8	C, I, M	S-A
AD1378	16	17	M	S-A
AD1170	18	1000	C	I

(Courtesy of Analog Devices)

resolution than the DAC integral to the conversion process (Geiger et al. 1990, pp.664–5). There are other methods of self calibration, backed by software support. Flash converters reach the highest throughput rates, and Table 6.20 contains a selection with 6-bit to 12-bit resolution. The characteristics of some static (non-sampling) ADCs are given in Table 6.21. The AD1170, the only converter operating in the integrating mode, differs in several other respects. It is a complete microcomputer-based measurement system, consisting of a charge-balancing integrating ADC, a single-chip microcomputer, and a CMOS controller chip, all in one package. It offers, unlike a dual-slope converter, one of several programmable integration times from 1 ms to 350 ms. Alternatively an arbitrary integration period may be set by a command on the interface bus. The ADC is fully auto-zeroed. All conversions are made to 22 bits and the resolution is programmable from 7 bits to 18 bits, the effective maximum due to measurement errors and noise. Self-calibration can be performed at any time with the aid of an external reference voltage.

Converter resolution is limited by quantization noise, and is clearly linked with the SNR. An expression relating SNR and resolution will now be derived using eqn (6.108). The peak magnitude of a noise voltage for a Gaussian distribution is three times its rms value (Franco 1988, pp.594–5, Sheingold 1986, pp.66–7). This relationship has been used previously in another context (see the comments below eqn (2.18)). The peak value

Table 6.22 DAC characteristics

Type	Resolution (bits)	Settling time (μs)	Update rate (MSPS)	Temperature range	Reference voltage	Supply voltage(s) (V)
DAC650	12		500	C	I	±15, 5, −5.2
AD9720	10		300	C, M	I	−5.2
AD9768	8		100	C, M	I	5, −5.2
DAC-10	10	0.085		C, M	E	±15
AD568	12	0.035		C, M	I	±15
AD7534	14	1.500		C, I, M	E	±15
DAC-16	16	0.500		I, M	E	5, −15

All types are Analog Devices elements, with the exception of the Burr-Brown DAC650.
Reference voltage: I= internal, E = external
(Courtesy of Burr-Brown Corporation and Analog Devices)

of the noise voltage, $\hat{v}_n$, is accordingly

$$\hat{v}_n = \frac{3K_v V_{REF}}{2^n \sqrt{12}} = \frac{K_v V_{REF}}{2^n \times 1.1547} \tag{6.122}$$

The SNR for maximum input ($K_V V_{REF}$) therefore comes to

$$\text{SNR} = 1.1547 \times 2^n = (1.25 + 6.02n) \text{ dB} \tag{6.123}$$

A different approach, taken in Sheingold (1986, p.542), is based on calculating the SNR for a sinusoidal input signal and the saw tooth quantization noise voltage shown in Fig. 6.59(b). That noise voltage has a peak amplitude of one LSB: $\hat{v}_n$ equals $K_V V_{REF}/2^n$. For a triangular pulse $\hat{v}_n/\bar{v}_n$ equals $\sqrt{3}$ and $\bar{v}_n$ comes to $K_V V_{REF}/(2^n\sqrt{3})$. The maximum rms input equals $K_V V_{REF}/\sqrt{2}$ leading to

$$\text{SNR} = 2^n \sqrt{\frac{3}{2}} = 1.2247 \times 2^n = (1.76 + 6.02n) \text{ dB} \tag{6.124}$$

Both approaches give very similar SNRs; the difference between them is only six per cent.

Data bearing on resolution sometimes includes one or more of the following: the signal-to-quantization noise ratio (S/N), which equals our SNR, the total harmonic distortion (THD), the intermodulation distortion (ID), and the signal-to-noise plus distortion ratio (S/(N + D)), which embraces all noise components (not just quantization noise) and distortion. Harmonic distortion is specifically quoted for the assessment of audio applications. The effective resolution allowing for all noise contributions and distortion cannot be deduced directly from the S/(N+D) ratio, because noise behaviour is complex. Transistor and resistor noise is largely independent of frequency and does not impact on resolution like quantization noise. For instance the AD7884 in Table 6.19 has a guaranteed resolution of 16 bits with no missing bits, making S/N equal to 97.5 dB according to eqn (6.123), whereas $S/(N+D)$ is specified to be 88 dB.

DAC characteristics are contained in Table 6.22. The first three DACs have ECL-compatible inputs and maximum output currents of (10–20) mA, capable of driving 50 Ω transmission lines. For the DAC650, the output is ±1 V (open circuit load) or ±20 mA (short-circuit termination). The ultrafast rate of the DAC650 is due to a combination of bipolar and GaAs technologies. The other converters have TTL/CMOS compatible inputs, and that also holds for the AD9721 which in other respects is identical to the AD9720, but has a reduced update rate of 100 MSPS. The maximum output currents of the DACs other than the first three are of the order (1–5) mA. The exact value is governed by the reference voltage and is consequently variable for the three multiplying DACs with an external V_{REF}.

An alternative to DACs with current output are DACs with voltage output, obtained by including the op amp and R_f in Figs. 6.71 to 6.74 within the chip. These, less numerous than current-output DACs, are considerably slower for a given resolution because of the op amp response, which reduces the speed of conversion by a faction between 5 and 10. The speed will of course also be reduced by an external op amp. However the user can choose a fast or ultrafast op amp—at extra cost— to minimize the loss in speed, and can also choose R_f for the desired output range.

6.5.6 Oversampling sigma–delta converters

Great strides have been made in digital signal processing. These advances have called for higher resolution in data converters. Attention has been drawn in Sections 6.5.3 and 6.5.4 to the limits imposed on resolution by noise, the accuracy of passive components (resistors and switched capacitors) and their temperature coefficients, and the balance of transistor, op amp, and comparator characteristics. The last decade has witnessed the emergence of sigma–delta (Σ-Δ) converters, which improve on the performance of conventional converters in a number of ways, and which have become firmly established. The methods of conversion are a mix of analogue and digital techniques. The digital aspects, pulse modulation and digital filtering, are outside the scope of this book. This does not preclude the inclusion of Σ-Δ conversion, but influences its treatment, which is in outline and in which mathematics has been kept to an absolute minimum. The brief introduction given here bristles with oversimplifications. It seeks to put forward qualitative concepts aimed at imparting a broad appreciation of the subject matter backed up by pointers to references for a deeper pursuit, if desired.

The crux of the conversion process, taking analogue-to-digital conversion first, is delta modulation, first proposed by Deloraine *et al.* (1946). In delta modulation the digital output does not represent the absolute value of the analogue input. Instead it is a one-bit quantum indicating the *difference* between successive samples of the input. The sampling frequency f_s is very much higher than the Nyquist frequency $2f_m$, now designated f_N, of eqn (6.112). The oversampling ratio OSR is defined by

$$\text{OSR} \equiv \frac{f_s}{f_N} \tag{6.125}$$

The principle of delta modulation is illustrated in Fig. 6.75. A differential increase is represented by 1, a decrease by 0. The integrator is effectively a one-bit DAC—sometimes named as such—whose output is subtracted from the analogue input $x(t)$. The difference is fed to the one-bit quantizer, a comparator which gives an output of 1 or 0, a positive or negative impulse according to the polarity of $x_e(t)$ (Fig. 6.75(c)). The integrator, driven by the impulse outputs of the quantizer, incre-

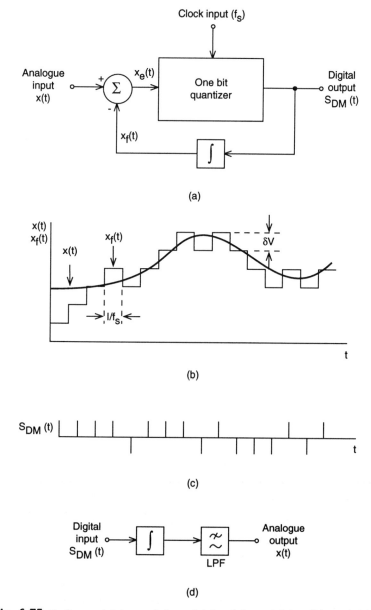

Fig. 6.75 Delta modulator and demodulator (a) modulator (b) modulator input waveform (c) modulator output waveform (d) demodulator

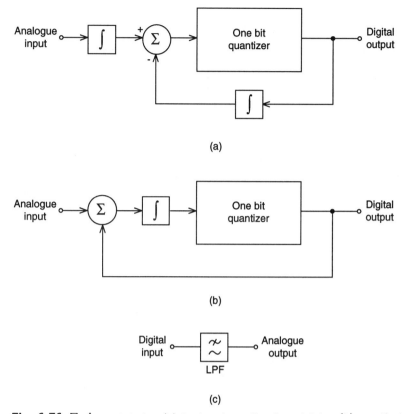

Fig. 6.76 Σ–Δ modulation (a) basic schematic of modulator (b) practical implementation of (a) (c) demodulator

ments or decrements $x_f(t)$ by δV accordingly for the next sample. The schematic for the delta demodulator is shown in Fig. 6.75(d). It consists of an integrator which reconstitutes an input like $S_{DM}(t)$ into a signal like $x_f(t)$, and is followed by a low pass filter for smoothing the output. Integration is a linear process, allowing the alternative of moving the integrator of the demodulator in Fig. 6.75(d) into the signal path of the modulator in Fig. 6.76(a). The functions of both integrators can be combined, leading to the practical implementations of a Σ–Δ modulator and demodulator in Figs. 6.76(b) and 6.76(c) (Park 1993; Marven and Ewers 1994, p.57). Σ–Δ modulation is a modification of delta modulation, proposal by Inose et al. (1962), and has only become a practical proposition with the arrival of VLSI. Σ–Δ modulation is inherently free from differential non-linearity and is more economic on hardware than delta modulation. It encodes the integral of the signal, not the signal itself as in delta modulation, before feeding it to the quantizer, thereby smoothing it out and making the encoding insensitive to its rate of change. An outstanding advantage of one-bit over n-bit conversion is that the need for relatively accurate components has been eliminated. Lastly the quantization noise is, in contrast to delta modulation, frequency dependent and this property is advantageous, allowing an improvement in the signal-to-noise ratio which will be explained presently. The analogue to

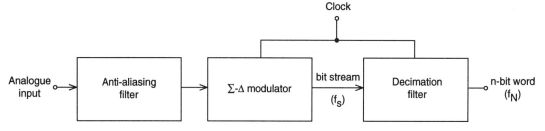

Fig. 6.77 Basic Σ–Δ ADC

digital conversion is completed by changing the single-bit word output of the modulator at f_s into a multiple n-bit word at f_N, the format of a traditional ADC output. The translation, called *decimation*, is made by means of a filtering process which downsamples the input from f_s to f_N and averages the bit-stream into n-bit words. The term decimation implies a reduction, but is in no way related to a decimal factor like 10. Decimation is almost universally carried out with digital filters, which take up less area than SC filters for the various signal processing steps in submicron CMOS (Friedman 1990). The decimation filtering, consisting of one of more filters, follows. The anti-aliasing filter precedes the Σ–Δ modulator (of the type shown in Fig. 6.76(b)) in the schematic of a basic Σ–Δ ADC contained in Fig. 6.77. The action is controlled by a high speed clock. The decimation filter fulfils the dual role of downsampling and reducing the quantization noise which, in Σ–Δ modulation, is spread over a frequency band $\pm f_s/2$ (Fig. 6.78), so that only a small proportion of the noise lies within the signal band. If this noise is white noise as in delta modulation, occupying the band $\pm f_s/2$, the noise power P_n in the signal band is given by (Candy and Temes 1992, p.3)

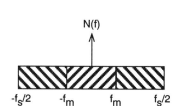

Fig. 6.78 Noise spectrum of Σ–Δ ADC

$$P_n = \frac{\bar{v}_n^2}{\text{OSR}} \quad (6.126)$$

where $\bar{v}_n$ is formulated in eqn (6.108). Equation (6.126) signifies that doubling f_s reduces the noise power by 3 dB and thereby increases the maximum resolution by half a bit. A four-fold increase in f_s is necessary to gain one bit in resolution and that is not easy. To raise the update rate of an ADC by a factor of four is a major undertaking. However for the first order Σ–Δ modulator in Fig. 6.76(b) (Candy and Temes 1992, p.3)

$$P_n = \frac{\bar{v}_n^2 \pi^2}{3(\text{OSR})^3} \quad (6.127)$$

which leads to a far greater reduction of noise with oversampling. Doubling f_s now reduces the noise power by 9 dB and provides 1.5 bits of extra resolution. The realization of an SNR close to the ideal expressed by eqn (6.127) is largely determined by the noise shaping of the decimation filter in an ADC, and the interpolation filter—to be described presently—in a DAC.

It is instructive to compare the conversion techniques of traditional and Σ–Δ ADCs and to enumerate the relative merits of the latter. A conventional ADC produces a pulse code modulation (PCM) output of n-bit words at f_N (or slightly higher), expressing the amplitudes of the

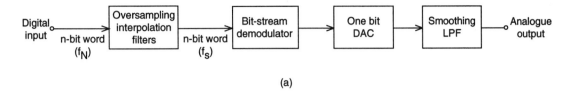

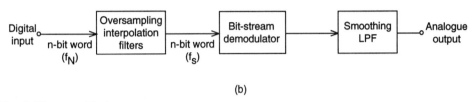

Fig. 6.79 Basic Σ–Δ DACs (a) PDM input (b) PWM input

sampled input at intervals of $1/f_N$. The Σ–Δ ADC produces the same output in a very different manner. The input is oversampled at f_s with $f_s \gg f_N$, rendering a sample-and-hold stage unnecessary. The modulator has a one-bit resolution. Its output gives the difference between successive samples, indicating only whether there has been an increase or decrease. Resolution has been traded for speed, and the technique requires nowhere near the accuracy of passive components (resistors and switched capacitors) and the matched transistor characteristics of traditional ADCs. This is a tremendous advantage for resolution of 16 bits or higher. Multi-bit resolution is obtained by passing the serial bit-stream through a decimation filter which downsamples the input and averages it into an n-bit word at f_N, n being chosen to give the highest resolution of which the converter is capable. Resolution increases with oversampling, which reduces signal noise more with a Σ–Δ than an ordinary ADC, and that makes the oversampled Σ–Δ converter capable of a higher resolution than a conventional ADC sampled at f_N. The specification of the anti-aliasing filter, which has to have a linear phase characteristic (constant group delay) in order to avoid distortion, is greatly relaxed, making for an easier design. However the Σ–Δ ADC requires much faster circuits and more complex processing.

The demodulators in Figs. 6.75(d) and 6.76(c) are for decoding a bit-stream. The other types to be considered are DACs which convert n-bit Nyquist words into an analogue signal. Two basic and very similar schematics of oversampled Σ–Δ DACs are shown in Fig. 6.79. The roles of the digital and analogue section in an ADC are reversed. The n-bit input at f_N is raised to f_s by an oversampling technique with *interpolation filters*, and is then changed into a bit-stream by a bit-stream demodulator, which is sketched in Fig. 6.80 and whose architecture is very similar to the Σ–Δ modulator in Fig. 6.76(b) (Marven and Ewers 1994, p.61). Oversampling is carried out by a repeated entry of each word into the interpolation filter, which inserts new words between the existing words by a process of linear interpolation. The oversampling ratio f_o/f_i where f_o and f_i are the output and input frequencies, is obtained by enter-

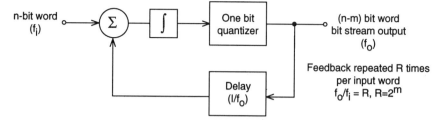

Fig. 6.80 Bit stream demodulator

ing each word (f_o/f_i) times. The filter adds $\{(f_o/f_i) - 1\}$ new words between each pair thus entered, giving an output an output rate of f_o. Interpolation is, like decimation, spread over two or more stages.

A direct conversion of an analogue signal with a moderate bandwidth into a one-bit stream is not possible, because the bit rate would be excessive, witness the following example. The standard Nyquist rate for compact disk (CD) audio signals is 44.1 kHz. This, for 16-bit words, gives a bit-stream rate of $2^{16} \times 44.1\text{kHz} = 2.9$ GHz. Such a rate is prohibitively high even for bipolar submicron VLSI, bearing in mind that the integrators in Figs 6.75 and 6.76 must be driven by inputs with rise and fall times very much less than the reciprocal of the bit rate.

The two established bit-stream modulations are pulse density modulation (PDM) and pulse width modulation (PWM), with PDM being the more popular of the two. In PDM pulses are modulated in both frequency and width, in PWM they are modulated in width at a constant rate. The output of the PDM bit-stream in Fig. 6.79(a) has to be fed to a 1-bit DAC, which is a simple switched capacitor circuit, for conversion into an analogue signal (Marven and Ewers 1994, p.62; Naus et al. 1987). For the PWM transmission in Fig. 6.79(b), the output of the bit-stream demodulator can be converted directly into an analogue signal by passing it through an LPF. The schematic of a bit-stream demodulator is shown in Fig. 6.80. For simple conversion with f_o equal to f_i and no change in resolution, the delay is set to $1/f_i$. Alternatively the bit-stream demodulator can be made to output the signal at a higher rate by setting the delay to $1/f_o$ and entering each input sample f_o/f_i times. This step of oversampling decreases the resolution from n to (n–m) bits in accordance with

$$\frac{f_o}{f_i} = 2^m \qquad (6.128)$$

As in A-to-D conversion, speed is traded for accuracy. The higher the output rate, the smaller is the signal noise and the simpler is the design of the LPF.

Interpolation is, like decimation, carried out with digital filters in the majority of cases, reflecting the general trend to maximize digital signal processing. Nevertheless analogue filters occupy an important place in signal processing, including Σ–Δ converters. Depending on the specification, SC filters can be faster, simpler, and less demanding on power than digital filters (Baher 1991). Digital and SC filters operate side by side, together with other SC circuits, in Σ–Δ converters and are likely to do so for the time being.

Table 6.23 Oversampled Σ–Δ ADCs

Type	Resolution (bits)	f_N (kHz)	OSR	Input bandwidth (kHz)	Temperature range	Typical applications
AD776[1]	16	100	64	50	I, M	Signal processing, audio.
	12	400	64	50		
AD1878/79[1]	16	44	64	20	−25 to 70 °C	Digital tape recorders. CD recording.
	18	44	64	20		
ADS1210[2]	up to 24	— see text —			I	DC and very low frequency measurements.

[1] Analog Devices [2] Burr-Brown
(Courtesy of Analog Devices, and Burr-Brown Corporation)

This section concludes with a brief reflection on the types of Σ–Δ converters on the market, their performance, and general issues relating to such converters. The converters have, until now, found their main outlet in the niche market of digital audio, catering for CD players, tape decks, etc. Other significant applications are voice telephony and dc or very low frequency measurements involving weigh scales, thermocouples, chromatograph, etc. Converters are usually application specific standard products (ASSPs) with standard values of f_N and fixed f_s; f_N equals 8 kHz and 44.1 kHz for voice telephony and the standard CD audio systems respectively. Alternatively they can function over a range of f_s and f_N at a fixed OSR by means of programmable filters. Digital and SC filters are programmable in various ways. One of these is the choice of the clock frequency, a specified multiple of f_s, supplied in this case externally. Another is software for programming the filter characteristics. Standard Σ–Δ converters are beginning to come on stream with a bigger variety of ADCs than DACs, an understandable situation because there are many applications in the field of measurements which require an ADC but not a DAC. The performance of some standard ADCs is summarized in Table 6.23. The technologies maximize on CMOS. The AD776 and AD1878/79 are BiCMOS, the ADS1210 is CMOS. The AD1878/AD1879 contains two dies in a single package, one (BiCMOS) containing the modulator and voltage reference, the other (CMOS) the digital decimation filters and the output interface. The ADS1210 is a second order Σ–Δ ADC, based on a differential switched capacitor architecture. A unique feature, called *turbo mode*, allows external control of f_s, which is normally at 20 kHz with a 10 MHz clock; f_s can be programmed to 40 kHz, 80 kHz, 160 kHz, and 320 kHz to improve the performance at higher frequencies. The effective resolution is listed together with the sample rate and the 3 dB bandwidth in Table 6.24. An increase in f_s increases the effective resolution at the cost of higher power consumption. The effective resolution at a data rate of 1000 Hz changes from 10 dB to 19 dB when f_s is raised from 20 kHz to 320 kHz.

Table 6.24 Performance of ADS1210 (System clock = 10 MHz, $f_s = 20$ kHz)

Data rate (Hz)	−3 dB frequency (Hz)	Effective resolution (bits)
10	2.62	22
30	7.86	21
60	15.72	20
100	26.20	18
250	65.50	15
1000	262.00	10

©Burr-Brown Corporation. Reproduced with the permission of Burr-Brown Corporation.

Two Σ–Δ DACs are included in this section; their schematics are shown in Fig. 6.81. Both are designed for CD players and digital audio tape recorders. Their performance is similar, but they differ in their structures. Both converters have 16-bit resolution, operate from a single 5 V supply, and have two output channels. The Philips TDA1311 and TDA1312 DACs (Fig. 6.81(a)) follow very closely the architecture described by Naus et al. (1987) and Philips (1992). The 16-bit input ($f_N = 44.1$ kHz) is oversampled to $4f_N$ in the first filter and by a factor 64 in the second filter, a 32X oversampling filter and a 2X oversampling sample-and-hold stage. The amplitude of the signal is increased by *dither*, a pseudo-random out-of-band signal added to the LSB in order to prevent audible idling of the noise shaper, a bit-stream demodulator which includes a limiter and a noise shaping filter. (Naus et al.; Marven and Ewers, 1994, pp.63–4). The increase in amplitude caused by the dither is accommodated by raising the word length to 17 bits. The bit-stream demodulator converts the 17-bit word at f_s (11.2896 MHz) into a bit-stream at the same rate for conversion into an analogue signal by the 1-bit DAC and the smoothing LPF.

The Burr-Brown PCM1710U, whose schematic is shown in Fig. 6.81(b), has an 8X oversampling interpolator at the input. It caters for 16-bit input words at f_N equal to 44.1 kHz. The bit-stream demodulator converts the input with further oversampling to either $32f_N$ (1.4112 MHz) or $48f_N$ (2.1168 MHz) depending on the choice of clock frequency. Its five-level quantizer, an innovative feature, gives a four-bit output, which is converted by a four-bit DAC into analogue form. A multi-level quantizer has been selected, because one-bit quantizers are prone to instability arising from loop feedback. The multi-level quantizer is stabilized by a phase compensation technique. It greatly reduces the effects of system clock jitter and rf interference.

The specification of the PCM1710U includes a dynamic range of 98 dB, an S/N ratio of 110 dB, and a total harmonic plus noise ($THD+N$) distortion 0.0025 per cent, corresponding to $S/(N+D)$ of 92 dB. The *dynamic range* is the effective resolution in dB, and the magnitude

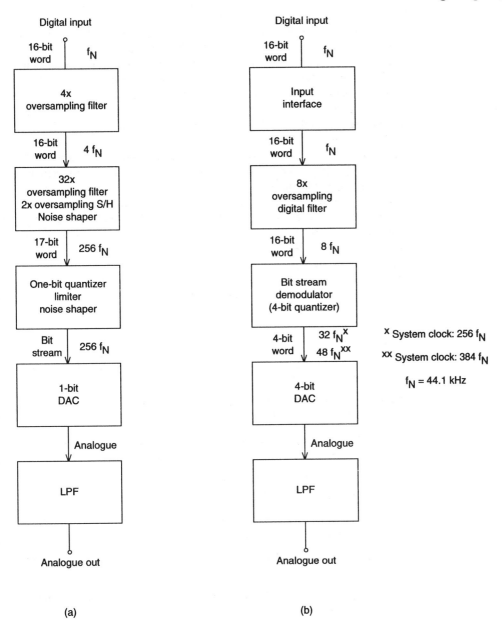

Fig. 6.81 Schematics of oversampling Σ–Δ DACs (a) Philips SA7322/7323 (b) Burr Brown PCM1710U (Courtesy of Philips Semiconductors and Burr Brown Corporation)

of 98 dB agrees with 16-bit resolution which, using eqn (6.123), makes S/N equal to 97.5 dB. The resolutions for S/N of 110 dB and $S/(N+D)$ of 92 dB equal 18 bits and 15 bits respectively. The intermediate value of 16 dB is in line with the observations made in Section 6.5.5 regarding the specification of the AD7884 ADC. Vendors provide a realistic estimate of the resolution, which depends partly on the application of the converter.

The main Σ–Δ products are 16-bit converters which are ASSPs designed for a specific application, in this case digital audio. The quality of reproduction with equipments like CD players and digital audio tapes is governed first and foremost by the DAC. There is a tendency to think that the one-bit Σ-Δ DAC is indisputably superior to conventional multi-bit converters. A searching evaluation of performance has shown that this is not the case, either in theory or in practice. The quality of the output depends on various factors, notably the total harmonic distortion, and a multiple-bit DAC with a conventional design or the Σ-Δ technique of the PCM1710U with its five-level four-bit quantization can give better results than one-bit conversion. The design and performance of the DAC have to be looked at in detail before making a judgement.

Σ-Δ converters have an exciting future, but it is not a foregone conclusion that they are to be preferred over multi-bit converters in general. They can achieve greater accuracy, and in applications calling for resolutions of 18 bits or higher they are likely to be preferred. At 16-bit resolution they will probably prove to be an increasingly cost-effective alternative to multi-bit converters because of the accumulating design experience, the elimination of high grade analogue components, and the ease of incorporating on-chip digital filters.

References

Allen, P.E. and Holberg, D.R. (1987). *CMOS analog circuit design*. Holt, Rinehart, and Winston, USA.

Allen, P.E. and Sanchez-Sinencio, E. (1984). *Switched capacitor circuits*. Van Nostrand Reinhold, USA.

Baher, H. (1991). Microelectronic switched-capacitor filters. *IEEE Circuits and Devices Magazine*, **7(1)**, 33–36.

Burns, S.G. and Bond, P.R. (1987). *Principles of electronic circuits*, pp.249–50. West Publishing Co., USA.

Candy, J.C. and Temes, G.C. (eds) (1992). *Oversampling delta–sigma data converters*. IEEE Press, USA.

Deloraine, E.M., Van Miero, S., and Derjavitch, B. (1946). French patent No. 932140.

Faulkenberry, L.M. (1977). *An introduction to operational amplifiers*. p.166. Wiley, U.S.A.

Franco, S. (1988). *Design with operational amplifiers and analog integrated circuits*. McGraw-Hill, USA.

Friedman, V. (1990). Oversampled data conversion techniques. *IEEE Circuits and Devices Magazine*, **6**, (6), 39–45.

Geiger, R.L., Allen, P.E., and Strader, N. (1990). *VLSI design techniques for analog and digital circuits*. McGraw-Hill, USA.

Gilbert, B. (1968). A precise four-quadrant multiplier with subnanosecond response. *IEEE Journal of Solid-State Circuits*, **3**, 365–73.

Gray, P.R. and Meyer, R.G. (1993). *Analysis and design of analog integrated circuits*, (3rd edn). Wiley, USA.

Huelsman, L.P. (1993). *Active and passive analog filter design*. McGraw-Hill, USA.

Inose, H., Yasuda, Y., and Murakami, J. (1962). A telemetring system by code modulation, delta–sigma modulation. *IRE Transactions on Space Electronics Telemetry*, **8**, 204–09.

Irvine, R.G.I. (1981). *Operational amplifier characteristics and applications*, pp.116–23. Prentice-Hall, USA.

Jacob, J.M. (1982). *Applications and design with analog integrated circuits*, pp.184-234, Reston, Prentice-Hall, USA.

Lynn, P.A. and Fuerst, W. (1990). *Introductory digital signal processing*, p.139. Wiley, England.

Malhi, S. and Chatterjee, P. (1994). 1-V Microsystems: scaling on schedule for personal communications. *IEEE Circuits and Devices Magazine*, **10**, (2), 13–17.

Marven, C. and Ewers, G. (1994). *A simple approach to digital signal processing*. Texas Instruments, England.

Menniti, P. and Storti, S. (1984). Low drop regulator with overvoltage protection and reset function for automotive environment. *IEEE Journal of Solid-State Circuits*, **19**, 442–8.

National Semiconductor (1993). *Operational amplifiers data book*. National Semiconductor Corporation. USA

Naus, P.J.A., Dijkmans, E.C., Stikvoort, E.F., McKnight, Andrew J., Holland, D.J., and Bradinal, W. (1987). A CMOS stereo 16-bit D/A converter for digital audio. *IEEE Journal of Solid-State Circuits*, **22**, 390–395.

Park S. (1993). *Principles of sigma–delta modulation for analog-to-digital converters*. Motorola, USA.

Philips (1992). *ICs for digital audio*. Publication SCB2, Philips Semiconductors, Holland.

Prince, B. (1991). *Semiconductor memories*, (2nd edn), p.74. Wiley, USA.

Sheingold, D.H.(ed.) (1986). *Analog-digital conversion handbook*, (3rd edn), p.486. Prentice-Hall, USA.

Sze, S.M. (1981). *Physics of semiconductor devices*, (2nd edn). Wiley, USA.

Widlar, R.J. (1965). Some circuit design techniques for linear integrated circuits. *IEEE Transactions on Circuit Theory*, **12**, 586–90.

Widlar, R.J. (1969). Design techniques for monolithic operational amplifiers. *IEEE Journal of Solid-State Circuits*, **4**, 184–91.

Widlar, R.J. (1971). New developments in IC voltage regulators. *IEEE Journal of Solid-State Circuits*, **9**, 2–7.

Further reading

Allen, P.E. and Sanchez-Sinencio, E. (1984). *Switched capacitor circuits*. Van Nostrand Reinhold, USA.

Candy, J.C. and Temes, G.C. (eds.) (1992). *Oversampling delta-sigma data converters*. IEEE Press, USA.

Franco, S. (1988). *Design with operational amplifiers and analog integrated circuits*. McGraw-Hill, USA.

Geiger, R.L., Allen, P.E., and Strader, N. (1990). *VLSI design techniques for analog and digital circuits*. McGraw-Hill, USA.

Gray, P.R. and Meyer, R.G. (1993). *Analysis and design of analog integrated circuits*, (3rd edn). Wiley, USA.

Huelsman, L.P. (1993). *Active and passive analog filter design*. McGraw-Hill, USA.

Marven, C. and Ewers, G. (1994). *A simple approach to digital signal processing*. Texas Instruments, England.

7
Semiconductor memories

7.1 Introduction

The importance of semiconductor memories has already been stressed in Chapter 1. Memories account for about one third of the total annual global consumption of MOS ICs in terms of revenue, and MOS ICs constitute about 70 per cent of all ICs. The biggest outlet for memories is naturally in data processing, which is still expanding rapidly. Some data processing is inherent in nearly all electronic system, from the smallest, like the chip for a quartz watch and a simple calculator, to the largest. Other expanding fields demanding substantial data storage are multimedia technology, telecommunications, and data communications.

Semiconductor memories can be grouped into two categories, RAMs and ROMs. The RAM allows entry and extraction at random, as the name implies. Data is written into a desired location, overwriting the information it contains. Data extraction (readout) is nondestructive i.e. the data at the location being read is retained. A RAM is volatile and all its information is lost when the power is removed or fails, unless special steps like battery backup are taken to maintain power. The ROM is a fixed content store and information is extracted nondestructively. Unlike a RAM it is nonvolatile.

The categorization of IC memories into RAMs and ROMs leads to a broad interpretation of their general use, an interpretation, let it be stressed, only put forward to explain their relative consumption. ROMs store programs and numerical data like constants and look-up tables. RAMs contain programs and are used for their execution, storing the intermediate and final results. A customary sequence is to enter the program, either in total or sequentially in sections, into RAM from a fixed store which might be a disk, diskette, or ROM. The architecture of a personal computer (PC) helps to illustrate the functions of the memories it contains. PCs (late 1994) come typically with (170–250) Mb hard disk and (4–8) Mb RAM storage—these are expandable. The hard disk contains the disk operating system (DOS) and software for alternative operating systems like Windows, together with numerous packages acquired by the user for specific tasks like word processing, games, simulation (like SPICE), electronic mail, computer-aided design (CAD), etc. Dedicated data processing hardware on a smaller scale includes ROMs which store the operating system and other specific programs, and a relatively small amount of RAM for program execution. Examples of such hardware are microprocessors, microcomputers, and microcontrollers.

ROMs compete with disk storage, which is too slow for program execution. In terms of volume, about 50 per cent of the data storage market is held by disks. Semiconductor memories account for 40 per cent, tapes for 10 per cent of the remainder. Within semiconductor memories, RAMs have about 70 per cent, mask- and user-programmable ROMs about 30 per cent of the market.

The memory remains the driving force in the thrust for higher transistor counts per chip and faster operation. It spearheads development and sets the pace for VLSI in general. Storage capacity increased dramatically with the adaptation of dynamic circuit techniques in the late 1960s. Until then RAMs were static. The dynamic RAM (DRAM) reduces the transistor count from four to six for an SRAM cell to one per cell (ignoring earlier extinct multi-transistor versions). Dynamic storage relies, like all dynamic logic—for example dynamic CMOS (Section 5.4)—on temporary charge storage across a nodal capacitor. The charge leaks away with time when the transistor driving the node is off. In synchronous logic such leakage, even at a moderate clock rate of 1 MHz, let alone at the clock rates of late 1994 data processing, 30 MHz or higher, is negligible. In a DRAM the interval between writing and reading information into and out of a cell is random, and the leakage demands a *refresh* operation, carried out for the DRAM at fixed intervals of about 8 ms with current technology. Because of the one-transistor cell, DRAMs exceed the storage capacity of SRAMs by a factor of between 4 and 8 for identical die size; their cost per bit (of storage) is consequently much less. SRAMs must not however be discounted. Originally, in the days of bipolar technology, they had a high power consumption, which has been greatly reduced by CMOS. Their superior speed combined with the absence of overhead circuits and refresh management ensures their continued existence. They make up about a quarter of the total RAM consumption.

The term ROM on its own stands for a mask (mask-programmed) ROM. Mask ROMs are suited to equipment manufactured on a large scale and being operated with fixed programs, like big computers. Equipments with smaller volumes of production and operated with software undergoing development and changes are satisfied far more economically with *user-programmable (field-alterable)* ROMs. The various categories of such ROMs are contained in Table 7.1. User-programmable ROMs enjoy a far high consumption than mask-programmable ROMs, by a factor of about three, because they are required by many more customers.

An overview of semiconductor memory performance is contained in Tables 7.2 and 7.3. The storages reflect the top end of the range for each category. Memory organization is expressed in the form (number of locations) × (bit length of word stored at each location), and is rounded off to figures like 64 K, 256 K, 1 M, which signify storages of 65 536 (2^{16}), 262 144 (2^{18}), and 1 048 576 (2^{20}) bits respectively. Similarly the 16 K (2 K × 8) PROM in Table 7.3 has a capacity of (2048 × 8) = 16 384 bits. The key parameter expressing speed is access time, the interval between the arrival of an address and the presentation of the data stored at the addressed location on the output pins. The information in Tables 7.2 and 7.3 is indicative rather than precise, because semiconductor memory characteristics change rapidly with time, more so than many other IC categories. The data are confined to chips established in full production. That explains the listing of 16 Mb DRAM in Table 7.2 compared with

Table 7.1 Categories of user-programmable ROMs

Type	Technology	Operation
EPROM	CMOS	Electrically programmable. UV light erasable.
OTP EPROM (OTP PROM)	CMOS	EPROM without erase facility
PROM	Bipolar	Electrically programmable. Cannot be erased.
EEPROM (E^2PROM)	CMOS	Electrically programmable and erasable. Usually erasable byte-by-byte
Flash EPROM	CMOS	EEPROM which cannot be erased byte-by-byte, but only by erasing the entire chip, or large sections thereof.
EAROM	CMOS	Electrically alterable (erasable and programmable) ROM. An earlier, now obsolete designation for EEPROM.

Table 7.2 RAM characteristics

Category	Technology	Capacity (b)	Organization (b)	Supply voltage (V)	Access time (ns)
DRAM	CMOS	16M	4M × 4	5	60.0
SRAM	CMOS	4M	1M × 4	5	25.0
SRAM	BiCMOS	4M	2M × 2	−5.2	10.0
SRAM	Bipolar—TTL	256K	256K × 1	−5.2	15.0
SRAM	Bipolar—ECL	16K	4K × 4	−5.2	4.5

the 64 Mb DRAM at the laboratory stage in 1991 (Table 1.2).

Some general deductions can be made from Tables 7.2 and 7.3. DRAMs maintain the highest RAM storage, but the SRAM is decidely faster. A striking feature is the performance of the BiCMOS RAM which achieves the same speed as bipolar–TTL SRAMs and yet has a higher storage level. At lower capacities bipolar ECL RAMs are decidedly faster; a 16 Kb SRAM has been included in Table 7.2 to confirm that. Table 7.3 shows that mask ROMs and flash EEPROMs have reached the storage capacity of DRAMs. The standard supply voltages are 3.3 V and 5 V.

7.2 Organization and operation

The basic schematic of a RAM is shown in Fig. 7.1, which is readily adapted for a ROM by replacing the bidirectional with a unidirectional

264 Semiconductor memories

Table 7.3 ROM characteristics (5 V supply voltage)

Category	Technology	Capacity (b)	Organization (b)	Access time (ns)
ROM	CMOS	16M	1M × 16	200
PROM	Bipolar	16K	2K × 8	25
EPROM	CMOS	4M	512K × 8	100
EEPROM	CMOS	256K	32K × 8	200
Flash EEPROM	CMOS	16M	2M × 8	80

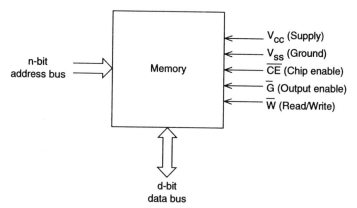

Fig. 7.1 Basic RAM schematic

data bus and omitting the $\overline{W}$ input. The memory has a storage of $2^n d$ bits. A location is accessed by the n-bit address input and the state of the $\overline{W}$ (write), alternatively $\overline{WE}$ (write enable) input, (formerly called $R/\overline{W}$ (read/write)) determines whether data is entered ($\overline{W}$ low) or extracted ($\overline{W}$ high). The logic high and low levels are V_{CC} and ground for CMOS. Vendors documentation retains V_{CC} for the supply voltage, although most memories are CMOS. A memory assembly made up of several chips is organized so that those memories not in use for a particular interrogation are disabled with $\overline{CE}$ (alternatively $\overline{E}$) in the high state. In that condition their *standby* input power is only a small fraction of the normal input power. The $\overline{G}$ (or $\overline{OE}$) control forces the output into the third state when it is set high. Most memories have a common I/O data bus, a practice which economizes on the number of pins, and which became firmly established when minimizing the pin count was important. Typical SRAM read and write cycle timings are shown in Fig. 7.2; the parameters therein are defined in Table 7.4 (Motorola 1992). Figure 7.2, although it applies to an SRAM, is a guide for semiconductor memories in general. A read cycle (Fig. 7.2(a)) commences with setting up the address, shown—like the data output—bivalent because it contains 1s and 0s. The $\overline{CE}$ input comes next, followed closely by $\overline{G}$. The output is enabled after an interval t_{GLQX}, reckoned from the $\overline{G}$ input, but does not take up the valid state until t_{AVQV}, which is the access time quoted in the memory specification. Evidently the cycle time is at least equal to the maximum access time. Read cycle and access times can be very similar if t_{ELQX} and t_{GLQX} are very small (which is usually the case),

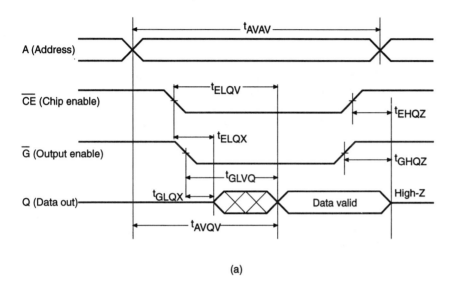

(a)

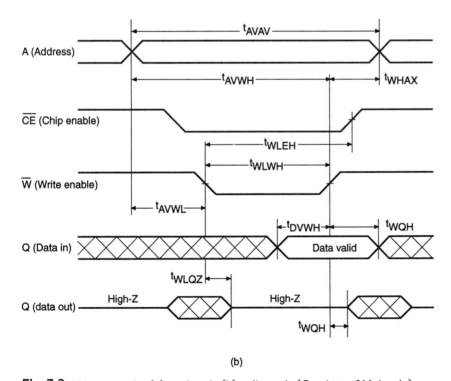

(b)

Fig. 7.2 Memory cycles (a) read cycle (b) write cycle (Courtesy of Motorola)

Table 7.4 Timing parameters (Fig. 7.2)

Parameter	Symbol	
	Standard	Alternative
Read cycle		
Read cycle time	t_{AVAV}	t_{RC}
Address access time	t_{AVQV}	t_{AA}
Chip enable access time	t_{ELQV}	t_{ACS}
Output enable access time	t_{GLQV}	t_{OE}
Chip enable low to output active	t_{ELQX}	t_{LZ}
Chip enable high to output high Z	t_{EHQZ}	t_{HZ}
Output enable low to output active	t_{GLQX}	t_{LZ}
Output enable high to output high Z	t_{GHQZ}	t_{HZ}
Write cycle		
Write cycle time	t_{AVAV}	t_{WC}
Address set up time	t_{AVWL}	t_{AS}
Address valid to end of write	t_{AVWH}	t_{AW}
Write pulse width	t_{WLWH}	t_{WP}
Data valid to end of write	t_{DVWH}	t_{DW}
Data hold time	t_{WHDX}	t_{DH}
Write low to data high Z	t_{WLQZ}	t_{WZ}
Write high to output active	t_{WHQX}	t_{OW}
Write recovery time	t_{WHAX}	t_{WR}

In the read cycle, identical alternative symbols, t_{LZ} and t_{HZ}, are used for different parameters. These, however, have much the same values in practice.
(Courtesy of Motorola)

and if the $\overline{CE}$ and $\overline{G}$ inputs are initiated soon after the address input. On the other hand control and address inputs have to be maintained to include set-up and hold times, which allow for worst case conditions with adequate margins for proper operation. In practice the cycle time may be up to twice as long as the access time. The speed of a memory tends to be judged in terms of its access time, but system assessment must allow for cycle time also.

The write cycle in Fig. 7.2(b) starts like the read cycle with address and $\overline{CE}$ inputs, followed by $\overline{W}$, which has to be maintained for a time t_{WLWH}. The data input has to become valid not later than t_{DVWH} and is maintained for a short time after the termination of $\overline{W}$. Read and write cycle times are either identical, or very nearly so, in practice.

The $\overline{G}$ input (Figs. 7.1 and 7.2(a)) is not a universal feature of semiconductor memories. Its function can to some extent be fulfilled by the $\overline{CE}$ control, which will in this case set the output into its third state (high Z in Table 7.4) when $\overline{CE}$ is high. The separate control of the output by $\overline{G}$ enhances system logic and virtually eliminates bus contention problems.

Bus contention arises in bus oriented systems when two or more devices try to output opposite logic levels on a common bus line. High

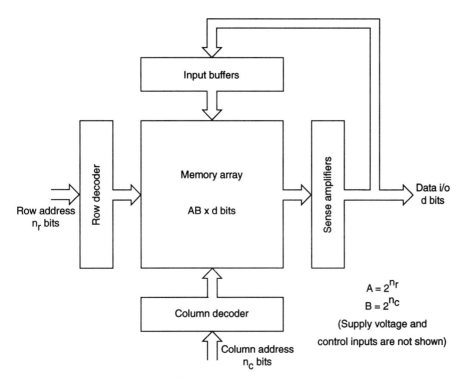

Fig. 7.3 SRAM architecture

speed RAMs with a common I/O data bus are highly susceptible to this malfunction, which occurs most frequently when switching from the read to the write mode and vice versa. Depending on the timing, the output of a RAM may not have turned off from the low level to the high impedance (third) state by the time another device is turned on and outputs a high level signal. Bus contention can be eliminated by judicious timing of the $\overline{W}$ and $\overline{CE}$ signals, better still by the $\overline{G}$ control if available. Some RAMs have separate I/O data, which helps with this type of bus contention and with contention of another kind, when a number of separate units wish to access the memory simultaneously. Finally a few SRAMs have a dual-port architecture with two independent I/O ports which are addressed separately. This allows simultaneous read or write at two separate locations.

7.3 SRAMs

The SRAMs architecture shown in Fig. 7.3 is an expansion of the schematic in Fig. 7.1, and is easily modified for a DRAM or a ROM. The shape of the array is a square or a rectangle with similar length and width. The architecture of a 256K × 4 memory might be 512 (rows) × 512 (columns) × 4, that of a 1M × 1 memory 512 (rows) × 2048 (columns) × 1, etc. Memory design is a complex combination of technology, architecture with specific attention to floor planning and layout, and circuit design. Circuit consideration of SRAMs and DRAMs is confined to the basic storage cells and sense amplifiers; these are the key components. CMOS is the preferred technology, complemented by BiCMOS for extra speed and bipolar ECL for the fastest operation obtainable.

268 Semiconductor memories

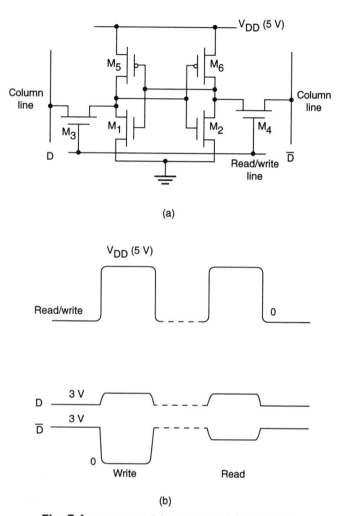

Fig. 7.4 SRAM cell (a) schematic (b) waveforms

The schematic of a six-transistor cell is shown in Fig. 7.4. It consists of a cross coupled latch driven by two gating pass transistors (M_3 and M_4). When the read/write line is at ground, the pass transistors are off; when it is at V_{DD}, they are on and the cell is active. The two column lines carry the data and its complement. A simplified explanation will now be given of the action, followed by a fuller account. Data is entered when the read/write line is pulsed to V_{DD}, enabling M_3 and M_4 and passing D and $\overline{D}$ to the gates of M_2, M_6, and M_1, M_5 respectively. The latching action locks the drains of M_1 and M_2 to D and $\overline{D}$, and they remain in that state when the read/write line returns to zero. Readout takes place when a subsequent pulse on the read/write line again enables M_3 and M_4, passing the information to the column lines.

The actual operation is modified by the *precharging* of the column lines. The column lines and other lines of SRAMs and DRAMs are usually precharged to a voltage in the range ($V_{DD}/2$) to V_{DD}. In this case assume a level of $(2/3)V_{DD}$ (3 V). The entry of a '1' ($D = 5$ V, $\overline{D} = 0$ V) into the cell is as follows. With the read/write line at 5 V, current flows from V_{DD} to ground via M_6 and M_4, pulling the gate of M_1 below V_T, provided M_4 has a much smaller resistance than M_6. The latching action sets M_1 drain to '1' (V_{DD}) and M_2 drain to '0' (0 V). D is raised slightly above 3 V by the current flowing through M_3 and M_5. For readout of a '1' the column lines are again precharged to 3 V. With the read/write line at 5 V, D will increase slightly as in 'write', and $\overline{D}$ will decrease by virtue of the current flow through M_2 and M_4; these transistors must be dimensioned to keep the drain at M_2 below V_T. The waveforms for the write and read actions are shown in Fig. 7.4(b).

The term *precharging a line* signifies that a line (wire) is set to a voltage V^+, which is stored on its capacitance C_1, by ϕ_1 until it is required for interaction with a receiver by closing ϕ_2 (Fig. 7.5) . Precharged lines represent a dynamic circuit technique which is employed for several reasons. First the power dissipation is very low, because there is no direct path from the supply voltage to ground. Second the switching is very fast. Depending on the condition of the receiver when ϕ_2 is closed, the line voltage will not change (or change only slightly), or the line may be discharged to zero. The discharge transistor of the receiver is dimensioned for drawing a large current to give a fast discharge, but it consumes no static power. The precharge of the column lines in Fig. 7.4 takes place before the final read/write steps and can be relatively slow without affecting cycle speed. The power saved and the improvement in speed make precharged lines vital parts of MOS SRAM and DRAM circuits.

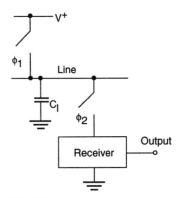

Fig. 7.5 Precharged line

The two preferred levels of precharge are V_{DD} and ($V_{DD}/2$). Precharge to V_{DD} is simply via a pull-up transistor, whereas precharge to ($V_{DD}/2$) requires a generator circuit. (Prince 1991, pp.251-60., and Bakoglu 1990). The ($V_{DD}/2$) precharge places the line potential close to the receiver threshold, and only a small change in input voltage will initiate the action when ϕ_2 is closed. The ($V_{DD}/2$) level imposes less stress on the cell capacitors—this is especially important for the DRAM cell described in the next section—and on interconnections over thin oxide. Last it economizes on power, because the $C\delta V^2 f$ product is smaller.

The differential voltage between D and $\overline{D}$ is restored by a sense amplifier to the full logic level. The sense amplifier is central to the performance of the memory. It largely determines the access time and a major part of the design effort is devoted to optimizing its performance. The circuit schematics of two widely used sense amplifiers are given in Figs. 7.6 and 7.7. The paired current mirror amplifier, the older of the two, is functionally identical to the amplifier in Fig. 6.16, but doubled up in order to give a differential output and constructed with MOSFETs instead of BJTs. The pMOS cross-coupled amplifier in Fig. 7.7 is faster because of the latching feedback. Both amplifiers have separate I/O column lines, and operate with a typical differential input voltage of 100 mV. They are controlled by an *activating* (enabling) pulse and an *equalization pulse*. The latter ensures correct operation and speeds up the sense detection, closing ϕ for a short time ($\sim$1 ns) during the leading edge transition of the inputs D_i and $\overline{D_i}$ (Sasaki *et al.* 1989). Great attention must be paid to the timing of the activation and enabling pulses and the layout of the bit lines in order to optimize performance.

Semiconductor memories operate from a single supply voltage. The

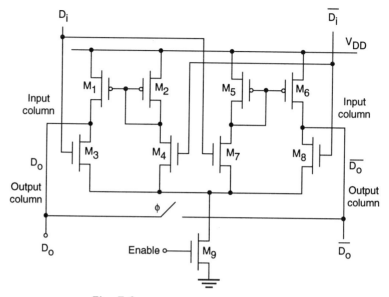

Fig. 7.6 Paired current mirror amplifier

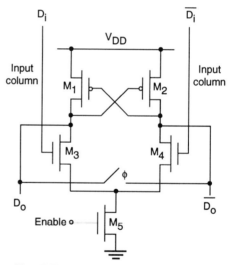

Fig. 7.7 pMOS cross-coupled amplifier

preferred standards are 3.3 V and 5 V, with 5 V memories at present in the great majority. A negative substrate bias voltage is advantageous for semiconductor memories and other MOS ICs. It reduces junction capacitance and makes V_T less susceptible to the *body effect*, the dependence of V_T on the source-substrate bias (Section 4.3.1). The early nMOS RAMs required a second supply voltage for that purpose. CMOS has decreased but not entirely eliminated the use of substrate back bias. When such a bias is required, it is generated on chip with a charge pump circuit, in which an oscillator generates a voltage which is stored by the relatively large substrate capacitance and rectified (Dillinger 1988; Glasser and Dobberpuhl 1985, pp.301–8). A charge pump circuit can generate a positive or negative voltage in excess of the supply voltage. One of its uses is to generate an internal supply voltage of ~3 V to power a high speed section with a geometry of ~0.5 μm contained within a 5 V CMOS RAM with a feature size of ~0.8 μm.

SRAMs and DRAMs are prone to *soft errors*, which set a cell into the wrong state and which come to light on readout. A soft error can cause a change of '1' to '0', or '0' to '1', is single, non-recurring, and causes no damage to the cell. *Hard errors* in contrast are repeated failures of a location (or a number of locations) to give the correct readout, and are attributable to structural defects.

The significance of soft errors was first positively identified in a classical paper by May and Woods (1979). Soft errors are caused by the radiation of particles which have sufficient charge to set cell data into the wrong state. The vast majority of soft errors in memories are due to alpha particles originating from materials within the device package (Section 2.6.3), and their consideration is confined to this cause. Soft errors arise from alpha particle bits on the bit lines, the sense amplifier, or the storage cell. DRAMs are particularly vulnerable, because they rely on temporary data storage across a capacitor, and it is this capacitor which is prone to be switched into the wrong state by an alpha particle. The probability of an error is closely related to the charge held by the capacitor at the time of impact, and this matter is examined in the next section. The soft error rate is greatly reduced by applying a protective coating like polyimide to the die. Another great improvement is the sapphire substrate of SOS CMOS (Section 3.2), but this is an expensive technology reserved for special applications.

The minimization of soft errors is an important factor in the design of the memory structure, including substrate doping, line composition and dimensions, and the storage cell. The development of RAMs with higher storage and faster cycle time has increased the danger of soft errors, which rise sharply with decreasing cycle time. Improvement in the soft error rate is one of the major issues in SRAM design (Sasaki 1989) and that holds for DRAMs.

A very small number of structural failures, even one failure only, will cause the die to be rejected unless special steps are taken to prevent this happening. Redundancy is often incorporated in chips which have a small raw yield. It consists of a few redundant rows and columns which are inserted in place of the faulty sections identified by probe tests prior to encapsulation. The redundant rows and columns are inserted in place of faulty rows and columns by one of three methods: current blown fuses, laser blown fuses, and laser annealed resistor connections (Prince 1991, p.127). The first method demands extra silicon estate for the high current transistors which blow the fuses, but has the advantage of not disturbing the passivation of the die. Replacement of faulty rows or columns with the aid of laser fusing or annealing has to be carried out in a clean room, and must be followed by a re-passivation of the substrate surface. Redundancy is employed for the initial production runs of memories with a low yield, and is phased out when the yield becomes adequate.

7.4 DRAMs

The schematic of a one-transistor DRAM cell is shown in Fig. 7.8. The

272 Semiconductor memories

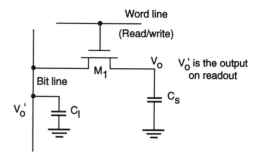

Fig. 7.8 One-transistor DRAM cell

cell contains no power supply and the transistor drain has no dc connection, being terminated by a capacitor. The circuit is strikingly simple, whilst its detailed design is highly complex. It is immediately apparent that the DRAM surpasses the storage capacity of an SRAM, requiring only one transistor per bit compared with six in an SRAM cell such as Fig. 7.4. Overhead circuits reduce that advantage, but for a given chip size DRAM exceeds SRAM storage by a factor of about four.

Information is written into the cell by raising the word line to V_{DD}, thereby enabling the access transistor and charging C_s to '1' or '0'. For V_{DD} and V_T equal to 5 V and 1.5 V respectively, V_o is either 3.5 V or 0 V, assuming CMOS overhead circuits. V_o is retained by C_s when the word line returns to zero until required for readout. V_o', the readout voltage developed across the bit line when the word line goes high, is related to V_o by

$$V_o' = \frac{C_s}{C_s + C_l} V_o \qquad (7.1)$$

C_l, the line capacitance, by far exceeds C_s, largely because of the many transistors connected to it. Typically $C_l \simeq 20 C_s$, making $V_o' \simeq (1/20)V_o$. The voltage affected by this attenuation is V_{OH}', the logic high output, which is only 167 mV, about one tenth of V_T. The readout is destructive, moreover V_{OH}' is in the wrong logic state. It is however sufficiently high to be identified as representing '1' by a sense amplifier to be described shortly. The smaller C_s the larger will be the attenuation from V_o to V_o'. The higher C_s the greater will be its charge and its ability to withstand an alpha particle hit. The intrinsic drain output capacitance is not large enough to satisfy these requirements, and C_s is increased by constructing an MOS capacitor. Let us estimate the capacitance needed to give a good immunity from soft errors. It has been found that the critical electron count to do so is $\sim$(1–2) million, an indicative estimate. The severity of an alpha particle hit depends also on other factors like the angle of incidence and the collection area of the capacitor. The critical charge Q_{crit} and the critical count N_{crit} are related by

$$Q_{crit} = q N_{crit} \qquad (7.2)$$

where $q = 1.60 \times 10^{-19} C$. $C_{s(crit)}$ is obtained from

$$C_{s(crit)} = \frac{Q_{crit}}{V_{oH}} \qquad (7.3)$$

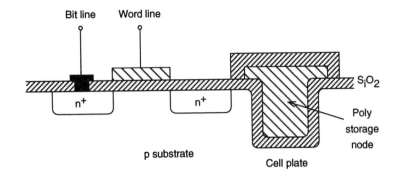

Fig. 7.9 Trench capacitor—DRAM cell

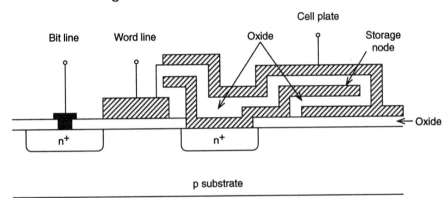

Fig. 7.10 Stack capacitor–DRAM cell

For $N_{crit} = 1.5 \times 10^6$ and $V_{OH} = 3.5$ V, $C_{s(crit)}$ equals $\sim$70 fF. In practice C_s is $\sim$(20–50) fF for DRAMs with a storage of 4 Mb or higher.

The two established structures for C_s are the trench and stack capacitors. The former is an adaptation of trench oxide isolation (Figs. 3.3 and 3.4(b)), and is illustrated in its basic form in Fig. 7.9. The polysilicon fill-in is the storage node, the part of the substrate surrounding the trench oxide is the capacitor plate. There are various different structures for such capacitors. The large sidewall area determines the capacitance and makes little demand on the silicon estate. The technology has matured thanks to the extensive use of trench oxide isolation. (Prince 1991, pp.137-8, 262-5).

The stack capacitor, whose structure is sketched in Fig. 7.10, benefits from the well established technology of interconnections above the gate oxide. The plates are separated by thin oxide and can rest on the various interconnect layers. Trench and stack capacitors achieve similar capacitances for the same silicon area. The trench capacitance is readily increased by deepening the trench, the stack capacitance is raised by multiple stacking. Submicron lithography is not suited for this type of structure above the silicon surface and separate processing steps may be needed, adding to the production cost. The limitations of the trench capacitor are leakage currents between trenches. These interfere with the stored charges which reside in the substrate in the case of trench structures whose polysilicon fill forms the capacitor plate taken to a fixed

Table 7.5 Typical DRAM cell areas and geometries

Memory capacity (Mb)	Cell area (μm^2)	Geometry (μm)
1	30	1.0–1.4
4	12	0.8–1.0
16	5	0.5–0.7
64	2	0.3–0.4

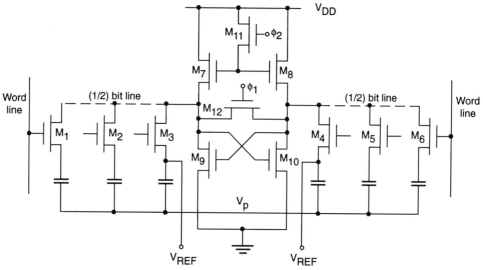

Fig. 7.11 DRAM cell readout

potential. This phenomenon becomes more serious for deep submicron geometries. Another problem with the trench capacitor is the exposure of the storage area within the substrate to alpha radiation. Both types of capacitor continue to be used, but there is a preference for the stack capacitor for DRAMs with storages of 16 Mb and higher. Geometries and cell dimensions for DRAMs with storages from 1 Mb upwards are contained in Table 7.5.

A schematic for cell readout is given in Fig. 7.11, in which the sense amplifier is located in the centre of a split bit line which has equal numbers of cells attached to it on either side. The clocking action of ϕ_1 and ϕ_2 is similar to the operation of the ϕ and enable signals in Figs 7.6 and 7.7. The bit lines are balanced as much as possible to have equal line capacitances. One favoured layout is folding the lines to be in parallel in order to obtain good symmetry. Each line includes a dummy cell (M_3 and M_4) which is precharged to V_{REF}, a reference voltage set halfway between the logic levels, i.e.

$$V_{REF} = \frac{1}{2}(V_{oH} + V_{oL}) \qquad (7.4)$$

Assuming perfect matching of cell and line capacitances, the outputs of

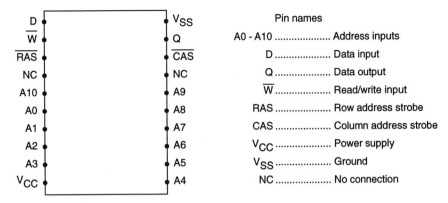

Fig. 7.12 Standard pin arrangement—4M × 1 DRAM

storage and dummy cells will be attenuated equally in accordance with eqn (7.1). For readout the selected cell is compared with the dummy cell on the other side of the sense amplifier. Precharging of the bit lines and the V_p rail is not shown (V_p is put at zero) in order to keep this explanation simple. In practice, these lines are likely to be precharged, probably to ($V_{DD}/2$) and V_{DD} respectively. With C_l/C_s in Fig. 7.8 equal to 20, and V_{OH} equal to 3.5 V (V_{oL} is zero), the inputs to the sense amplifier are 167 mV and 83 mV respectively. The sense amplifier has adequate sensitivity to detect that differential voltage and the latching action restores the bit line to the full logic level, which is entered back into the cell by an internal write operation following on immediately from the read cycle. Differential is far superior to single-ended operation and is established practice for DRAMs and SRAMs (see Figs 7.4, 7.6, and 7.7). The actual reference voltage differs a little from V_{REF} in eqn (7.4). Allowance has to be made for the inequalities in storage and dummy cell capacitances. The dummy cell has an additional area of contact for the V_{REF} input, applied via a coupling capacitor (not shown).

The full complexities of sense amplifiers and their detailed clocking are, like the intricacies of the DRAM cell, beyond the scope of this text. It has to be stressed that the technology and structure of the DRAM cell, the sense amplifier design, and the clocking are some of the most complex and ingenious features of VLSI.

The inevitable charge leakage from the cell capacitor C_s in Fig. 7.8 necessitates a periodic refresh at fixed intervals, which for current DRAMs are from 8 to 16 ms. The refresh operation consists of reading out the entire memory: the readout automatically restores the cell contents to their full value in accordance with the readout mechanism described for Fig. 7.11. The refresh modes are:

(i) $\overline{RAS}$ only refresh;
(ii) $\overline{CAS}$ refresh;
(iii) hidden refresh;
(iv) self refresh.

Read and write cycles are initiated by the address input and the setting of $\overline{RAS}$ and $\overline{CAS}$. The address input is time multiplexed. Taking a 4M × 1 CMOS DRAM with the pin count given in Fig. 7.12, two 11 bit address inputs are entered sequentially, the first with $\overline{RAS}$, the second

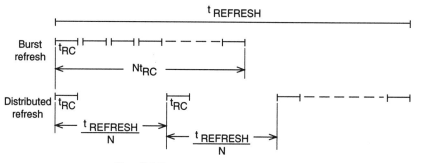

Fig. 7.13 Burst and distributed refresh

with $\overline{CAS}$ going low.

An $\overline{RAS}$ only refresh is carried out by entering addresses A_o to A_9 and switching $\overline{RAS}$ low. Address bit A_{10} is ignored in this mode, because it identifies which half of the array is to be accessed. $\overline{CAS}$ is held high throughout the operation, which is terminated by setting $\overline{RAS}$ high, and repeated for the next row. This refresh requires an external counter and externally-supplied row address. $\overline{CAS}$ before $\overline{RAS}$ refresh, also known by the acronym CBR (column before row) refresh, is carried out by first switching $\overline{CAS}$, and then $\overline{RAS}$ from high to low. This, the reverse order for a read or write cycle, initiates an internal row counter which generates the address for refresh. A hidden refresh is a $\overline{CAS}$ before $\overline{RAS}$ refresh initiated during a read or write cycle, which would normally be terminated by switching $\overline{CAS}$ and subsequently $\overline{RAS}$ high. Now $\overline{RAS}$ is switched to high and $\overline{CAS}$ is held low, concluding the $\overline{RAS}$ cycle. $\overline{RAS}$ is switched low again, beginning another $\overline{RAS}$ cycle and initially the refresh action. As long as $\overline{CAS}$ is held low, the valid data remains at the output whilst the refresh operation takes place. The refresh operation is hidden within what is an elongated read cycle. Self-refresh will be explained after dealing with refresh execution.

In *burst* refresh, the memory is refreshed in a burst of consecutive cycles, each cycle refreshing a row (Fig. 7.13). The time taken to refresh a row is the write cycle time t_{RC} (Table 7.4), regardless of the refresh mode, leading to a total refresh time Nt_{RC} where N is the number of rows. The proportion P_{REF} of the operating time taken up by the refresh is given by

$$P_{REF} = \frac{Nt_{RC}}{t_{REF}} \qquad (7.5)$$

P_{REF} comes to about one per cent in practice, witness the following example. A typical 4M × 1 CMOS DRAM has $t_{REFRESH}$ and t_{RC} equal to 16 ms and 130 ns respectively. The refresh is organized into 1024 cycles refreshing 4096 bits each. The total refresh time is 1024 × 130 ns = 133.1 μs and P_{REF} comes to 0.83 per cent. In distributed refresh the cycles are spaced by intervals of $t_{REFRESH}/N$. The total refresh time is Nt_{RC} but the inhibition of the memory is distributed instead of being a quantum of Nt_{RC}.

The reduction of input power continues to be a leading objective in DRAM design. Many DRAMs are now being offered in standard and low power versions, which draw far less current in standby. The active power dissipations of both versions are equal. The low standby current reflects the leakage currents, which determine the loss of cell charge in dynamic storage. The low-power versions have an extended refresh time, up to eight times the normal refresh interval. A further development incorporated in the low-power versions is the battery backup (BBU) mode, which includes an extended self-refresh (Konishi et al. 1990). The command for such a self-refresh is similar to the $\overline{CAS}$ before $\overline{RAS}$ operation, but these strobes are now held low for a much longer time, $\sim$ (100-300) μs. The BBU includes an external circuit incorporating a battery which automatically takes over from the dc power supply in case of failure (Oats et al. 1993). The power failure detection circuit initiates the self-refresh, which is very similar to the $\overline{CAS}$ before $\overline{RAS}$ mode except that these strobes are held low for much longer, as has just been mentioned. In self-refresh a new row is refreshed every (100–300) μs until the refresh is completed. Although the refresh peak current is high ($\sim$100 mA), its mean value is very small because of the extended cycle time. The low leakage in the BBU mode has made this extended cycle time, which is from 8 to 16 times the normal refresh period, possible (Konishi et al 1990). BBU is solely for data retention in case of power failure, and the extended self-refresh mode should only be used for that purpose.

7.5 ROMs

A general representation of a ROM array is shown in Fig. 7.14. PLA and PAL arrays are very similar (Chapter 9). The AND array is fixed in accordance with the binary address input, the OR array is programmed to contain the desired data at each location. The simplified representation of the AND and OR gates in Fig. 7.14 is explained in Fig. 7.15. A memory array is decomposed into subsections. It would be impracticable to have a 20-bit address decoded with AND gates having 2^{21} input lines (true and inverse address bits). The OR gate for each would need to have about a million (2^{20}) input lines! The simplified schematic does however illustrate the principle of ROM array architecture with its fixed AND and programmable OR planes. An example of an 8 × 4b ROM is shown in Fig. 7.14(c), in which the locations 0 to 7 have outputs of 1001, 101, 110, 1101, 1011, 1110, 1001, and 1100 respectively.

User-programmable (field-programmable) ROMs are electrically programmed. Erasure, if provided, is either electrical or optical. ROM storage is usually implemented with one transistor per bit. A customary ROM structure, the OR array in Fig. 7.16, is in effect a hardware implementation of the OR plane in Fig. 7.14. One of the rows, selected by the decoder, is at '1', all others are at '0'. A transistor at the intersection of a data column and an activated row will pull the data column to zero ('0'). The inverters at the end of the data columns change the output logic from NOR to OR.

The earliest PROMs were bipolar. Programming of a bipolar PROM is carried out by a fuse incorporated in each cell. The fuse metals are nickel chromium, titanium, tungsten, platinum silicide, and doped silicon. They are deposited on the silicon dioxide surface and are blown by passing a current in the range 5 to 25 mA, obtained from a supply voltage of 10 to 15 V. There was some regrowth of blown fuses in the early stages of this technology, which has now been mastered to give high reliability.

278 Semiconductor memories

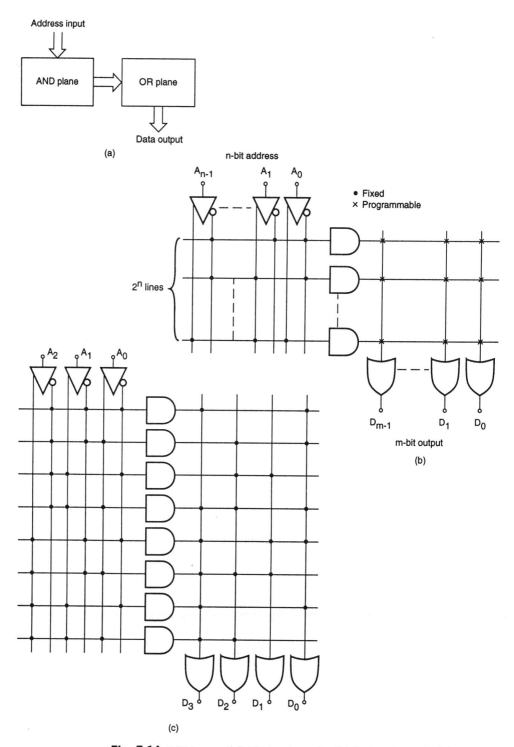

Fig. 7.14 ROM array (a) block schematic (b) logic schematic (c) example

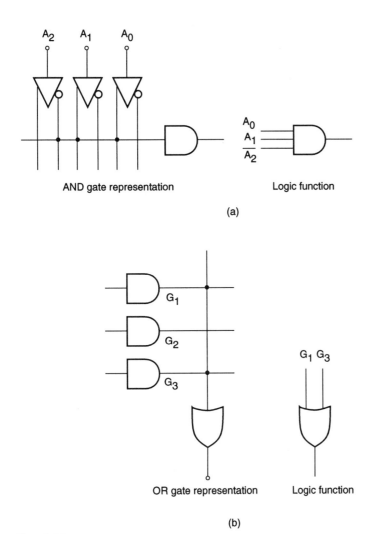

Fig. 7.15 Gate representation in Fig. 7.14 (a) AND gate (b) OR gate

An alternative technique has emerged, the antifuse. (Green et al. 1993; Hamdy et al. 1988). The antifuse is an open-circuit path rendered closed-circuit by applying a voltage in excess of the normal supply voltage across its terminals. The antifuse technology was first used in FPGAs, but is now being extended to PROMs. The structure of the PLICE antifuse is shown in Fig. 7.17. PLICE, a proprietary trademark of Actel Corporation, stands for Programmable Low Impedance Circuit Element. The PLICE antifuse consists of two conductors, an n^+ deposited polysilicon and an n^+ diffused layer, separated by an oxide-nitrogen-oxide (ONO) dielectric, about 10 nm (100 Å) thick. Conduction between these layers is achieved by applying a 16 V programming pulse for about 1 ns across them. The pulse melts the dielectric and establishes conduction, which is limited to $\sim$5 mA. The PLICE antifuse occupies 1.2×1.2 μm^2, a far smaller area than that taken up by a conventional fuse, and has a total capacitance, including the contributions of the two

280 Semiconductor memories

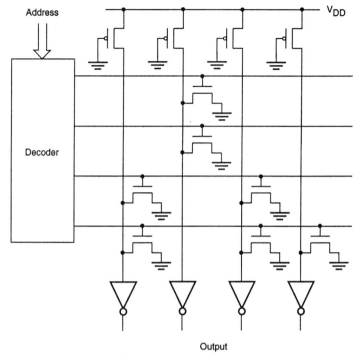

Fig. 7.16 ROM OR array

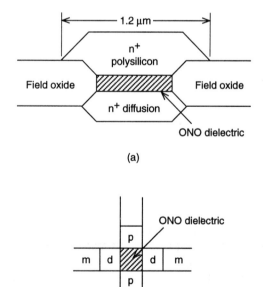

Fig. 7.17 PLICE antifuse (a) profile (b) top view (By permission of Actel Corporation)

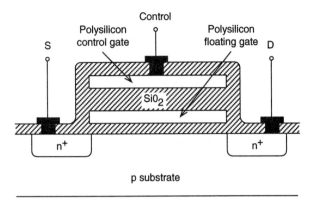

Fig. 7.18 FAMOS transistor

n^+ layers and the metal lines connected to them, of ~5 fF. The resistance in the off state exceeds 100 MΩ, and is about ~500 Ω when the antifuse is blown. The PLICE antifuse is already firmly established in FPGA technology, and is a highly attractive structure for PROMs because of its small dimensions and low capacitance. Its expected lifetime is over 40 years at 125 °C.

The usefulness of field-programming is greatly enhanced by the ability to erase and program repeatedly. This permits modifications to develop a computer program and subsequent changes of proven software to allow for alterations in the specification etc. The mature technology for that purpose is the EPROM. The profile of the MOSFET for such a memory, the floating gate avalanche (FAMOS) transistor, is shown in Fig. 7.18. The modifications of the standard MOSFET are a floating, i.e. unconnected, gate and a second control gate. The thickness of the oxide layers below the floating gate and between the two gates are identical, and equal to the thickness of the gate oxide (thin oxide) for an nMOSFET. The transistor is programmed to be permanently off or to be operated by the control gate like an ordinary MOSFET. The programming consists of grounding the source and applying a high positive voltage (~25 V) to the control gate and the drain. The high electric field causes avalanche breakdown in the channel and produces an abundance of hot electrons (Section 4.3.4), some of which have sufficient energy to penetrate the thin oxide and to be attracted to the floating gate by its positive potential, surmounting the 3.2 eV energy barrier between the substrate and the gate oxide in order to do so. The floating gate acquires a negative charge which it retains on removal of the programming voltage with the result that the transistor is non-conducting even when the control gate is taken to V_{DD}. If the programming step is omitted, the transistor behaves like a standard MOSFET. The EPROM has a one-transistor storage cell, and the cells are organized in an array like that in Fig. 7.16 or a similar arrangement, except that there is a transistor for each cell with its gate connected to the word line. The charge retention on the floating gate is usually guaranteed for a period of 10 years or longer. Erasure is effected by exposing the die to UV light under closely defined conditions. The chip, encased in a package with a transparent quartz window over the cell array, is moved from its mounting, and

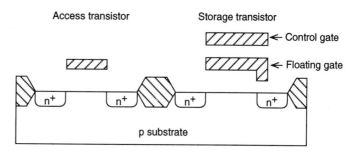

Fig. 7.19 EEPROM storage cell

placed ∼2.5 cm away from a 12 W/cm² UV lamp (wavelength 2537 Å). Erasure takes ∼20 mins. Unwanted erasure at a much slower rate occurs when the chip is exposed to strong light or standard fluorescent light and the transparent lid should be covered with an opaque label, which is removed for erasure. A modified version of the EPROM has become available in the form of the one-time programmable (OTP) EPROM (alternatively OTP PROM). The only change from the EPROM is in the package which has no transparent window, is therefore much cheaper, and can be plastic. Vendors offer the alternative of an EPROM or OTP EPROM for many of their products.

UV erasure is slow and the next move in field-programming was the EEPROM, in which programming and erasure are both electrical. The structure of the EEPROM storage transistor is shown, together with the select (access) transistor which makes up a one-bit cell, in Fig. 7.19. The vital difference between the structure of the FAMOS and the EEPROM transistors is the extension of the floating gate over the drain with a protrusion which gives an ultra-thin *tunnel* oxide region of 5 to 10 nm. The floating gate is charged and discharged by Fowler–Nordheim (F–N) tunnelling of *cold* electrons between drain and gate. These electrons tunnel through the oxide, provided it is thin enough, without surmounting the 3.2 eV barrier overcome by hot electrons. The structure of this storage transistor is known by the name of floating gate tunnel oxide technology (FLOTOX), a widely used Intel trademark. The transistor is programmed by grounding the control gate and connecting the drain to a high positive voltage. Electrons tunnel from the floating gate to the drain, leaving the floating gate positively charged and putting the transistor permanently on. The cell is erased by applying a positive voltage to the control gate and grounding the drain. F–N tunnelling now attracts electrons to the floating gate, putting the transistor permanently off. The control gate serves only for programming and erasure; it has no function for readout, which is obtained via the select transistor. The placement of EPROM and EEPROM cells within an array is shown in Fig. 7.20. The EEPROM cell has two transistors per bit compared with one for the EPROM cell, and this leads to a smaller storage capacity. The critical step in the fabrication of an EEPROM transistor cell is the formation of the high quality tunnel oxide with a thickness of 5 to 10 nm. The advances in submicron MOS processing have now reached a stage which gives high reliability with a charge retention of 10 years and 10^5

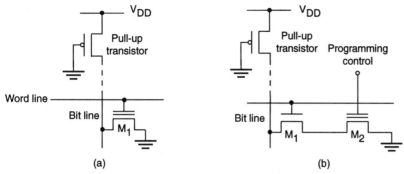

Fig. 7.20 EPROM and EEPROM cell placement (a) EPROM (Fig. 7.17) (b) EEPROM (Fig. 7.19)

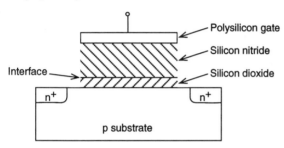

Fig. 7.21 MNOS storage transistor

program–erase cycles.

The metal-nitride-oxide semiconductor (MNOS) transistor for an EEPROM is an alternative to the charge storage on a floating gate. The gate dielectric of the MNOS transistor, shown in Fig. 7.21, consists of a very thin layer of silicon dioxide ($\sim$5 nm) and a much thicker layer of silicon nitride ($\sim$50 nm). The silicon dioxide/silicon nitride interface is a storage reservoir, trapping the charge induced in it by the tunnelling of cold carriers (electrons or holes), because it is surrounded by insulators of high quality. In the absence of such a charge, the transistor functions in the normal enhancement mode. It is programmed by applying a high positive voltage (15 to 25 V) to the gate with source and drain grounded. Tunnelling cold electrons establish a negative charge in the interface and put the transistor in the off state. The transistor is erased by applying a similar negative voltage to the gate, returning the electrons to the substrate. The MNOS cell demands, like the EEPROM cell in Fig. 7.20, a select transistor for readout. The MNOS transistor was the core of the first electrically programmable and erasable ROM cell, and the memory was called electrically alterable ROM (EAROM), a term which has been replaced by EEPROM.

The two-transistor EEPROM cells in Figs. 7.19 and 7.20 are complemented by the single-transistor cell for flash EEPROMs in Fig. 7.22. The split gate has a dual role. It functions like a conventional gate over the channel region, and controls the erasure of the floating gate. The cell behaves like two transistors. It is programmed by hot-electron injection and is erased through cold electron tunnelling from the floating gate to the drain. The reduction from two transistors to one transistor per bit increases the storage capability of flash EEPROMs built with this cell

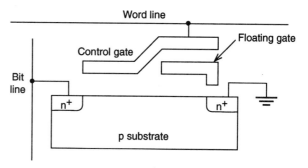

Fig. 7.22 Split gate EEPROM cell

to the magnitudes attainable with EPROMs. The absence of a control gate makes this cell only suitable for flash erasure of the entire memory or large blocks thereof.

In hot electron injection the electronic charge which builds up on the floating gate increasingly repels electrons in the oxide injected towards it and the process is self-limiting. It is very hard to overprogram an EPROM. Hot electron injection and UV erasure are both self-limiting mechanisms, F–N tunnelling of cold electrons is not. The floating gate of the storage transistors in Figs. 7.19 and 7.22 can and often is over-erased and thereby positively charged, but the cell in Fig. 7.19 is kept off until required for readout by the select transistor M_1 in Fig. 7.20(b), and the cell in Fig. 7.22 is similarly inhibited by the section of the control gate directly over the channel.

Chronologically bipolar PROMs and (UV) EPROMs were the first field-programmable nonvolatile memories. The F–N tunnelling of cold electrons and holes ushered in the EEPROM and the flash EEPROM. Tunnelling, which is an operating mode in all EEPROMs, is not as reliable as avalanche injection or UV erasure because of the stress across tunnel oxide. F–N transport calls for electric fields of ~ 10 MV/cm and these can only operate reliably with very high quality tunnelling oxide. MNOS EEPROMs continue to exist but are overshadowed by the FLOTOX types. The MNOS cell excels in radiation hardness and has very good endurance, but suffers from data degradation with write/erase cycling. The reliability of the FLOTOX EEPROM cell is governed by the ability of the tunnel oxide to withstand the stress of repeated program/erase cycling.

The EPROM maintains its leading position. At one time it was thought that the EEPROM would surpass its market share, but this has not happened. The proven reliability and technology of the EPROM have maintained its dominant place. The flash EEPROM, which builds on EPROM technology and has a similar storage capability (thanks to the one-transistor cell) is likely to become the major alternative and may indeed take more of the market than the EPROM in the long term.

7.6 Characterization and performance

Semiconductor memory storage continues to expand with DRAMs indisputably in the lead. Current DRAM capacities range from about 1 Mb to 16 Mb. Lower storage is catered for by SRAMs with some overlap between the two types up to the highest SRAM storage of 4 Mb. DRAMs

Table 7.6 Typical DRAM organizations

Memory capacity (b)	Organization alternatives (locations) × (word length)			
1 M	1M × 1	256K × 4	64K × 16	
4 M	4M × 1	1M × 4	512K × 8	256K × 16
16 M	16M × 1	4M × 4	2M × 18	

Table 7.7 CMOS DRAMS (The second entry in each category has self-refresh for BBU)

Device	Storage (b)	Access time (ns)	Refresh interval (ns)	Supply voltage (V)	Input power Active/Standby (mW)
MCM411000[1]	1 M	70	8	5.0	440/5.5
MCM41l1000[1]	1 M	70	64	5.0	440/1.5
TMS46100[2]	4 M	70	16	3.3	200/1.7
TMS46100P[2]	4 M	70	128	3.3	200/1.0
μPD4216400L[3]	16 M	60	64	3.3	265/2
μPD42516400L[3]	16 M	60	128	3.3	265/0.5
TMS416800[2]	16 M	60	64	5.0	450/5
TMS416800P[2]	16 M	60	512	5.0	450/2.5

[1] Motorola [2] Texas Instruments [3] NEC
(Courtesy of Motorola, Texas Instruments, and NEC)

emerged originally with a 1-bit word. That practice continues, but alternative organizations, which also apply to SRAMs, are available and the most common patterns are given in Table 7.6. The quest for speed is manifested by the drive to increase system bus bandwidth (SBW), expressed in eqn (5.48) and signifying the number of bytes which can be transferred per second on the system bus. SBW is increased by widening the bus and raising memory word length correspondingly. Memory word length can be increased by paralleling chips with common address and control signal inputs. It is however more efficient to increased the word length of the memory chip itself. Other considerations apart, this leads to an access time with less spread than that of an equivalent multi-chip combination.

A selection of DRAMs is listed in Table 7.7, which contains the standard and low standby-power self refresh versions in each range. Laboratory and prototype samples are of course ahead. A 64 Mb DRAM was reported in 1991 (Table 1.2) and 256 Mb DRAMs are at an advanced stage of development. Memory capacity can be increased by using memory modules, multichip modules (MCMs) which consist of chips mounted on a substrate together with a decoupling capacitor mounted under each chip. The modules are single sided or, when the number of chips becomes large, double sided with chips mounted on both sides of the substrate.

Table 7.8 CMOS DRAM modules

Range	Organization (b)					
Min	256K × 8	256K × 9	256K × 32	256K × 36	256K × 40	1M × 72
Max	16M × 8	16M × 9	8M × 32	8M × 36	8M × 40	8M × 72

(Courtesy of Motorola)

Table 7.9 Fast CMOS SRAMs

Device	Storage (b)	Organization (b)	Access time (ns)	Supply voltage (V)	Input power Active/Standby (mW)
MCM6206C-15[1]	256 K	32K × 8	15	5	825/100
MCM62V06D-25[1]	256 K	32K × 8	25	3.3	200/3
CYC107-A[2]	1 M	1M × 1	12	5	825/275
MCM6249-25[1]	1 M	1M × 4	25	5	750/250

[1] Motorola [2] Cypress Semiconductor
(Courtesy of Motorola and Cypress Semiconductor)

The range of DRAM modules available is indicated in Table 7.8. Some memories have word lengths of 9 bits or multiples thereof, for the purpose of parity checking. The examples given in Tables 7.6 (2M × 18) and 7.8 (16M × 9) draw attention to the availability of the limited number of DRAMs and SRAMs with such provision.

SRAMs come in various technologies. CMOS, the preferred technology, is followed by the faster BiCMOS, and by bipolar memories for the fastest speeds. CMOS SRAMs are divided into two categories: one with fast memory, the other with low-power memory and reduced speed. The speed–power trade-off is evident from Tables 7.9 and 7.10. The improvement in speed of BiCMOS over CMOS SRAMs stands out from Table 7.11. The fastest performance is naturally obtained with bipolar RAMs (see Table 7.12), but the difference in access times between them and BiCMOS RAMs is small, and their superior speed is confined to

Table 7.10 Low-power CMOS SRAMs

Device	Storage (b)	Organization (b)	Access time (ns)	Supply voltage (V)	Input power Active/Standby (mW)
μPD43256B-B12	256 K	32K × 8	120	3.3	150/0.2
μPD43256B	256 K	32K × 8	55	5.0	250/10
μPD431000A-B15	1 M	128K × 8	150	3.3	230/0.2
μPD431000A	1 M	128K × 8	70	5.0	350/10
μPD434000	4 M	512K × 8	70	5.0	375/10

(Courtesy of NEC)

Table 7.11 BiCMOS SRAMs

Device	Storage (b)	Organization (b)	Access time (ns)	Supply) voltage (V)	Input power Active/Standby (mW)
μPD46258L[1]	256 K	32K × 8	9	3.3	430/66
CY7B1094[2]	256 K	64K × 4	6	5.0	900/100
μPD461016L[1]	1 M	64K × 16	10	3.3	860/230
MCM6728A-5[3]	1 M	256K × 4	8	5.0	875/100
MCM101524-10[3]	4 M	1M × 4	10	−5.2	950/700

[1] NEC [2] Cypress Semiconductor [3] Motorola
(Courtesy of Cypress Semiconductor, Motorola and NEC)

Table 7.12 Bipolar ECL SRAMs (Supply voltage −5.2 V)

Device	Storage (b)	Organization (b)	Access time (ns)	Input power (W)
μPB1076LL	4 K	1K × 4	3	1.82
μPB10484A	16 K	4K × 4	6	1.66

(Courtesy of NEC)

storages of 64 Kb or less. SRAMs are, like DRAMs, available in modular assemblies, typified by the data in Table 7.13.

By and large semiconductor memories are much slower than logic gates made with the same process, a state of affairs which is understandable considering their function and architecture. The address decoder, the read/write organization, and the routing of data are far more time consuming than the operation of a logic gate. A similar observation holds for the system speed of a VLSI chip like a microprocessor or microcontroller.

Soft errors caused by alpha radiation originating within the package have been referred to in Sections 7.3 and 7.4. The error rate is very rarely specified, but there is a tacit aim among semiconductor houses to contain soft error failures in RAMs to a maximum of one in a million hours. This corresponds to 1000 FITs; a FIT (failure in time) signifies one failure in 10^9 hours. Two other parameters are involved in the assessment of soft errors, the SER and the alpha flux. The soft error rate (SER) expresses

Table 7.13 CMOS SRAM modules

Range	Organization (b)		
Minimum	256K × 8	64K × 16	16K × 32
Maximum	2M × 8	64K × 16	1M × 32

(Courtsey of Cypress Semiconductor and Motorola)

the failure rate in terms of incident alpha particles: an SER of 1×10^{-4} means one error per 10 000 alpha particles. The alpha flux F is the number of incident particles per cm^2 of die area per hour ($F = \alpha/\text{cm}^2 h$), α being the number of particles). The alpha flux with a die of area A cm^2 is FA and

$$FIT = SER.F.A.10^9 \qquad (7.6)$$

Equation (7.6) expresses the fact that the failure rate is proportional to the alpha flux, a proportionately which holds over eight decades of flux (May and Woods 1979). The SER is too low for measurement on an acceptable time scale. Instead it is obtained by extrapolation from a measurement of the error rate with the die exposed to an alpha particle source whose flux is several orders of magnitude higher. The flux typically extends from 0.001 to 0.01 α/cm^2h, but can approach 0.05 α/cm^2h. (Glasser and Dobberpuhl 1985, p.234). In order to achieve a failure rate of 1000 FITs, a RAM with a chip area of 1 cm^2 and a flux of 0.01 α/cm^2h must have an SER of 1×10^{-4}. Soft errors are a major issue in RAM design. Originally far more attention was paid to soft errors in DRAMs than in SRAMs, because memory expansion was dominated strongly by DRAM development when soft errors were first positively identified. The importance of soft errors in SRAMs is now being fully appreciated (Sasaki *et al. 1989)*. Indeed SRAMs may even have a higher error rate than DRAMs (Carter and Wilkins 1987). The on-going shrinkage to deep submicron geometries calls for special efforts in memory design in order to keep the error rate at a maximum level of 1000 FITs. Another avenue being explored is a novel use of a soft error correcting circuit (Mazumder 1992).

GaAs is the fastest digital logic. To realize its speed advantage to the full, a GaAs chip will need to contain some GaAs high speed memory. There have been ongoing developments in GaAs DRAMs and SRAMs. The DRAM storage capacitor consists of a reverse-biased pn junction utilizing a JFET, MESFET, or heterojunction bipolar transistor. Extensive developments have resulted in storage capacitors with acceptable charge retention and refresh times (Cooper 1993). The easier pursuit of GaAS SRAM development has progressed to yielding a 1 Kb laboratory model with an access time between 1 and 3 ns (Chandna and Brown 1994). GaAs DRAMs are still at an early experimental stage, and are not likely to make a significant inroad into the silicon memory market.

In the fixed-storage sector, mask ROMs hold a minor share of the market, which is dominated by UV EPROMs and flash EEPROMs. OTP EPROMS come next, and EEPROMs last. Table 7.14, which lists mask ROMs, illustrates the general tendency of ROMs to be slower than RAMs. A lower speed of operation is tolerable because data can be transferred, in total or in part, from ROM to RAM for faster execution if necessary. EPROMs are listed in Table 7.15, which reflects the growing practice of offering such a memory for UV erasure or in an opaque and much cheaper package for one-time programming. The separate supply voltage for programming is from 12 to 13 V. Flash EEPROMs, listed in Table 7.16, have a similar programming voltage. Their data retention is normally guaranteed for 10 years and the program/erase endurance

Table 7.14 Mask CMOS ROMs (Supply voltage 5 V)

Device	Storage (b)	Organization (b)	Access time (ns)	Input power (mW)
μPD23C1001EA	1 M	128K × 8	150	200
μPD23C40005	4 M	512K × 8	100	500
μPD23C16000J	16 M	2M × 8	120	350
μPD23C32140†	32 M	2M × 16	120	1000

† (Under development)
(Courtesy of NEC)

Table 7.15 CMOS EPROMs and OTP EPROMS (The second entry for each category is the OTP version)

Device	Storage (b)	Organization (b)	Access time (ns)	Supply voltage (V)	Programming voltage (V)
TMS27C256-10 TMS27PC256-10	256 K	32K × 8	100	5	13
TMS27LV010A-20† TMS27LV010A-20	1 M	128K × 8	200	3.3	12.75 ($V_{cc} = 5$ V)
TMS27C210A-1 TMS27PC210A-12	1 M	64K × 16	120	5	13
TMS27C240-10 TMS27PC240-10	4 M	256K × 16	100	5	13
μPD27C8000†‡ μPD27C8000‡	8 M	1M × 8/ 512K × 16	150	5	12.5

† Identical designations are used for the UV and OTP versions in two cases
(Courtesy of Texas instruments and NEC‡)

Table 7.16 Flash EEPROMs (Supply voltage 5 V)

Device	Storage (b)	Organization (b)	Access time (ns)	Programming voltage (V)
TMS28F512-10	256 K	64K × 8	100	12
TMS28F210-10	1 M	64K × 16	100	12
TMS28F040	4 M	512K × 8	100	12
NM29N16†	16 M	2M × 8	80	—

(Courtesy of Texas Instruments and National Semiconductor†)

Table 7.17 CMOS EEPROMs (Supply voltage 5 V)

Device	Storage (b)	Organization (b)	Access time (ns)
μPD28C04	4 K	512 × 8	200
μPD28C64a	64 K	8K × 8	150
μPD28C256	256 K	32K × 8	200

(Courtesy of NEC)

is 10^4 cycles as a rule. The Texas Instruments flash EEPROMs in Table 7.16 are offered in three versions with endurances of 100, 1000, and 10 000 cycles respectively, allowing the choice of cheaper chips with smaller endurance. The NM29N16 flash EEPROM in Table 7.16, and all EEPROMs in Table 7.17, dispense with a separate supply voltage for programming. The separate voltage for programming is generated on chip from the supply voltage by the charge-pump process (Section 7.3). The attractive practice of operating with a single supply voltage is on the increase.

References

Bakoglu, H.B. (1990). *Circuits, interconnections, and packaging for VLSI*, pp.175-6. Addison-Wesley, USA.

Carter, P.M. and Wilkins, B.R. (1987). Influences of soft error rates in static RAMs. *IEEE Journal of Solid-State Circuits*, **22**, 430-6.

Chandna, A. and Brown, R.B. (1994). An asynchronous GaAs MESFET static RAM using a new current mirror memory cell. *IEEE Journal of Solid-State Circuits*, **29**, 1270-6.

Cooper, J.A. (1993). Recent advances in GaAs dynamic memories. In *Advances in electronics and electron physics*, **86**, (ed. P.W. Hawkes). pp.1-79. Academic Press, USA.

Dillinger, T.E. (1988). *VLSI engineering*, pp.490-501. Prentice-Hall, USA.

Glasser, L.A. and Dobberpuhl, D.W. (1985). *The design and analysis of VLSI circuits*. Addison-Wesley, USA.

Green, J., Hamdy, E., and Beal, S. (1993). Antifuse programmable gate arrays. *Proceedings of the IEEE*, **81**, 1042-56.

Hamdy, E., McCollum, J., Chen, S., Chiang, S., Eltoukhy, S., Chang, S. et al. (1988). Dielectric based antifuse for logic and memory ICs. Technical Digest, IEEE International Electron Devices Meeting, pp.786-9.

Konishi, Y., Dosaka, K., Komatsu, T., Inove, Y., Kumanoya, M., Tobita, Y. et al. (1990). A 38-ns 4-Mb DRAM with a battery-backup (BBU) mode. *IEEE Journal of Solid State-Circuits*, **25**, 1112-17.

May, T. and Woods, M. (1979). *Alpha-particle-induced soft errors in dynamic memories.* IEEE Transactions on Electron Devices, **26**, 2-9.

Mazumder, P. (1992). On-chip ECC circuit for correcting soft errors in DRAMs with trench capacitors. *IEEE Journal of Solid-State Circuits*, **27**, 1623-33.

Motorola (1992). *Fast static RAM*. Publication DL156D, pp.8-7 to 8-10. Motorola, USA.

Oats, P.A., Hansen, J.P. and Polansky, P.J. (1993). *Battery backup of self-refreshing dynamic random access memory.* Application Note 1202, Dynamic RAMs and memory modules, Publication DL155D, pp.9-67 to 9-70. Motorola, USA.

Prince, B. (1991). *Semiconductor memories*, (2nd edn). Wiley, USA.

Sasaki, K., Ishibashi, K., Yamanaka, T., Hashimoto, N., Nishida, T., and Shimohigashi, K. (1989). A 9-ns 1-Mbit CMOS SRAM. *IEEE Journal of Solid-State Circuits*, **24**, 1219-25.

Further reading

Hu, C. (ed.) (1991). *Nonvolatile semiconductor memories.* IEEE Press, USA.

Prince, B. (1991). *Semiconductor memories,* (2nd edn). Wiley, USA.

8
ASIC design styles

8.1 Introduction

The case for ASICs was argued in Section 1.3.5. It follows on from the case for VLSI presented in Section 1.3.1. VLSI ASICs are an extension of standard VLSI, embodying architectures and design practices tailored to individual user requirements. The coverage of ASICs in this and the following three chapters is of value for a general understanding of VLSI design: the design techniques for standard and ASIC chips have much in common. Some ASICs fall within LSI. Those that do have a device count closer to VLSI than to MSI; their architecture and design tend to follow VLSI practice.

Recapitulating very briefly, electronic systems of substance are rarely satisfied by standard (V)LSI, but also contain a considerable amount of SSI and MSI. This *glue*, to quote an accepted generic term, can be mopped up entirely, or very nearly so, by ASICs, which make the equipment more economic and give it superior performance. There are of course many permutations. It could be that one application is best met by combining standard (V)LSI with an ASIC designed solely for replacing glue. (The generic term (V)LSI embraces both LSI and VLSI in connection with subject matter which applies to both). A more likely situation is that optimization comes about with one or more ASICs which replace standard (V)LSI devices and mop up the MSI and SSI chips. Yet another combination might consist of several different ASICs and standard (V)LSI.

The rapid development of ASICs has been accompanied by an increasing variety of design styles. An understanding of these is a prerequisite for appreciating the capability of ASICs, their design styles, and the design process. This chapter gives an overview of design styles in terms of their architecture and relative position. The continuing increase in the transistor count of VLSI chips has fertilized ASIC developments which have brought about new architectures. At the same time many PLDs at LSI level continue to fulfil a vital role in electronic equipment. The range of ASICs available caters for smaller systems satisfied at LSI level and for very large systems based on standard and/or ASIC VLSI. Digital ICs are far more numerous than analogue ICs in the majority of electronic equipments, but mixed-signal and analogue ASICs are increasingly replacing many standard linear ICs.

8.2 Categories

ASICs can be divided into the following categories:

(i) Custom (full custom, handcrafted) ASICs. These are designed and processed like standard products with a full set of masks for all fabrication layers.

(ii) Semicustom ASICs. These are either partly or fully *prefabricated*, or they are structured like standard and custom ICs, but with largely *predesigned* mask patterns.

The semicustom ASICs are:

(i) Gate arrays. The gate array is prefabricated with the exception of the metallization. It is completed by the design and fabrication of the metal masks, which establish the required interconnections.

(ii) Standard cells. A standard cell is processed like a standard IC and requires a full set of masks. The mask patterns are predesigned for a large variety of logic functions, which range from logic gates and bistables to the more complex MSI functions like counters and shift registers and may extend to complex VLSI megacells like microprocessors, RAMs, etc.

(iii) Cell-based ASICs. These hybrids contain a mixture of standard cell and gate array architecture. The term 'cell-based' is usually reserved for ASICs which contain some very large cells like those mentioned in (ii). A cell-based ASIC may also incorporate a full custom design.

(iv) Mixed-signal and analogue ASICs. The two established techniques are mixed-signal (digital–analogue) standard cells and linear (analogue) arrays. The linear array is similar to a gate array, prefabricated with the exception of the metallization masks, which determine the analogue function.

(v) Programmable logic devices (PLDs), which are completely prefabricated. The desired logic function is obtained by programming of the interconnections; the technique is in principle similar to the programming of field-programmable ROMs.

The technology of the ASICs in this chapter is assumed to be CMOS, unless stated otherwise. Figure 8.1 shows a tree of the various categories, which will now be described.

8.3 Gate arrays

The architecture of a channelled gate array is sketched in Fig. 8.2(a). The core is occupied by rows of identical *array elements* with spaces, the *routing channels*, kept free for the interconnections, which pass over the oxide covering the substrate. The substrate and the oxide layer contain all the electrodes, the gates being within the oxide. A typical array element usually consists of two p- and two nMOSFETs, but is sometimes larger. The functional power of a chip is expressed by either the transistor count or the number of *equivalent gates* (the word equivalent tends to be omitted). The transistor count is the parameter quoted more for standard ICs, the equivalent gate count is preferred for ASICs. What constitutes an equivalent gate is a matter of interpretation. By a broad concensus it is generally taken to be a 2-input CMOS gate. That makes the transistor count equal to four times the gate count. That interpretation is adopted for all categories of ASICs. The core array is

Gate arrays

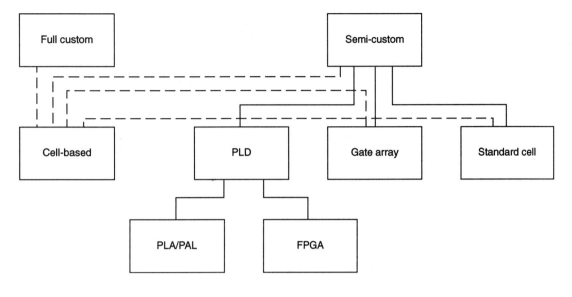

Fig. 8.1 ASIC categories

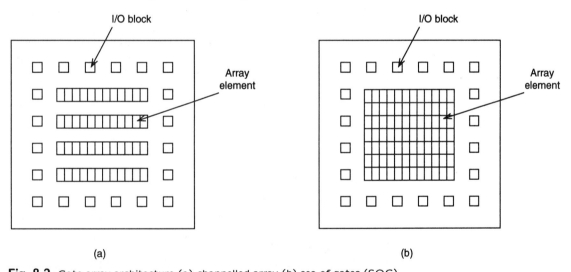

Fig. 8.2 Gate array architecture (a) channelled array (b) sea of gates (SOG)

supplemented by the peripheral I/O blocks, which can be configured into input, output, or bidirectional buffers.

Alternative designations for 'array element' are 'primary cell', 'core cell', and the original still widely used term, 'basic cell'. The use of the word cell for the array element and the I/O block is highly confusing, because of the generic term *standard cell*, which identifies a very different type of ASIC (see Section 8.2). Adhering to the descriptions 'array element' and 'I/O block' allows the label 'cell' to be reserved for an interconnected group of components which form a circuit. This definition will be retained throughout the coverage of ASICs, and will help to maintain the important distinction between cell and (uncommitted) array structures. The chip function is implemented by the metallization, which interconnects selected array elements. The vendor predesigns metal mask patterns for a comprehensive variety of macrocells (macros) which are circuits of varying complexity. These cover a wide spectrum of SSI and MSI logic, and usually include some macrocells with LSI and VLSI functions. The VSLI macros are, as has been pointed out, designated megacells. The metallization may be single- or multi-layer (Fig. 1.6). The smaller macrocells are formed by interconnections which are largely local. Macros with increasing function complexity call on larger numbers of array elements and are formed with local interconnections and global interconnections in the routing channels.

An alternative architecture, the *sea-of-gates* (SOG) or *channelless* array, occupies the entire core of the chip (Fig. 8.2(b)). All interconnections now pass over array elements, inhibiting some from being used, because the space occupied by an array element permits local but only a moderate amount of global interconnections. On the other hand the number of array elements has been increased by a factor of about two. The net result is that the SOG architecture allows a larger number of array elements to be utilized than a channelled structure for the same size of die. The proportion used depends on the chip function and is limited by connectivity, the ability to make the interconnections. It is typically between 50 and 90 per cent, but may be higher. The design process of *place* and *route* (*placement* and *routing*) establishes the locations (placement) of the used array elements and the pattern (routing) of their interconnections. The architectural trend is towards SOG architecture, but channelled gate arrays are still in full production.

A gate array family is supplied in a range of chips with different die areas and equivalent gate counts. The customer aims at the smallest die which is likely to meet the chip function. If the design process shows that this cannot be done, the next larger die is evaluated and the process is repeated until the specified function is met.

8.4 Standard cells

The standard cell is fabricated with a full set of masks. The design support includes a library of macrocells similar in the range of functions to the macros for a gate array. The macros contain the patterns for all the masks, and the design is completed by the placement and routing, which determines the location of the macros and their interconnections. A standard cell is more efficient than a gate array, requiring less silicon for a given function because the macros are more compact. This advantage has to be set against the extra cost of producing a full set of masks. The architecture of a standard cell in Fig. 8.3 contains a symbolic representation of macrocells, which can fulfil digital or analogue functions. Some ASIC vendors offer standard cells and gate arrays with libraries which contain functionally identical macros. It is important to be clear about the different nature of gate array and standard cell macros. A gate array macro contains the data for the metal masks, a standard cell macro contains the patterns for the full set of masks.

The macros which have been described in this and the previous section are hardwired (hard): they have a fixed *relative* placement and fixed routing (interconnect pattern). A software (soft) macro is a combination

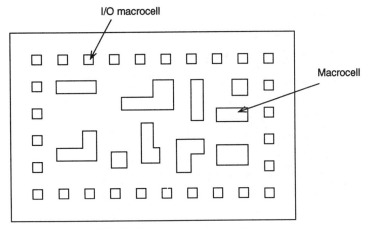

Fig. 8.3 Standard cell architecture

of hard (and possibly other soft) macros, and evidently presents a much more complex function than a hard macro. The soft macro identifies the macros it is composed of and their interconnections, but does not predetermine placement and routing. The above distinction between hard and soft macros holds for gate arrays, standard cells, and FPGAs. Vendors' ASIC libraries for standard cells and gate arrays usually contain hard and soft macros. The function of a soft macro can of course be produced in the design process, but a soft macro has the advantage of giving the user a predesigned complex circuit function, thereby easing the overall design. Moreover a soft macro can easily be modified to suit user requirements.

The standard cell is not offered in a range of different chip dimensions, but the die is restricted to a maximum area. The die size is chosen to contain the chip function within the smallest possible area and the utilization of the macros is 100 per cent. The macrocells are not, like the array elements of a gate array, uniformly placed (Fig. 8.2). Figure 8.3 reflects the architecture with its far more irregular spacing of macros and routing areas than for channelled gate arrays. There is a move towards a SOG-type architecture in order to improve efficiency.

It is vital to be clear about the different nature of gate array and standard cell ASICs. The fundamental distinction is between the uncommitted *prefabricated* array elements of a gate array and the *predesigned* circuit macros of a standard cell. The array elements are turned into macros by the metallization, whereas the macrocells of a standard cell are produced with a full set of masks which contain predesigned patterns. To emphasize this distinction, the word cell is in this and the following chapters on ASICs reserved for a *completed circuit function*.

8.5 Cell-based ASICs

Cell-based ASICs implement large systems which include *megacells* like microcontrollers, memories, etc. They are hybrid combinations of standard cell and gate array (usually SOG) architectures. The standard cell blocks are *embedded* within the gate array, i.e. they are surrounded by a gate array structure (Fig. 8.4). The embedded cells are likely to have been predesigned by the vendor for general use in ASICs. Alterna-

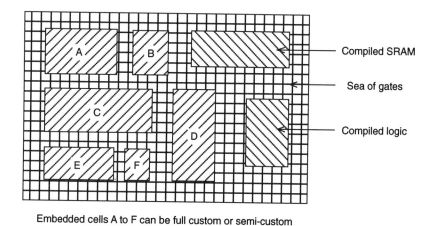

Embedded cells A to F can be full custom or semi-custom megacells, or semi-custom standard cell blocks

Fig. 8.4 Embedded cell architecture

tively an embedded cell may be fully custom designed (Fig. 8.1). The term 'embedded' distinguishes a standard cell macro (for example an embedded microcontroller) from a gate array macro. Another nomenclature labels these cells 'diffused' (for standard-cell embedded macros) and 'metallized' (for gate array macros or megacells).

The attraction of the cell-based ASIC, also called *embedded array*, is that it allows complete prefabrication of the substrate, which contains the predesigned megacells (and possibly smaller embedded cells) and the array elements. The ASIC is finalized to customer specification by designing the metal masks. The interconnect patterns for the embedded cells are already available and the design, which is quite fast, consists of compiling the pattern for the gate array logic and for the global interconnection of the entire system.

8.6 Mixed-mode and analogue ASICs

Mixed-mode (mixed-signal) ASICs contain digital and analogue circuits. The analogue section of the ASIC covers functions like operational amplifiers, comparators, voltage references, and data converters, to mention some of the more prominent applications. Analogue circuits can be constructed in standard-cell format, in which case analogue and digital macros are side by side in a mixed-mode ASIC. Alternatively they are formed by interconnecting a linear array which contains an assembly of bipolar transistors, resistors, and MOS capacitors. Some arrays are in CMOS or BiCMOS technology; BiCMOS includes bipolar and MOS transistors and polysilicon resistors. The analogue circuits are formed, like the digital circuits of a gate array, by the interconnections patterned in accordance with the data for the macros, supplied by the vendor.

8.7 PLDs

8.7.1 PLA and PAL

All PLDs are fully prefabricated and are configured by blowing fuses (including antifuses), or by programming transistors which act like fusible links in the interconnections. These transistors are either MOS pass transistors (Fig. 5.8) or EPROM/EEPROM-type MOSFETs (Figs. 7.18 and

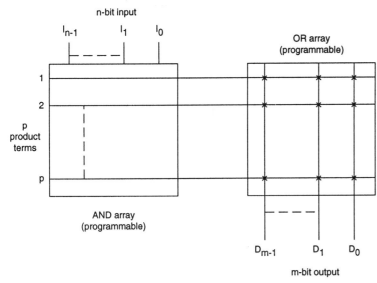

Fig. 8.5 PLA architecture

7.19). The principle underlying PLA and PAL architecture is that any logic function can be expressed in sum-of-products form. For example, the sum (S) and carry (C_H) outputs of a full adder are given by

$$S = ABC_L + A\overline{B}\,\overline{C}_L + \overline{A}\,\overline{B}\,C_L + \overline{A}B\overline{C}_L \tag{8.1}$$

$$C_H = AB + AC_L + BC_L \tag{8.2}$$

where A, B, and C_L are the inputs, C_L being the carry forward from a lower order. The PLA and PAL structures are adaptations of the ROM array in Fig. 7.14. The PLA, the first PLD on the market, was established long before the PAL. The earliest PLAs were mask-programmable and the field-programmable PLAs were at first labelled FPLAs to distinguish the one from the other. Mask-programmable PLAs are now obsolete, and the prefix F has been dropped. The PLA schematic in Fig. 8.5 is for an n-bit input AND array which generates p product terms. That number is very much less than for a ROM array where an n-bit input leads to 2^n product terms i.e. $p \ll 2^n$. The product terms feed into the OR array. The AND and OR arrays are fully programmable, and every product term can be used repeatedly, if necessary, for any of the output lines. This generous choice is inevitably demanding on hardware, and is modified in a PAL by retaining the programmable AND array (Fig. 8.6). The AND plane product terms are arranged into more or less equally sized groups, whose outputs are connected to the PAL output lines. A PAL with 56 product terms and an 8-bit output might have 7 product terms in each group. These outputs are fed via OR gates to the data output lines (Fig. 8.6). The amount of hardware is far less than for a comparable PLA. The reduction is reflected in the number of fusible links, which are $p(2n + m)$ and 2 np for the PLA and PAL structures in Figs. 8.5 and 8.6 respectively. A PAL product term can only be used once and is connected to a specific output line. Should such a product be

300 ASIC design styles

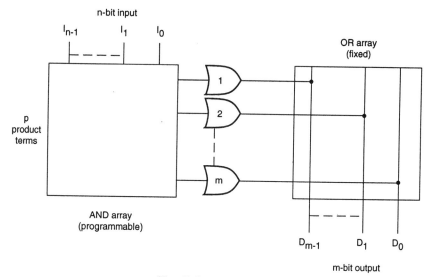

Fig. 8.6 PAL architecture

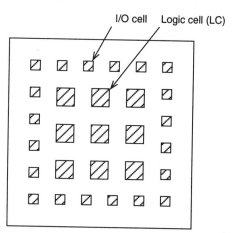

Fig. 8.7 Typical FPGA architecture

needed for another output bit, it has to be generated again in another group. Only a few functions need a repeat of product terms and the PAL has become the far more popular choice of the PLA-PAL family.

8.7.2 Field programmable gate arrays

The field programmable gate array (FPGA) consists of an array of logic cells (blocks, modules), and programmable interconnections. FPGAs are, like PLAs and PALs, completely prefabricated. A schematic of a typical FPGA architecture is shown in Fig. 8.7, but some FPGAs have a structure, more like that shown in Fig. 8.2. The composition of a logic cell varies widely between the products of different vendors. It consists of logic gate combinations with a preference for multiplexer-based logic. A multiplexer (MUX), used in communications for selecting one of several inputs for the output path, is used in logic for generating sum-of-product terms. Figure 8.8 contains an MUX in its simplest 2-input form. The output Z is given by

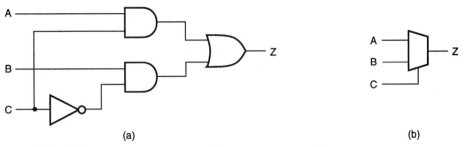

Fig. 8.8 Two-input multiplexer (a) logic schematic (b) symbol

$$Z = AC + B\overline{C} \qquad (8.3)$$

The MUX has generated two product terms. Unlike the array element of a gate array, the logic module is a complete circuit by virtue of fixed interconnections which establish a basic logic entity. It is configured to generate a specified function by *programmable interconnections*. There is a vital difference between this action and the formation of gate array and standard cell interconnections by metal masks. The FPGA is fabricated with an enormous number of programmable interconnections, very much larger than the final number which establishes the chip function. The interconnections have, like those in a gate array, the dual role of forming macrocells and interconnecting them to obtain the required logic.

There are two techniques for establishing the interconnections.

(i) Fusible links are blown, opening a conventional fuse or shorting an antifuse (Section 7.5).
(ii) nMOS pass transistors or EPROM/EEPROM cell transistors are programmed to open- or short-circuit the links.

These alternatives apply equally to FPGAs. The pass transistors (Fig. 5.8) are in effect programmable links which take the place of fuses.

8.8 Overview

The growth and development of ASICs was spearheaded by gate arrays and standard cells. These still hold the major share of the ASIC market, and have kept pace with the increase in VLSI component density due to the reduction in feature size. Currently ASICs have a maximum gate count of about 500 000. One way of using such a capacity to advantage is the hybrid structure of the cell-based embedded array, in which megacells perform the megafunctions, supplemented by smaller cells if needed. The cell-based section is embedded within a gate array which serves for the speedy completion of the ASIC.

Analogue circuits, implemented sparsely at first, are now commonplace. Vendors are offering more and more analogue macros for mixed-mode standard cells in their libraries. All-analogue ASICs are based on linear arrays of transistors, resistors, and capacitors which are turned into analogue circuits by the interconnections.

Mask-programming is eliminated in PLDs, which are completely prefabricated, user-programmable and free from non-recurrent engineering (NRE) development costs. The PLA and the PAL, who ushered in the era of ASICs before that term had been coined, continue to satisfy a significant demand for ASICs with modest gate counts. The developments of VLSI ASICs with the highest possible number of gates understandably steal the limelight, but the need for small ASICs must not be overlooked. The FPGA has experienced a meteoric rise since its introduction in 1985. Combining the attraction of field-programming with a versatile and powerful logic cell structure, the FPGA has established a niche for itself in many applications which had hitherto been met by a gate array or a standard cell.

The design styles have now progressed to a growing variety and a multitude of hybrid combinations. A few of these have been mentioned: there are many others. For example an embedded array might include a PLA megacell. CMOS has become the technology of choice for ASICs. BiCMOS and bipolar ASICs are marketed on a smaller scale for high speed applications. The original PLA and PAL families were all bipolar and many of them still are, but CMOS is replacing most of the ranges. With CMOS has come another change, this time in programming. The fuse is giving way to the transistor link which can be an nMOSFET or an EPROM/EEPROM cell transistor. ASIC design styles have been outlined in this chapter in order to portray the various alternatives open to the user, before embarking on an investigation of their relative merits and their design. These matters are covered in the chapters which follow.

Further reading

Brown, S.D., Francis, R.J., Rose, J., and Vranesic, Z.G. (1992). *Field programmable gate arrays*. Kluwer Academic Publishers, USA.

Coli, V.J. (1989). VLSI programmable devices. In *VLSI Handbook* (ed. J.D. Giacomo), pp.4.1–4.39. USA.

Einspruch, N.G. and Hilbert, J.L. (1991). *ASIC technology*. Academic Press, USA.

Hill, F.J. and Peterson, G.R. (1981). *Introduction to switching theory and logic design*, (3rd edn), pp.196–209. Wiley, USA.

Huber, J.P. and Rosneck, M.W. (1991). *Successful ASIC design the first time through*, pp.1–16. Van Nostrand Reinhold, USA.

9
ASICs—Programmable logic devices

9.1 Overview

In terms of function density, PLDs occupy the lower range of ASICs. PLAs and PALs have been established longest and satisfy a demand for combining a modest amount of SSI and MSI in a single chip which is easily programmed. The PAL is far more widespread than the PLA. Originated by Advanced Micro Devices (AMD), who coined the trade mark PAL, it is a standard off-the-shelf IC and is produced in a mature range of chips which are second sourced by a good number of vendors. To label them 'standard' is not in contradiction to their ASIC role. They are ASICs, but are field programmed in contrast to other types of ASICs, which are mask programmed. PALs have not been overtaken by events. Within their function capability, which has changed little over the last decade, they are a highly efficient implementation of SSI and MSI, combining a wide choice of standard TTL/CMOS functions.

Proven design tools and programming hardware (which is an adaptation of ROM programming hardware) make PAL design fast and economic. The main criteria for design style selection are that the ASIC has the required density (meaning function density, often shortened to density) and that it is, allowing for production and development costs, the most economic solution. PALs have gate counts which extend from $\sim(100\text{--}1000)$ equivalent gates. An 'equivalent gate', defined in Chapter 8 and simply called 'gate' in the context of chip density, corresponds to four MOSFETs. PAL architecture has improved and families have been introduced which supplement the original TTL PALs. These, however continue to remain in full production. Many of the advances are embodied in the general array logic (GAL) series, which were originated by Lattice and are, like the standard PAL families, second sourced. There are also more advanced PLDs which incorporate the AND–OR array of a PAL but whose architecture makes them very similar to FPGAs. These are however very small in number and are mainly produced by vendors who concentrate on PAL-type PLDs and top up the upper end of their range with such products.

If PAL architecture has remained relatively static since the inception of these devices over fifteen years ago, the technology has changed. Bipolar are making way for CMOS ICs wherever possible, and the original TTL PALs have been overtaken by CMOS families which are equally fast, consume less power, and allow reprogramming. EPROM (ECMOS) came first, and is being overtaken by EEPROM (EECMOS) techniques for reprogrammable PALs.

The outstanding PLD development has been the FPGA, which started with modest gate counts of a few hundred and has increased in density to an upper limit—still going up— around 25 000 gates. It stands side by side with gate arrays, which are often termed *mask-programmable gate arrays* (MPGAs) to distinguish them from FPGAs. CMOS is the technology of choice. Programming techniques include ECMOS, EECMOS, the antifuse and SRAM cells.

PALs achieve maximum flip-flop clock rates of around 100 MHz and system speeds, which depend on the application, are in the range (50–75) MHz. FPGAs do not by and large reach the speeds of the fastest PALs. Their system speeds are less on aggregate, roughly (40–60) MHz.

The FPGA is still on a growth curve. Its attractions are field programmability, the absence of NRE costs, and an architectural flexibility smilar to that of MPGAs. They are standard off-the-shelf chips which allow users to meet specific needs by electrical programming. These features, combined with the speed of development and the ease of producing pilot quantities, have given FPGAs a significant and growing share of the ASIC market. The very large number of interconnections, only a small percentage of which are used, constrains the FPGA to a much lower density than the MPGA. Nevertheless it covers a function range for which there is a large demand, and in which it is likely to be the first choice for moderate quantities of production.

Good design tools are prerequisites for ASIC engineering. PALs are backed by long-established proven design tools. FPGA design tools are similar to the design support for MPGAs. All FPGAs are strongly supported by design software and are programmed with hardware similar to PROM programmers. PAL design tools are outlined after descriptions of PAL-type PLDs and FPGAs. The design tools for FPGAs are covered in Chapter 11, because of their affinity with MPGA computer-aided design and engineering (CAD/CAE).

9.2 PAL-based PLDs

9.2.1 Structures

PLDs with a PLA/PAL architecture have been outlined in Section 8.7.1, where it was pointed out that the PAL is the more popular structure of the two. By far the greater majority of devices in the PLA/PAL families are PALs and this section describes their architectures, which have undergone significant modifications since they appeared in their original forms in 1975 (PLA) and 1978 (PAL). The basic version of the PAL still exists, but numerous refinements have been added to later families, enhancing their capability.

The first of these, leading to the *registered* PAL, is shown in Fig. 9.1. The diagram illustrates the PALCE22V10 family output logic macrocell (OLMC) . The cell includes a register, the D flip-flop. The four possible output configurations are governed by the S_o, S_1 inputs to the 4-to-1 multiplexer (MUX). The two multiplexers in Fig. 9.1 are for routing and not for function generation. The rectangular shape of the multiplexer symbol differs from the more common trapezoid for the MUX in Fig. 8.8. It has been adopted here in keeping with vendors' diagrams

PAL-based PLDs 305

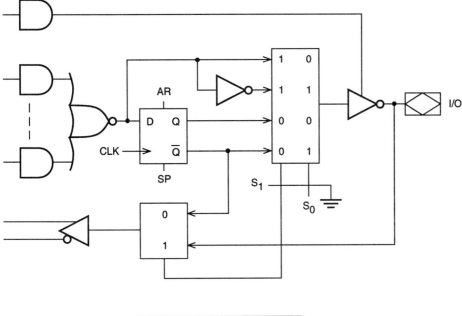

S1	S0	Output configuration
0	0	Registered/active low
0	1	Registered/active high
1	1	Combinatorial/active low
1	0	Combinatorial/active high

0 = Programmed EE bit
1 = Erased (charged) EE bit

Fig. 9.1 Output logic schematic—PALCE22V10 (Courtesy of AMD)

for PLD logic. The PALCE22V10 is an EECMOS PAL, whose links are programmable with EEPROM-type storage transistors (Fig. 7.19). The intersection of the horizontal ground line with the vertical S_o and S_1 lines signifies that they can be programmed to give any one of the four combinations tabulated in Fig. 9.1. A schematic to illustrate PLA/PAL programming in general is shown in Fig. 9.2. The output Z is given by

$$Z = (A + \overline{f}_1)(\overline{A} + \overline{f}_2)(\overline{B} + \overline{f}_3)(B + \overline{f}_4) \qquad (9.1)$$

with f equals 0 for a blown fuse, and 1 for an intact fuse. (The opposite conditions hold for an antifuse). Programming with EECMOS gives $f = 0$ with the transistor programmed, and $f = 1$ with the transistor erased. The AND–OR input logic driving the D flip-flop and the 4-to-1 MUX is based on the schematic of Fig. 8.6. The number of AND gates feeding the OR gate in Fig. 9.1 equals the number of product terms applied to it from the AND array. The three-state output inverter, controlled by a product term, allows the I/O port to be configured for the input or output. This provision has become general practice for PALs and the more advanced PAL-based PLDs. It greatly enhances the variety of logic configurations possible, and aids optimization to suit each application. The output, whether registered or combinatorial, can be set

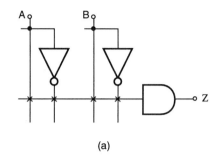

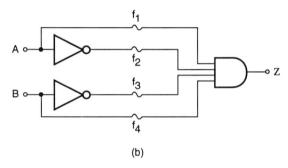

Unprogrammed PLA-PAL : all fuses intact
Fuse blown : f=0
Fuse intact : f=1

Fig. 9.2 PLA–PAL programming (a) equivalent schematic (b) logic schematic

to active high or active low. The registered output may be fed back, if desired, from $\overline{Q}$ to the AND array via the 2-to-1 MUX. The logic is arranged for synchronous operation (clock and synchronous preset), but permits asynchronous reset. The PALCE22V10 family can fulfil functions like counters and shift registers, and is particularly well suited for state machines. D latches are preferred for the registers, and it is usual to have the Q output active low. The choice of combinatorial or registered output at each I/O port allows the user to synchronize combinational logic with the clock input to the registers.

The logic schematic for the PALCE22V10 is shown in Fig. 9.3. The AND array capacity is expressed by $(2i \times p)$ where i and p are the numbers of inputs and product terms respectively. The term $2i$ arises because each input must have true and inverse lines in the AND array. In this case there is a maximum of 22 inputs and 132 product terms. These are divided among the 10 outputs, which have between 8 and 16 product terms each. The device designation PALCE22V10 signifies maxima of 22 inputs and 10 outputs. The allocation of inputs and outputs is flexible. Eleven inputs are *dedicated*, whereas all the 10 I/O ports can be used for inputs and outputs. If any of the ports are used for an input, the three-state buffer control sets the inverters feeding them into the open circuit output state. The maximum of 22 inputs cannot be obtained in practice, because the PAL would then have no output. What the

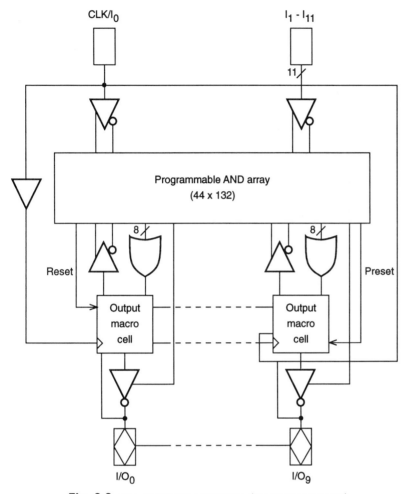

Fig. 9.3 PALCE22V10 schematic (Courtesy of AMD)

provision does mean is that any one of the I/O ports may be selected for an input applied to any of the undedicated lines of the AND array. (Bostock 1993, pp.96–103; Pellerin and Holley 1991, pp.39–44.)

Another refinement of the basic PAL is the addition of an exclusive–OR gate in the data path between the AND array and the I/O port. Figure 9.4 gives a basic schematic for combinatorial and registered outputs driven via exclusive–OR gates. The use of such gates in the output macrocells is not confined to setting the output polarity. They can help to configure different functions. One such example is a macrocell embodying a D flip-flop and an exclusive–OR gate together with some multiplexers. Such a cell can be programmed for six different flip-flops (J-K, $\overline{J}$-$\overline{K}$, $\overline{J}$-K, J-$\overline{K}$, T and $\overline{T}$). The exclusive–OR gate reduces the amount of logic for many applications, counters in particular. (Coli 1989, pp.4.16–4.18.)

The registers in the macrocells of Figs 9.1 and 9.3 are for synchronous operation. Combinatorial outputs are asynchronous, but can be operated synchronously. Some applications call for registers with asynchronous capability, and a typical output macrocell for asynchronous operation is shown in Fig. 9.5. The register, a D flip-flop, includes an optional preload

308   ASICs—Programmable logic devices

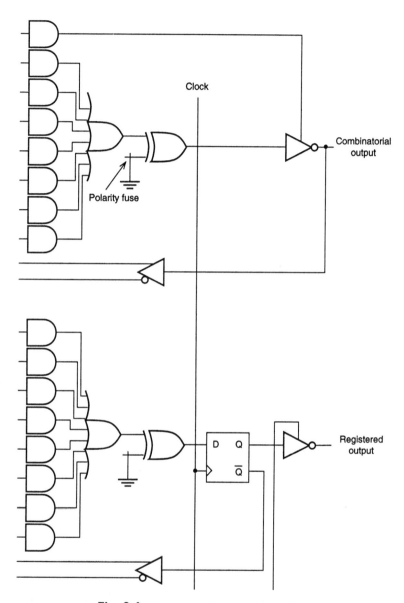

Fig. 9.4 Programmable output polarity

(PL) for presetting the registers of all macrocells simultaneously. This facility is especially useful for functional testing of sequential designs. Separate product terms supply the inputs for the clock, preset (AP) and reset (AR) signals in the asynchronous mode. Alternatively the register can be set or preset from the I/O port by applying a logic low to the PL input. It will then take the logic level of the I/O port, which reaches the P (sometimes called PD) input of the register via an inverter. Another feature of the register is the reset during power-up, which puts the flip-flop output in the high state irrespective of the input; this is especially

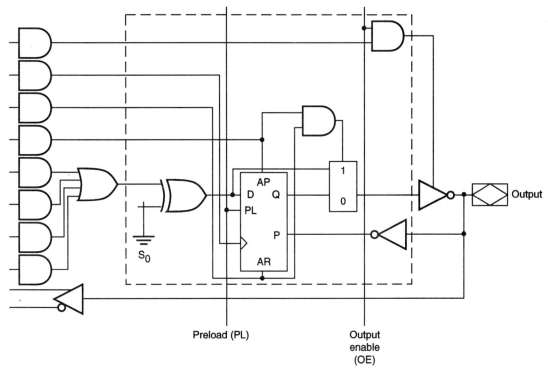

Fig. 9.5 Asynchronous macrocell schematic—PALCE20RA10 (Courtesy of AMD)

valuable in state machine initialization. The user can choose to have a resistered or combinatorial output. If both preset (AP) and reset (AR) are logic high, the flip-flop is bypassed and the output is combinatorial.

The registers in Figs. 9.1 and 9.5 feed I/O ports or can be inhibited from doing so. An alternative arrangement is to have registers which cannot be accessed for providing an output. Such registers are said to be *buried*, and their outputs are usually fed back into the AND array. The PLD outputs are in such cases a function of the buried register output and other inputs to the AND array. Buried registers are advantageous for complex machine design.

Another enhancement of PAL architecture is the folded feedback structure illustrated in Fig. 9.6, which contains a functional block schematic and an extract of the logic schematic diagrams for the Philips programmable macrologic device PLH5501. The principle of the feedback logic will be explained first, its purpose afterwards. The feedback logic is based on an AND–NAND array. Some of the NAND gates have folded outputs which are fed back into the array. The remaining NAND and AND gates, only two of which (NAND gates) are shown in the simplified schematic, drive the output ports. Figure 9.6(b) contains:

(i) NAND gates whose outputs are fed back (folded) into the array;
(ii) two NAND gates which drive the output ports via buffers.

Figure 9.7 illustrates that a NAND–NAND combination can generate Boolean sum-of-products terms like an AND–OR combination. The output Z is given by

310 ASICs—Programmable logic devices

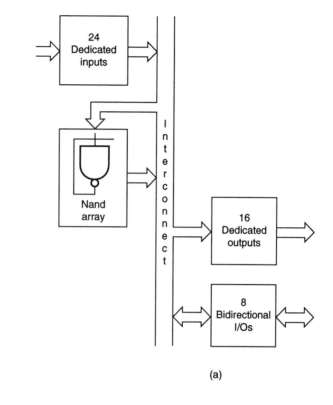

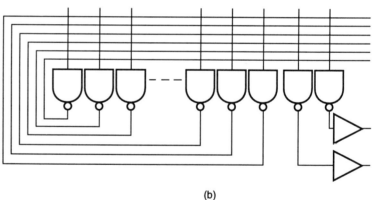

Fig. 9.6 Folded product line architecture—PLH5501 (a) schematic (b) logic schematic of folded section (Courtesy of Philips Semiconductors)

$$Z = \overline{\overline{P_1}\,\overline{P_2}\cdots\overline{P_n}} = P_1 + P_2 \cdots P_n \qquad (9.2)$$

P_1 to P_n being the individual product terms fed into gates 1 to n. Sum-of-product terms can be generated equally well with an AND array driving NAND gates i.e. AND gates in place of the NAND gates 1 to n. In this case the output is given

$$Z = \overline{P_1 P_2 \cdots P_n} = \overline{P}_1 + \overline{P}_2 \cdots + \overline{P}_n \qquad (9.3)$$

The P terms in eqn (9.2) can be made equal to the $\overline{P}$ terms in eqn (9.3),

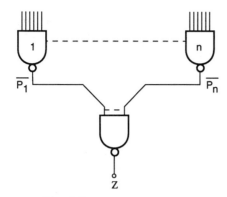

Fig. 9.7 NAND–NAND array

because input variables are available in true and inverse form. It is possible to have folded feedback with NOR–NOR gates in place of NAND–NAND gates, and a few PLDs have this logic. The folded feedback in Fig. 9.6 has several attractions:

(i) A single product term can be realized with one gate.
(ii) It is possible to design many macrofunctions economically.
(iii) It is possible to construct multi-level logic. The traditional AND-OR array of a PAL restricts the user to two-level logic. The folded feedback permits the generation of as many logic levels as required. To generate comparable logic with a standard PAL would demand far more additional hardware and would cause a serious, possibly unacceptable loss of speed.

The programmable logic sequencer (PLS) is a complex PLA (note: not a PAL) whose outputs are all registered. There are however a few PLS devices with some I/O ports. These have combinatorial outputs; the dedicated outputs are always registered. The registers are R-S or J-K flip-flops in practice. The latter are preferred because they can toggle, an important asset in counters. The structure overcomes an inherent limitation of the PAL, namely the limited number of product terms per output bit, typically eight or less. The PLS is fundamentally a state machine device. The registers are divided into two sections of similar bit-size, output registers and buried state registers, whose outputs are fed back as inputs to the AND array. The PLS is specially suited for applications like shift registers, synchronous counters, microcontrollers, etc.

The AMD MACH family is an example of a high density PLD, based on blocks with an AND array and macrocells, interconnected by a switch matrix. The EECMOS MACH family spans a range from 900 to 3600 gates, and extends the logic power of PAL devices up to the capacity of small gate arrays. A block schematic of the MACH architecture is shown in Fig. 9.8. Output macrocells are available in all families, buried macrocells are confined to the MACH2 series. The outputs can be registered or combinatorial with programmable polarity. The routing of the output macrocells is similar to the logic in Figs 9.1 and 9.4. The output macros may be turned into buried cells, in which case the register output is fed back and the I/O port can be used for an input. The *dedicated* buried

312 ASICs—Programmable logic devices

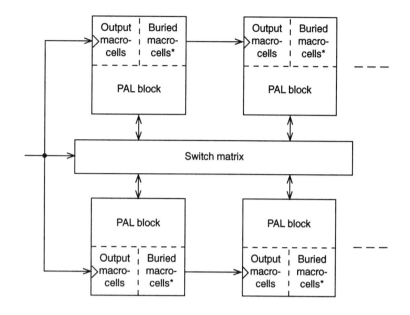

* Buried macrocell available on MACH 2 devices only

Fig. 9.8 MACH 1 and 2 architecture (Courtesy of AMD)

macrocells can provide input registers. In that role they are useful for synchronization.

One feature, which is not shown in any of the schematics, is the *security fuse* (or *security cell*). Incorporated in virtually all PLDs, it prevents identification of the device program and protects proprietary designs from competitors. Without a security fuse the interconnect pattern can be read like a PROM, or can be copied by a programmer. For transistor-programmed links (ECMOS and EECMOS) protection is ensured by a security cell. Once this cell has been programmed, the array will read as if all its links had been activated. The security cell can only be erased by reprogramming the PLD. It is a more effective protection than the security fuse, because it is easy to remove the package and to ascertain the pattern of fused interconnections by inspection with a microscope. Such detection is virtually impossible for transistor-programmed links (Pellerin and Holley 1991, p.79).

Before going on to look at PLD characteristics, let us briefly recapitulate on the main families:

(i) Basic PAL/PLA

The output polarity is usually programmable, active low being the preferred standard. Many or all of the I/O ports can be used for inputs or outputs;

(ii) Universal PAL

Outputs may be either registered or combinatorial with programmable output polarity. GALs have a similar architecture, but their output macrocells have superior logic power;

(iii) PLS (alternatively FPLS)

This is a PLA in which some or all the outputs are registered. Buried

registers, if available, feed their data back into the AND array;
(iv) Asynchronous PAL

The PLDs in (i) to (iii) are synchronous. An asynchronous PAL has outputs each of which can operate asynchronously in relation to the other outputs;

(v) Folded array PAL/PLA

The array architecture is an AND - NAND/AND in place of an AND-OR combination. Part of the array is folded: some of the NAND gate outputs are fed back to the input lines.

9.2.2 PAL characteristics

The characteristics of PALs are good indications of the performance of PAL-type PLDs in general, so that the information in this section broadly portrays the capabilities of various PLDs. The performance of a PAL is characterized by the following main parameters:

(i) the number of inputs and outputs;
(ii) the number of product terms;
(iii) speed;
(iv) power consumption;
(v) programming and reprogramming techniques.

The parameters under (i) and (ii) above have been standardized, and current PAL families closely follow the original devices. PALs have, in common with (V)LSI in general, moved from bipolar to CMOS, and occasionally to BiCMOS technology. The CMOS transistor links are ECMOS and EECMOS, with EECMOS being the favourite choice. Fuse links are used in bipolar and BiCMOS chips. The CMOS families present the second generation, which has as one of its aims the minimization of power consumption. The two conditions to be considered are the power demands in the operating and quiescent states. For a fixed supply voltage (the present standard for PALs is 5 V) the power is proportional to the supply current I_{CC}. There is the customary speed–power trade-off, an example of which has been given for VLSI with the data for fast high-power and slow low-power CMOS SRAMS in Tables 7.9 and 7.10. Power consumption has been greatly decreased relative to what it was by going from bipolar to CMOS technology. Were all the circuits of a PAL CMOS, the quiescent power would be zero, but this is not the case. The PAL core, the array, is nMOS: CMOS is confined to the peripheral I/O circuits. In consequence there is a substantial consumption of current in the steady state. *Zero-power* PALs are available with near-zero I_{CC} in the standby mode. They incorporate special circuit techniques which entail—directly or indirectly—detecting an input transition and which inevitably incur a speed penalty. One such technique, called input transition detection (ITD) is taken over from memory design (Philips Semiconductors, 1993). Advantage has been taken of the micron and submicron dimensions to produce PALs which consume far less power but have much the same speed as the original geometrically larger TTL families. This matter will be discussed in more detail shortly after the tabulations summarizing PAL performance.

The key parameters expressing speed are :

(i) t_{PD}, the delay between input or feedback to combinatorial output;

314 ASICs—Programmable logic devices

Table 9.1 PAL characteristics (All devices are synchronous)

Device	Technology	t_{PD} (ns)	I_{CC} (mA)	$f_{max(F)}$ (MHz)	$f_{max(NF)}$ (MHz)
PAL16L8-4	TTL	4.5	210	117	125
AmPAL22P10B	TTL	15.0	180	—	—
PALCE16V8H-5	EECMOS	5.0	125	143	166
PALCE26VH12-7	EECMOS	7.5	115	105	125

All values are worst case
(Courtesy of Advanced Micro Devices (AMD))

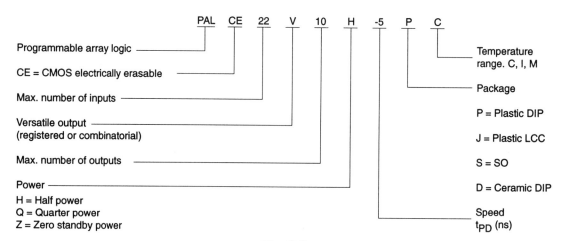

Fig. 9.9 Standard PAL device code

(ii) t_s the set up time from input or feedback to clock;
(iii) t_{CO}, the delay between clock and combinatorial output;
(iv) t_{WL}, the clock width (low);
(v) t_{WH}, the clock width (high).

The maximum exercizing frequencies are given by

$$f_{MAX(F)} = \frac{1}{t_s + t_{co}} \quad (9.4)$$

$$f_{MAX(NF)} = \frac{1}{t_{WH} + t_{WL}} \quad (9.5)$$

where the qualifications (F) and (NF) apply to operation with external feedback and without feedback respectively. Vendors' data occasionally quotes f_{MAX} equal to $1/t_{PD}$, but this figure is over-optimistic on its own, although t_{PD} is a useful initial guide. The expressions in eqns (9.4) and (9.5) are to be preferred if data is available for their computation.

The characteristics of some standard TTL and CMOS PALs in Table 9.1 outline the type of performance obtained with current chips. The code for device designation is given in Fig. 9.9, which applies universally with minor changes and holds for Table 9.1 with the exception of the AmPAL22P10B. There the letter B is a code for t_{PD} in the range

Table 9.2 Characteristics of EECMOS PALs

Device	Package pin count	t_{PD} (ns)	I_{CC} (mA)	Comments
PALCE16V8H5	20	5.0	125	
PALCE16V8Q-25	20	25.0	55	
PALCE20V8H-5	24	5.0	125	
PALCE20V8Q-25	24	25.0	55	
PALLV18V8-10	20	10.0	55	3.3 V supply
PALCE22V10H-5	24	5.0	115	
PALLV22V10-10	24	10.0	55	3.3 V supply
PALCE26V12H-7	28	7.5	115	
PALCE26V12H-20	28	20.0	105	
PALCE20RA10H-7	24	7.5	100	Asynchronous
PALCEM16H-25	24	25.0	100	
PALCE29MA16H-25	24	25.0	100	Asynchronous

All values are worst case
(Courtesy of AMD)

Table 9.3 PALs—AND array organization

Device	Product lines × Product terms
PALCE16V8	32 × 64
PALCE22V10	44 × 132
PALCE24V10	48 × 80
PALCE26V12	52 × 150
PALCE29M16H	58 × 188
PALCE29MA16H	58 × 178

(Courtesy of AMD)

(15–25) ns. The superior speed power product of the CMOS chips is immediately apparent. Most TTL and all EECMOS families are available with registered outputs. These, in CMOS PALs can, it will be recalled, be by-passed in favour of combinatorial outputs. Like semiconductor memories, PALs have a modest pin count (see Table 9.2). The preferred package alternatives are the DIP, PLCC, and SOIC encapsulations.

Table 9.2 gives a cross section of EECMOS PALs, including some devices for asynchronous operation. The entries for each family represent the devices with the smallest and largest propagation delay. The device designation gives the maximum number of inputs and outputs. The letter V (for versatile) signifies that the outputs can be either registered or combinatorial. The PALCE22V10H family, for instance, has 12 dedicated inputs and 10 I/O ports, giving a maximum of 22 inputs. The logic capability is governed by the AND array, typical compositions of which are given in Table 9.3. The number of product lines is simply twice the number of inputs; each input gives rise to true and inverse lines for the array. Inspite of that obvious deduction, it has been quoted because the AND array composition is usually documented this way (see Table 9.3).

Table 9.4 PLA and PLS characteristics

Type	Device	Organization†	$t_{PD}(max)$ (ns)	f_{MAX} (MHz)	Power (mW)
PLA	PLUS 153-10	18 × 42 × 10	10		825
PLA	PLUS 173-10	22 × 42 × 10	10		850
PLS	PLUS 105-10	16 × 48 × 8		45	800
PLS	PLUS 105-70	16 × 48 × 8		70	800
PLS	PLC 415-16	17 × 68 × 8		16	80

† Inputs × Product terms × outputs
(Courtesy of Philips Semiconductors)

Reference has been made to standard PAL families. The word 'standard' is at first sight a contradiction in terms for a PLD which, after all, is an ASIC. However the standard PALs are foursquare ASICs, configurable by fuse or transistor programming to individual requirements.

The CMOS PAL and GAL families include half-power and quarter-power chips, which have been designed to draw those proportions of the power for the original TTL PALs, operating at approximately the same speed. Quarter-power versions are confined to t_{PD} from 10 ns upwards, half-power versions have t_{PD} of 10 ns or less; there are no 'full-power' PALs. The currents of zero-power and half-power PALs with identical t_{PD} are approximately equal. The variation of I_{CC} with frequency is shown for a variety of CMOS PALs in Fig. 9.10. The half-power represents the second, the quarter-power the third generation of PALs. Maintaining the speed of TTL PALs with half or quarter the original current has come about by the appropriate dimensioning of the CMOS successors with their greatly reduced micron and submicron geometries. But who is going to purchase a half-power PAL, when an equally fast quarter-power PAL with the same speed is available? The answer is that the quarter-power PALs are at present slightly more expensive and a little slower than the half-power versions. Speed is not simply governed by t_{PD}. It depends on the application and is a function of various parameters, the most important of which are contained in eqns (9.4) and (9.5). The price differential will gradually erode. In the meantime the half-power and quarter-power versions will continue to be marketed side-by-side, just like the various TTL families in Table 5.6 will be maintained for some time to come. Note the similarity between Figs 5.38 and 9.10(b) due to the presence of 'non-CMOS' components, the bipolar transistors in the BiCMOS gate and the nMOS array in the PAL.

Although PLAs are outweighed by PALs in volume of sales and variety, they have, together with PLS devices, an established place in the market. A range of PLA and PLS chips is shown in Table 9.4. These are all bipolar with the exception of the ECMOS PLC415-16. The number of product terms is more modest than for PALs (see Table 9.3), but the hardware is nonetheless similarly substantial, because each PLA product term may be used repeatedly for any of the outputs. The product terms of the devices in Table 9.4 are organized into logic and control groups, 32 and 10 respectively for the PLUS 153-10; this is a representative

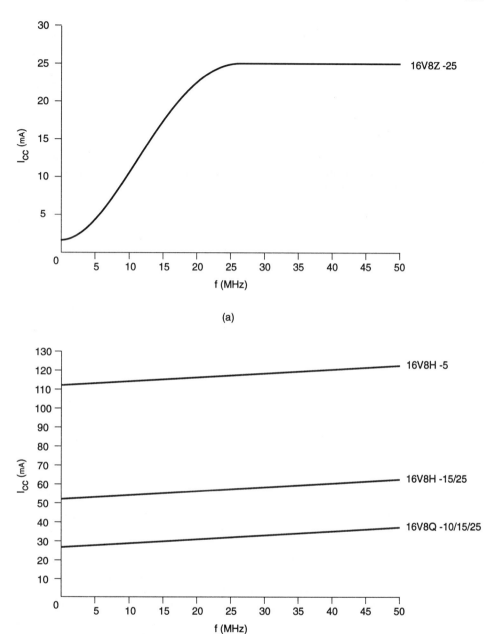

Fig. 9.10 I_{CC} vs frequency—PAL16V8 series (a) zero power (b) quarter and half power (Courtesy of AMD)

distribution.

GAL devices are similar to the PALs which they can emulate, but have more powerful OLMCs and surpass bipolar PALs in speed whilst consuming significantly less power. Indeed their power consumption is lower than that of the corresponding CMOS PALs. Lattice GAL16VP8 and GAL20VP8 are high-power versions which can sink and source

64 mA and 32 mA respectively like the BiCMOS family listed in Table 5.16. The BiCMOS technology has improved on CMOS performance and two 3.3 V GAL-type PLDs, the Philips LVT16V8-7 and LVT20V8-7, have a typical t_{PD} and an I_{CC} of 7.5 ns and 91 mA respectively.

9.3 FPGAs

9.3.1 Introduction

The definition of an FPGA, given in Section 8.7.2, needs to be elaborated now that a variety of architectures are being considered, and in order to differentiate between an FPGA and other types of PLD. It is probably best to start not by defining what an FPGA is, but by pointing out what it is not. The danger of confusing it with a gate array has been eliminated by renaming a gate array *mask-programmable* gate array (MPGA), a term which will be used frequently from now on. The terms FPGA and MPGA distinguish the two types of programming, but leave the ambiguity about the meaning of 'gate array'. Neither the MPGA nor the FPGA are gate arrays in the proper sense of that term. The MPGA is composed of uncommitted array elements (it was called uncommitted logic array in the early years of the ASIC era), the FPGA is based on configurable logic cells. It is easy enough to distinguish between an FPGA and an MPGA; the distinction between an FPGA and other types of PLDs is far more difficult.

The FPGA can arguably be defined to consist of programmable logic configurable cells (combinations of gates, bistables and multiplexers), in-interconnected by a programmable wiring network. The other PLDs have AND, in the case of PALs and PLAs also OR arrays, and macrocells in the data path between the AND array and the I/O ports. These macros have a more prescribed function than the macros of an FPGA. On the other hand the high density PAL-type PLDs with their profusion of macrocells have a strong architectural affinity with low density FPGAs. The likeness is striking for FPGAs with a PLD cell structure, signifying macrocells which include an AND array. Talking of density, the logic power of a PLD is expressed by its array capacity (product lines x product terms) or, in the case of a FPGA, by either the PLD equivalent gate count, alternatively the 'gate array' equivalent gate count, defined in Section 8.3. PLD programmable transistor links consist of one or two transistors and as a general rule the EPLD gate count is from two to three times the equivalent gate ('gate.array' equivalent gate) count. It is more usual to find the capacity defined in terms of PLD gates; occasionally both PLD and gate array figures are provided. The Actel A1225A FPGA, to take a specific example, has a capacity of 2500 'gate array' equivalent gates or alternatively of 6250 PLD equivalent gates.

FPGA architecture is based on logic blocks which contain some hardware in addition to gates and bistables. Two such components are multiplexers and look-up tables (LUTs). The MUX was introduced in its simplest form in Section 8.7.2. A MUX for a n-bit Boolean function is shown in Fig. 9.11(a) (Blakeslee 1979; Hill and Peterson 1981). It has $2^n - 1$ inputs and (n-1) address lines. One of the variables goes to the input lines, the remaining (n-1) variables are applied to the address lines. Figure 9.11(b) contains the schematic of such a MUX and Fig. 9.11(c)

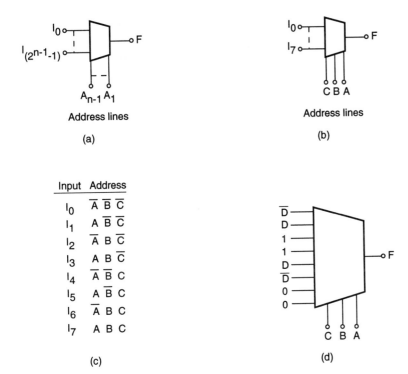

Fig. 9.11 MUX function generator (a) n input MUX (b) 8-input MUX (c) tabulation for (b) (d) solution for example 9.1

gives its truth table.

Example 9.1

Generate the function

$$F = \overline{A}\,\overline{B}\,\overline{C}\,\overline{D} + \overline{A}\,\overline{B}\,C\,D + A\,B\,\overline{C}\,D + \overline{A}\,B\,\overline{C}\,D$$
$$+ A\,\overline{B}\,\overline{C}\,D + A\,\overline{B}\,C\,\overline{D} + \overline{A}\,B\,\overline{C}\,D + A\,B\,\overline{C}\,\overline{D}$$

using the MUX in Fig. 9.11.

The solution, obtained with the aid of Table 9.5, is shown in Fig. 9.11(d).

The architectures of different FPGAs place varying emphasis on AND array logic, multiplexers, and LUTs, which will be explained in the next section.

9.3.2 Selected families

The FPGAs of four leading vendors in this field are outlined in this section, and give a fair representation of current practice. They are:

(i) Xilinx XC2000/3100A/4000E Logic Cell Arrays;
(ii) Actel ACT 1/2/3 FPGAs;
(iii) Lattice pLSI AND ispLSI 1000/2000/3000 families. pLSI stands for programmable Large Scale Integration; i in ispLSI signifies in-system programming;

Table 9.5 Truth table—Example 9.1

Input line	Address	D states	Input
I_0	$\overline{A}\,\overline{B}\,\overline{C}$	$\overline{D}$	$\overline{D}$
I_1	$A\,\overline{B}\,\overline{C}$	D	D
I_2	$\overline{A}\,B\,\overline{C}$	$D+\overline{D}$	1
I_3	$A\,B\,\overline{C}$	$\overline{D}+D$	1
I_4	$\overline{A}\,\overline{B}\,C$	D	D
I_5	$A\,\overline{B}\,C$	$\overline{D}$	$\overline{D}$
I_6	$\overline{A}\,B\,C$	—	0
I_7	$A\,B\,C$	—	0

Table 9.6 FPGA all structure and programming technology (All devices are CMOS)

Company	Cell structure	Programming technology
Xilinx	Look-up table	SRAM
Actel	Multiplexer	Antifuse
Lattice	PLD	EECMOS
Altera	PLD	ECMOS, EECMOS, and SRAM

(iv) Altera Max 5000 and Max 7000 EPLDs.

It speaks for the variety of generic labels that only one of the above four families, all of which are CMOS, is actually called FPGA. A summary of logic structures and programming technologies is given in Table 9.6 (Brown et al. 1992). The entries under 'cell structure' state the *distinguishing* feature of the logic block. Multiplexers are used freely in all cells of the the FPGAs listed.

The architecture of Xilinx FPGAs is shown in Fig. 9.12 (Brown et al. 1992), which resembles Fig. 8.7 and typifies the general practice of having orthogonal routing, although not always in the completely symmetrical form shown. The schematic of the configurable logic block (CLB) of the XC2000 logic cell array (LCA) family is contained in Fig. 9.13. The configurable logic blocks of Xilinx FPGAs consist of one or more LUTs, which are function generators based on a ROM, in this case with 16×1 storage. A 16-bit ROM can generate any logic function up to four input variables (A to D in Fig. 9.13). The LUT has two outputs F and G, which are independent functions of three variables. The output of the flip-flop can be fed back to become an input in place of D, and this mode is an efficient technique for constructing counters and state machines.

FPGAs are, like MPGAs, fabricated in a range of different sizes. The Xilinx FPGAs are SRAM programmable. The Xilinx LCA devices (LCA is a Xilinx trade mark for their FPGA architecture) are configured by programming *configuration* memory cells, which establish the function of the CLBs, I/O blocks, LUTs and the interconnections. The config-

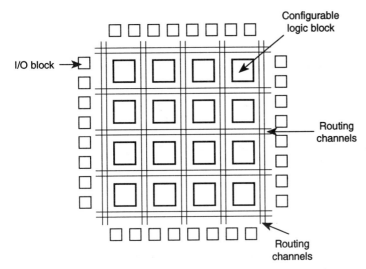

Fig. 9.12 Architecture of Xilinx FPGAs (Reproduced by permission of Kluwer Academic Publishers)

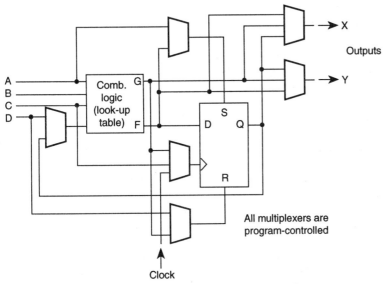

Fig. 9.13 Configurable logic block—Xilinx XC2000 LCA (Courtesy of Xilinx, Inc. 1994. All rights reserved)

uration program is loaded into the configuration memory cells when the chip is powered up. It can reside in ROM (EPROM, EEPROM) on the PCB or on disk. The configuration program sets the LUTs into the desired mode for generating the required functions, controls all the multiplexers, and establishes the interconnections. There is no RAM as such on the chip. The memory cells are one-bit stores attached to the nodes they control. The CMOS five-transistor configuration memory cell in Fig. 9.14 is a latch made up of two inverters and driven by a gating transistor. It is written in once when the chip is powered up; its power consumption in operation is zero. The speed of the configuration

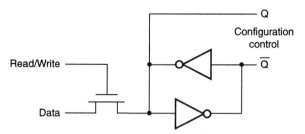

Fig. 9.14 Configuration memory cell—Xilinx XC2000 LCA (Courtesy of Xilinx, Inc. 1994. All rights reserved)

memory cells is relatively unimportant, because they are only used on initialization. The configuration program is entered serially—it is a serial bit-stream—and the time taken to do so, ~30 ms at the most, presents no problem. The configuration memory cells are very reliable; their FIT rate is ~1000 (Xilinx 1991). The short interval for program entry allows real-time changes to be made during operation, because the chip can be reset and reloaded in-circuit with a modified configuration program.

The XC4000 family allows the LUTs to be configured as a user RAM. The LUT was labelled a ROM in the description of it earlier on, a designation favoured in the literature on this subject. It does act in that manner, because after its cells have been set by data from the configuration program its outputs are an unalterable function of the data inputs. If instead of setting the input to the LUT by the configuration program, data can be entered during operation, the LUT is turned into a user RAM. That alternative significantly enhances the system capability of the device for functions like FIFOs, register files, and RAM-based shift registers. The configuration of LUTs into user RAMs inevitably curtail the logic functionality of the CLB. A compromise is usually arrived at, because in many application the RAM requirement can be satisfied by using only a proportion of the CLBs available. The maximum RAM capacity for the XC4000E family extends from 2 Kb to 32 Kb.

The ease of reprogrammability is the outstanding advantage of SRAM programming. Modifications can be entered into the FPGA without removing the chip from the board. Furthermore SRAM programming permits *readback*, the readout of the program, the state of the flip-flops and the memory cells, for proving the design and debugging. Reprogramming can be undertaken without extra cost and quickly. An obvious disadvantage is the SRAM volatility; the FPGA has to be reprogrammed when it is powered up. That process is however quite fast and is set in train by an automatic initialization. A summary of performance for the XC LCA families is given in Table 9.7. The system speed depends on the application and the figures quoted are only indications. ASICs are configured to individual requirements and the system speed is likely to vary widely from case to case. The XC2000 and XC3100A LCA families are available in low voltage (3.3 V) versions, which incur an unavoidable loss of speed (~30 per cent) relative to the 5 V chips. The higher function density of the XC3100A and XC4000 series is not merely a matter of increasing the number of CLBs. At the lower end of the range each of the three families in Table 9.7 contains 64 such blocks. The higher density, reflected by the

Table 9.7 Performance of Xilinx XC LCA families

Parameter	XC2064	XC2018	XC3020A	XC3195A	XC4002A	XC4025E
Equivalent gates (approx)	800	1500	1000	8000	2000	25 000
CLBs	64	100	64	484	64	1024
I/O (max)	58	74	64	176	64	256
Typical system speed (MHz)	30	30	40	85	40	65
Configuration program (bits)	12 038	17 878	14 779	94 984	31 628	422 128
User-RAM (Max bits)	—	—	—	—	2048	32 768

(Courtesy of Xilinx, Inc. 1994. All rights reserved)

Table 9.8 Time constants of FPGA programming technologies

Technology	R on-resistance (Ω)	C (fF)	RC (ps)
nMOS pass transistor (SRAM)	1500	15.0	23
ECMOS	3000	15.0	45
EECMOS	3000	15.0	45
PLICE antifuse	500	5.0	2.5
ViaLink antifuse	70	1.2	0.084

equivalent gate count of the XC3100A and XC4000E series, is due to the more powerful CLBs of these second and third generation families.

The Actel FPGAs are fuse programmed by means of the PLICE antifuse described in Section 7.5. The performance of these devices is greatly influenced by the characteristics and reliability of the antifuse. Extensive studies indicate a lifetime of well over 40 years operation at 5.5 V and 125 °C for the PLICE antifuse of the Actel FPGAs (Greene et al. 1993). Fuse-programmed PLDs can be configured once only, regardless of whether fuses or antifuses are being used. That is their disadvantage vis-à-vis the other programming technologies in Table 9.6. Programming is simpler for the antifuse than for a fuse or transistor-link technique, because only those antifuses which have to establish links, typically two per cent of the total, have to be blown (Greene et al. 1993). An outstanding advantage of the antifuse is the small area it occupies and its consequent small self-capacitance. An alternative, the ViaLink antifuse of Quicklogic (ViaLink is a trade mark of the Quicklogic Corporation) has a smaller on-resistance and self-capacitance than the PLICE antifuse. Both types have a far smaller intrinsic time constant than the other programming technologies, see Table 9.8. The data in Table 9.8 is based on data from Rose et al. (1993) and vendors' FPGA data books.

324 ASICs—Programmable logic devices

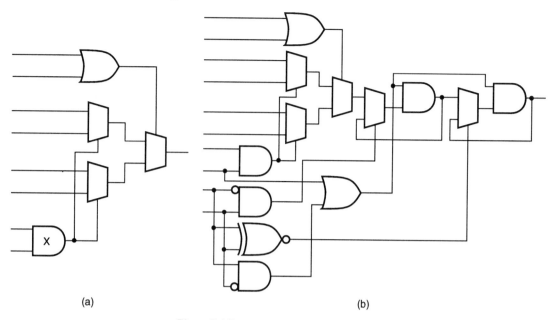

Fig. 9.15 ACT 2 and ACT 3 module architecture (a) C module—combinational (b) S module—sequential (Courtesy of Texas Instruments)

Evidently the antifuse minimizes routing delays. Furthermore it is small enough to fit in with the width of the interconnection and occupies far less space than a transistor–programmable link.

The Actel FPGA floor plan is similar to the row-based layout of an MPGA shown in Fig. 8.2(a), with logic modules in place of array elements. The series is produced in three families, ACT 1, ACT 2, and ACT 3. The ACT 1 family, which was first on the scene, is based on a combinational (C) module, augmented by a serial (S) module, which includes a flip-flop, for the ACT 2 family. A more powerful S module has been added to the C module for the ACT 3 family, introduced in 1993. Block and architectural schematics of the C and S modules are given in Figs 9.15 and 9.16. The detailed logic schematic of Fig. 9.15 was obtained from Texas Instruments (TI) data; TI second source these FPGAs. The only difference between the C modules of the ACT 1 and ACT 2 devices is the addition of the AND gate in Fig. 9.15 marked X for the latter. On a point of detail, logic capability is related to the flip-flop density, and device data usually includes the number of dedicated flip-flops. For the ACT 3 family this is one flip-flop per S, and two per I/O module. Additional flip-flops can be configured from two C modules, allowing the flip-flop density to be raised above the sum total of the S and I/O modules if necessary. C and S modules are placed side by side in the module rows, and are very nearly equal in number, as can be seen from Table 9.9. The PLD gate count is from two to three times the equivalent gate count, a ratio to be expected from comments on this subject in Section 9.3.1.

The very large number of PLICE antifuses illustrate the vast choice of interconnections available. Quotations of speed in terms of a single

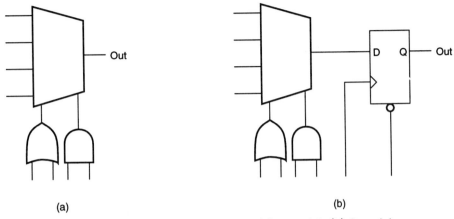

Fig. 9.16 ACT 2 and ACT 3 module schematic (a) C module (b) S module (By permission of Actel Corporation)

Table 9.9 Characteristics of Actel FPGAs

Parameter	ACT 1		ACT 2		ACT 3	
	A1010B	A1020B	A1225A	A1280A	A1415A	A14100A
Equivalent gates	1200	2000	2500	8000	1500	10 000
PLD equivalent gates	3000	6000	6250	20 000	3750	25 000
C modules	295	547	220	608	96	698
S modules	—	—	231	624	104	697
I/O modules	57	69	83	140	80	228
F_{max} (MHz) 16-b prescaled counter	~70	~70	105	85	145	115
PLICE antifuses	112 000	186 000	250 000	750 000		

(Courtesy of Actel Corporation)

value per device are of limited value. The speed of a (V)LSI ASIC depends on the nature of its function, whereas the transient response of logic gates and the toggle (count) rates of flip-flops can be stated precisely. Indications of system speed extending over a frequency range are more appropriate. Comments to that effect have been made before in relation to ASICs. The performance of the ACT 3 family for three different functions is a good example of ASIC behaviour. The worst-case maximum frequencies of 16-bit accumulators, loadable counters, and prescaled loadable counters are 47 MHz, 82 MHz, and 150 MHz respectively.

The move towards ECMOS and EECMOS programming is exemplified by the EECMOS FPGAs of Lattice, made in two versions aptly named programmable large scale integration (pLSI) and in-system programmable large scale integration (ipLSI). The ispLSI devices have the unique facility of on-board erasure and programming with the standard 5 V supply voltage. The pLSI and ipLSI ranges are absolutely identical otherwise, and are produced in three families, the 1000, 2000, and 3000 series. The on-board programming alternative is also available for one of the Lattice GALs, the ispGAL22V10.

326 ASICs—Programmable logic devices

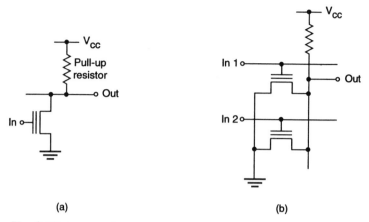

Fig. 9.17 ECMOS/EECMOS programming (a) inverter (b) two-input NOR gate

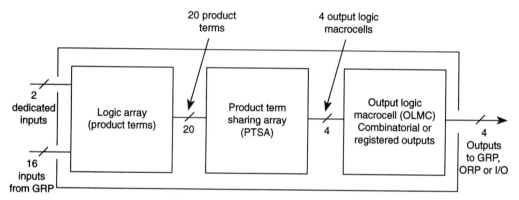

Fig. 9.18 Generic logic block—Lattice FPGA (Courtesy of Lattice Semiconductor Corporation)

ECMOS and EECMOS programming links are unlike the serial SRAM and antifuse links, which connect two wires. Instead they are pull-down devices for logic components like inverters, gates, and multiplexers. The transistor M_1 in Fig. 9.17(a) will, if left unprogrammed, pull the output to zero for a '1' on the input. If programmed, the transistor will be permanently open circuit and the output line will be high. A programmable two-input NOR gate is shown in Fig. 9.17(b).

The floor plan of the pLSI/ispLSI families differs from those described so far, and the architecture of the logic blocks contains a number of innovations. The devices are based on powerful generic logic blocks (GLBs). The block schematic of the GLB for the 1000 and 2000 series is reproduced in Fig. 9.18. GLBs are linked to the global routing pool (GRP) and the output routing pool (ORP), with the GRP occupying the core of the die (Fig. 9.19). The GRP connects the entire internal logic and allows it to be shared among all the GLBs. The logic array of the GLB, an array of product terms, has 16 inputs from the GRP and two dedicated inputs. Its twenty product terms feed the product terms

FPGAs

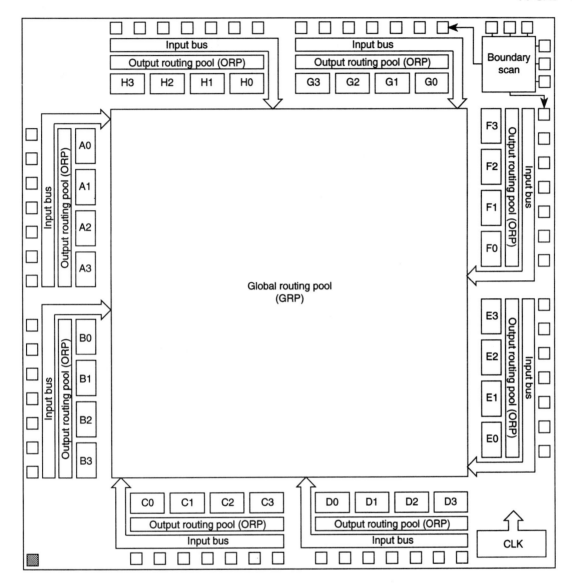

Fig. 9.19 Architecture of Lattice FPGA (Courtesy of Lattice Semiconductor Corporation)

sharing array (PTSA), where they can be shared with any of the four GLB outputs. The four outputs of the PTSA drive the four OLMCs, which offer combinatorial or registered outputs. The 3000 series has a 'twin' GLB in which 24 inputs to the logic array yield two groups of 20 product terms which are fed to two PTSAs. These supply 2×4 outputs to the groups of four OLMCs, leading to a total of eight outputs. The facility of feeding inputs from the GRP into the GLBs, combined

Table 9.10 Characteristics of Lattice FPGAs (The information applies equally to the ispLSI families)

Parameter	pLSI 1000		pLSI 2000		pLSI 3000	
	1016	1048	2032	2096	3192	3320
PLD gates	2000	8000	1000	4000	8000	14 000
f_{MAX} (MHz)	80	80	137	110	110	80
GLBs	16	48	32	96	192	320
I/O (max)	36	106	34	102	96	160

f_{MAX} applies to a 16-bit loadable counter.
(Courtesy of Lattice Semiconductor Corporation)

with the nature of the PTSA, gives the pLSI/ispLSI families the ability to implement complex logic. The performance of these FPGAs is summarized in Table 9.10. The pLSI/ispLSI 1000 family implements high integration functions and achieves typically 80 MHz system speed. The pLSI/ispLSI 2000 series is distinguished by a high I/O to logic (PLD gates) ratio and is faster, achieving system speed of up to 137 MHz. The highest density is available with the pLSI/ispLSI 3000 family, which includes a boundary scan, a provision for testing to the IEEE standard 1149.1, which is described in Chapter 10. The typical system speed is down to ∼100 MHz. An increase in density means an increase in the number of wires, and some loss of speed is to be expected.

The last of the four vendors listed in Table 9.5, Altera, markets CMOS FPGAs with ECMOS, EECMOS, and SRAM technologies. Altera divide their Erasable Programmable Logic Devices (EPLDs) according to their architecture into two categories, the 'classical' types with the traditional AND array and a PAL-type architecture, and the multiple array matrix (MAX) 5000 and 7000 EPLDs, which are justifiably classed FPGAs. The floor plan of the MAX 7000 family, sketched in Fig. 9.20, is similar to the layout in Figs 8.7 and 9.12. The FPGA is composed of logic array blocks (LABs), interconnected by a programmable interconnect array (PIA). The LAB is made up of a macrocell array (a combination of eight or sixteen macrocells), an expander product term array, and an I/O block (Fig. 9.21). The macrocell includes an AND array, a product term array, an exclusive-OR gate, and a D flip-flop. It is the prime resource for generating the required logic. Its power is enhanced by the nature of the AND array, which contains dedicated inputs, programmable interconnect signals, and an expander product-term array, consisting of a group of free-standing inverted product terms.

The ECMOS MAX 5000 family extends from the EPM5016, which contains a single LAB, to the EPM5192 made up of 12 LABs (192 macrocells). The characteristics of the MAX 7000 family are given in Table 9.11. This second generation of MAX EPLDs is largely in EECMOS, although a few of the devices are ECMOS, an interim state of affairs. The quotation of available and usable gates in Table 9.11—equivalent gates are implied—is the first of its kind in this text. It occurs again in Table 9.12 and for MPGA characteristics in Chapter 10. The figures in Table 9.11 imply a 50 per cent utilization. The explanation for this

FPGAs 329

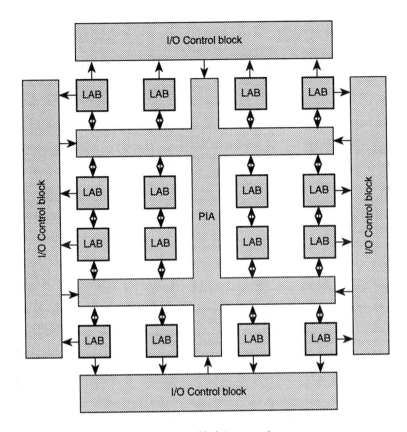

PIA - Programmable interconnect array

Fig. 9.20 Global architecture for Altera MAX 5000/7000 FPGAs (Courtesy of Altera Corporation)

Table 9.11 Characteristics of Altera MAX 7000 EPLDs

Parameter	EPM7032	EPM7256
Available gates	1200	10 000
Usable gates	600	5000
Macrocells	32	256
I/O (max)	36	164
f_{MAX} (MHz) (16-bit loadable counter)	100	63

(By permission of Altera Corporation)

factor is best understood by referring to the MPGA architectures in Section 8.3. In the channelled gate array about half the silicon is reserved for routing, and in the SOG array practically the entire die area is taken up by array elements (equivalent gates for this purpose). Gate utilization is not quoted for channelled arrays. It depends strongly on the application and is likely to be somewhere between 50 and 90 per cent. It is however becoming usual to find a fixed quotation of gate utilization (fixed in the sense of not being tied to an application) for SOG architectures. That

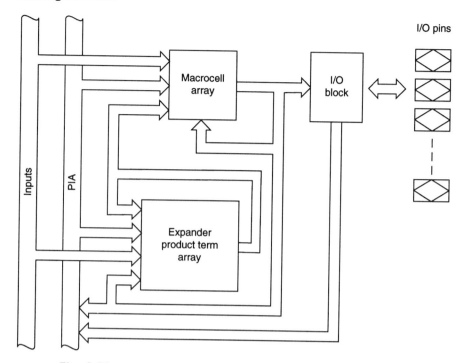

Fig. 9.21 Logic array block—Altera MAX 500 series (Courtesy of Altera Corporation)

Table 9.12 Characteristics of Altera FLEX 8000 EPLDs

Parameter	EPF8452	EPF81188
Available gates	8000	24 000
Usable gates	4000	12 000
Logic elements	336	1008
I/O (max)	120	184

(By permission of Altera Corporation)

utilization is close to 50 per cent in nearly all cases and this applies to the FPGAs in Tables 9.11 and 9.12. The implication is that with an SOG architecture or, in this case, the architecture of the MAX FPGA families, a 50 per cent utilization due to the constraints of connectivity can be virtually guaranteed independent of the application. A similar figure may hold for other FPGAs with a related architecture. The 50 per cent utilization leads to a higher (equivalent) gate count for an SOG than for a channelled MPGA. A channelled gate array of equal size will have fewer used gates, because the total number available is about half that for the SOG array.

The latest addition to the Altera FPGAs is the Flexible Logic Element Matrix (FLEX) 8000 family, fabricated in CMOS SRAM technology. The architecture is similar to Fig. 9.20, but the composition of the LABs is very different. The FLEX 8000 LAB is made up of eight logic elements (LEs) each of which contains a four-input LUT and a programmable flip-flop, permitting a registered or combinatorial LE output. In this respect the compositions of the LE and the Xilinx CLB (Fig. 9.13) are similar. The LE includes two fast data paths, the carry and cascade chains. These connect all LEs in a LAB and all LABs in the same row. The FLEX 8000 operates like the Xilinx SRAM FPGAs with regard to the loading of a configuration program on power-up and in-circuit modification. The characteristics of the FLEX 8000 family are given in Table 9.12.

The four FPGA families which have been described represent a fair cross-section of current products. The tabulations indicate their densities with an upper limit arount 25 000 equivalent gates and system speeds which are in the range (25–75) MHz. ASIC design style selection, embracing all PLDs and the other two categories of semicustom circuits, MPGAs and standard cells, is covered in Chapter 10. Some issues appertaining to the relative merits of PLDs and the other semicustom ASICs can be identified at this stage. The PLD, being completely prefabricated, eliminates NRE and mask fabrication costs. PLD post-design configuration is done with proven software and hardware tools, based on established techniques used for PROM programming. These practices are easily applied to fuse, SRAM, and ECMOS/EECMOS links. The disadvantages of PLDs are the higher cost of quantity production (each PLD has to be programmed individually), the lower density, and the slower speed. A high proportion of the silicon in an FPGA is taken up by the overhead circuits for configuration. The structure includes far more potential interconnections than are actually used in order to offer a good choice for optimizing links. Only a few per cent of the available interconnections are utilized.

The speed of the FPGAs is limited by the time constants of the programmed interconnections, where antifuses have a decided advantage over other technologies (Table 9.8). Propagation delay increases with wire length and system speed is reduced for devices with a high density. The extent of the overhead circuitry for establishing links is evident from the figures for configurable bits and PLICE antifuses of the Xilinx and Actel families in Tables 9.7 and 9.9.

The FPGA was originally developed to advance PAL/PLA density. Subsequent advances have made FPGAs an alternative to MPGAs at densities up to about 30 000 equivalent gates. That figure, like the gate count of MPGAs, continues to increase. FPGA architecture and the supporting design tools are undergoing extensive development in order to raise density and accelerate the design process. FPGA architecture is characterized by its *granularity* i.e. the complexity of the basic logic module (Trimberger 1994). The logic blocks are either *fine* or *coarse grain* (Rose et al. 1993; Greene et al. 1993). Fine-grain logic blocks consist of simple cells like the MPGA array elements, which are equivalent to a two-input NAND or NOR gate. Coarse-grain logic includes more complex circuits like multiplexers and flip-flops. The Xilinx CLB (Fig. 9.13) and the Actel S and C modules (Figs 9.15 and 9.16) are examples of coarse-grain logic elements. There are varying degrees of granularity falling in between the array element of an MPGA and the Xilinx CLB in Fig. 9.13. Fine-grain modules occupy a small space and have a small internal delay. They are easily compounded for higher logic functions, but this puts a strain on the routing architecture: there is a danger of exceeding the capacity of the local connectivity. Coarse-grain (usually muliplexer-based) modules have a higher functionality per transistor count. They require more inputs and a correspondingly large amount of routing. The granularity of FPGA modules is a compromise to achieve the best performance in terms of speed and density. The optimization depends somewhat on the nature of the applications envisaged.

It was pointed out in Section 9.3 that advanced PLDs with PAL-type arrays can be classified as FPGAs, even if they are not given that title in their generic description. By and large FPGAs are targeted at the higher densities of the PLD ranges. The Altera MAX EPLDs can and are deservedly placed in the FPGA category. The same applies to the Cypress MAX EPLDs, which adopt the MAX architecture of Altera with a family going up to 192-macrocell chips. Philips likewise have programmable macro logic (PML) devices based on a folded NAND array (Fig. 9.6) and designated FPGAs in the preface of their data handbook (Philips Semiconductors 1993). These observations have been made to reinforce the advice given in Section 9.3.1 when explaining the architecture of an FPGA in juxtaposition to other PLDs. Placing the FPGA into the category of PLDs makes it easier to gain an overall appreciation of PLDs and their capabilities.

9.4 Design outline

An ASIC is only as good as the design tools which support it. These and the design process itself are described in the next two chapters. The design outline which now follows briefly discusses some general concepts and describes PAL/PLA design.

The first factor in ASIC choice is a realistic estimate of the complexity in terms of equivalent gates to fulfil a specified function. The task might be a straightforward combination of standard TTL/CMOS SSI and MSI—mopping up glue—into a single chip. At the other end of the spectrum one or more ASICs may be required to implement a microcontroller. These two examples are representative of the bottom-up (mopping up glue) and top-down (microcontroller) design routes. The bottom-up technique is the easier and more straightforward of the two. The ASIC function is precisely defined in terms of TTL/CMOS SSI and MSI. The term CMOS will be dropped from now on when referring to SSI and MSI in relation to ASICs. TTL is understood to signify the SSI and MSI functions of the standard TTL and CMOS series; most of these overlap. The combination of standard TTL SSI and MSI into a single chip was and still is a major application for PALs and PLAs. It is a fact that FPGAs were also heavily used for that purpose in the early years of their development. They can still fulfil that role, but are intended for more complex applications calling for top-down design. Returning to the design of a microcontroller or another system with similar complexity, this could be done by a bottom-up process. A logic schematic of the system can be built up laboriously by designing the various building blocks like counters, shift registers, ALUs etc. and detailing these in terms of the basic gates and bistables. Such a method would take far too long, and a complex design can only be completed on an acceptable time scale by a top-down approach, commencing with a behavioural description and carried out with the aid of electronic design automation (EDA). The design, whether top-down or bottom-up, leads to a complete logic schematic (confining consideration to digital designs for the present), formulated in a netlist which contains all the logic components and their interconnections. (Alternative design entries germane to PLD ASICs will be mentioned presently). A bottom-up approach simply combines the SSI and MSI elements into the required logic schematic, translated as before into a netlist. Once the schematic and the netlist are to hand, the next phase of the design can commence.

The above introduction holds for PLDs and mask-programmable ASICs alike. The subsequent design procedures for these two categories vary, although they have a good deal in common. A key process in FPGA design is the selection of the cells to be used, their configuration and their interconnection. The utilization, i.e. the proportion of logic cells used, is less—sometimes far less—than 100 per cent, being constrained by the feasibility of achieving the interconnections. The final step in the design, the placement (location) of the used cells and their interconnection, is known by the name of *place* and *route* (alternatively *placement* and *routing*). An important distinction between FPGA and MPGA is the selection from a large number of possible interconnections for FPGA routing, in contrast to mask design for MPGA routing. FPGA design entry can be either in the form of a logic schematic, or in sum-of-products Boolean expressions, or in a state machine formulation with a truth table representation, in which case no graphical display is involved.

The step following on from the schematic capture (or in the case of an FPGA, possibly from one of the alternative forms of design entry mentioned above), is *technology mapping*, which signifies the most efficient conversion of the technology-independent components of the netlist derived from the schematic capture into a netlist of components obtainable with the logic blocks of the FPGA. This is followed by placement and routing, design verification, and calculation of circuit performance. When all these checks have been made, the FPGA is ready to be configured. Having presented this outline, we leave FPGA design because it is described in more detail together with MPGA and standard cell design in the next two chapters in a unified manner. The only comment added at this stage is about the macro support. FPGAs and mask-programmed ASICs are backed by substantial design support with hard and soft macros, which were defined in Sections 8.3 and 8.4. The four FPGA families in Section 9.3 enjoy such design backing. The Xilinx XC4000 series contains ~300 macros. The macro totals for the Actel families are ~250, ~350, and ~500 for the ACT 1, ACT 2, and ACT 3 series respectively. Over 275 macros are available for the Lattice pLSI/ispLSI families. Lastly there are ~300 macros for the MAX 7000 family of Altera. In all cases these totals are combinations of hard and soft macros. Many of the macros emulate TTL SSI and MSI.

PAL design is supported by proven software design tools and programming hardware, which operates with individualized circuits (*personality cards*) to blow fuses/antifuses or program ECMOS/EECMOS links of specified standard PALs. The design entry for programming a PAL can be in one of several formats. It is either a schematic capture or

334 ASICs—Programmable logic devices

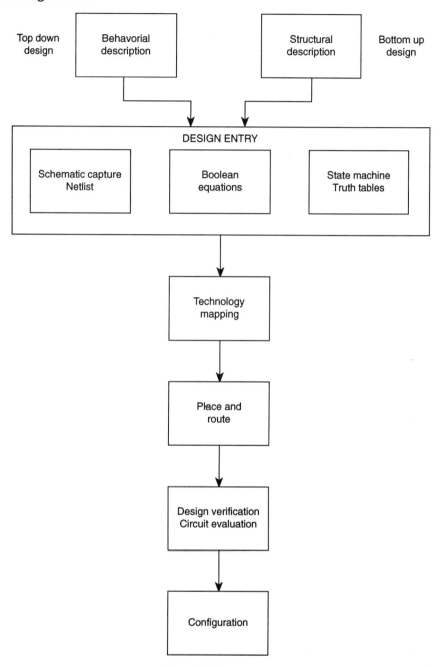

Fig. 9.22 PLD design flow

a netlist of the logic schematic. Alternatively it can be a set of Boolean equations, or a state machine entry with truth tables (Fig. 9.22). The design flow applies to an FPGA or a PAL. The customary route for a PAL is bottom-up design. The initial cycle of the design flow is similar for ASICs in general. Technology mapping translates the design entry into configured circuits of the PAL and decides on the proportion of PAL cells to be used. The place and route software determines the location

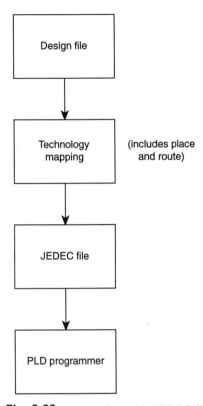

Fig. 9.23 PLD design with JEDEC file

of the used cells (placement is alternatively called *annealing*) and their interconnections. The design data has to be in a format acceptable to the programming hardware. A firmly established standard first proposed in 1980, JEDEC standard 3, has been drawn up for that purpose. The JEDEC file defines all fuse links of a PLD in a form recognized by the programming hardware designed to that standard. In addition to programming a JEDEC file includes test vectors. The design flow using a JEDEC file is shown in Fig. 9.23.

PLD design tools can be grouped into three categories

(i) assemblers;

(ii) compilers;

(iii) architecture-independent design tools.

Assemblers convert a Boolean equation design entry into programming data. PAL design is mainly carried out with assemblers. Compilers support various high level design entries and incorporate logic synthesis to produce design data in several formats. They require that the type of PLD to be used is specified in advance, a restriction absent in assemblers. Architecture-independent design tools are more advanced and are suitable for the complex architectures of FPGAs and mask-programmed ASICs.

PALASM, which stands for PAL assembler, is a Fortran IV computer program pioneered by AMD. It converts design descriptions in sum-of-products Boolean equations directly into JEDEC files for downloading into the programmer. The restrictions of the sum-of-products format limits the usefulness of PALASM, because the design entries nowadays are no longer confined, like they were in the early days of the PAL, to Boolean equations. PALASM supports only a limited number of devices, generally PALs marketed by AMD. These are however strongly second sourced. A comprehensive description of PAL design based on PALASM is contained in a text by the technical staff of Monolithic Memories (1981), the forerunner of AMD. A recent new version of PALASM, PALASM 2, supports state machine in addition to Boolean equation entries, and includes a logic minimization module for a restricted number of PALs.

PAL vendors offer comprehensive in-house design tools. The Altera A+ PLUS, PLCAD-SUPREME, and PLS-SUPREME programmable logic systems for their classic EPLDs are examples of advanced, comprehensive design support. The two configurations are:

(i) PLCAD-SUPREME

A complete development system embracing software and programming hardware.

(ii) PLS-SUPREME

The complete software suite of PLCAD-SUPREME, but no hardware.

The A+ PLUS software supports the following design entries:

(i) LogiCaps schematic capture;
(ii) Boolean equations;
(iii) state machine and truth table.

The LogiCaps software includes two libraries, which may be supplemented by user-created libraries:

(i) the A+ PLUS Primitive Library containing macros for the basic logic gates and flip-flops;
(ii) the A+ PLUS Macrofunction Library containing over 120 macros for TTL functions, including MSI like counters, etc.

The Altera Design Processor (ADP) transforms the design into a JEDEC file for programming.

References

Blakeslee, T.R. (1979). *Digital design with standard MSI and LSI*, pp.65–76. Wiley, USA.

Bostock G. (1993). *Programmable logic handbook*, (2nd edn). Newnes, England.

Brown, S.D., Francis, R.J., Rose, J., and Vranesic, Z. (1992). *Field programmable gate arrays*. Kluwer Academic Publishers, USA.

Coli,V.J. (1989). VLSI programmable devices. In *VLSI Handbook*, (ed. J. DiGiacomo), pp.4.16–4.18, McGraw-Hill, USA.

Greene, J., Handy, E., and Beal, S. (1993). Antifuse field programmable gate arrays. *Proceedings of the IEEE*, **81**, 1042–56.

Hill, F.J. and Peterson, G.R. (1981). *Introduction to switching theory and logical design*, (3rd edn), pp.196–209. Wiley, USA.

Monolithic Memories (1981). *Designing with programmable array logic*, (2nd edn). Monolithic Memories and McGraw-Hill, USA.

Pellerin, D. and Holley, M. (1991). *Practical design using programmable logic*. Prentice Hall, USA.

Philips Semiconductors (1993). *Programmable logic devices data handbook*, pp.673–5. Philips Semiconductors, USA.

Rose, J., Gamal, A.E., and Sangiovanni-Vincentelli, A. (1993). Architecture of field-programmable gate arrays. *Proceedings of the IEEE*, **81**, 1013–29.

Trimberger, S.M. (ed.)(1994). *Field-programmable gate array technology*. Kluwer Academic Publishers, USA.

Xilinx (1991). *Static memory technology overview*. Xilinx Technical Brief, Xilinx Inc., USA.

Further reading

Bostock G. (1993). *Programmable logic handbook*, (2nd edn). Newnes, England.

Brown, S.D., Francis, R.J., Rose, J., and Vranesic, Z. (1992). *Field-programmable gate arrays*. Kluwer Academic Publishers, USA.

Pellerin, D. and Holley, M. (1991). *Practical design using programmable logic*. Prentice Hall, USA.

Trimberger, S.M. (ed.)(1994). *Field-programmable gate array technology*. Kluwer Academic Publishers, USA.

10
ASICs—Characteristics and design issues

10.1 Introduction

The choice of standard ICs is a matter of selecting the best component for a given application without undertaking any IC design. The choice of an ASIC is the selection of an IC which has to be completed with a substantial design effort. That distinction stamps the difference between merely selecting a component and engaging in its design. Whether this effort, which depends on the nature of the application, is large or small, the task of transforming the raw die into a finished product is formidable. The design style selection is inextricably linked to the design of the ASIC and is affected by this. The user has to take a critical look at not only at the specifications, but also at the vendors of the various ASICs and at the design effort entailed.

The ASIC(s) must be justified on grounds of cost. Before embarking on ASIC design, the customer must know, even if only in round terms, the types of chips, their functionality, and their equivalent gate count. Such a preliminary estimate is followed by a more detailed assessment of the system, leading to an improved quantification of the requirements. The initial estimate allows a critical appraisal of the economics and determines whether to go ahead with ASICs or not. The major factors to be considered are NRE cost, chip production costs, and system production costs, which are linked to the quantities of ASICs required. The term ASIC design is in the singular: ASICs are designed to an individual specification one at a time. A system on the other hand may require several ASICs and the initial appraisal will establish the number and types of ASICs, possibly combined with standard ICs, to implement the system. There may be subsequent slight changes during the detailed design, but the specifications emerging at this stage should be pretty near the mark.

The task of designing an ASIC is a formidable undertaking. One only has to examine the data sheets of a simple logic gate in TTL/CMOS, or of an operational amplifier to realize the plethora of parameters which have been specified in order to allow prediction of circuit behaviour under varying conditions (changes in temperature, supply voltage, fan-in, fan-out, etc.). Close liaison between vendor and customer is essential for successful design of an ASIC composed, as it is, of thousands, possibly many thousands, of such circuit 'bricks'. The vendor specifies the performance of the macros and supplies adequate circuit and layout data for various checks and simulations. ASIC design is based on the vendor's macro library, augmented occasionally—but not all that frequently—by user-defined macros.

ASIC design proceeds along one of the following lines:

(i) The turnkey route. The customer undertakes the initial system design, selects the design style, and designs the logic schematics of the ASICs. At that stage the customer hands over the design to the vendor. The execution of the design includes fabrication of the dies, their encapsulation and testing.

(ii) The foundry route. The customer undertakes the entire design and the vendor's involvement is limited to fabrication of the dies. The vendor is a silicon foundry (or GaAS foundry; some ASICs are GaAs), whose services usually include packaging and testing.

The turnkey route is by far the more widely used of the two. The design procedures for both routes are essentially identical and are only distinguished by the differences in involvement on part of the vendor and the client. Newcomers to ASIC design and many users without substantial design experience will opt for the turnkey solution. The foundry alternative is followed by those who have amassed comprehensive design acumen, have acquired all the necessary design tools, and are confident of proceeding on their own.

The flow initiating ASIC design is illustrated in Fig. 10.1. The system requirement leads to a system specification, and at this stage the customer will have a good idea about the types of ICs required in terms of their functions, and equivalent gate and pin counts.

Even the turnkey route demands probably some design effort, and at the very least a good understanding of the design process, on the part of the user. In the initial design phase, most of which is in the user's court, of Fig. 10.2 there will be close cooperation between user and vendor and this should be maintained throughout the design. The vendor's expertise and know-how does not absolve the user from contributing to the design. Co-operation between them is highly desirable at all stages. The foundry and turnkey routes are not clear-cut compartmentalized entities. The user will be (or should be) heavily involved with the vendor in the design, whatever path is followed.

Having arrived at the system specification in Fig. 10.1, the economic factors are carefully weighed up by the customer in the light of the chip specification(s), the cost of development and production, and the cost of the design tools, if these represent an initial outlay. The cost is linked to the design style selected and to the projected IC quantities. It was pointed out in Chapter 9 that for modest numbers of chips, the FPGA is a far more cost effective solution than the gate array with a similar density. If, weighing up the various factors, the decision is taken to use ASICs, design style selection including the ASIC manufacturer comes next, and ASIC design can commence.

Design for testability is a vital part of the design process. In the turnkey process it is the responsibility of the customer, but the testing of the chips is undertaken by the vendor. This part of the design is a major part of the total effort and demands a significant portion of the silicon estate.

The design style selected determines the design tools required. The

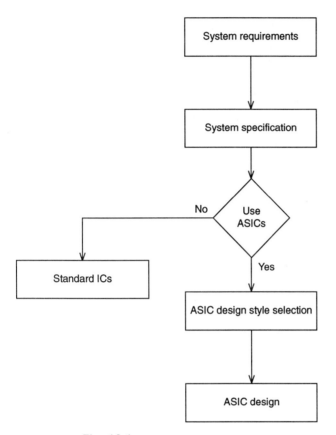

Fig. 10.1 ASIC design approach

choice of the best ASIC for a given application calls for a searching examination of ASIC characteristics. The information on selected ASIC families in Section 10.5 is more detailed than the data for standard digital and analogue ICs in Chapters 5 and 6, because the ASIC user has to delve into considerable detail in arriving at a choice and taking part in the design. Digital ASICs naturally come first to mind. They paved the way of the ASIC revolution and far outnumber mixed-mode (analogue and digital) ASICs. However there has been a pointed change. With the impact of multimedia technology and its manifestation in data and telecommunications, and with the universal expansion of communications in general, analogue circuits have gained in importance within the ASIC sector. Mixed-mode ASICs have been in existence on a small scale ever since ASICs appeared on the market. The number of vendors offering such chips has greatly increased during the last few years as has the variety of analogue circuits within such ASICs. A major factor in marketing mixed-mode ASICs has been the reduction in the testing costs of the analogue section. These costs have now come down sufficiently to make mixed-mode ASICs economically attractive.

The decision of going ahead with ASICs for a specific project is a tripartite selection of the design style, the vendor, and at least some, if not all, of the design tools. The information in this chapter is aimed

342 ASICs—Characteristics and design issues

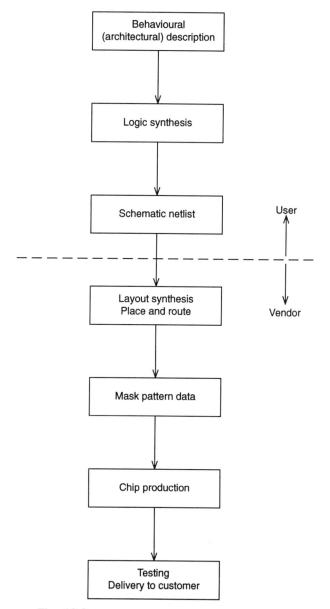

Fig. 10.2 ASIC design sequence—turnkey route

at enabling the user to choose the right ASICs and to be aware of the design process, which is described fully in the next chapter.

10.2 Design methodology and design tools

The design flow in Fig. 10.2 starts where the design initiation in Fig. 10.1 left off. The last block in Fig. 10.1, ASIC design, is equivalent to the entire schematic in Fig. 10.2. The design sequence, which applies to top-down design, starts with a behavioural or architectural high-level description. The first step is its translation by means of logic synthesis into a structural schematic complete with interconnections. The schematic, any part of which may be displayed on the screen of the workstation, is stored in the form of a netlist, which contains all the logic elements and their interconnections. The next move is layout synthesis (place and route) which transforms the schematic into the mask layout, composed of the selected macros and their interconnections. Chip production, testing, and delivery to the customer follow. That overview of the design flow points to the nature of the design tools. The hardware core is the *workstation*, first introduced in 1982.

A workstation is a specialized stand-alone computer for handling a specific task, in this case ASIC design. It is dedicated to a small number of users. Its main facilities are networking and interactive graphics. To engage in ASIC design, it has to have a very large database. The database holds substantial software for aiding in the design, namely the design information (the behavioural description), the netlist, the layout, the mask designs, and the results of the various checks and simulations. Alphanumeric and raster window displays are expected facilities as is a multiwindow display which allows simultaneous viewing of schematics, simulation waveforms, and alphanumeric data, etc. The workstation is located in an office unlike the mainframe computer before it, which was probably located in an air-conditioned room. Workstations are a key component of CAD/CAE. The workstation is central to concurrent design, in which several engineers are engaged simultaneously on different parts of the design. Networking makes it possible for these engineers to operate from different locations. The chief features of a workstation are:

(i) the operating system, which enables the software tools (logic synthesis, layout synthesis, simulation, etc.) to function;
(ii) networking—simultaneous access from terminals linked to the workstation;
(iii) interactive graphics for alphanumeric and raster displays of behavioural, functional, and structural models, simulation results etc. and multiwindow displays;
(iv) a database for holding software packages, results of all design steps, including checks and simulation.

The software design tools consist of the following groups:

(i) design. Tools which translate a behavioural description into the hardware structure, and then translate that structure into the layout of the circuit.
(ii) netlist, schematic capture, and mask pattern generations;
(iii) test procedure—test vectors and test organization;
(iv) simulations, extending from behavioural (top) level to component level;
(v) checks to ensure that the design rules (layout geometry) and the electrical rules (fan-out), unconnected nodes etc.) have not been violated and additional checks to ensure that the mask patterns will yield the ASIC correctly.

The term *silicon compiler* is often used in relation to software tools. A silicon compiler transforms a behavioural description, expressed in a hardware description language (HDL) into mask layouts, expressed in a geometrical specification language. Several standard HDLs and geometrical specification languages have become established, and some vendors have their own in-house language for that purpose. The silicon compiler, taking the definition given above, is a Utopian concept. The stage has not yet been reached in ASIC design where software produces a mask layout directly from a high-level description. It is better to think in terms of silicon compilation, an expression signifying an assembly of software packages which transform a behavioural description into a mask layout in stages. There will be two such stages according to the schematic in Fig. 10.2, logic and layout synthesis.

The software design tools are a formidable collection of packages for converting the behavioural description into the mask patterns, and for the simulations and checks to confirm the design. One might add another design tool, a hardware accelerator which, although optional, is not a luxury. The hardware accelerator greatly speeds up the simulations it can undertake. All in all the combination of hardware and software tools is a sophisticated CAD/CAE assembly. The term CAD/CAE is used in preference to CAD or CAE on their own, because these two generic descriptions tend to merge, sometimes imperceptibly, into one another. An approach for arriving at distinction between them is to compare design with engineering. CAD applies more to designing a component, for instance an ASIC, and CAE is more apt for system design. In CAD, the designer has to think and make intelligent use of the design tools (Hollis 1987, p.14). CAE can be regarded as the implementation of a CAD product. The workstation is the vehicle for CAE. That said, the difference between implementation and engineering design is slight: the use of CAD/CAE embraces both activities.

The workstation has ousted the main frame computer for IC design, and much of general engineering design. It can be implemented with modern PCs, which, with CD ROM-storage backup, have adequate power for ASIC design. Their attraction is that they are likely to be available for most design engineers.

10.3 Design for testability

Testing ICs is an integral part of ASIC production, which culminates in the delivery of the tested chips to the customer. Testing does not end there. It is followed by in-circuit (on-board) tests with the ICs, which might be a mix of ASICs and standard components, mounted on the PCB, and often by functional board tests of the entire PCB assembly. The testing of ICs has undergone significant changes with the establishment of (V)LSI. SSI and MSI was and still is tested by applying stimuli in parallel to the input pins, and comparing the measured performance at the output pins with the expected behaviour. The chips are mounted in automatic test equipment (ATE) and the testing is computer-aided. On-board testing of SSI and MSI has been undertaken for many years with the *bed-of-nails* technique, which is only applicable for through-hole mounting. The bed-of-nails test is carried out by means of a board of 'nails', pins with needle-shaped points, patterned to establish contact with selected pins of the ICs under test, and connected to the ATE, which applies the test vectors and measures the response. The ICs are mounted on one side, the bed-of-nails make contact with the IC pins on the other side of the PCB (Fig. 10.3). The method allows in-situ testing

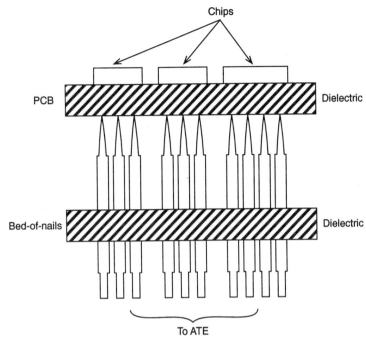

Fig. 10.3 Bed-of-nails test

of the individual chips and their interconnections, but is impracticable for VLSI for the following reasons:

(i) Surface mounting technology (SMT) is taking over from through-hole mounting in general, and is mandatory for VLSI. With SMT components are mounted on both sides of the PCB;

(ii) To accomodate VLSI chips with their high pin counts, the pitch of PCB wires has decreased. The long-established standard of 100 mil (2.54 mm) is being supplemented by 25 mil (0.63 mm), with a yet smaller spacing of 10 mil (0.3 mm) being practicable and coming into use (Section 2.6.5).

(iii) The gate-to-pin ratio for ASICs is very high. ASIC chips with about 20 000 and 260 000 usable equivalent gates have typical gate-to-pin ratios of 100 and 400 respectively. Such architectures allow only a vanishingly small proportion of the circuit to be tested by the traditional method.

The situation brought about by VLSI and SMT called for a new approach to test design. In 1985 a number of European electronic equipment manufacturers and IC vendors set up an *ad hoc* Joint Test Action Group (JTAG). Rapid progress was made by the end of 1986, when the membership of JTAG had increased and widened to include representatives from both European and North American companies. An initial proposal of a boundary scan test, made earlier that year, underwent several refinements and was subsequently submitted to the IEEE for approval, which was granted in February 1990, when it became the IEEE Standard (Std) 1149.1. (Maunder and Tulloss 1990, pp.23–9).

The basic architecture of the IEEE Std 1149.1 boundary scan (frequently just called IEEE 1149.1 boundary scan) is shown in Fig. 10.4.

346 ASICs—Characteristics and design issues

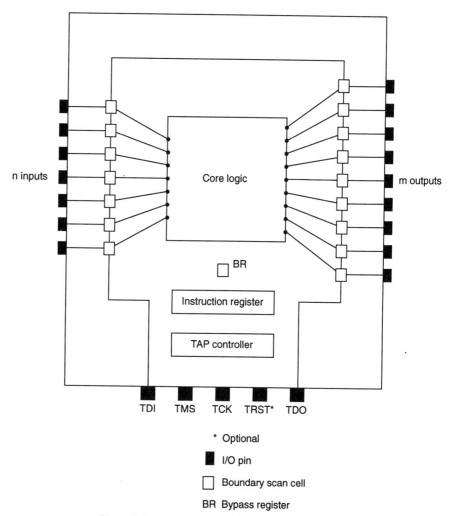

Fig. 10.4 Basic schematic of IEEE 1149.1 boundary scan

The description of the boundary scan in this section applies to the IEEE Std 1149.1 throughout. The principle of operation is based on the addition of a storage cell, the boundary scan cell (BSC), for each I/O pin, and a configurable connection of all these storage cells into a serial shift register, the boundary shift register (BSR). Test data is entered via the test data in (TDI) and extracted via the test data out (TDO) pins of the test access port (TAP), which consists of five pins, one of these being optional. The other pins are for the test clock (TCK), test mode select (TMS), and the test reset (TRST) entries; the last of these is the optional facility. The test data is entered and extracted serially, being scanned through the BSR. An alternative transmission path is from TDI to TDO via the one-bit bypass register (BR). This mode is used for quick transmission of data when the chip is not being tested. Serial data transmission, in order to economize on the number of pins, has been encountered before in the entry of the configuration into the Xilinx FPGAs (Section 9.3.2). The two main tests are:

(i) An external on-board test (EXTEST), verifying the correctness of interconnections between chips by checking for faults like open circuit, short circuit of a wire to ground, short due to solder spill between adjacent wires etc.

(ii) an internal test (INTEST) for checking the core logic. This test can be carried out before and after insertion of the chip on the PCB.

The boundary scan overcomes the limitations of the bed-of-nails technique. The test vectors can have any desired length and the requirement for test points is limited to four, possibly five additional pins (the TAP) per chip. The boundary scan test (BST), exemplified by the IEEE Std 1149.1, has been adopted worldwide by virtually all manufacturers of PCB electronic assemblies. It saves a good deal of time compared with traditional testing, and greatly reduces test costs by not requiring special ATE and hardware for each application. The saving in cost is substantial; it can amount to between 50 and 70 per cent (Bleeker et al. 1993, p.xv).

The versatility of testing is linked to the options of configuring the BSC. Remember that each I/O pin is supported by such a cell. The various modes of cell operation are illustrated in Fig. 10.5. Figure 10.5(a) shows the four switches of each cell. The parallel-in parallel-out routing in Fig. 10.5(b) is for normal operation: the cell is transparent to the user. The connections in Fig. 10.5(c) configure the BSC to be part of a serial shift register for scan transmission. Figures 10.5(d) and 10.5(e) illustrate parallel-in serial-out and serial-in parallel-out operation, the former for extracting, the latter for entering test data. The parallel-series modes are in practice applied to a group of adjacent BSCs. For instance the status of all input pins can be captured by the BSC group in a parallel operation in one test clock cycle and the information can then be transmitted serially for evaluation via the TDO pin. Similarly a test vector can be input to the core logic in Fig. 10.4 by parallel-routing of all the boundary scan cells in one clock cycle. The result is placed simultaneously on the m-bit section of the boundary by parallel routing and is then extracted by serial shifting. An interval of more than one clock cycle may have to be allowed to ensure adequate test time.

Examples of interconnect and component tests are illustrated in Figs 10.6 and 10.7 (Maunder and Tulloss 1990, pp.15–17). The test in Fig. 10.6 can be extended to a larger number of chips. The test vector length is equal to N bits, where N is the total number of I/O pins being tested (in this case 12). The test shows up two postulated faults, a short to ground in one of the interconnections between IC1 and IC3, and a short between two adjacent interconnections emanating from IC3. The tabulated inputs and outputs list the correct no-fault and the actual response. The test in Fig. 10.7 verifies the correct operation of a two-input NAND gate. It is a simple example which does not obtain in practice, because a VLSI ASIC contains several thousands of primitives like gates and bistables. Figure 10.7 highlights the constraint on testing imposed by the limited number of I/O pins, which only allows a very small fraction of the core logic to be tested; more about that shortly.

Figure 10.8 is a general illustration of test data being fed through the scan path. The diagram gives no information about the nature of

348 ASICs—Characteristics and design issues

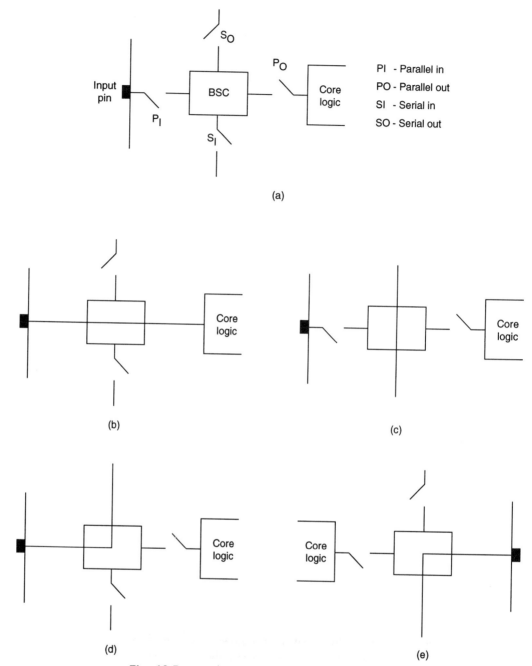

Fig. 10.5 Boundary scan routing (a) switches (b) parallel-in– parallel-out (c) serial-in–serial-out (d) parallel-in–serial-out (e) serial-in–parallel-out

the tests or the configuration of the BSCs. The tests would be external (for interconnections) or internal (for core logic) or a mixture of both. The B registers facilitate the quick passage of test data for the situation illustrated in Fig. 10.9, where only one of the chips, IC3, is being tested. The hypothetical layout of the PCB allows access to IC3 via the other two ICs only. The B registers on IC1 and IC2 enable the

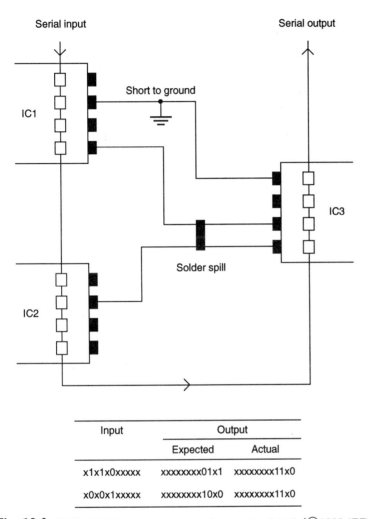

Fig. 10.6 IEEE 1149.1 boundary scan—interconnect test (©1990 IEEE)

Input	Output	
	Expected	Actual
x1x1x0xxxxx	xxxxxxxx01x1	xxxxxxxx11x0
x0x0x1xxxxx	xxxxxxxx10x0	xxxxxxxx11x0

test data to reach IC3 far more quickly than threading them through via the boundary scan registers. The B register exists purely for such a speed up. The miscellany of test modes in Fig. 10.10 has IC1 connected for an external, and IC3 for an internal test; the other component, IC2, is being bypassed. Figure 10.10 contains the entries for the instruction registers, which govern the test mode (including the bypass action). The IRs are set by a serial input stream via TDI; the serial transmission of test data–also via TDI—follows later. The TMS and TCK inputs are always applied to all chips in parallel.

The test operations will now be explained in more detail with reference to the architecture of the IEEE boundary scan in Fig. 10.11 (Maunder and Tulloss 1990; Bleeker et al. 1993). This schematic enlarges on the simpler diagram of Fig. 10.4 by including additional registers and the basic support logic. The TDI input is routed to the BSR for test data entry, to the BR for bypassing the chip, and to the IR for setting up the commands which control the data registers. The test mode is determined by the TMS input, which activates the TAP controller. The

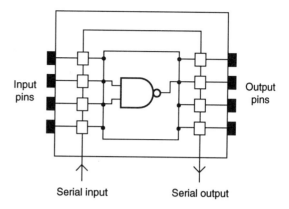

Input	Correct output
x10xxxxx	xxxxx1xx
x01xxxxx	xxxxx1xx
x11xxxxx	xxxxx0xx

Fig. 10.7 IEEE 1149.1 boundary scan—test of core logic (©1990 IEEE)

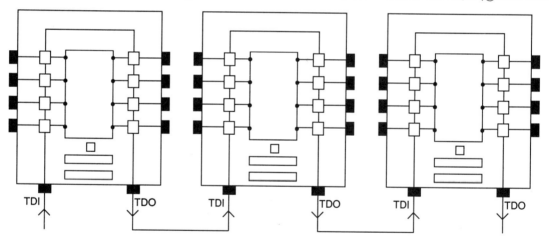

Fig. 10.8 IEEE 1149.1 boundary scan—test of all boundary scan registers (©1990 IEEE)

two alternative serial inputs to TDI are test data and data for setting the IR. The TAP controller determines the route for the TDI input via the select switch, and MUX1 selects one of the test data registers for the scan path. The operation of the other logic elements in the scan path will be explained in relation to the various inputs and registers described below.

(i) The TMS input controls the operation of the test logic via the TAP controller.

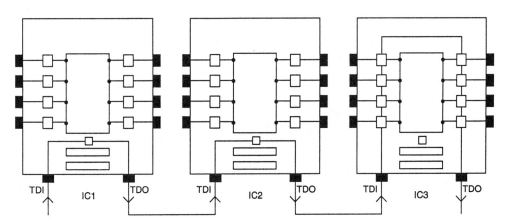

Fig. 10.9 IEEE 1149.1 boundary scan—B registers and BSR on scan path

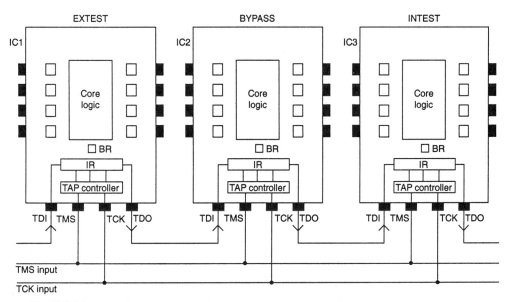

Fig. 10.10 IEEE boundary scan—various tests governed by IR settings

(ii) The TAP controller generates the clock and control signals required for the operation of the IR and the test data registers from the TMS input. It is a 16-state finite state machine (Maunder and Tulloss 1990, pp.37–43). Half the states govern the operation of the test logic, the other half control logic settings when the test is suspended. When no data is being scanned, the buffer driving the TDO pin is set into the third state.

(iii) The test clock input TCK is independent of the system clock(s) for the ICs, enabling test operation to be sychronized for all the chips on the PCB. Signals are entered at the input pins on the rising edge and are shifted out via the D bistable to TDO on the falling edge of the TCK signal.

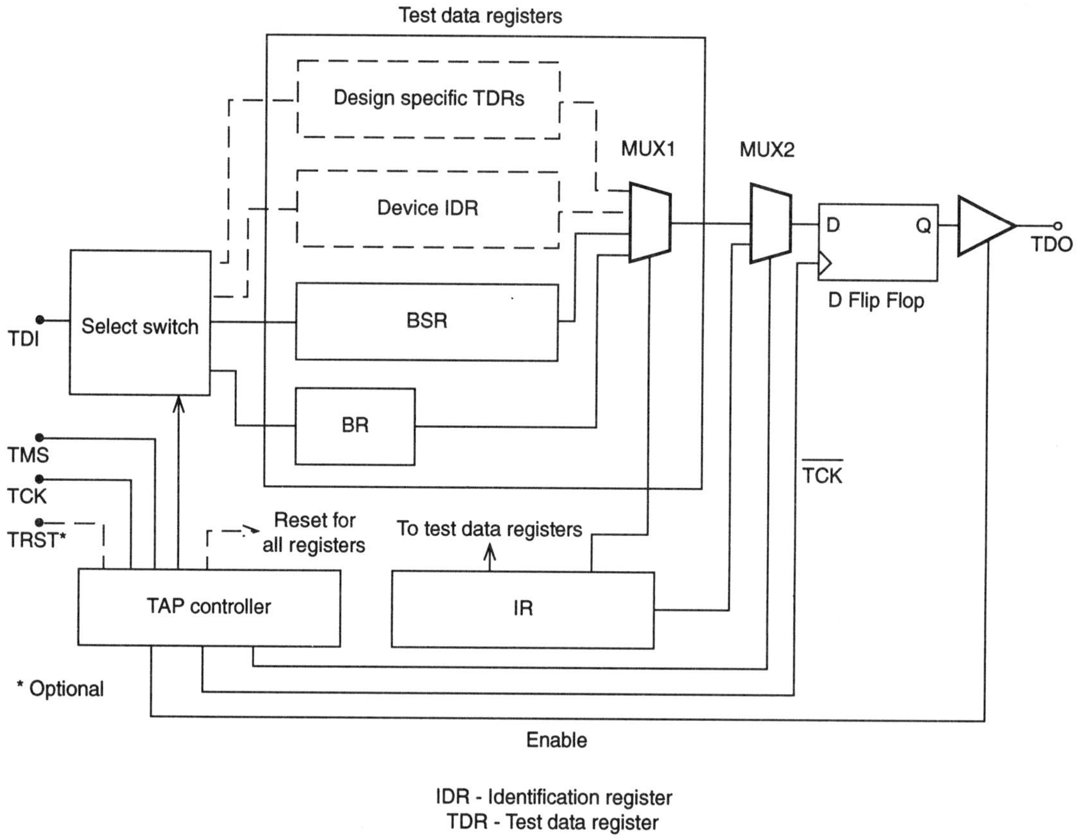

352 ASICs—Characteristics and design issues

Fig. 10.11 IEEE 1149.1 boundary scan architecture (Reproduced by permission of Kluwer Academic Publishers)

IDR - Identification register
TDR - Test data register

(iv) The data input TDI is fed to either the instruction register or one of the data registers, depending on the state of the TAP controller and hence the previous TMS input.

(v) The data output from either the IR or one of the test data registers is fed to TDO via the D flip-flop, depending on the previous TMS input, which selects the output of the TAP controller.

(vi) The optional TRST input forces the test logic *asynchronously* into its reset mode. It is optional because the test logic must be designed so that the reset state can be obtained quickly by the control of the TMS and TCK inputs.

(vii) The IR is a shift register whose instructions specify the address of the data test register chosen and the test to be carried out. It allows test instructions to be entered into each chip included in the boundary-scan path. The IRs are daisy chained for that purpose so that identical or different test instructions can be loaded into the chips on the scan path (Fig. 10.10). The IR must be 2 bits wide at least. Its length is determined by the number of test instructions provided; the maximum is not defined.

(viii) Test data registers.

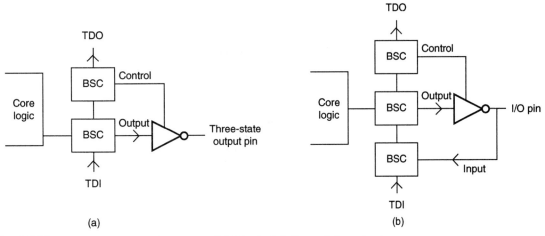

Fig. 10.12 IEEE 1149.1 boundary scan—BSC logic (a) three-state output pin (b) I/O pin

The two mandatory registers are the BSR and the BR. The BSR is series shift register configured from all the BSCs. The standard BSC is based on D flip-flops and multiplexers (Maunder and Tulloss 1990, pp.71–3; Bleeker et al. 1993, pp.28–9, 37). The pattern of one BSC per I/O pin has to be modified for three-state outputs and for pins which are not dedicated, but can be for either input or output (a practice already met repeatedly in PLDs; see Chapter 9). The arrangements for the above two cases are shown in Fig. 10.12. The bypass register is, as has already been pointed out, a one-bit register for a quick transmission of data not required for that chip.

The optional registers are the device identification register (device IDR) and the design-specific test data registers (Fig. 10.11). The device IDR contains information about the vendor's name, the component designation, and the variant (if any) of the component. Some of its applications are to verify that the correct IC has been entered, and to adjust the test program for a variant replacement. Design-specific test data registers are optional (there may only be one). They contain data generated by the manufacturer for his own design, and might be used for in-house testing only. Alternatively they may be made available to the customer.

Two other aspects of testing are included in this section, the level sensitive scan design (LSSD) and the built-in self-test (BIST). The LSSD is a technique for the shift register operation in the boundary scan. It was pioneered by IBM and the schematic of an LSSD cell, the shift register latch (SRL) which can be used for the BSC, is sketched in Fig. 10.13. (William and Parker 1983, pp.103–4; Bennetts 1984, pp.53–62). The SRL deals with timing constraints which arise from the parallel transfer of data into the BSCs. The inputs traverse different path lengths and incur different delays. The operation of the standard BSCs is stongly dependent on the rise and fall times of the clock pulses, propagation delays in the paths to and from the cells, and on the variations in response between the cells. The SRL in Fig. 10.13 overcomes those timing hazards, which can lead to possible oscillation and race conditions, by having two

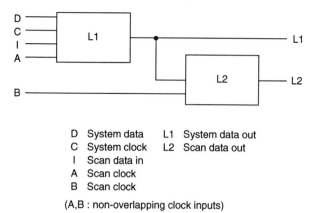

D System data   L1 System data out
C System clock   L2 Scan data out
I Scan data in
A Scan clock
B Scan clock

(A,B : non-overlapping clock inputs)

Fig. 10.13 Shift register latch—LSSD

level-sensitive latches in place of edge-triggered flip-flops. The D and C inputs are the system data and clock entries for parallel transfer like in Figs 10.5(d) and 10.5(e). The scan path function is obtained with the aid of two non-overlapping scan clock signals A and B, derived from the TCK input. The output of the previous SRL is the input I for L1, and L2 drives the input of the next stage.

LSSD has undergone considerable development since its origination in 1978. (Eichelberger and Williams 1978). The schematic in Fig. 10.13 is a mature version favoured for the IEEE 1149.1 boundary scan. Computer, semiconductor, and electronic equipment manufacturers sometimes have their own in-house designs, which do not follow the IEEE standard precisely but are fairly close to it and achieve much the same. They exercise that freedom to modify the architecture, including LSSD cells. One such development is an LSSD technique by IBM for testing of ASICs with high pin counts using a relatively inexpensive ATE with a reduced pin count. (Basset et al. 1990).

The component (circuit) test, the test of the core logic in Fig. 10.7, is limited to the number of input pins, which is very small compared with the gate count of a VLSI chip. The built-in self-test (BIST) examines a far larger proportion of the core logic. To test a VLSI chip for all possible logic combinations is not on. Consider a purely combinational network with n inputs. A complete functional test will have 2^n combinations. Added to this m latches, which are part of the logic, and 2^{n+m} inputs are needed for a complete functional test (Williams and Parker 1983, p.100). If we take a VLSI chip with a modest total of 50 for n and m, the total number of test patterns comes to $2^{100} = 1.27 \times 10^{30}$. Assuming that each test occupies 100 ns, the total test time comes to $\sim 4 \times 10^{15}$ years, well and truly beyond the Psalmist's allotted lifespan of threescore and ten years! Naturally a test, disregarding the enormous timescale, need not be that comprehensive. There is no point in testing flip-flops by trying to set and reset them simultaneously, and there will be plenty of combinations which are pointless. All the same testing has to extend over far more logic elements than those accessed by the core logic inputs to give an adequate fault coverage. The combination of BIST and boundary scan testing, known by the acronym of BBS, presents the most promising solution to testing at component and board level (Ballew and Streb 1992;

Rosquist 1991). The core logic of the chip is grouped into sequential and combinational blocks. The sequential components (flip-flops, counters, registers, etc.) are designed to fulfil a dual role. They perform their normal logic function but can also be configured into a serial shift register which is incorporated into the boundary scan. The combinational block is connected to these storage elements and is tested rather like the core logic in the boundary scan. The parallel loading of this shift register, followed by serial transmission, is very similar to the operation of the BSR in a boundary scan test. The BIST is easily accomodated within the boundary scan architecture (Sedmak 1989). The self-test is initialized by an instruction via the TAP. The test signal is usually a pseudo random generator and the test output is extracted from the TDO pin.

The tests so far have been based on an IEEE Std 1149.1 architecture for all chips. PCB assemblies are likely to include ICs which do not incorporate bounday scan provision. These may be chips with old designs, or ASICs with low densities (PLDs), etc. The four (or five) pins needed for the TAP are a small price to pay for ASICs with 50 pins or more, but become an important consideration for ICs with a smaller number of pins, where they will cause a significant increase in package area and cost. The testing of PCBs with such a mixed composition will be inevitably less efficient. Special techniques have been developed to optimize on-board tests embracing chips with and without boundary scan architecture. Ballew and Streb (1992) describe a method in which an ASIC specifically designed for tests under such conditions leads to greater efficiency.

A comprehensive coverage of testing is outside the scope of this text. The material presented in this section has the limited aim of indicating the main issues of design for testability with special reference to the IEEE 1149.1 boundary scan, and of outlining the nature of the design effort. The salient factors are summarized below.

(i) A testing strategy is vital. It must come at the beginning of the system design and not as an afterthought at the end.

(ii) The tests should extend to component level, board level, and system level.

(iii) The IEEE Std 1149.1 boundary scan has become the established test method for ASICs.

(iv) The boundary scan is far more economic than the bed-of-nails test, which has become impracticable because of the fine pitch between the wires of PCBs and VLSI chips. The boundary scan is independent of packaging technology.

(v) Economic considerations are not confined to the cost of ATE and test design, but extend to the speed with which the test can be carried out. The time to market is of vital importance and boundary scan testing is far faster than previous methods. Design cycle times have decreased whilst test program development times have increased, largely because of the higher gate counts. Test design has assumed more importance and now exerts a dominant influence over the total design effort.

(vi) Costs per fault are estimated at $0.30, $3, $30, and $300 for component, board, system, and field failures respectively (Williams and

Parker 1983, p.100). These figures will no doubt have increased since they were published, but the relative increase by a factor of about ten with progression to the next higher level is likely to hold.

(vii) The bounday scan can be adapted to include the BIST, which is the preferred method for testing the core logic with adequate probability of fault detection. The combination of BIST and boundary scan (BBS) is deemed to be the best test strategy.

(viii) The additional number of pins per package for the IEEE Std 1149.1 architecture is easily absorbed in chips with pin counts of 50 or higher. It can however lead to a significant increase in package cost for ICs with a small number of pins.

(ix) The inherent serial mode of the boundary scan may, depending on its length, be too slow for DRAMs. These should be tested by other means.

(x) A typical boundary scan for a 10 000 gate ASIC with 40 I/O pins demands about 900 gates i.e. the overhead gate cost is nine per cent (Maunder and Tulloss 1990, pp.141–9). The number of test gates goes up considerably when a BIST is included, and the test section of an ASIC can easily demand 15 to 20 per cent of the silicon estate (Huber and Rosneck 1991, p.156).

(xi) The continuing rise in ASIC complexity puts a severe strain on test development, which can dominate the design effort for chips with gate counts in excess of 100 000 (Rosqvist 1991, p.131).

10.4 Economics

The case for VLSI on purely economic grounds was presented initially in Section 1.3.1, where a simple calculation highlighted the tremendous saving in equipment cost made possible by replacing an all-SSI implementation with VLSI. The cost of the VLSI version was found to be between 1 and 5 per cent of the cost for the SSI version.

ASIC design is preceded by a critical comparison between ASIC and standard ICs. The initial appraisal will show whether ASICs, possibly combined with standard components, are a cost-effective solution for the equipment. The estimate has to include the production and non-recurrent development costs. The designer will think in terms of overall cost rather than the price of individual ICs. The cost of ASICs has to allow for the NRE of chip design, including design for testability, and mask manufacture. These NRE costs are additive to the chip fabrication costs (die production, packaging, and testing). The price charged by the ASIC vendor will depend on the number of equipments and hence ASICs to be produced, because the NRE expenditure, which is a major outlay, has to be recovered from the sale of equipment. The cost estimate which will now be advanced is in terms of two key components, the cost of the ASICs and the production cost of the equipment. The various parameters are:

Table 10.1 Design alternatives for a specified equipment

Parameter	TTL/SSI	ASIC gate array	ASIC full custom
Number of ICs	1667	4	1
C_P^* ($)	0.45	8.55	12.83
C_{OH} ($)	2.43	17.13	25.69
Number of PCBs	33	1	1
Pins per IC	18	84	128
Total pin count	30 006	336	128
D_{NRE} ($)	—	67 500	30 000
D_{EQ} ($)	142 500	9000	4500

* includes profit margin
(Courtesy of GEC Plessey Semiconductors)

C_P = raw production cost of chip
C_{OH} = overhead added to C_P to allow for equipment *production* cost (but not for development cost)
D_{NRE} = non-recurrent *engineering development* cost to produce the chip
C_P' = cost of chip, amortizing D_{NRE}
D_{EQ} = non-recurrent *engineering development* cost to produce the equipment (mainly PCB development cost)
C_{PROD} = production cost of equipment, amortizing D_{EQ}
n = number of ASICs per equipment
m = number of equipments to be produced

The definitive equations linking the above parameters are

$$C_P' = C_P + \frac{D_{NRE}}{mn} \qquad (10.1)$$

$$C_{PROD} = n(C_P' + C_{OH}) + \frac{D_{EQ}}{m} \qquad (10.2)$$

Production costs will now be given for an equipment built with either SSI, gate arrays, or a full custom ASIC. (The information has been kindly supplied by GEC Plessey Semiconductors.) Using the information in Table 10.1 with eqns (10.1) and (10.2), C_{PROD} has been calculated for four production quantities in Table 10.2. This example of an equipment produced with three different design styles quantifies the economic advantage of going over from standard chips to two categories of ASICs, a gate array and a full custom design. The solution for the gate array was obtained with four different ASICs, but there are many applications where a single semicustom ASIC will suffice. Large systems will contain a greater number of ICs, and these may be a blend of standard and ASIC components. There is however a trend to go for all-ASIC solutions in such cases. Cell-based ASICs, possibly combined with some smaller ASICs, are a means of implementing large systems very economically.

The production cost of an ASIC has to be distinguished from its NRE development cost, which includes the design and fabrication of the

Table 10.2 C'_P and C_{PROD} for equipment in Table 10.1 (1989 prices)

IC category	C'_P ($)			C_{PROD} ($)		
	m			m		
	500	10 000	50 000	500	10 000	50 000
TTL/SSI	0.45	0.45	0.45	5086	4815	4804
Gate array	42	10	8.9	250	110	104
Full custom ASIC	73	15.8	13.4	108	42	39

(C'_P is based on C^*_P, which includes a profit margin, in Table 10.1)

Table 10.3 VLSI manufacturing prices (1989 prices)

Wafer cost ($)	1050
Die size (mil^2)	150 × 150
Dies per wafer	1144
Expected yield (%)	80
Good dies	915
Assembly yield (%)	85
Test yield (%)	90
Final chip count	700
Cost per chip ($)	2.55
Typical sales price ($)	8.25
Profit per wafer ($)	3990

(Courtesy of GEC Plessey Semiconductors)

masks. The production cost itself is quite modest, witness the breakdown for a typical production run in Table 10.3. The data in Table 10.3 gives an initial yield of 1144 dies. The area of the 6 in wafer allows for 1256 dies. The square die images imposed on the circular wafer and the space between dies lead to perimeter and interspace losses, resulting in a yield of 1144 dies. Further yield losses due to packaging and testing give a final total of 700 chips. The cost per chip covers die production, probe tests, packaging and final tests. The sales price is typical for a VLSI chip in the middle of the range with an upper limit around 500 × 500 mil^2. The purpose of Table 10.3 is to bring home the comparatively modest cost of chip production. The non-recurrent ASIC engineering and equipment development costs linked to the production schedule are the decisive factors.

Example 10.1

An electronic equipment is constructed with five different ASICs mounted on a PCB. Each ASIC costs $10.80 to produce. The ASIC development cost and the equipment development cost are $75 000 and $12 000 respectively. The production cost per equipment (excluding the cost of the ASICs) is $112.50. A total of 8000 equipments are to be produced. The vendor supplies the ASICs to the customer at a price which yields a 15 per cent profit margin over the chip production, and a 20 per cent profit margin over the ASIC development cost. Calculate the sales price per equipment, allowing for a 20 per cent profit by the customer

Chip production cost + profit = $10.80 \times 1.15 = $12.42
D_{NRE} + profit = $75 000 \times 1.2 = $90 000

There is a total of 40 000 ASICs, adding $(90 000/40 000) = $2.25 to the cost of each chip and making the cost of chip to the customer $(12.42 + 2.25) = $14.67. To recover the equipment development cost, $(12 000/8000) = $1.50 has to be added to the price per equipment. The total equipment cost comes to $112.20 + $5 \times 14.67 (ICs) + $1.50 = $187.35. The final price for a profit margin of 20 per cent comes to $187.35 \times 1.2 = $224.82.

Chip cost is sometimes considered in isolation from equipment cost. Doing so leads to a useful but limited initial assessment for design style selection. In general terms a full custom ASIC occupies less silicon than a standard cell, which in turn demands less silicon than a gate array for the same function. The relative D_{NRE} cost will be in reverse order, being highest for the full custom design. The quantity for which ASICs with different design styles break even may be obtained from eqn (10.1), bearing in mind that C_P excludes any profit margin. Rearranging eqn (10.1), the chip costs for ASICs with design styles 1 and 2 will balance at a quantity M given by

$$M = \frac{D_{NRE(1)} - D_{NRE((2))}}{C_{P(2)} - C_{P(1)}} \qquad (10.3)$$

Consider a standard cell (1) and a gate array (2) which are functionally equivalent. Suppose that the standard cell occupies half the area of a gate array. C_P is to a close approximation, like the yield of chips per wafer, proportional to the die area: $C_{P(1)}$ and $C_{P(2)}$ are put at $2 and $4 respectively. Assume 12 mask levels for the ASICs (in line with the observations in Section 2.5.2). All masks for the standard cells have to be designed and fabricated. The gate array will only require the design and fabrication of *four additional masks*, stipulating two-level metallization for both gate array and standard cell. The cost of the other gate array masks is excluded, because the dies are prefabricated in large quantities. A very small addition to the chip production cost is sufficient to recover the cost. The typical cost of producing an optical mask is $3 000, leading to mask costs of $36 000 and $12 000 for the standard cell and the gate array respectively. The NRE design costs for the standard cell and the gate array are put at $75 000 and $37 000 respectively. These are the amounts which might be expected for ASICs with about 20 000 gates. Actual charges depend on the nature of the system, the design tools and the allocation of tasks between vendor and customer, the development schedule etc. Adding the development and mask costs, $D_{NRE(1)}$

= \$111 000 AND $D_{NRE(2)}$ = \$49 000. Substituting these values and the previously given figures for $C_{P(1)}$ and $C_{P(2)}$ in eqn (10.3), M comes to 31 050, a result in line with practice where the break-even point is frequently in the range (20 000–100 000).

The example just given is one of the approaches to design style selection economics, which has to take into account other factors explained in or inferred from material in this chapter. Take the alternative of two-level or three-level metallization. The higher the number of metallizations, the easier it will be to achieve the necessary connectivity, but the higher will be the cost of mask fabrication. It could well be that the size of the ASIC die chosen can meet the specification with fewer (two, rather than three) metal layers and can therefore be produced cheaper. Alternatively a larger die with fewer metal layers may prove to be more economic than a smaller die with more metal layers.

Another consideration is the role of the FPGA vis-à-vis the gate array and the standard cell. D_{NRE} is virtually nil provided the user undertakes design and testing, a practice strongly encouraged by FPGA vendors. On the other hand C_P must include the cost of programming the chip. The bottleneck for quantity production is the time it takes, because FPGAs have to be programmed individually one at a time. The FPGA, confining ourselves purely to economics, is suitable for quantities up to several hundred, a situation often met in small equipments marketed in low volume.

A comprehensive treatment of ASIC economics has been undertaken by Fey and Paraskevopoulos (1987), who have elaborated on their pioneering approach in a VLSI handbook frequently referred to in this text (Fey and Paraskevopoulos 1989). They set the following criteria for the economics of design style selection:

(i) the development schedule, defined to be the interval between completion of the chip specification(s) and the successful testing of the prototypes (fabrication times are excluded);
(ii) the chip function density;
(iii) the chip cost per gate;
(iv) risk.

Their assessment includes a comparison of ASICs with standard ICs and a comparison between the various ASIC design styles. The analysis is based on a key parameter, the level of integration (LOI), defined by

$$\text{LOI} = \frac{(\text{no. of gates}) \times (\text{degree of usage})}{\text{no. of pins}} \qquad (10.4)$$

for a single chip and

$$\text{LOI} = \frac{\Sigma\{(\text{no. of gates}) \times (\text{degree of usage})\}}{\Sigma(\text{no. of pins})} \qquad (10.5)$$

for a PCB assembly (PCBA).

The number of pins in eqns (10.4) and (10.5) holds for signal (I/O) plus control pins. It is shown that the level of integration largely determines the total *development* cost (production cost is excluded). In

Table 10.4 LOI composition for a PCBA

Category	Quantity	No. of gates	Usable gates (%)	No. of pins	LOI
Gate array	1	30 000	70	200	105
LSI	1	2000	100	40	50
MSI	8	240	100	144	1.67
Total		†23 240		384	60.5

† Usable gates

ASIC parlance the degree of usage is the proportion of available gates being used. This will be the order 50 per cent (Table 9.11) in FPGAs and SOG arrays. A higher proportion, sometimes approaching 90 per cent and very exceptionally higher still, may be expected for channelled gate arrays which however have only about half the gate density of SOG arrays. The LOI is obviously highest for standard and full custom VLSI, decreasing for standard cells, becoming yet smaller for gate arrays, and being smallest for PLDs. SSI and MSI have an LOI below 5. PALs are higher, but their LOI is likely to be below 10. The LOI of FPGAs is, according to Tables 9.9 to 9.12, in the range $\sim(20\text{--}45)$.

Gate arrays with 30 000 or more usable gates have LOIs from 100 upwards; the value approaches 500 for chips with about 300 000 usable gates. Taking a specific standard VLSI chip, the Intel 486 microprocessor, the 1.2 million transistors are equivalent to 300 000 gates. Leaving out the numerous supply line and ground pins, there are about 100 I/O pins and LOI $\simeq 3\,000$. The total cost of a product is governed not by the LOI of the chips, but by the PCBA LOI, which is easily reduced by combining VLSI with SSI/MSI and LSI. The LOI in the example of Table 10.4 is lowered from 105 for the gate array to 60.5 for the PCBA.

The analysis based on the LOI and the estimate of relative performance made previously lead to similar conclusions. Fey and Paraskevopoulos (1987, 1989) find that the system development schedule is more important than the ASIC design schedule for overall economic evaluation. The advantage in system redesign derives from the compaction of a PCB when changing over from SSI/MSI to ASICs. The change can be quite dramatic. A system based on a 10×10 in^2 PCBA and containing 80 SSI/MSI ICs can easily be replaced by a single ASIC mounted on a PCB measuring 2.5×2.5 in^2.

The fourth of the main selection criteria advanced by Fey and Paraskevopoulos (1987, 1989) is the risk factor. Profit is related to the development and production schedules. The latter is the more predictable of the two, the development schedule is vulnerable. It may, and often does slip because of a larger number of design iterations than estimated, and because of consequent redesigns arising from errors which come to light during the simulations. The mask patterns are only finalized when extensive simulations have confirmed the validity of the design. An error discovered subsequent to mask manufacture would be far more serious and costly. The economic viability is dependent on meeting the develop-

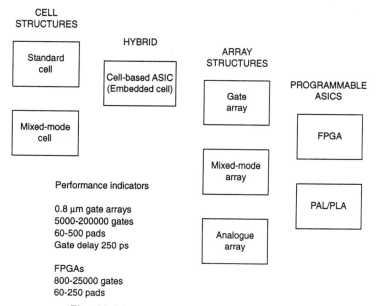

Fig. 10.14 Semicustom ASIC design styles

ment schedule, which is kept to a minimum with possibly some allowance for slippage. Fey and Paraskevopoulos (1989, p.256) quote figures provided by Reinertsen (1983) on the cost penalty for failing to complete engineering projects of this nature on target. According to this estimate, a six-months delay causes the profit to decrease by 34 per cent, assuming a 20 per cent market growth rate, a 12 per cent annual price erosion, and a five-year product life. A ten-month production delay might, making certain stipulated assumptions, reduce the profit to zero. These estimates, although advanced in 1983, are still freely quoted in recent technical literature and commercial publications (Texas 1991). Lengthening the development time in order to allow for slippage might seem a way forward to guard against such loss. However there is a premium on minimizing development time. In this age of rapid technological change, the life of equipment is fairly short. The five-year span mentioned above is typical, and the time to market has to be minimized. Getting the ASIC right first time, or with single rework at the most is of paramount importance. There is a fair chance of doing so given the right design tools, close cooperation between vendor and customer, and some ASIC design experience of the customer's part.

10.5 Characteristics and performance

10.5.1 Design styles

Building on the foundation established in Chapter 8, the various design styles will now be discussed before going on to describe their characteristics and performance. The schematic for that purpose, Fig. 10.14, has much in common with Fig. 8.1. The reason for the minor differences between those diagrams will become evident in the course of explaining the design styles.

The gate array and the standard cell are the mainstream of ASICs. The feature size of a CMOS ASIC is sometimes expressed in the form 0.7 µm (0.8 µm drawn), signifying a lithographical *drawn* gate length of 0.8 µm and an *effective* channel length, reduced by the lateral spread of source and drain layers, of 0.7 µm. If only one figure is quoted, it signifies the drawn gate length. The functionality of a gate array is specified by the total number of (equivalent) gates, qualified generally by the number of usable gates for an SOG architecture; the specified proportion of usable gates is typically half the total. Only the total gate count is specified for a channelled gate array. The proportion of gates used in such an array depends, more so than in an SOG architecture, on the particular application and is typically in the range (60–90) per cent. Coming back to the SOG array, the figure quoted for usable gates is only a guide; it too depends on the application. The characteristics for a standard cell do not include a gate count except an estimate of the upper limit for the maximum die size available. Below that level, the size of die is adjusted to accommodate the function required.

The logic capability of a gate array or standard cell can be ascertained from the macro library which supports it. The information for each macro includes its electrical characteristics and the mask patterns defining its geometry. For the same size of die, the standard cell has a considerably higher density and a higher system speed. Typical values for an 0.7 µm (0.8 µm drawn) standard cell are an upper limit of ∼350 000 gates and a gate delay of ∼200 ps.

The usage of cell-based ASICs (embedded cells) is expanding with the ongoing increase in chip density and equipment complexity. These ASICs incorporate megacells for RAMs, ROMs, microcontrollers etc. and are in many cases a more cost effective solution than a standard cell or a gate array above. The vertical positions of the ASICs in Fig. 10.14 have quantitative significance: the standard cell has the highest function density. However the cell-based ASIC may, depending on the application, be cheaper overall by combining the major part of the system, pre-designed cells, with metallized (array) logic to meet the customer's specification. Photographs of the layout for three types of 0.8 µm ASICs are shown in Figs 10.15 to 10.17. Figure 10.15 illustrates an embedded cell with ∼130 000 gates. It contains three different types of diffused RAM; the remainder of the array is an uncommitted SOG. The die of the compiled array shown in Fig. 10.16, some of which is taken up by compiled RAM and ROM and a coder, has an SOG architecture. Figure 10.17 illustrates a channelled gate array with about 20 000 gates.

FPGAs are now achieving equivalent gate counts which place them well into the capacity of gate arrays and standard cells at low and medium densities. They are experiencing one of the highest growth rates in the IC market. Since full information on FPGAs and other PLDs is contained in Chapter 9, these devices will not be described any further, but are referred to in the overview concluding this chapter.

Figure 10.14 enlarges on ASICs in Fig. 8.1 by including mixed-mode and analogue chips. These mixed-mode semicustom circuits, taken together with some all-analogue arrays, are one of the most rapidly expanding IC sectors.

The categories of analogue ASICs are:

Fig. 10.15 Embedded cell layout (Copyright ©Fujitsu Limited, Japan)

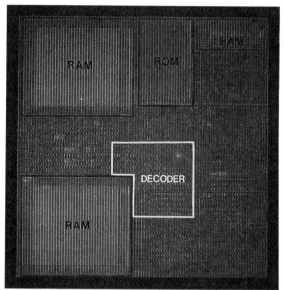

Fig. 10.16 Compiled RAM and ROM array layout (Copyright ©Fujitsu Limited, Japan)

(i) Mixed-mode (CMOS) cells. Their composition is a blend of digital and analogue diffused macrocells. The term standard cell usually signifies a digital ASIC. It is possible to construct an all-analogue standard cell, but such semicustom circuits are very rare, if they exist at all. The usual practice is a mixed-mode cell with a relatively modest digital section in immediate support of the analogue function. The total number of transistors is generally below the count at the bottom end of gate arrays, and the same can be said for mixed-mode analogue arrays.

(ii) Mixed-mode arrays. These assemblies of bipolar array elements can

Fig. 10.17 Channelled gate array layout (Copyright ©Fujitsu Limited, Japan)

be configured to give digital and analogue functions.
(iii) analogue arrays. A few vendors offer bipolar analogue arrays include resistors and capacitors. The macros for these ICs are confined almost entirely to analogue functions.

It should be clear by now that macrocells characterize the functional capability of an ASIC. The data in the descriptions which follow is based on information from various vendors supplied from late 1992 onwards. Although the development of ASICs continues apace, the characteristics of devices which have been on the market for two or three years are fairly representative of current practice. Moreover information on mature products tends to be fuller than recent and probably provisional data. The practice of looking at selected products which represent what is currently on the market was adopted in Chapter 9. It is being continued here and, in respect of design tools, in Chapter 11. Analogue ASICs are described in more detail than digital ASICs, whose composition is relatively prescribed. The description of analogue ASICs is aimed at incorporating enough information for appreciating the main factors in their design.

10.5.2 Gate arrays

The key parameters of a CMOS gate array are:
 (i) the equivalent gate count;
 (ii) the number of programmable pads—most of these serve for I/O pins;
 (iii) the range of dies, specified in terms of the gate count for each die;
 (iv) the macros available;
 (v) the number of metal layers—sometimes there is a choice;
 (vi) the electrical characteristics of the macros.

Table 10.5 CLA80000 double-layer metal arrays (high density pads)

Array type	Array elements	Usable gates	Total programmable pads
CLA81XXX	2816	1400	64
CLA82XXX	8736	4260	88
CLA83XXX	17 920	8400	112
CLA84XXX	30 784	13 600	136
CLA85XXX	54 720	22 000	168
CLA86XXX	100 048	30 000	216
CLA87XXX	157 872	48 000	264

Table 10.6 CLA80000 triple-layer metal arrays (high density pads)

Array type	Array elements	Usable gates	Total programmable pads
CLT81XXX	2816	1680	64
CLT82XXX	8736	5200	88
CLT83XXX	17 920	10 700	112
CLT84XXX	30 784	18 000	136
CLT85XXX	54 720	32 500	168
CLT86XXX	100 048	58 000	216
CLT87XXX	157 872	90 000	264
CLT88XXX	307 568	170 000	360
CLT89XXX	513 136	260 000	456

Another key factor, although not a parameter, is the design support available. The GEC Plessey Semiconductors (GPS) CLA80000 SOG series will be used to illustrate the performance of CMOS gate arrays. It has an 0.7 μm (0.8 μm drawn) geometry, a typical gate delay of 210 ps, and operates with a supply voltage in the range (2.7–5.5) V. The series is produced in three versions, a double layer metal series with a high pad density, a similar three layer metal series, and a series with a smaller number of pads. The families are tabulated in Tables 10.5 to 10.7. The three layer metal array has a density of 2000 gates/mm^2.

The programmable pads of all three families can be configured as power, ground, input, output, or bidirectional pads. All the arrays have eight fixed pads which are permanently wired to the power rail or ground. The library supporting the CLA80000 arrays includes over 100 macrocells, mainly logic macros (GPS call these microcells). It also includes some generic cells which can be parameterized into many configurations. These are in effect megacells (GPS call these paracells). Typical applications are RAMs and ROMs.

The cell library comprises the following groups:

Table 10.7 CLA80000 standard density pad arrays (MLA arrays have two, MLT arrays three layers of metal)

Array type	Array elements	Usable gates	Total programmable pads
MLA85XXX	54 720	22 000	144
MLA87XXX	157 872	48 000	224
MLT85XXX	54 720	32 500	144
MLT87XXX	157 872	90 000	224
MLT88XXX	307 568	170 000	304
MLT89XXX	513 136	260 000	376

(The information in Tables 10.5 to 10.7, and other data of the CLA80000 series are reproduced by kind permission of GEC Plessey Semiconductors)

Buffers and inverters
AND, NAND, OR, NOR gates
Exclusive OR and adders
Clock drivers
Latches
Registers (D flip-flops)
Output driver cells

Multiplexers
Input protection cells
Complex gates
Paracells (RAM, ROM etc.)
Tristate drivers
Level shifters
JTAG (IEEE Std 1149.1) cells

The preferred parameter for transient response is the delay time t_P between the input and output of the cell. It is expressed by

$$t_P = K_1 + K_2.LOAD + K_3.EDGE - \frac{K_4.EDGE}{\exp(K_5.LOAD/EDGE)} \quad (10.6)$$

where $LOAD$ = capacitive load across the output, normalized to one $LOAD$ unit (LU), which equals 49 fF, the notional input capacitance of a gate. To give an example, $LOAD = 72/49 = 1.47$ LU for a capacitive load of 72 fF;.
$EDGE$ = edge speed (transition time) of the input;
K_1 = intrinsic delay i.e. the delay with the load and the input edge speed set to zero;
K_2 = delay sensitivity to load;
K_3 = delay sensitivity to input edge speed;
K_4 and K_5 = coefficients which reduce the effect of edge speed for light output loads.

The approximate form of eqn (10.6) is

$$t_P \simeq K_1 + K_2.LOAD \quad (10.7)$$

$LOAD$ in eqns (10.6) and (10.7) is given by

$$LOAD = \frac{(C_L + C_W)\,(\text{fF})}{49}\;\text{LU} \quad (10.8)$$

C_L and C_W being the total load and wire capacitances across the output. Equation (10.7) is a useful approximate guide; accurate delays are calculated by the GPS *Universal Delay Compiler (UDC)* using eqn (10.6).

The nature of most categories in the cell library is self-evident. JTAG (IEEE Std. 1149.1) cells are a near-universal provision of ASIC vendors for boundary scan testing. We have referred to this by the name of IEEE 1149.1. Alternative acronyms for JTAG (IEEE Std. 1149.1) cells are JTAG cells, or simply JTAG (boundary scan) on its own. Clock grid drivers are listed separately. They include buffers with a very high output capacitance, which enables them to drive $\sim$(1000–3000) LUs, whereas clock drivers have an upper capacity of about 50 LUs. The level shifter cells include multiple output drivers, TTL input drivers with 3.3 V and 5 V supplies, Schmitt triggers, and a 3 V to 5 V interface.

The data for three cells, given below, contains general information. Fuller particulars are supplied by the vendor to the customer for accurate analysis and simulation of dc and transient behaviour.

Cell NAND2—2-input NAND gate

Inputs: A,B Output: F
Switching characteristics:
(See eqn (10.7) for definitions of t_i and K; one LU = 49 fF)

t_{PHL} (A to F) $K_1 = 113$ ps $K_2 = 42$ ps/LU
t_{PLH} (A to F) $K_1 = 195$ ps $K_2 = 68$ ps/LU
t_{PHL} (B to F) $K_1 = 77$ ps $K_2 = 42$ ps/LU
t_{PLH} (B to F) $K_1 = 238$ ps $K_2 = 60$ ps/LU

Cell parameters
  Input loading   (pins A and B)   1 LU
  Drive capability (pin F)   21 LUs
  Cell size   1 array element

Cell DFRS—Master–slave D flip-flop with set and reset

The schematic and truth table are given in Fig. 10.18
Switching characteristics
(One LU = 49 fF)

t_{PHL} (CKT, CKTI to Q) $K_1 = 421$ ps $K_2 = 45$ ps/LU
t_{PLH} (CKT, CKTI to Q) $K_1 = 360$ ps $K_2 = 71$ ps/LU

Cell parameters
  Input loading   CKT pin   2.6 LUs
                     CKI pin   2.5 LUs
                     D pin   3.1 LUs
  Drive capability   Q pin   22 LUs
                     QBAR pin   20 LUs
  Cell size   8 array elements

OPT6—Standard output driver

The schematic and truth table are contained in Fig. 10.19
Switching characteristics

t_{PHL} (N to OP) $K_1 = 139$ ps $K_2 = 11$ ps/LU
t_{PLH} (P to OP) $K_1 = 251$ ps $K_2 = 27$ ps/LU

Cell parameters

Characteristics and performance

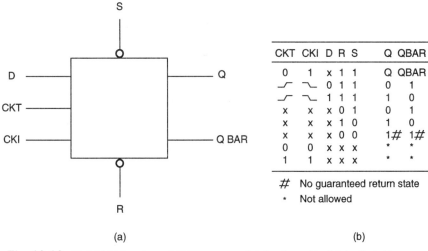

Fig. 10.18 CLA80000 series—DFRS macrocell (a) schematic (b) truth table (Courtesy of GEC Plessey semiconductors)

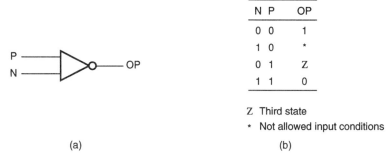

Fig. 10.19 CLA80000 series—OPT6 macrocell (a) schematic (b) truth table (Courtesy of GEC Plessey semiconductors)

Input loading (N)   68.2 LUs
Input loading (P)   64.6 LUs
Drive capability   6 mA

The parameters for CLA80000 power dissipation (5 V) are:

(i) Gate dissipation = 4.1 μW/MHz
   This holds for a fan-out of two standard loads.
(ii) Output buffer dissipation = 25 C_L μW/MHz
   where C_L is the load capacitance (pF).

Example 10.2

A CLA80000 ASIC in a certain application has 100 048 array elements and 85 output buffers. The maximum dissipation of the package is 2 W. The utilization is 63 per cent, and the average PCB load per buffer is 42 pF. Find the maximum system clock rate if 15 per cent of the chip is exercised at any one time.

Number of gates utilized = 100 048 × 0.63 = 63 030

370 ASICs—Characteristics and design issues

Table 10.8 Characteristics of VSC470 series

Parameter	Value
Drawn gate length (μm)	0.8
Effective gate length (μm)	0.7
Flip-flop toggle frequency (MHz)	> 330
Gate delay (ps)	195
Max. usable gates	> 350 000
Typical gate power (μW/MHz)	1.9
Macrocells available	~400

(Courtesy of VLSI Technology, Inc.)

Total gate power dissipation $= 63\,030 \times 4.1 \times 0.15$ μW/MHz
$= 38.76$ mW/MHz
Total buffer dissipation $= 85 \times 25 \times 42 \times 0.15$ μW/MHz $= 13.39$ mW/MHz
Total dynamic power dissipation $= 52.15$ mW/MHz
Hence the maximum system clock rate $= 2000/52.15 = 38.4$ MHz

ASIC vendors offer a wide choice of packages and the CLA80000 series is no exception. The preferred packages for the CLA and CLT families in Tables 10.5 and 10.6 are: metric quad flat pack (MQFP), fine pitch quad flat pack (FQFP), thin quad flat pack (TQFP), plastic J leaded chip carrier (PLCC), and small outline (SO). The recommended options for prototyping are the CERQUAD, LDCC, and ceramic SO packages. (Most of these packages are listed in Table 2.13). Various types of ceramic leaded chip carriers and ceramic pin grid arrays are recommended for the standard density pad arrays in Table 10.7. A vendor can usually satisfy a customer's request by providing a package which is not in the preferred list.

10.5.3 Standard cells

The standard cell is documented in terms of the macrocell library. The macros broadly cover the same functions as the macros for gate arrays. Indeed the documentation for gate and standard cell macros is very similar. The series chosen for illustration is the VLSI Technology VSC470 series. Now there are no prefabricated dies of different sizes. Instead the standard cell is designed to have the smallest area for achieving the specified function. The characteristics of the VSC470 series are listed in Table 10.8.

The VSC470 Standard Cell Library has two distinguishing features. First it is—like some other standard cell and gate array libraries of VLSI Technology—portable: the library contains cells which can be implemented in either standard cell or gate array designs in a number of technologies. In this case the design can be undertaken equally well in an 0.8 μm CMOS standard cell or gate array. The second special feature is the complementary VSC470L library, in which all the macros of the VSC470 (5 V) library are documented for the lower supply voltage of 3.3 V.

The documentation for standard cell and gate array macros is very similar. The data for the two macros given below should be compared with the data for the corresponding microcells in Section 10.5.2.

VSC470 and VSC470L Standard Cell Libraries

Macro NDO2DO—2-input NAND gate

	5 V supply	3.3 V supply
t_{PLH} (ns)	1.13	2.17
t_{PHL} (ns)	0.93	1.64
Gate power (μW/MHz)	0.80	0.30

The delay times apply to two standard (std) loads
Std load = 21 fF
Input capacitance = 15 fF
Output capacitance = 19 fF

Macro DFBTNB—D flip-flop

	5 V supply	3.3 V supply
t_{PLH} (ns) (CP to Q)	2.23	4.76
t_{PHL} (ns) (CP to Q)	2.63	5.52
Power dissipation (μW/MHz)	10.2	4.4
Gate equivalents	9	9

The delays apply to two standard (std) loads

Std load = 21 fF
    D-input capacitance 19 fF   Clock output capacitance 20 fF
    Preset input capacitance 52 fF   Clear output capacitance 53 fF
    Q output capacitance 27 fF $\overline{Q}$ output capacitance 27 fF
(Data reproduced by courtesy of VLSI Technology, Inc.)

Further library support for the VSC470 standard cells includes a cell computer library, a datapath compiler library, and a pad library. The cell compiler library has macros for RAMs, ROMs, and multipliers. The datapath compiler contains designs for high-speed ALUs, multipliers, comparators, etc. I/O pad macros are covered by the pad library.

The data for the macros includes their gate equivalents, which are also supplied for gate array macros (see Section 10.5.2). Such information allows an estimate of the silicon estate needed. For a gate array it helps in the initial selection of the die size. In the case of a standard cell it indicates the die size likely to suffice. The choice of packages available for the VSC470 standard cells is shown in Table 10.9.

Table 10.9 and the listing of the packages for the CLA80000 gate array in Section 10.5.2 give ample evidence of the variety in packages offered by ASIC vendors. This situation applies to ASICs universally.

10.5.4 Cell-based ASICs

The design style for ASICs fulfilling complex functions with very high gate counts is swinging strongly towards the cell-based ASIC (embedded cell), because it proves to be the most economic solution in many such cases. These ASICs are a blend of cell and gate array macros (including megacells), with a preference for cells, which are usually more efficient by taking up less silicon than a gate array. Megacells for high density RAMs, and ROMs, processor cores extending from 8-bit microprocessors

Table 10.9 VSC470 series—packages

Package	Pin range
DIP	16–48
Ceramic sidebrazed DIP	8–64
SO	20–28
PLCC	20–84
LDCC	28–84
LCCC	28–84
PPGA	64–208
CPGA	68–180
High performance PPGA	223, 229
High performance CPGA	223, 299, 391
MQFP (formerly PQFP)	80–240

(Courtesy of VLSI Technology, Inc.)

to 32-bit RISC architectures, and microcontrollers are typical of what is available. An embedded cell is often based on established designs, and is completed by the fast designs of the metal layers for the gate array section, composed mainly of macros available for that purpose. It would be overoptimistic to expect that an embedded cell can always be assembled entirely with pre-designed macros, but in most cases it will be largely constructed with these, possibly supplemented by some full custom macrocells which can subsequently be used for other designs. There are also embedded *arrays*, made up entirely of metallized megafunctions and macros on a smaller scale. Figure 10.16 is one such example. The cell-based ASIC and the embedded array are becoming the preferred design styles for ASICs with the highest density.

10.5.5 Mixed-mode and analogue ASICs

Mixed-mode and analogue semicustom circuits represent the area of extremely rapid growth in the IC market, a situation which mirrors the expansion of data and telecommunications. Attention has been drawn to this in the preface and in Chapter 6. Mixed-mode ASICs have been around since the origination of ASICs themselves, but have until recently been produced on a very small scale. Their emergence in earnest owes more to the development of analogue design tools than to the technology and circuit techniques for linear ICs, which are well established. The result is the existence of mixed-mode and all-analogue ASICs which fully complement digital semicustom circuits. Because of the rapid progress in this field and because of its inherently diverse nature, it is covered with some attention to detail.

The CMOS mixed-mode cell, set to become the preferred ASIC for combining analogue and digital functions, is made up of analogue and digital standard cells. The term 'standard' is often dropped; standard cells are simply called cells. Their structure is emphasized by alternative designations like diffused cells, diffused RAM, and embedded cell. The term embedded is also used for a gate array structure with descriptions like embedded array, embedded metallized ROM etc. These

Table 10.10 AMS mixed-mode CMOS cells—operational amplifiers

Cell name	Gain	CBP	Slew rate	Load	AMS process		
	(dB)	(MHz)	(V/μs)		CBH	CAE	CZE
OPEA01	94	1.7	2.2	1 MΩ/1 nF	*	*	*
OPO2B	81	3.9	4.6	100 kΩ/50 pF	*	*	*
OPO3B	85	10.8	20.0	1 MΩ/40 pF	*	*	*
OPVIDO	55	50.0	35.0	75 Ω/25 pF		*	*

* available

Geometries of the CBH, CAE, and CZE processes are contained in Table 10.13

(The data in Tables 10.10 to 10.13 and the data on analogue cells, following on from Table 10.12, are reproduced by kind permission of Austria Mikro Systeme International AG., Austria)

Table 10.11 AMS mixed-mode CMOS cells—ADCs

Cell name	Mode	Resolution	Sample rate	Power consumption	Process		
		(bits)	(Hz)	(mW)	CBH	CAE	CZE
Aquarell	Σ–Δ	20	1	1.8	*		
ADC02	Success. approx.	8	10^4	2.5	*	*	*
SADC13	Success. approx.	13	5×10^4	20	*		

* available

All ADCs have a maximum differential non-linearity of 0.5 LSB

comments, which reinforce previous explanations and definitions, have been interposed to 'deconfuse' the reader in an area characterized by a bewildering technology. The products chosen for illustration are the mixed-mode CMOS cells of Austria Mikro Systeme (AMS), who specialize in this facet of the ASIC art. The AMS mixed-mode families are available in 1.0, 1.2, and 2.0 μm geometries. A comprehensive selection of the analogue cells is given in Tables 10.10 to 10.12. Concise details of some other analogue cells are:

(i) Cell BGREF
   Bandgap voltage reference, 1.25 V, supply current 100 μA;
(ii) Cell OSC_DD2
   Crystal oscillator, (32–40) kHz, supply current 600 nA at 1.2 V;
(iii) Cell OSC1B
   Crystal oscillator, 200 kHz to 20 MHz, supply current 200 μA at 5 V;
(iv) Cells RC41, RC42, and RC43
   Filters, RC or switched capacitor, cut off frequency $\sim$(20–40) kHz.

Table 10.12 AMS mixed-mode CMOS cells—DACs

Cell name	Mode	Resolution	Sample rate	Power consumption	Process		
					CBH	CAE	CZE
		(bits)	(kHz)	(mW)			
DACS 1	Σ-Δ	10	1	2.5	*		
DACST 1	Stochastic	8	3.8	2.0	*	*	*
VDAC 8	Flash	8	25	24		*	*

* available
All DACs have a maximum differential non-linearity of 0.5 LSB

Table 10.13 AMS mixed-mode CMOS—general parameters

Parameter	Process		
	CBH	CAE	CZE
Feature size (μm)	2.0	1.2	1.0
Effective channel length (μm)	1.25	1.1	0.85
Gate delay (ps)	660	400	240
D flip-flop Max. toggle frequency (MHz)	200	300	400

Some of the CMOS circuits operate in weak inversion (Section 4.3.1) to save power. Considerable effort was spent in their design and in the development of accurate device models. The existence of all these analogue cells in CMOS shows how far CMOS analogue techniques have advanced. The supply voltage for the CBH process can go up to 10 V, which is an advantage for circuits like op amps when a large output signal is needed. The standard voltage for the analogue cells made with the CAE and CZE processes is 5 V. The supply voltage for the digital sections extends from 2.2 V to 5.5 V. The key parameters for digital cells are given in Table 10.13. The AMS digital cell library contains about 30 hard and 50 soft cells. The hard cells form logic gates, latches, flip-flops, multiplexers, etc., the soft cells fulfil more complex functions like shift registers, adders, and counters.

Getting an ASIC design right the first time through is a noble ideal. It can be and has been attained, but in reality a second layout and hence a second production of the full mask set are likely to be required for fine tuning, if not for major modifications. In order to reduce mask costs, AMS have adopted the practice of multiple die reticles for shared wafer production, in place of the standard single die image (Section 2.5.2, Fig. 2.16). The 5X multiple-die reticle serves for producing the different dies on a single wafer. The quantity obtained for each die will naturally be smaller than for the traditional single die image, but will still be more than adequate for prototype evaluation. With this multi-product wafer service, mask production charges are shared and the cost falling on an individual customer is greatly reduced.

The uncommitted logic array (ULA) mixed signal families of GEC Plessey Semiconductors (GPS) are a benchmark for bipolar ASICs which combine analogue and digital functions. The information on these products is reproduced by kind permission of GEC Plessey Semiconductors. The six families are:

(i) DA series
   A general purpose family;
(ii) DS series
   A high performance array for 100 MHz digital systems;
(iii) DF series
   The logic section of the DS series is retained, but the analogue capabilities are greatly enhanced;
(iv) DT/DV series
   A high performance family with digital system speeds and analogue capacity up to 200 MHz;
(v) DX series
   The fastest family with digital system speeds up to 600 MHz.

All the series adopt a common architecture with a central core of digital array elements surrounded on all sides by analogue array elements. The generic term *array element* is being retained in preference to *cell* (the data book designation) in order to emphasize that the D series cells are formed by metallizing array elements. Two of the families, the DA and DF series, will now be described.

The DA series is produced in eight dies. The smallest of these contains 20 digital and 6 analogue array elements, the largest is made up of 1428 digital and 40 analogue array elements. The digital core of the DA series has an SOG architecture. Each digital array element consists of six transistors and four resistors, and can be configured as a CML 2-input NOR gate: CML is the logic for the DA family. The array element consequently is equivalent to two gates. The analogue array element is composed of nine npn and four pnp transistors, 21 resistors, and a bond pad, which can be used as a low value capacitor (~ 9 pF). The analogue cell library contains the following:

(i) input cells;
(ii) output cells;
(iii) comparators;
(iv) op amps, amplifiers;
(v) oscillators and timing circuits;
(vi) voltage regulators and voltage references;
(vii) application specific circuits.

All dies in the DA series have four special array elements, placed at the corners of the die and called *corner cells*. These are:

(i) TLC cell (top left corner)
   Bandgap core components, level shift components, pass elements. Main application: bandgap reference, 1.25 V, temperature coefficient of 50 ppm/$°C$;
(ii) TRC cell (top right corner)

Two large capacitors (nominally 35 pF), resistors, npn and pnp transistors.
Main application of capacitors: frequency compensation in amplifiers;

(iii) BLC cell (bottom left corner)
Four 60 mA npn transistors, resistors, supply voltage regulator for CML logic.
Main applications: high current outputs for triac drivers *etc*. Stable supply voltage for CML;

(iv) BRC cell (bottom right corner)
Four banks of eight 5 kΩ resistors, matched to better than 0.5 per cent, npn and pnp transistors, four 200 kΩ resistors.
Main application for resistor banks: data converters.

The input and output cells are buffers with facilities like TTL compatible input, three-state output, and open collector output. Seven comparators are available. Their specific attributes are general purpose, Darlington input, low power, hysteresis, ground and rail sensing (meaning sensing the proximity of the input to ground or rail (supply) voltage), and window comparison. The six oscillator and timing cells include RC and crystal oscillators, an RC monostable, and an on-chip monostable which is configured with a TRC cell and requires no external components. The seven operational amplifiers include two operational transconductance amplifiers (OTAs, see Section 6.2.2). Two of these are *ground sensing* op amps, signifying that the high common mode rejection ratio holds at ground levels. Another cell furnishes a voltage follower. Five of the nine regulator and reference cells are 5 V regulators operating with input voltages in the range (6.5–18) V. They are configured with the aid of a TLC cell, except for the 5 V standby regulator which requires one analogue array element (a standard requirement for most of the analogue cells) and has an output current of 50 μA. Two reference cells output the 1.25 V bandgap voltage, another gives an accurate divide-by-two of the input voltage.

Application specific cells are rather like application specific standard products (ASSPs), referred to in Section 1.3.5. They incorporate application-specific circuits *within* the ASIC. Examples of application-specific analogue cells are:

(i) an open loop triac phase angle controller;
(ii) a 5 V shunt regulator and bandgap reference;
(iii) a high voltage piezo oscillator;
(iv) a high voltage piezo driver;
(v) a 4-bit ADC;
(vi) a zero crossing detector with a programmable threshold voltage.

The ULA DF family is available in six sizes. The smallest die has 224 digital and 32 analogue array elements. It is much faster than the DA family, achieving a digital system speed and analogue capability of 100 MHz. The digital core is channelled, with rows of array elements separated by routing channels. A digital array element consists of eight transistors and two resistors, and is equivalent to two 2-input gates. It can be configured into one of four basic gate structures:

(i) NOR (CML);
(ii) RNOR (ECL);
(iii) PNOR (active pull-up and pull-down output, similar to TTL output—Fig. 5.24);
(iv) WRNOR (a derivative of the RNOR gate with twice the tail current and having twice the RNOR differential logic swing).

The types of analogue circuits contained in the DA and DF families are similar. The same can be said of the other D series; the distinction is in their performance. The DF input buffer has a typical t_P of 3 ns compared with 90 ns for the emitter follower macro in the DA series. The video amplifier has a gain of 20 dB with a 60 MHz bandwidth. Oscillator macros include a crystal oscillator going up to 20 MHz and an overtone crystal oscillator going up to 100 MHz.

The *tile array* series of Micro Linear is a contrast to the mixed-mode CMOS cells of AMS and the bipolar arrays of GPS. It is all-analogue—at least very nearly so—and is largely in bipolar technology. A tile array is composed of *mini tiles*, the basic analogue array elements which are made up transistors, diodes, resistors, and capacitors. The mini tiles are in effect micro array elements, which are compounded in various combinations into macro array elements (but not cells). The three general purpose array families are:

(i) the FC3500—5 V, BiCMOS;
(ii) the FB3600—12 V, bipolar;
(iii) the FB3400—±15 V, bipolar.

The Micro Linear data book does not list analogue macros as such (Micro Linear Corporation 1993). Micro Linear's standard products are built with their proprietary *tile array* technology. These products are very easily incorporated in their semicustom analogue ASICs. What it amounts to is that Micro Linear can satisfy many requests by drawing on their standard products, which can be modified to meet a specific requirement outside the standard specification.

The FC3500 BiCMOS tile array has, in round figures, 360 npn and 100 pnp transistors, 70 nMOSFETs and 70 pMOSFETs, resistors totalling 2.62 MΩ, MOS capacitors totalling 30 pF, and 28 bondpads. The die size is 70 × 88 mil^2. It has a 1.5 μm geometry and gives the highest performance of all the families.

The FB3600 family contains the widest range of general purpose arrays. Each of the nine tile arrays is a selected combination chosen from 19 mini tiles and targeted at a specific function. The composition of selected tile arrays is given, together with a statement of their functions and the nature of the mini tiles, in Table 10.14.

Here is a concise summary of the mini tiles. T1 (npn, pnp, resistors; npn, pnp are used as shortened notations for npn, pnp transistors) is the most frequently used of the (mini) tiles. T2 (npn, pnp, capacitor) has the next highest usage. T3, T4, T12, T15, and T16 are made up exclusively of transistors and diodes (including Schottky devices) for special applications like high frequency, power, and Schottky-based circuits. T6 (npn, resistors) is for npn intensive circuits, including ECL, which however is mainly implemented with T9, That tile (npn, diodes, resistors) can be

Table 10.14 Selection of FB3600 tile arrays

Array Description	FB3605 SHF	FB3610 SGP	FB3621 MHF	FB3623 MHP	FB3631 LMAD	FB3635 LMAD
T1 General	10	48	48	64	92	36
T1A General	10					
T2 Specialized	2	6	8	12	12	4
T2A Specialized	2					
T3 Power	2	4	2	4	4	2
T4 Low noise		4	4	2		
T5 Precision				4		2
T6 NPN intensive			8			8
T7 High frequency			12		4	12
T8 Schottky core						4
T9 ECL logic	10				22	42
T10 ECL bias					1	1
T11 TTL output	8		4		8	8
T12 Schottky peripheral						2
T13 High frequency	4					
T14 High power npn				4		
T15 Medium pnp				4		
T17 General					1	

SHF = Small high frequency, SGP = Small general purpose, MHF = Medium high frequency, MHP = Medium high power, LMAD = Large mixed analogue/digital

Numbers apply to the tiles used for a particular tile array

(The information for Tables 10.14, 10.15, and the other data on the FC3500, FB3600, and FB3400 tile arrays is contained in the Micro Linear Semi-Standard Analog 1993 Data Book. It is reproduced with permission of Micro Linear Corporation, U.S.A.)

Table 10.15 FB3600 tile arrays die—area and function complexity

Tile array	Die area (mil^2)	Analogue complexity	Digital complexity
FB3605	70×110	4	28
FB3610	82×102	6	
FB3620	102×115	12	
FB3621	102×115	8	
FB3622	112×125	12	
FB3623	115×122	12	
FB3630	131×150	24	
FB3631	192×156	12	62
FB3635	131×150	9	130

Analogue complexity is in terms of 741 op amps (see Table 6.2)
Digital complexity is in terms of 2-input NAND gates

configured into three 2-input ECL gates. T10 (npn, pnp, resistors, capacitor) provides the necessary temperature-compensated bias voltage for ECL logic (similar to V_{BB} in Fig. 5.17). T7 (npn, resistors) is aimed at (60—100) MHz amplifiers. T8 (Schottky npn, Schottky diodes, resistors, capacitor) is intended for analog and digital designs. T11 (npn, Schottky npn, resistors) contains components for TTL- or CMOS-level output stages and for logic level conversions. T13 (npn, pnp, resistors) is a general purpose tile similar to T1.

The variety of active and passive tile components leads to high quality analogue circuits covering a wide range of applications. Transistors can be selected for particular characteristics (like high frequency, power) unlike transistors in most ICs, where the only variation allowed is the transistor area; the analogue circuits built with the mini tiles will have a superior performance. The tiles also contain a variety of resistors, base resistors (225 Ω), implant resistors (4 kΩ and 8 kΩ) and *precision* resistors (850 Ω, matched to 0.5 per cent, in T5). The high component density of the tiles meets one of the main objectives in ASICs, the construction of a circuit with the minimum amount of silicon. The die area for the FB3600 series is shown, together with the function density, now called complexity in Table 10.15. The diminutive size of the dies is evident when these are compared with a typical VLSI die. The areas of the smallest and largest dies in the FB3600 series are 5 per cent and 13 per cent of that for a 1×1 cm^2 die.

The general purpose tile arrays are supplemented by *application focussed* tile arrays. These, selected assemblies of mini tiles, in the FB3400/FC3500/FB3600 series, form the foundation for applications of a prescribed nature. Examples of application focussed tile arrays are:

(i) The FB3651 local area network (LAN) transceiver tile array, which is optimized for building the circuits for a LAN transceiver. Nine independent timer functions can be arranged and the array can yield up to 18 analogue blocks and 150 ECL gates.

(ii) The FC3560 read channel tile array. This BiCMOS array has optimized blocks for the following circuits: a pulse detector, servo demodulator, data separator, frequency synthesizer, write compensation, two crystal oscillators, and a bandgap reference. There is an additional logic capability equivalent to 800 gates.

Coming very briefly to the FB3400 tile array family, this is designed for ± 15 V standard operation with a maximum of ± 18 V, or 36 V for single polarity operation. The family consists of three general purpose tile arrays, made up from five mini tiles and having an analogue complexity of 4, 12, and 16 respectively. The types of functional blocks which can be implemented are op amps, comparators, voltage references, video amplifiers, modulators, demodulators, etc.

The three categories of ASICs described in this section give evidence of the progress made in the analogue field. The advances are not merely in technology, but also in the analogue design tools for macro formation and, what is very important, for ASIC testing. The analogue macros are not confined to those in the data books which, in this respect, are far more comprehensive for digital ASICs. They are being continually augmented by modified and new designs. GPS, AMS, and Micro Linear all have substantial experience in analogue techniques incorporated within standard ICs or ASSPs, and have large macro libraries available. Existing macros can easily be modified at small extra cost to meet a customer's specification. CMOS analogue cells have established a strong niche for themselves. All the same bipolar techniques afford more flexibility for analogue design and tend to result in functions with superior performance, because the circuits can be optimized by choosing from a variety of transistors. Mixed-mode CMOS will remain highly popular, and is likely to be preferred in general, but will continue to be complemented by bipolar ASICs, which can outperform them in some applications and have a wider coverage of the analogue field.

10.6 Overview

This chapter could have been concluded with a section on design style selection. Instead this section contains an overview which is intended to help in the selection process by looking at other issues than the mere comparison of ASIC design styles in isolation. For standard ICs the choice of the right components for the application is clear-cut. The chips have to satisfy specified electrical and electro-mechanical characteristics (packaging, environmental properties, *etc.*) and should preferably be second-sourced. Choosing an ASIC is a very different matter. It is not just the selection of the design style, but also the selection of the design tools and the ASIC and design tool vendors. There is no second-sourcing in the accepted sense of that term. The design is carried out with sophisticated tools which may sometimes call for a choice between the proprietary tools of the ASIC vendor or the third party tools of an EDA vendor. Another consideration is linked to the nature of the customers, who span the spectrum from the big players with a need for large quantities of ASICs, more likely than not for sophisticated systems, to manufacturers of equipment produced in small quantities and calling for a correspondingly small numbers of ASICs.

The first parameter which comes to mind when categorizing ASICs is their function density. At the high end we have the gate array, the standard cell, the embedded cell (cell-based ASIC), and the embedded array. At the low end the gate array and the standard cell compete with the FPGA. Lower still, the PAL and PLA are attractive means of mopping up a modest amount of glue. The ASICs considered so far have all been CMOS with the exception of some bipolar PAL/PLA families. BiCMOS, ECL, and GaAs are alternative technologies and ASICs are available in all of these, albeit with a much smaller choice and volume of production than CMOS chips.

BiCMOS ASICs are few and far between. BiCMOS gate arrays tend to be constructed with gates having CMOS and ECL logic levels on the same chip; typical gate delays are 170 ps and 140 ps respectively. The maximum gate count is ~130 000 (CMOS, with about 5 per cent ECL). The speed advantage of BiCMOS over CMOS becomes highly marginal for 0.5 μm structures, which are beginning to come on stream. ECL gate arrays have gate delays of ~(40–60) ps and an upper gate count of ~150 000. GaAs arrays go up to ~250 000 gates. They have gate delays of ~70 ps, but consume only about one third of the gate power for ECL and cost about 50 per cent more. Their superior radiation resistance can be a telling advantage in some applications. The role of VLSI GaAs remains uncertain. Is it standing in the wings to burst forth in earnest and take a substantial slice of the ASIC market, or will it remain rather specialist with a niche in that part of a system which is beyond the speed of CMOS? For the moment it looks as if GaAs and ECL fall into the latter category. They are assured of a place at the front end of telecommunication systems for processing incoming gigahertz signals and for passing the output at (100–200) MHz to 0.5 μm CMOS.

The superior density of standard cells over gate arrays remains their forte. The differential in density between these ASICs has been reduced but not eliminated with the SOG architecture. If density were the only criterion, the standard cell would win every time. The narrowing of the gap in gate count, combined with the far quicker design time of the gate array and the reduced cost of metal masks compared with diffusion masks, lead to a preference for gate arrays in many cases. The premium put on the time to market has been stressed repeatedly. It can dominate over an analysis purely in terms of cost without regard to the development and production schedule. Collett (1991) reinforces the estimates of loss in profitability due to slippage in production, based on the reference of Reinertsen (1983) cited earlier.

The combination of standard cell and gate array architectures in the cell-based ASIC illustrates their complementarity. Thanks to the maturity of the ASIC art, vendors have an abundant and ever increasing library of embedded cells, extending to megafunctions and including analogue cells. The array section can be quickly designed to complete the logic for the function required. The embedded cells are sometimes combinations of existing (semicustom) and new (full custom) standard cells; the full custom become semicustom cells for subsequent designs. At the other end of the spectrum the FPGA has helped to place ASICs within the reach of customers who need them only in small quantities, where NRE costs make gate arrays and standard cells uneconomic. One way of exploiting the advantages of the various ASIC design styles is *migration*, the transfer of a proven design from one architecture to another. The most widely used migration is from an FPGA to a gate array or a standard cell. The cost effectiveness of the FPGA for small quantities is thereby transferred to one of the aforementioned alternatives for high volume production. Most FPGA vendors have design tools available for that purpose.

The advances in EDA design tools, which are available from numerous suppliers, have made ASIC chip design universally available instead of letting it remain an esoteric art. The turnkey design route does not absolve the customer from partaking in that aspect of the design which is the prime responsibility of the vendor. The design is an interactive process and success calls for co-operation between the two sides at all stages. Moreover the customer is responsible for the initial design phase (Fig. 10.2) and has to possess the necessary tools for that purpose. The outlay on hardware and software tools (the hardware is likely to be usable for purposes other than just ASIC design) imposes only a modest burden on NRE costs in the long run for the likely situation that the customer will have an ongoing involvement with ASIC design.

The factors to be weighed when choosing an ASIC manufacturer are matters like production capability, reliability, NRE and production costs, dynamic characteristics (speed and power), design support, and compatibility of design tools. Mask-programmable ASICs are offered in the main by semiconductor houses whose products include other VLSI, either in the form of standard components or ASSPs. Some establishments are ASIC vendors first and foremost, but generally also offer ASSPs. Programmable products like FPGAs and other PLDs are, in contrast to mask-programmable ASICs, the preserve of vendors exclusively engaged in their manufacture.

VLSI is moving towards a reduction of the standard feature size from ~ 0.8 μm (mid 1995) to ~ 0.5 μm, likely to be achieved by 1997. The cost of establishing foundries for 0.5 μm geometries is enormous. It is of the order \$500 million to \$1.5 billion and can only be undertaken by IC manufacturers in the top league. ASIC specialists will probably reach agreements with these foundries for their services should they wish to adopt such a geometry. It is not absolutely essential for them to follow suit immediately. Many applications can and will continue to be satisfied with ASICs having a feature size of (0.8–1.2) μm or even higher. The larger the geometry, the less critical is the design process and the less expensive is the product. In some analogue circuits, large transistors have several advantages over smaller devices. The majority of ASICs will however have to adopt the new feature size of standard VLSI in the long term.

The design of ASICs is a complex undertaking. The cost is tied up with the NRE component, chip production and testing. It has to be related to system cost for a meaningful economic appraisal. The time to market greatly influences revenue, and the possibility of a second layout (and hence a second mask set) for fine tuning or the elimination of errors should be allowed for. The choice of ASIC and EDA tools vendors is weighed in the light of factors some of which are subjective. Objective judgements should be maximized, but every decision involves an irreducible subjective element. It will not be taken on quantitative data alone, and ASIC selection is not likely to be a purely reductionist matter.

The major developments during the last few years have been:

(i) The improvement in EDA design tools, which has greatly increased the prospect of an ASIC functioning when inserted in the target system the first time through. The design tools will however have to be developed further to match the growing complexity of ASICs due to the ongoing shrinkage of VLSI geometries. Technology is ahead of design, and the reduction of that gap is of paramount importance.

(ii) The production of mixed-mode and analogue ASICs, matched by design tools and test equipment, which have made them affordable even for small firms requiring only moderate quantities.

(iii) The growth of the FPGA, which competes with other semicustom chips at the low end of their functionality. The FPGA can also be a quickly designed prototype which is turned into a mask-programmable ASIC by migration.

(iv) The spread of cell-based ASICs, which permit the rapid design of highly complex functions.

References

Ballew, W.D. and Streb, L.M. (1992). Board-level boundary scan: programming observability with an additional IC. *IEEE Transactions on computer-aided design*, **11**, 68–75.

Basset, R.W., Turner, E.M., Panner, J.H., Gillis, P.S., Oakland, F.S., and Stout, D.W. (1990). Boundary-scan principles for efficient LSSD ASIC testing. *IBM Journal of Research and Development*, **34**, 339–53.

Bennetts, R.G. (1984). *Design of testable logic circuits*. Addison-Wesley, England.

Bleeker, H., Eijnden, P. van den., and Jong, Frans de. (1993). *Boundary-scan test*. Kluwer Academic Publishers, USA.

Collett, R. (1991). In *ASIC technology* (eds. N.G. Einspruch and J.L. Hilbert), pp.7–24. Academic Press, USA.

Eichelberger, E. and Williams, T. (1978). A logic design structure for LSI testability. *Journal of Design Automation and Fault Tolerant Computing*, **7**, 167–78.

Fey, C.F. and Paraskevopoulos, D.E. (1987). A techno-economic assessment of application-specific circuits: current status and future trends. *Proceedings of the IEEE*, **75**, 829–41.

Fey, C.F. and Paraskevopoulos, D.E. (1989). In *VLSI handbook*, (ed. J.D. Giacomo), pp.25.3–26.21. McGraw Hill, USA.

Hollis, E.E. (1987). *Design of VLSI gate array ICs*. Prentice-Hall, USA.

Huber, J.P. and Rosneck, M.W. (1991). *Successful ASIC design the first time through*, p.156. Van Nostrand Reinhold, USA.

Maunder, C.M. and Tulloss, R.E. (1990). *The test access port and boundary scan architecture*. IEEE Computer Society Press, USA.

Micro Linear Corporation (1993). *Semi-standard analog 1993 data book*. Micro Linear, USA.

Reinertsen, G. (1983). Whodonit? The search for the new product killers. *Electronic Business*, **9**, 62–6.

Rosqvist, L. (1991). In *ASIC technology*, (eds. J.L. Hilbert and N.G. Einspruch), pp.221–41. Academic Press, USA.

Sedmak, R.M. (1989). In *VLSI handbook*, (ed. J.D. Giacomo), pp. 9.1–9.29. McGraw Hill, USA.

Texas (1991). *Testability primer*. Semiconductor Group, Texas Instruments, USA.

Williams, T. and Parker, K.P. (1983). Design for testability—a survey. *Proc IEEE*, **71**, 98–112.

Further reading

Bleeker, H., Eijnden, P. van den., and Jong, Frans de. (1993). *Boundary-scan test*. Kluwer Academic Publishers, USA.

Einspruch, N.G. and Hilbert, J.L. (eds.) 1991. *ASIC technology*. Academic Press, USA.

Hollis, E.E. (1987). *Design of VLSI gate array ICs*. Prentice-Hall, USA.

Huber, J.P. and Rosneck, M.W. (1991). *Successful ASIC design the first time through*. Van Nostrand Reinhold, USA.

Maunder, C.M. and Tullos, R.E. (1990). *The test access port and boundary scan architecture*. IEEE Computer Society Press, USA.

11
ASICs—Design techniques

11.1 Overview

The outline of ASIC design in Section 10.2 (Fig. 10.2) and, to a lesser extent, the design outline for PLDs—primarily FPGAs—in Section 9.4 (Fig. 9.22) contain preparatory material for the treatment of design techniques in this chapter. Their placement in Chapters 9 and 10 served to familiarize the reader with the nature of the ASIC design process in order to decide on the ASIC design style and on the choice, or at least the assessment of the design tools. Portraying the design issues in outline was not putting the cart before the horse. Decisions like design style selection often have to be made prior to being familiar with the subject of design to the extent with which it is pursued here.

This overview follows on from the aforementioned material, and outlines the design sequence by a concise description of the various procedures in the design flow. These and related design facets are the subjects of the following sections. There is some repetition in the subject matter within this chapter. This is a deliberate tactic to drive home the design techniques in a way which prompts a flexible approach for understanding the differences between the design tools of various vendors, and at the same time appreciating the common factors underlying ASIC design. Such differences may loom large on first acquaintance with a particular design procedure, but the essential design core is much the same in all cases.

The top–down design flow in Fig. 11.1 blocks begins with a behavioural description of the system function, represented by the 'building' blocks in Fig. 11.1(a). That begs two questions. First, what is the nature of the behavioural description and second, what type of function do the behavioural blocks represent? The second question can be answered by referring to a yet higher level, an architectural level. Suppose it is intended to design a microprocessor with a specification which cannot be met by a standard product. The blocks in Fig. 11.1(a) would then be units like an arithmetic logic unit (ALU), a microprogram ROM, an instruction decoder, etc. The designer has already decomposed the architectural construct, the microprocessor, to arrive at these behavioural blocks. It is safe to say that as a rule top–down design *starts* with such a decomposition, and there is little point in a diagrammatic or descriptive representation of an architectural block without any indication whatever of how the function it represents is executed. The answer to the other question is that the behavioural description is couched in a hardware description language (HDL). Two standard HDLs have emerged. It was not all that long ago that ASIC vendors relied on their own in-house HDL.

386 ASICs—Design techniques

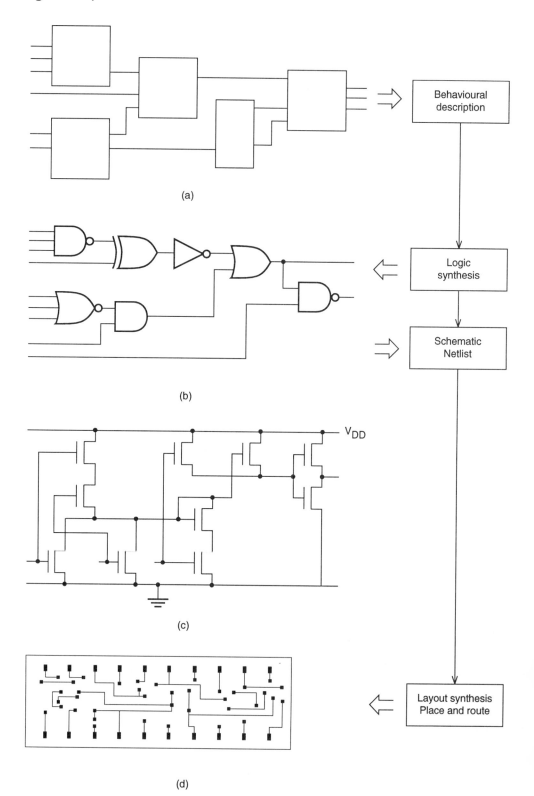

Fig. 11.1 Top–down design (a) behavioural (functional) blocks (b) logic schematic (c) circuit schematic (d) mask pattern

Many of them still retain their own version, but the two standard HDLs are strongly preferred and in-house HDLs are likely to disappear. HDLs describe a function at various levels of abstraction. The *behavioural* description becomes the specification for the lower level, the *structural* description of the implementing hardware. The transformation from the behavioural description of the blocks in Fig. 11.1(a) to their logic implementation at gate level in Fig. 11.11(b), a purely symbolic schematic, is the process of logic synthesis carried out by a compiler specially developed for that purpose. The logic schematic of Fig. 11.1(b) is expressed in terms of the macros contained in the ASIC library and is stored in a *netlist* containing all the logic elements and their interconnections.

The netlist becomes the specification for the final transformation via the circuit schematic of Fig. 11.1(c) into the mask patterns of Fig. 11.1(d) by means of the layout synthesis (place and route), which selects the array elements to be used (gate array) or places the macrocells in the desired locations (standard cell). This process, it will be recalled from Chapter 10, is called *annealing*. The second step of place and route is the formation of the interconnections. The number of mask patterns required depends on the nature of the ASIC. It is a full set of masks for a standard cell, and a set of metal masks for a gate array. For an FPGA place and route configures the hardware programmer or, in the case of SRAM FPGAs, the SRAM.

Simulations and other checks have been omitted—they are covered later—in order to emphasize the central role of the two transformations, logic and layout synthesis. Figure 11.1 is at the heart of ASIC design and can easily be adapted for the bottom–up route, which has a much simpler approach by starting with the structural composition, effectively the logic schematic at gate level. In bottom–up design, logic synthesis is omitted.

The design flow is described in Section 11.2, followed by an outline of hardware descriptive languages in Section 11.3. Simulation and checking are the themes of Section 11.4, and commercial design tools are the subject of Section 11.5. Whilst the design techniques for mask-programmed semicustom circuits and FPGAs have much in common, there are sufficient significant differences to justify a separate Section, 11.6, for the latter. Finally Section 11.7 contains some reflections on the state of ASIC design in general.

11.2 Design flow and methodology

Figures 10.2 and 11.1 may have given the impression that the design process is quite simple, whereas it is a highly complex undertaking calling for considerable skill and a good appreciation of the ASIC art. Figure 11.2 puts flesh on the structure of Fig. 11.1. It has been compiled after consulting numerous schematics of data flow, found mainly in the literature of ASIC and EDA design tool vendors. The design flow it portrays is not dogmatically definitive, but can vary for a number of reasons. A chip with a very high density may call for more sophisticated tools than an ASIC at the low end of the range. Different design tools may have different deployments in the flow sequence. The simulation steps and the procedures for testability are two further potential variants. All the same Fig. 11.2 does give a representative flow diagram for ASIC design

388 ASICs—Design techniques

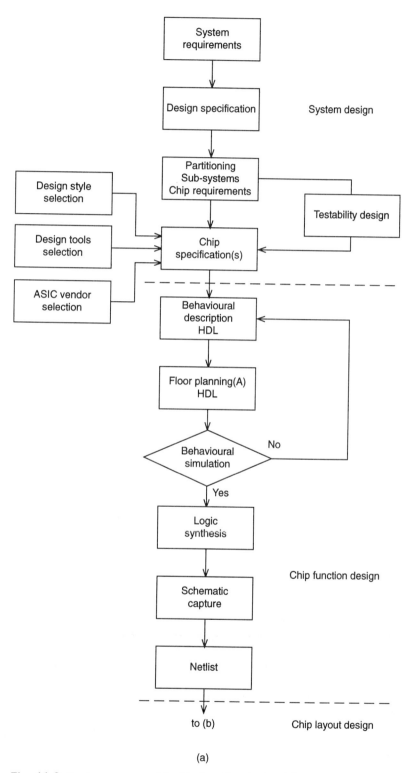

Fig. 11.2 Design sequence (a) chip function design (b) chip layout design

Design flow and methodology 389

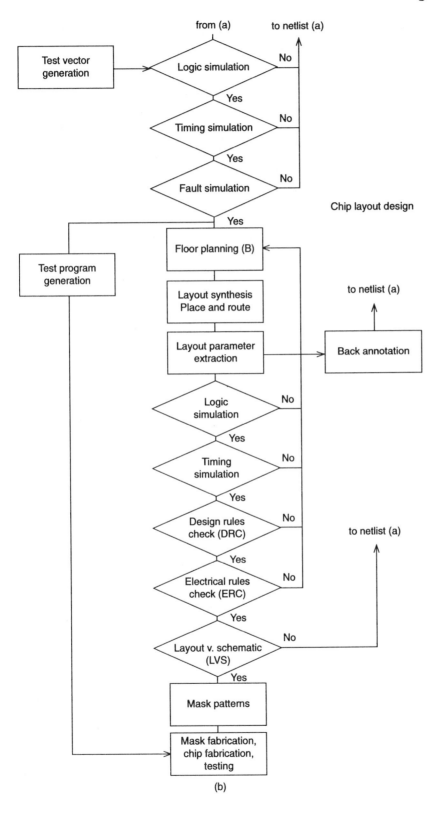

(b)

and for appreciating the various steps and their sequence in the design process. It lays a foundation for the design strategy regardless of minutiae in the order of executing the various steps, or for that matter in the steps themselves.

Figure 11.2(a) covers the design up to and including the netlist, which represents the functional specification of the chip. Figure 11.2(b) takes over at that point and covers the design of the physical layout. The design specification, which corresponds to the system specification in Fig. 10.1, provides the basis on which to decide whether to go for standard ICs or ASICs. It is one of the most important steps in the whole of ASIC design and is prone to be underestimated by a novice in this field. Much effort and skill are called for to draw up a clear, unambiguous, and complete design specification of acceptable quality. The design specification has not only to define the function of the chip on its own, but must guarantee its correct interaction with the other chips mounted on the PCB. Attention to detail is required and that tends to be overlooked. It is worth browsing through the data book for a microprocessor in order to appreciate the detail in and the quality of the design specification, including testability and debugging support (Intel 1992). At a lower level, familiarization with the data of TTL/CMOS MSI like the 181 4-bit ALU and the 160–163 synchronous counters is recommended for the same reason (RCA 1986; Texas 1984).

Having completed the design specification the system is, if necessary, partitioned into subsystems and the chip requirements become clear. The stage is set for selecting the design style, the ASIC supplier, and the design tools. These decisions may have been made right at the start if the customer already has substantial design experience. In that case, some, possibly all the design tools needed will be available. The drawing up of chip requirements marks the initiation of the design for testability, which is part of the specification. The design of the individual ASIC(s) now begins. If the chip requirements call for a number of different ASICs, the design process starting with the block *chip specification* in Fig. 11.2(a) has to be undertaken for each individual ASIC separately. The logic synthesis is based on the behavioural description in an HDL. The synthesis tool usually includes or operates in conjunction with a *floorplan*, which will be explained presently. The first objective is the transformation of the behavioural description into a structured hardware implementation by logic synthesis. Figure 11.2 applies to digital ASICs, but is readily adapted to mixed-mode semicustom circuits. The input description is in behavioural HDL.

Many vendors have their own in-house HDL, developed before the two industry standards, Verilog and VHDL, became available. The majority of ASIC designs is now undertaken with these HDLs. The highest level of an HDL is *behavioural*, an algorithmic description which contains no information about the structure of the hardware. Next down in the hierarchy comes the *register transfer level* (*RTL*), also called *data flow*. RTL is a behavioural description with architectural significance, leading to yet another designation 'behavioural/RTL'. The *structural* or *gate level* is the third tier. There are some lower levels which are of no interest at this stage. One of these, the switch level, is used in circuit simulation (Section 11.4). The nature and significance of the various levels is dealt with in Section 11.3. The RTL level is by far the most widely used HDL description for entry to the logic synthesizer for reasons given later. The compiler is the heart of the logic synthesis block in Fig. 11.2(a), transforming the input into structural HDL in terms of the ASIC macros.

Two steps are interposed between the behavioural description and the logic synthesis in Fig. 11.2(a). The behavioural simulation is the first of the simulations and checks undertaken in the design process. It verifies the validity of the HDL input. Floorplanning(A) has been included in keeping with the aim of presenting a comprehensive picture of ASIC design that contains some steps which are not universal practice; floorplanning(A) is one of these. Floorplanning is concerned with the layout of the die. We shall now proceed to logic synthesis and shall return to floorplanning presently.

The process of logic synthesis is illustrated in Fig. 11.3. The input to the synthesizer compilers is a behavioural HDL construct. There are two other inputs. One of these is the ASIC macrocell library, simply called ASIC library by most ASIC and EDA tool vendors. This library holds the full data for all macros. The other is a list of the constraints for controlling the output. These are factors like timing (the most important), area (of specific parts), power consumption, and testability. Figure 11.3(a) simply shows the conversion of a behavioural RTL into structural HDL.

The process of synthesis is illustrated in Fig. 11.3(b). The HDL input is translated into a Boolean description consisting of primitives like logic gates, latches, and flip-flops. That Boolean description is optimized by the algorithms of the compiler(s); it is likely that several compilers perform this task. One established and popular optimization technique is to convert the Boolean description to a PLA sum-of-products format (Section 8.7.1) and submit this to well established PLA optimization techniques for producing a Boolean description of the minimized hardware.This process of conversion is called *flattening*, because the output contains only two levels, the AND and OR planes. The description is mapped to the ASIC macros, drawing on the constraints and ASIC macrocell library inputs (Shankar and Fernandez 1989, pp.217–24; Perry 1994, pp.235–41). The resulting structural HDL output is used to compile the netlist and for the schematic capture, a graphical display for inspection and user-interactive modifications. The schematic capture is obtained via a standard graphics interchange format.

The techniques for optimization depend on the nature of the logic: separate approaches are taken for the datapath and the control logic of a digital system. The datapath, sometimes called execution unit, comprises the parts of a data processing system which operate on the data in contrast to the parts which control that operation (Geiger *et al.* 1990, p.856; Glasser and Dobberpuhl 1985). The datapath contains hardware like shift registers, ALUs, multipliers etc. Digital filters like the decimation filters for oversampling delta-sigma converters in a DSP (Section 6.5.6) are another example of a datapath. A broader interpretation classes a datapath to be any form of *structured* logic, which is characterized by regularity in layout and interconnections. The datapath in this case includes memory chips with their highly structured arrays (Geiger

392 ASICs—Design techniques

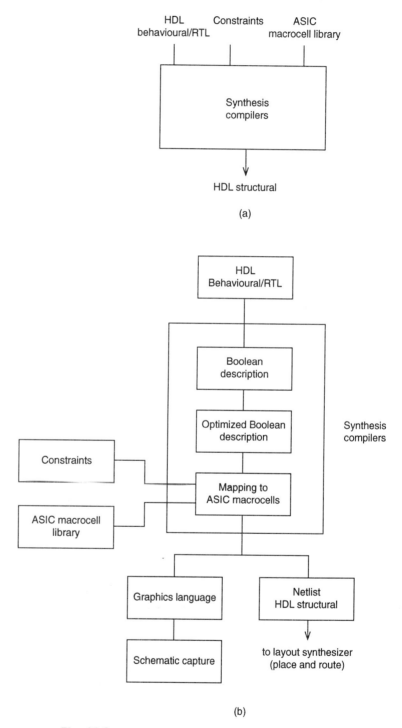

Fig. 11.3 Logic synthesis (a) schematic (b) design flow

et al. 1990, p.779). A digital system can broadly speaking be divided into datapath (structured) logic and control (random) logic, which consists of combinational logic and storage cells (latches and bistables) assembled with a random composition of ICs and with a layout to suit a particular application.

The best optimization of logic sythesis is achieved by having separate compilers for the datapath and the random logic. Many synthesizers automatically partition the HDL input into those two sections for that purpose. The synthesized output is, as had already been pointed out, available in two formats, a netlist and a graphical schematic. The netlist is the specification for the next transformation into a physical layout by layout synthesis. Its validity is established by three simulations. The logic and timing simulations verify the logic and check that the propagation delays are within prescribed limits. At this stage the timing simulation is a pre-layout approximation based on statistical estimates of wire lengths for the gate-level schematic. The fault simulation establishes that the test vectors are of adequate quality. When all simulations are satisfactory, after iterations if necessary, the netlist inputs its data for layout synthesis via the floorplanner(B). The two blocks *floorplanning(B)* and *layout synthesis/place and route* are likely to be represented by a single block in many schematics where place and route embraces floorplanning. The term layout synthesis will probably be omitted: *place and route* on its own is the norm. The term 'layout synthesis' has been added to stress the nature of the place and route operation. Floorplanning(B) is shown separately to emphasize its importance in the layout strategy. The floorplan arranges the physical layout by partitioning the schematic into blocks and placing these in the best possible position. Figure 11.4(a) shows an initial and Fig. 11.4(b) a final block layout. Manual interaction enables the designer to move the positions of the blocks.

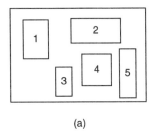

(a)

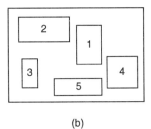

(b)

Fig. 11.4 Floorplan (a) initial layout (b) optimized layout

There are three types of macrocells, labelled *hard, soft,* and *firm*; the first two of these were mentioned in Section 8.4. The hard macro has fixed placement and routing and hence fixed timing. The soft macro has flexible placement and routing and variable timing. The firm macro has fixed placement, flexible routing, and variable timing. In relation to these definitions, each macro is an entity on its own. *Placement* refers to the relative position of the transistors which make up the macro, and *routing* to their interconnections.

The physical layout is governed by factors like:

(i) the specified location of the pins, especially I/O:
(ii) the placement of the blocks. These are located to even out the interconnections between them, so as to obtain an equitable distribution. This step eliminates or at least reduces local and potentially prohibitive congestions in connectivity;
(iii) timing constraints for specific signals, calling for selected routing in case of need.

The floorplanner in many cases is an initial place and route tool, first dividing the chip logic, expressed by the netlist, into blocks of selected macros, and arranging the place and route for each block. The blocks are placed and interconnected on the first pass (Fig. 11.4(a)) and their placement and interconnections are subsequently optimized paying attention to constraints for selected signals (11.4(b)). Layout synthesis tools frequently integrate floorplanning with place and route. Semicustom design does not demand that the designer deals with chip geometries. The only involvement with geometries is likely to be graphical interactive modification of place and route, and that is easily handled without requiring a detailed knowledge of the macro structures. The established routing in ASICs follows the Manhattan style, for which wires are horizontal or vertical. A common (but not universal) practice is to reserve one set of wires for datapath and the orthogonal set for control logic interconnections. Many vendors modify the Manhattan style by the addition of diagonal wires at 45° from the orthogonal interconnections.

The technique of arriving at the physical layout by floorplanning is becoming the method of choice, and has replaced earlier techniques for layout synthesis which operated in a far more random fashion. Floorplanning has many advantages but also has some drawbacks. One of these is a loss in chip area. The partitioning of the full schematic into logic blocks leaves some of the silicon unoccupied (except for interconnections). That loss will be partially offset by the superior routing within the blocks. We can now return to the step of floorplanning(A) in Fig. 11.2(a). Whether the division of the chip function into logic blocks is carried out integrally with the behavioural HDL description or as part of the layout synthesis depends on the set of design tools being used. If it is the former, floorplanning will still be part of the physical layout. Floorplanning(B) is not confined to partitioning of the logic into blocks, but also optimizes their positions. The action of floorplanning(A) is more limited. An important advantage of floorplanning high up in the hierarchical structure is that it can be used to obtain very realistic estimates of pre-layout timing. The timing of a logic system determines its ability to function correctly. In logic design race hazards have to be contained and proper functioning calls for the presence of two or more signals at specified nodes within a narrow time window. Floorplanning(A), by being placed upstream, permits realistic estimates of timing early in the design.

The design of the physical layout is followed by simulations and checks which may involve iterations in order to obtain an acceptable layout. The graphical layout display and the data for mask production are obtained by means of a graphical interchange format, a graphical language which, if it is only suitable for mask production, is called a geometrical specification language. Most graphical exchange formats can do both. One of these, which has HDL and graphical capability and comes up again in Section 11.3, is EDIF (Electronic Design Interchange Format). Established graphical exchange formats are CIF of Caltech (Caltech Intermediate Form), GDS II Stream (often simply called GDS II) of Calma, SLL (Symbolic Layout Language) of Cadence, and GDT of Mentor Graphics. EDIF and CIF, because they are in ASCII (American Standard Code for Information) have much larger files than the most popular industry standard, the binary GDS II Stream. The design is completed by chip fabrication, probe testing of the dies, and final testing of the encapsulated chips which are the prototypes passed on to the customer.

Let us briefly look at the post-layout checks and simulations. Taking the vital assessment of timing, the wire lengths (from which the timing parameters of the interconnections are calculated) are obtained with LPE (Layout Parameter Extraction) and lead to an accurate timing simulation. The logic simulation confirms that the circuit fulfils the correct function. DRC (Design Rule Check) verifies that the physical design rules (Section 3.7) are obeyed. ERC (Electrical Rule Check) examines conditions like maximum permitted fan-out, shorts of V_{DD} to ground, etc. Lastly LVS (Layout Versus Schematic) verifies that the logic derived from the physical layout agrees with the logic derived from the netlist.

Strictly speaking none of these post-layout steps, save possibly the timing simulation, should be necessary if all the software tools right from the start operated without error. They don't. The hardware and software tools are, like all of us, highly fallible. By imposing multiple checks at all stages of the design, the probability of errors slipping through is reduced to negligible proportions at, it has to be said, very great cost. Simulations and checks are the only means of establishing design validity; breadboards went out of the window years ago. A single mistake can ruin the entire design. Errors apart, the optimization of the design depends strongly on the quality of the EDA tools. The protracted checks and simulations serve to achieve the best possible performance. It is tempting to treat iterations arising from errors or moves to optimize the performance lightly. After all many HLL (high level language) programs contain algorithms involving iterations. However in ASIC design, a reiteration can be very time consuming and expensive. It is likely to mean reiterations of other checks and simulations and, may be, a rework of the synthesis. On the other hand it is frequently possible to localize a reiteration to that part of the ASIC which has failed a simulation or check.

The proceedings of Fig. 11.2 are generally adhered to, although there are differences in detail from vendor to vendor. The two schematics in Figs 11.5 and 11.6 are examples of design flow documentation provided by ASIC suppliers. Figure 11.5 is an illustration of the design flow for the CLA80000 gate array (Section 10.5.2). The customer's responsibility in turnkey design ends with the production of the design capture/compilation (schematic capture/netlist). The subsequent design is undertaken by the vendor (turnkey route) or the customer (foundry route). GPS, in common with many other ASIC houses, offer design tools of their own (PDS) or accept third party tools for most stages of the design. Figure 11.5 does not contain all the steps shown in Fig. 11.2, largely because it is a more concise representation.

The design flow is Fig. 11.6 for the VSC370 standard cell described in Section 10.5.3 is another example of industrial practice whose congruency with the flow in Fig. 11.2 is evident on close inspection. Both Figs 11.5 and 11.6 contain provision for design reviews between vendor and customer. The logic synthesis in Fig. 11.6 highlights the importance of the datapath. Interactive floorplanning corresponds to floorplanning(B) in Fig. 11.2. The interconnect delay extraction and physical design verification are the equivalents of LPE and LVS in Fig. 11.2(b).

Summing up, the design flow is based on the following processes:

(i) The customer specifies the system and partitions it, if necessary,

396 ASICs—Design techniques

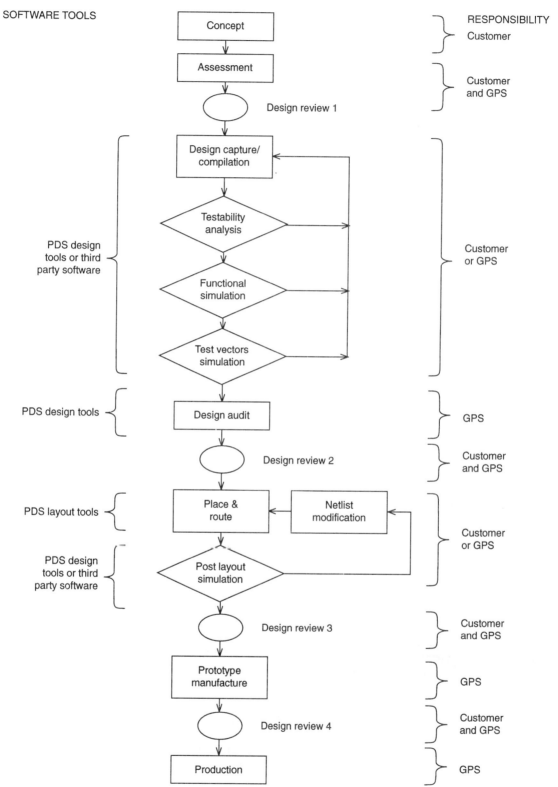

Fig. 11.5 Design flow—GPS CLA80000 gate array (Courtesy of GEC Plessey Semiconductors)

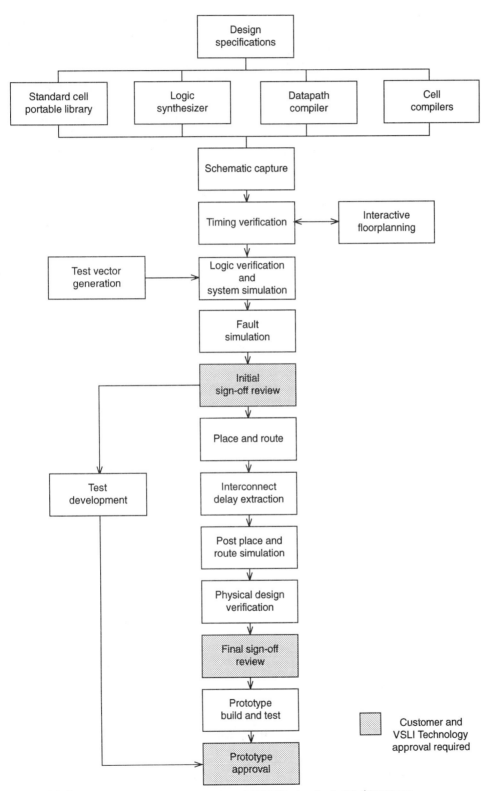

Fig. 11.6 Design flow—VLSI Technology VSC470 standard cell (Courtesy of VLSI Technology, Inc.)

into functional specifications of the individual chips.
(ii) Logic synthesis transforms the specification, expressed in behavioural or RTL HDL, into a structural hardware schematic, held in a netlist file in structural HDL;
(iii) The netlist is the input for place and route, the physical layout tool, whose output is available for graphical display with user interactive modification on the workstation, and for generating the mask patterns.
(iv) Extensive simulations and checks, some prior to, but most after, logic synthesis, verify the logic validity of the hardware, timing, check the electrical and design rules, and call for reiterations and modifications if necessary. Some of these changes are interactive and are manually inserted by the designer.
(v) The customer is responsible for testability and has to establish the quality of the test vectors by fault simulation. The ASIC vendor is responsible for test development and for testing the chips.

11.3   Hardware description languages

Top–down chip design starts with a description at behavioural level devoid of any pointer to the architecture, and progressively moves down to structural and circuit levels. All these levels are described in an HDL. (The SPICE syntax can be interpreted to be an HDL). VLSI spawned the creation of such languages, and ASIC design gave an intensive boost to their development. In the early 1980s ASIC manufacturers produced their own in-house HDLs. They had to: no industry standards were available. That situation has changed and three internationally recognized HDLs are available for design and interchange of design data: VHDL, Verilog, and EDIF. A description of their syntax is outside the coverage of this book. Two of these HDLs will now be outlined to help with the understanding of their roles in ASIC design.

VHDL is an acronym for very-high-speed integrated circuits (VHSIC) hardware description language. Its beginnings go back to 1981 when the US Department of Defense (DoD) called for a hardware language to describe digital hardware designs for military projects. The objective was the production of unambiguous documentation in a standard format, clearly understood by all interested parties, of completed projects. In 1983 the DoD issued a specification of VHDL and awarded a development contract to IBM, Texas Instruments, and Intermetrics. Military restrictions on VHDL development were lifted in 1985, and the IEEE obtained the copyright for further development and standardization. The resultant IEEE VHDL 1076-1987 standard was promulgated in December 1987 and gained rapid international acceptance. VHDL obtained a further boost in 1988, when the DoD required all completed ASIC designs to be documented in that language (Military standard MILSTD 454L). By then VHDL had moved away from the narrow confines of design documentation to the realm of ASIC design, and had already become comprehensive with constructs at behavioural level, register transfer level (RTL), also called data flow level, and structural (gate) level. VHDL is now first and foremost a VLSI/ASIC design language; documentation of the final design is of course still part of its function.

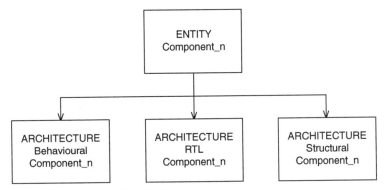

Fig. 11.7 VHDL architecture

Two further VHDL standards have been established. The IEEE 1164 standard/logic package, published in 1991, lays down a standard for describing the interconnection data used in VHDL modelling. It is reproduced in full by Perry (1994, pp.307–26). Second the IEEE revised standard of 1993, IEEE VHDL 1076-1993, supercedes the 1076-1987 standard; there are no drastic differences between these definitive documents (IEEE 1987, 1992, 1993).

VHDL has been structured to describe hardware for design, simulation, testing, and documentation. Now for a brief outline of its syntax. The information provided and the examples given are intended to portray the flavour of the language, which is not just machine-, but also human-readable, that is the designer can visualize what the program seeks to express. VHDL is formulated in terms of *entity* and *architecture* declarations. The entity contains the component's name and its input and output ports; physical and other parameters (signals, constants etc.) may be included. The architecture specifies the function of the component in terms of its inputs and the influence of physical and other parameters. All designs are expressed in the form of entities, which are the basic building blocks. The multi-level nature of VHDL comes about by the three different main architectures available for an entity, allowing it to be specified at behavioural (hierarchical), RTL (data flow), or structural (gate) level (see Fig. 11.7). The unbracketed designations will be used from now on, with an occasional inclusion of the bracketed alternatives. The behavioural and structural levels are more familiar concepts for newcomers to ASIC design than the RTL construct. RTL is ' a behavioural description with architectural significance', to quote a comment made in Section 11.2. It is of the utmost importance in ASIC design, because it is the standard HDL input for logic synthesis. RTL is a description of data flow which has been in existence since the mid-1960s, when it was called register transfer *language*, a term still retained in much of the literature on computer architecture (Bartee 1985).

A simple example of an RTL statement is

$$A \longrightarrow B \qquad (11.1)$$

meaning that the contents of register A are entered into register B, the registers being equally wide.

$$0 \longrightarrow D \tag{11.2}$$

sets D to 0. This could be a register being reset to 0, or—more likely—the D input of a D flip-flop being set to 0. RTL statements apply to a user-defined architecture and are unambiguous. Another statement,

$$R, S, : IC + 1 \longrightarrow IC \tag{11.3}$$

means that the IC register is incremented by 1 when R and S are both at 1. An unconditional increment has the form

$$IC + 1 \longrightarrow IC \tag{11.4}$$

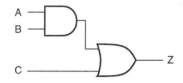

$$MB_{15-0} \longrightarrow MA \tag{11.5}$$

describes the transfer of bits in locations 0 to 15 of the MB register into (by implication) the 16-bit register MA. The RTL statements in eqns (11.6) and (11.7) are typical ADD and SUBTRACT instructions for the CPU of a microprocessor, with AC and BR being the accumulator and the buffer register respectively.

```
entity EXAMPLE_1 is
  port (A, B, C: in BIT)
end EXAMPLE_1;

architecture HL_1 of EXAMPLE_1 is
  begin
  Z <= (A AND B) OR C
  end HL_1;
```

Fig. 11.8 VHDL description of AND–OR combination

$$AC + BR \longrightarrow AC \tag{11.6}$$

$$AC - BR \longrightarrow AC \tag{11.7}$$

The AND and OR functions are represented by

$$A \wedge B \longrightarrow C \tag{11.8}$$

$$A \vee B \longrightarrow C \tag{11.9}$$

with HDL alternatives of

$$A \text{ AND } B \longrightarrow C \tag{11.10}$$

$$A \text{ OR } B \longrightarrow C \tag{11.11}$$

```
entity OR_3 is
  port (a, b, c: in BIT; out BIT);
end OR_3;
architecture RTL of OR_3 is
  begin
  d < = a OR b OR c;
  end RTL;
```

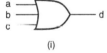
(i)

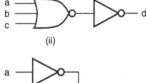
(ii)

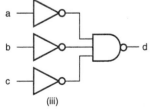
(iii)

Fig. 11.9 VHDL description of OR gate

In eqns (11.8) to (11.11) each bit of the A register is ANDed or ORed with the corresponding bit of the B register, and the result is entered into the C register. These equations might equally apply to single-bit operation.

The statements in eqns (11.1) to (11.11) are simple examples of data flow. Further descriptions of register transfer design that are helpful for appreciating the RTL construct of an HDL are contained in two books on hardware by Hill and Peterson (1978, 1984).

The first example of an RTL VHDL description is a program for the AND–OR combination in Fig. 11.8. Figure 11.9 contains a VHDL description of an OR gate. This RTL declaration can evidently be executed in a number of different ways as shown. The compiler may synthesize the VHDL input into the gate schematic of (ii) or (iii) rather than (i), possibly because inversion gives very much faster logic for most technologies.

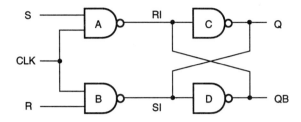

ENTITY declaration
entity NAND_FF is
   port (S, R, CLK: in BIT;
         Q, QB: out BIT)
end NAND_FF;

(a)

```
architecture RTL of NAND_FF is
   signal RI, SI: BIT;
begin
   RI <= S NAND CLK;
   SI <= R NAND CLK;
   Q <= RI NAND QB;
   QB <= SI NAND Q;
end RTL;
```

```
architecture GATE of NAND FF is
   component NAND GATE
      port (A, B: in BIT; Z: out BIT)
   end component
   signal RI, SI: BIT;
begin
   A : NAND GATE port (S, CLK, RI);
   B : NAND GATE port (R, CLK, SI);
   C : NAND GATE port (RI, QB, Q);
   D : NAND GATE port (SI, Q, QB);
end GATE;
```

(b)   (c)

Fig. 11.10 VHDL description of NAND flip-flop (a) schematic and entity declaration (b) RTL declaration (c) gate-level declaration

The VHDL descriptions for the NAND gate flip-flop in Fig. 11.10 are given at RTL and structural levels. The latter construct defines the gate implementation precisely, whereas the RTL program allows for alternatives of the kind given in Fig. 11.9. The VHDL description of the full adder in Fig. 11.11 includes a statement at behavioural as well as RTL and structural level. Ideally the compiler synthesizing the VHDL input should be able to produce the structural (gate level) hardware from a purely behavioural description. In practice this ideal has not been attained nor is it likely to materialize in the near future, although considerable research is being undertaken towards that end (Carlson 1991, pp.3–4).

The examples of VHDL descriptions in Figs 11.8 to 11.11 have been for simple logic, but ASIC design is far more complex. Turning the algo-

402 ASICs—Design techniques

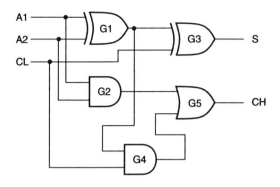

ENTITY declaration
entity FULL_ADDER is
  port (A1, A2: BIT; CL: in BIT;
    S: out BIT; CH: out BIT);
end FULL_ADDER;

(a)

architecture RTL_LEVEL of FULL_ADDER is
  signal SI, C1, C2: BIT;
  begin
    SI <= A1 XOR A2;
    CI <= A1 AND A2;
    C2 <= SI AND CL;
    S <= SI XOR CL;
    CH <= C1 OR C2;
  end RTL_LEVEL;

(c)

architecture BEHAVIOURAL of FULL_ADDER is
  begin
  process
    variable N: integer
    constant sum_vector: bit_vector (0 to 3): = '0101';
    constant carry_vector: bit_vector (0 to 3): = '0011';
  begin
    wait on A1, A2, CL;
    N: = 0
    if A1 = '1' then N: = N+1; end if;
    if A2 = '1' then N: = N+1; end if;
    if CL = '1' then N: = N+1; end if;
    Sum <= sum_vector (N) after 20 ns;
    CH <= carry_vector (N) after 30 ns;
  end process;
end BEHAVIOURAL;

(b)

architecture GATE_LEVEL of FULL_ADDER is
  component AND_GATE port (A, B,: in BIT; Z: out BIT);
  end component;
  component XOR_GATE port (A, B: in BIT; Z: out BIT);
  end component;
  component OR_GATE port (A, B: in BIT; Z: out BIT);
  end component;
  signal SI, C1, C2: BIT;
  begin
    G1: XOR_GATE port (A1, A2, SI);
    G2: AND_GATE port (A1, A2, CI);
    G3: XOR_GATE port (SI, CL, S);
    G4: AND_GATE port (SI, CL, C2);
    G5: OR_GATE port (C1, C2, CH);
  end GATE_LEVEL;

(d)

Fig. 11.11 VHDL description of full adder (a) schematic and entity declaration (b) behavioural declaration (c) RTL declaration (d) gate-level declaration

rithms of a behavioural construct for a complex function into gate-level hardware is very different from doing the same for a simple function like a 4-bit adder. Even at lower levels a purely behavioural (algorithmic) HDL entry may cause problems. Suppose it is required to synthesize an 18-bit up/down counter. The synthesizer, whose compiler is assumed to be technology-independent, will map this to its logic. If a 4-bit counter is part of this, five such counters could be used together with additional logic for suppressing counts above 18 bits. When this has been done another potential hurdle remains. Can the ASIC library help to successfully map the technology-independent compiler output to its macros? In practice the answer is likely to be yes, but this example has been presented to point out some potential difficulties which escalate with design complexity. The present situation is that by and large a behavioural description devoid of any architectural content is not acceptable. RTL with its structural pointers has an implied architecture and has become the standard level for HDL synthesis (Carlson 1991, pp.3-4).

The examples of VHDL descriptions given so far have been confined to logic operators. Various categories of VHDL operators are summarized in Fig. 11.12. A few more examples illustrate an *arithmetic operator* used for 4-bit addition (Fig. 11.13), the *wait* statement in a T (toggle) flip-flop (Fig. 11.14), and the *if–then–else* statement applied to a multiplexer (Fig. 11.15). The listing below gives the main features of VHDL.

(i) It is a comprehensive language for hardware design.
(ii) A function can be described at behavioural, RTL, or structural level.
(iii) Lower levels may be used, like the switch level, described in Section 11.4.
(iv) It is generic at structural level, allowing for technology processes, device size characteristics, timing, environment, etc.
(v) The design models behave like hardware.
(vi) It is technology-independent.
(vii) It is descriptive, being human—and machine—readable.
(viii) It is a leading edge standard which has obtained international recognition.
(ix) The execution of the program is concurrent: statements contained between *begin* and *end* in the architecture declaration are executed in such a way that they have to all intents and purposes been executed simultaneously. This mode reflects hardware operation, in which all parts of the system are active simultaneously. In contrast statements in a high level language program are executed sequentially.

Logic
NAND, AND, NOR, INV, XOR
Operand and result: Bit (Boolean)

Relational
=, /=, <, <=, >, >=

Operand: any result: Boolean

Arithmetic
+, -, *, /, **
Operand and result: real integer

Concatenation
&
Operand and result: array of any type

Fig. 11.12 VHDL operators

```
Example of using ARITHMETIC OPERATOR
entity ADD_INTEGERS IS
  port (A, B: in INTEGER range 0 to 15;
        C: out INTEGER range 0 to 15);
  end ADD_integers;

architecture ADDITION of ADD_INTEGERS is
  begin
    c<= A + B;
  end ADDITION;
```

Fig. 11.13 VHDL description of adding two 4-bit numbers

404 ASICs—Design techniques

```
Example of using WAIT statement
entity FLIP_FLOP is
  port (ENABLE: in BIT; CLOCK in BIT; CLOCK in BIT;
        TOGGLE: buffer BIT);
end FLIP_FLOP

architecture TOGGLE of FLIP-FLOP is
  begin
    process begin
      wait until CLOCK' event and CLOCK = '1';
      if (ENABLE = '1') then
      TOGGLE <= not TOGGLE;
    end if;
  end process;
end TOGGLE;
```

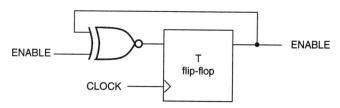

Fig. 11.14 VHDL description of toggle flip-flop

The texts by Navabi (1993), Perry (1994), and Lipsett et al. (1989) are recommended for gaining a background in VHDL description and design. Ott and Wilderotter (1994) have produced a comprehensive text on VHDL synthesis.

It was stated at the beginning of Section 9.4 than an ASIC is only as good as the design tools which support it. Likewise an HDL is only as good as the compilers and simulators which support it. This support is forthcoming and rapidly developing for VHDL and the other standard HDL, Verilog, which was created in the winter of 1983/84 by Gateway, who later merged with Cadence. Verilog is intended for top–down hierarchical design and, like VHDL, has behavioural RTL, structural and switch levels of abstraction. It is concurrent in the sense defined for VHDL. Verilog is based on the C language (C for short), leaning heavily on its syntax and semantics. C is becoming very popular with engineers. Many ASIC designers are likely to be familiar with it and will therefore find it easy to learn Verilog. Indeed it is easier to master Verilog than the far more comprehensive and consequently less user-friendly VHDL. And yet VHDL seems to have made all the running and to steal the limelight. The root cause is probably the backing it received by the DoD in the first place, combined with the mandatory obligation to use if for the documentation of all military ASIC designs. The development of VHDL has been accompanied by a bountiful supply of books on the subject. The number of EDA tool vendors (over forty) which support it is equally impressive (Maliniak 1994). It might be inferred that Verilog has been completely overshadowed by VHDL, but this is not so. VHDL enjoys a pre-eminence in Europe, yet about half of all ASIC designs in the US are undertaken with Verilog. Not only is Verilog easier to learn, but a comparable description can be up to 50 per cent shorter in Verilog than in VHDL (Sternheim et al. 1990, p.3). The literature on Verilog is relatively sparse; the author has come across only two definitive texts. Thomas and Moorby (1991, 1995) have followed their first with a substantially enlarged second edition. The book by Sternheim et al. (1990) contains detailed examples of system design. We are likely to witness a resurgence of Verilog, which is being evaluated by the IEEE for standardization and is being strongly supported by design tools. Most of the tool vendors who originally supplied tools using VHDL only, now offer their products for VHDL or Verilog. Taking an overview VHDL is likely to become the leading HDL worldwide, but Verilog will maintain a strong position and may even increase its share of ASIC designs. The two languages are very different and in the light of all the circumstances the introduction in this section of an HDL for ASIC design has been confined to VHDL.

Verilog and VHDL are accompanied by a third HDL, EDIF (electronic design interchange format). Interchange is the *motif* behind EDIF which, at the time of its inauguration in 1984, was intended to reach behavioural level. That idea was soon abandoned when the role of VHDL became evident (Shankar and Fernandez 1989). By the time EDIF 200 was adopted by ANSI (American National Standard Institute) in March 1988, its role and levels of abstraction had become crystallized. With a syntax similar to LISP (List Programming Language) EDIF extends from gate level down to physical presentation in either symbolic or graphical form (Fuccio and Hinstorff 1989; Geiger et al. 1990, pp.930–3). The capability of expressing a logic schematic, circuit description, and layout in symbolic *and* graphical form distinguishes EDIF from other interchange formats, which can only represent one or other of these categories. EDIF is used at all its levels for interchange of data, netlists, layout schematics, synthesizers and mask patterns. Typical examples of EDIF interchanges are:

(i) An ASIC engineer can use any commercial CAD system with its own software tools and generate an EDIF file at any design level (below RTL). That level will be acceptable on another CAD system and hence by another design group.

(ii) In a silicon foundry design, the customer can use EDIF to send the mask patterns to the foundry.

Surveying the roles of the various languages, VHDL and Verilog enter the behavioural or RTL description of the chip function into the logic synthesizer from which it emerges in VHDL, Verilog, or EDIF at the structural level for the netlist file. The schematic is captured with EDIF or one of the standard graphics interchange formats listed in Section 11.2. VHDL and Verilog have been extended for high-level description of analogue and mixed signal circuits. These AHDLs (analogue hardware description languages), called VHDL-A and Verilog-A, are already supported by some commercial design tools and moves are

IF-THEN-ELSE- statement
entity MUX is
  port (A, B, CT,: in BIT; D: out BIT);
end MUX;

process (A, B, CT) begin
  if (CT = '1') then
  D <= B
  else D <= A
  end process;
end MUX;

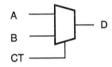

Fig. 11.15 VHDL description of multiplexer

being made to have them standardized.

11.4 Simulation and checking

Simulations and checks can be placed into three groups in accordance with the design flow in Fig. 11.2:

(i) behavioural simulation prior to logic synthesis;
(ii) pre-layout simulations;
(iii) post-layout simulations and checks.

The simulation under (i) verifies the correctness of the input for logic synthesis, described in Section 11.2. The preference for this input is to be in HDL at RTL level. The vendors favour a software package combining VHDL or Verilog synthesis and simulation. The categories of pre-layout and post-layout simulations and checks are:

(i) logic simulation;
(ii) timing simulation;
(iii) fault simulation;
(iv) switch-level simulation;
(v) ERC;
(vi) DRC;
(vii) LVS.

The first two simulations are common to pre-layout and post-layout processing. The pre-layout simulation confirms that the logic of the netlist is correct. Logic simulation signifies that timing is not included: logic states are assumed to change in zero time and propagation delays are completely ignored. Timing simulations are probably the most critical of all, because timing causes the biggest problems, presenting one of the greatest hazards in VLSI design. The timing constraints fed into the synthesizer in Fig. 11.3 have to be met. Apart from that simulation makes a critical assessment of timing in general. Here and in some other simulations the models should preferably be at board level, allowing for the connection of the ASIC to other ICs on the PCB.

Fault simulation ascertains the quality of the test vectors. Testability is a joint responsibility divided between customer and vendor. The customer undertakes testability design at the top level, where it is incorporated in the chip specification. The vendor is responsible for the test program, beginning with drawing up the vectors for that program, which will be executed by the automatic test equipment (ATE) at the end of the design when the prototypes are evaluated. The fault simulation establishes the fault coverage ie. the proportion of simulated faults identified by the test vectors. The fault coverage should lie in the range (90–95) per cent for high confidence in the test/program. (Huber and Rosneck 1991, p.134).

The post-layout simulations and checks are driven by the output of the layout synthesizer, and this presents no linguistic problem. The tool vendor provides compatible synthesis, simulation, and checking software packages for integral operation. Alternatively proprietary tools of the ASIC vendor or other third-party synthesis tools may integrate automatically into the EDA vendor's tools through software interfaces designed

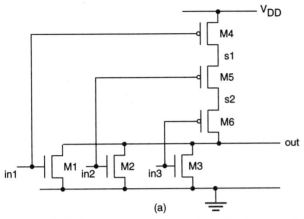

Fig. 11.16 Switch level description (a) schematic (b) description

for that purpose. The timing simulation improves on its pre-layout predecessor, which was an approximation generally based on a statistical estimate of wire lengths; now the exact lengths are available. The LPE and the back annotation feed back wire lengths, and the delay of the wires is usually expressed, depending on the length of wire, in the form of an RC time constant or a transmission line representation for each interconnection. It does not take much imagination to visualize the large data bank required for handling such a volume of information. The heavy demand on workstation storage capacity applies to simulations, checks, and synthesis in general. The LPE and annotation processes are good points for focusing on the enormous data storage demanded for ASIC design, where a hard-disk space of at least 100 Mbytes is specified for the workstation. The highest accuracy for circuit simulation is obtained with SPICE or a similar package; the great majority of EDA tools use SPICE. An analysis of the entire circuit with SPICE is prohibitively long. SPICE simulations are confined to selected parts of the circuit where timing and possibly the nature of the overswings and undershoots need to be looked at carefully. Switching simulation, not shown in Fig. 11.2 because it does not enjoy the universality of the other simulations, is a powerful technique of verifying the logic at a simplified circuit level. It can be carried out at the pre-layout or post-layout stage; usually the latter is chosen. The MOSFETs are represented as on/off switches. Timing simulation is sometimes included by modelling the switches with RC time constants. The models are often generated automatically from LPE data, in which case RC representation is easily incorporated in the model. A switch-level model of a 3-input NOR gate is shown in Fig. 11.16. The arbitrary syntax of the input netlist is representative of actual practice.

Simulations and checks are inevitably time consuming. For ASICs with 100 000 gates or more, a simulation run might take several days. Various steps have been taken to cut down on simulation time. These include:

(i) Mixed-mode simulation, which consists of simultaneous simulations at different levels. A typical combination is RTL and structural (gate-level) simulation, but there are other combinations.

(ii) Event-driven simulation, where only circuits whose inputs have been changed are evaluated.
(iii) Hardware accelerators.

A *hardware accelerator* (alternatively *hardware engine, hardware simulation engine, accelerator*, sometimes just *engine*) is a special purpose processor which undertakes simulations far more quickly than software tools. (Hollis 1987, pp.348–51; Huber and Rosneck 1991, pp.99–102). The hardware accelerator is normally a server to the workstation, which compiles netlists for the hardware simulations and the display of the results. Hardware accelerators have logic and fault simulation capability at least; some can do far more. The hardware accelerator has the same requirements as software simulators in respect of the following:

(i) netlist (simulation) input;
(ii) models;
(iii) stimulus (input vectors);
(iv) control.

Hardware accelerators can only handle their own primitives and these are less numerous than the primitives for traditional simulation. By and large hardware accelerators are most efficient for simpler models. One effective use of hardware accelerators is to apportion the simulation of the more complex models to the workstation, leaving the accelerator to undertake these simulations for which it is best. Factors to consider when comparing the speed of conventional and accelerator simulations are:

(i) The creation of netlists in the workstation.
(ii) The transfer of the netlists to the hardware accelerator.
(iii) Translation of the stimuli for software to hardware simulation, and their transfer to the accelerator.
(iv) The transfer of the simulation results to the workstation.

Simulation time is reduced by a factor of between 250 and 1000, and the extra cost of the hardware accelerator is easily jusified on economic grounds.

Mixed-signal simulation is, like mixed-signal synthesis, briefly referred to in the next section. Mixed-signal EDA tools are becoming widespread, but have not yet reached the maturity of the support for all-digital EDA.

The three post-layout checks are DRC, ERC, and LVS. The arrangement for LVS is illustrated in Fig. 11.17. The netlist derived from the layout by reverse computation is compared with the netlist containing the output of the logic synthesizer for logical equivalence. The DRC checks that the layout adheres to the design rules, the principles of which were outlined in Section 3.7. Special attention is paid to spacing and width, overlap rules, and details like metal surrounding a contact cut. The ERC checks for factors like:

(i) fan-out;
(ii) fan-in;
(iii) voltage drops across wires;
(iv) floating input/output pins;

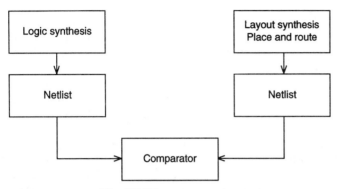

Fig. 11.17 LVS operation

(v) instances without pins like an I/O pad without an I/O pin;
(vi) shorts of V_{DD} to ground;
(vii) unconnected nodes.

11.5 Commercial design tools

The call for high quality ASIC EDA has driven development in several directions. The dominant advance may be ascribed to third-party tool vendors, who have taken the art of ASIC design from its early beginnings to a stage where highly complex chips can be designed and tested within tight and previously unattainable production schedules. When ASICs first came on the market, their manufacturers had to develop their own software tools. Now numerous tool vendors supply software ASIC design at all levels. Some ASIC houses retain their full in-house design tools, but times have changed and design nowadays is largely undertaken with third-party design tools and, of course, a workstation, the *hardware platform*. When an ASIC vendor's tools are used, they more likely than not support third-party vendor's tools which offer more variety and embrace a wider range of design processes. FPGA design tends to be based more on the manufacturer's tools because of the individualistic nature of its architecture. These tools support compatible third-party software, and the customer usually employs such a combination.

The expressions *workstation* and *hardware platform* are synonymous. The expression *hardware platform* is more generally used. Workstation stands, or at any rate used to stand for *dedicated* CAD/CAE hardware and was defined that way in Section 10.2. The role of a workstation is fulfilled not only by a dedicated computer specifically designed for that task, but is generally undertaken by a hardware platform which can be used for other purposes as well. The types of hardware platforms employed for ASIC design have settled down to a number of recognized computer systems. The most popular of these are:

(i) Hewlett-Packard (HP): Apollo DN3XXX, DN4XXX, 9000/300, Series 400, 700;
(ii) Digital Equipment Corporation (DEC): DEC Station, VAX/Station VMS;
(iii) SUN: SUN 3/XXX, SUN 4/XXX, SUN SPARC Stations.

386/486/Pentium IBM compatible PCs are also being used as workstations, provided they have the necessary memory back-up for that purpose. The RAM storage should be 32 Mbytes at least, combined with a far higher hard disk or CD-ROM back-up from 100 Mbytes upwards. Many software tools are written in the C language and will run under the UNIX operating system.

The composition of the tools can be deduced from the material in Sections 11.2 to 11.4. The software packages have to support logic and layout synthesis, simulations and checks. The ASIC vendor is at full stretch to keep pace with the rapid developments in hardware and relies increasingly on third-party design support. Semiconductor houses are likely to have substantial design experience for standard VLSI but ASICs, because of their architecture and customer involvement, need a different approach. The customer is strongly engaged in the design process, even when it follows the turnkey route, and has to acquire design tools which will be used in future, possibly for ASICs from a different vendor. ASIC manufacturers and their customers might have preferences to a point where products from different tool suppliers are preferred for specific parts of the design process. It is not uncommon to have tools from different sources for logic synthesis, layout synthesis, checks and simulation. Tool vendors are helping with such a situation by profuse support of third-party tools. The potential divergence of tools between vendor and customer is contained by extensive interfacing software to link packages of different suppliers. Compatibility is especially essential between simulations which are carried out by the customer, but are repeated independently by the ASIC vendor to confirm the performance prior to commencing production. Indispensable support of another kind is furnished by ASIC vendors, who gladly supply their library (of macros), sometimes called macro library, to EDA tool vendors. The more ASIC libraries a tool vendor holds, the more likely is he to attract custom by offering immediate service. The ASIC vendor will similarly benefit from being supported by CAD/CAE tools from many sources.

Two features deserve to be mentioned before describing commercial products, because they are offered by tool vendors in general. Logic synthesizers detect HDL source code which cannot be executed because of a syntax error or limitations in the compilers. The synthesizers incorporate a check for HDL compliance and contain debugging software. The second feature is multi-level design and multi-level simulation with multiple-window displays and user-interaction. To give an example, the multi-window display for simulation might contain gate-level logic, a graphical display of the waveforms, a list of the display, and the HDL code at RTL level.

The products of four vendors will now be briefly described in order to bring home the commercial aspects of ASIC design support. EDA has reached a stage where many software houses supply design tools of high quality. Some offer special support for MCM and PCB layout, microwave circuit and switched-capacitor design, etc. The accent in this section is on support tools for semicustom circuit design. The description avoids a mere repeat of giving the characteristics of the core tools from all four suppliers. Instead the packages have been selected to present a broad view of what is available. The four vendors selected are Cadence, Compass, Mentor Graphics and Synopsys.

Compass

Starting with Compass Design Automation, to give it its full name, this software house was originally a subsidiary of VLSI Technology, but become an independent company some years ago. The Compass *Navigator* series is for top–down design and consists of two parts, the *ASIC Navigator* and the *Silicon Navigator*. The *ASIC Navigator* undertakes the design steps in Fig. 11.2(a). The *Silicon Navigator* is responsible for the layout design (place and route) and the mask patterns. The *Navigator* series contains several design tools for logic and layout synthesis, floorplanning, simulation and checks. The tools include testability design and test vector generation. A full range of entry formats is available. The input to the ASIC navigator can be a graphical or textual description, including the alternatives given for PLD design in Fig. 9.22, but the recommended entry is in VHDL or Verilog. The top–down approach based on behavioural/RTL entry in VHDL or Verilog is comparatively recent, although it is unquestionably the way to go. ASIC vendors are still having to push this approach to hardware engineers who have been brought up on graphical design entry. The textual entry in an HDL is a swing from hardware to software and designers will have to master at least one of the two established HDLs and the esoteric skills in its use. Engineers will have to be trained that way and the onus for laying the appropriate foundation in degree courses that include hardware and system engineering rests with the universities. The hardware engineer has to master the software for a textual specification in VHDL or Verilog and for design management in that medium.

The initial step in top–down design is taken with the *ASIC Navigator* which partitions the system, decides on the number of ASICs required, and produces the ASIC specifications. At this stage the *ASIC Synthesizer* takes over. It synthesizes directly from VHDL, Verilog, state machine or datapath diagrams and some other formats. VHDL and Verilog entries are automatically partitioned into datapath and random logic. The *ASIC Synthesizer* generates an optimized gate-level netlist. This output is available for three optional tools, the *Datapath Synthesizer*, the *Logic Synthesizer*, and the *FPGA Optimizer*. The *Datapath Synthesizer* accepts VHDL or Verilog and has an extensive set of datapath libraries. It includes a *Datapath Compiler* for optimizing the output, which is a graphical specification of the datapath and the corresponding control logic. The *Logic Synthesizer* accepts VHDL, Verilog, or EDIF and writes the output in one of these languages at structural level, and in the graphics language SDF for schematic capture. The FPGA Optimizer supports the most popular FPGA architectures, including the products of Actel, Altera, and Xilinx. It accepts a VHDL or Verilog input from the ASIC Synthesizer or the Logic Synthesizer and maps the output to the FPGA architecture. The output formats include EDIF and XNF (a Xilinx format, see Section 11.6).

The design of the physical layout is carried out by the *Silicon Navigator*, which is based on the following tools:

(i) *Chip Planner*. A floorplanner for gate array and cell-based ASICs. It predicts timing early in the design cycle, thereby reducing design iterations;

(ii) *Gate Compiler*. Place and route for a two- or three-layer metal gate array;

(iii) *Chip Compiler*. Place and route for a two- or three-layer metal cell array;

(iv) *Rose*. A symbolic editor supporting layouts with Manhattan and 45° geometries;

(v) *Pathfinder*. A tool which integrates floorplanning with place and route.

Ample support tools are available for the post-layout simulation and the DRC, ERC, and LVS checks. A number of test tools are provided. *The Boundary Scan Solution* is for the automatic insertion of the complete IEEE 1149.1 boundary scan architecture. Its library includes TAP controller configurations to support boundary scan instructions like INTEST, EXTEST (Fig. 10.9) and BIST. A very similar design support, but one which is not tied to the IEEE 1149.1 standard and allows the user an architecture of his choice, is the *Scan Solution*. Automatic synthesis and insertion of the BIST structure, including linear feedback shift registers for pseudo-random pattern generation and signature analysis, is undertaken by the *BIST Solution*. Finally *Test Synthesis Solutions* integrates architectures like the boundary scan, BIST, and interval scans to suit the customer's specification.

Just like hardware platforms have an operating system like UNIX, EDA software tools require an in-house operating system, preferably one that supports the tools of other third parties and ASIC vendors. Integration of this kind is offered by most vendors, including Compass.

Mixed-signal synthesis and simulation is catered for by a number of tools. The *Analog Assistant* serves for mixed-mode entry. The analogue part can be described at various levels and the input is synthesized into a mixed-mode netlist and simulated by an analogue simulator of another company, *ELDO* from Anacad EES, supplemented by the Compass digital *QSIM/SIM* simulators.

ASIC designs are verified by extensive simulations at various levels of abstraction. *Formal verification* using *formal methods* is an alternative for verifying the lower-level implementation of a higher-level specification by logic manipulation, based on first-order, higher-order, and temporal logic (Yoeli 1990). It is a method of proving formally that the descriptions at different levels are in agreement. Taking ASIC design, formal methods might be used to verify that the RTL and structural levels of a VHDL description are equivalent, or that the mapping of a technology-independent output from a logic synthesizer to the ASIC macros is correct.

Traditional simulation is demanding and difficult. The models may allow some errors to remain undetected. The generation of test vectors with adequate quality calls for much effort and some of the results can be ambiguous. Conventional simulation is prone to errors, apart from being expensive and time consuming. Formal verification is a means of reinforcing traditional simulation and reducing the probability of errors. A successful pioneering application of formal methods was undertaken to design the VIPER, an 8-bit microprocessor for real-time control of safety-critical systems, developed by the Royal Signals and Radar Establishment (RSRE) of the British Ministry of Defence (Birtwistle and Subramanyam 1988). May *et al.* (1992) pioneered the application of formal methods to more complex logic in their design of the floating point unit, a datapath controlled by microcode, for the INMOS T800 transputer.

Compass have come out with an innovative tool, *VFORMAL*, which creates a formal mathematical proof that two synchronous VHDL descriptions are logically identical for every possible set of legal inputs and internal states; timing is ignored. VFORMAL does not require input test vectors or an expected response.

Formal verification is not a replacement but a complement for conventional simulation. It is an art which is still at an early stage and in need of further research and development before it is widely accepted in industry. The application of formal methods is not confined to verification where they are a valuable technique which, when combined with conventional simulation, greatly reduces the chance of undetected errors. Their use in the construct of an HDL is by itself an improvement on previous practice, and aids the reliability of conventional simulation.

Mentor Graphics

The core of Mentor Graphics EDA is the *Falcon Framework for Concurrent Design*, a design structure with the declared object of integrating Mentor Graphics, other third-party, and proprietary ASIC vendor's tools. In order to achieve that, a very strong software infrastructure had to be established. By designing such software with easy and virtually transparent integration of tools from a variety of sources, Mentor Graphics have produced a CAD/CAE system with a distinct appeal for customers who do not wish to confine themselves exclusively to EDA tools from one particular vendor.

User interfaces give access to the database and to all the design tools. Singling out a few of these, logic synthesis is carried out by *AutoLogic*, which is capable of synthesizing from both VHDL and graphical high-level descriptions. AutoLogic, used with other recommended packages, executes the top–down design. It can synthesize semicustom, full custom, and FPGA ASICs. *AutoLogic* is primed by the *Design Architect*, which can capture a design at architectural behavioural, gate, and circuit level. It has a VHDL editor for entry of the architectural/behavioural input. *Design Architect* runs on all industry-standard hardware platforms and integrates schematic, symbol, and text editors. The provision of high-level schematic entry, to be distinguished from schematic entry in bottom–up design, caters for many designers who visualize an architectural entry at top level graphically, and who do not think solely in terms of an HDL description. *AutoLogic Blocks* lets the user specify datapath elements like adders, converters, and mulipliers; these can be described efficiently in graphical format. Two tools complete VHDL synthesis and simulation. *QuickSim II* can perform logic simulation with and without timing. The tool which brings *AutoLogic* and *QuickSim II* into the VHDL orbit is *System-1076*, so called because it supports the standard IEEE 1076-1987 VHDL. (That standard applies to all commercial design tools, unless stated otherwise). The two arms of *System-1076* are the *VHDL Compiler* for synthesis, and the *VHDL Simulation Solver* for simulation, which operates in conjunction with *QuickSim II*. *System-1076* includes a checker of the source code, the *System-1076 Debugger*, which is integrated with *SimView*, the Mentor Graphics analysis environment. It allows *System-1076* to be used with *QuickSim II* for debugging by interactive graphics, and for trace displays and hierarchical traversal. The *System-1076 Debugger* sees to the correct syntax of the source code, and the *VHDL Simulation Solver* sees to its simulation. *System-1076* enables *QuickSim II* to simulate VHDL at virtually all levels of abstraction. The tools permit multi-level design and simulation, features that are available with most EDA software. Tools for mixed-signal design include *LSIM*, which can simulate at RTL, gate-level, switch level and SPICE circuit level. *Explorer LSIM* undertakes concurrent digital and analogue simulation, embracing the full range of *LSIM*.

Cadence

The logic synthesizer is Synergy 250/251 which accepts VHDL, Verilog, and Cadence's in-house HDL, SKILL. The logic simulators are the *Leapfrog VHDL Simulator* and the *Verilog-XL Simulation Family* consisting of the *Verilog-XL* and the *Verilog-XL Turbo*. The Verilog simulators support behavioural (algorithmic), RTL, gate, and switch levels. The *VHDL Mode Import* for the Verilog-XL Simulation Family brings VHDL and Verilog into a single design environment. VHDL source code can be debugged and simulated with the Verilog-XL simulators without having to master VHDL.

The physical layout is implemented with the *Gate Ensemble* for gate arrays and the *Cell Ensemble* for cell-based ASICs, both operating in conjunction with the *Preview* floorplanner. An optional post-placement synthesis tool, *Synergy PBS* (placement based synthesis) optimizes the layout by branching out in a new direction. Logic synthesis has so far been purely logic-based: the optimization is solely for logic. We are witnessing a change in design philosophy with a merger of the logic synthesis and layout synthesis algorithms. Reading the initial placement information the *Synergy PBS* optimizes the placement and routing to produce a revised netlist and physical layout (Katsioulas 1994). Hitherto the layout synthesizer operated on a fixed netlist. Now that netlist is modified by post-placement synthesis which is interactively coupled with logic synthesis.

A very similar action takes place in logic synthesis with the Compass ASIC Synthesizer supported by the *ChipPlanner*, a tool included here in order to place it in juxtaposition with the *Synergy PBS*. The *ChipPlanner* can generate a floorplan prior to logic synthesis (Floorplanning(A) in Fig. 11.2(a)), or—the situation being discussed now—after the production of the netlist (Floorplanning(B) in Fig. 11.2(b)). The timing constraints are passed to the floorplanner, and the timing delay information it generates after drawing up the floorplan is passed back to the ASIC synthesizer for re-analysis and re-optimization. The revised netlist is automatically passed on to the floorplanner for updating place and route. The approach is similar to the action of the *Synergy PBS*. In both cases the timing information obtained, either after placement (which is governed very much by the floorplan) in the Compass *Synergy PBS* , or prior to place and route in the Cadence *ChipPlanner*, is the driving force behind a placement-based synthesis operating on the logic synthesis. A purely logic synthesis without taking into account timing may lead to a netlist which cannot be routed in accordance with the timing constraints, a situation which is becoming more severe with the thrust to smaller submicron geometries. Multiple logic and layout iterations to meet the timing conditions could impose a prohibitive penalty on the design costs.

Dracula is a stand-alone verification which has become an industry standard. The latest version, *Dracula III*, has *HDRC*, *HERC*, *HLVS*, *HLPE*, and *HPRE* modules (H for high speed). *HPRE* (PRE for parasitic resistance extraction) reinforces *HLPE* with RC networks for the interconnects, changing from lumped to distributed (transmission line) representation for long wires in order to improve accuracy.

The prime tool for analogue and mixed-signal ASICs is the *Analog Artist Design System*, which is composed of the *Analog Artist Electrical Design System* and the *Analog Artist Physical Design System*. The former is for design capture and simulation, the latter for physical layout. There are two circuit simulators, *Spectre* and *Spectre HDL*. *Spectre* is written in C and uses the same basic algorithms as SPICE. These have however been substantially revised and improve on standard SPICE performance. *Spectre* is capable of simulating circuits in excess of 50 000 transistors. Its speed is at least three times faster than SPICE and its accuracy is equal to or better than SPICE. *Spectre HDL* is for *behavioural* analogue modelling and simulation, and supports the emerging analogue hardware description languages (AHDLs) VHDL-A and Verilog-A, mentioned in Section 11.3. The mixed-level modelling of *Spectre HDL* speeds up simulation by a factor of ten or more.

Synopsys

Synopsys tools concentrate on logic synthesis accepting descriptions at behavioural, RTL, or structural level, or at a mixed-level containing a combination of these modes. The synthesis is backed by simulation tools. Physical layout, its simulation, and the standard post-layout checks are obtained with tools from other vendors. The *Design Analyzer* controls and manages entry and analysis in a graphical environment. Synthesis is carried out by the *HDL Compiler Family* with an entry in VHDL or Verilog and feeding the *Designer Compiler Family*, which consists of the *DC Professional*, and *DC Expert* packages (DC for Design Compiler). Its output options include EDIF, VHDL, Verilog, and TDL (Tegas design language). Synopsys offer the option of optimizing the layout with post-layout placement synthesis. The tool for that purpose is the *DC Expert* which, operating together with the *Design Computer* and using back annotation of post-layout delays and parasitics, optimizes the layout. This is yet another example of combining logic with layout synthesis. By concentrating on synthesis, Synopsys have produced tools for that purpose at the cutting edge of the market, using industry standards whenever possible. Synopsys software is written in C and runs under UNIX. Synopsys products are supported by the following hardware platforms: Sun Microsystems, HP/Apollo, IBM RISC System/6000, MIPS, Digital Equipment DEC Station, and the Solbourne families of workstations. A minimum of 32 Mbytes RAM and a CD-ROM drive are required.

Synopsys integration with other tool vendors is exceptionally strong, because a complete design kit based on Synopsys high-level synthesis has to be combined with the tools of other vendors for place and route, and for post-layout simulation and checking. The Cadence to Synopsys interface links the Synopsys synthesis tools with the *Cadence Design Framework*, an architecture giving access to Cadence and third-party tools. The Mentor Graphics interface fully integrates the Synopsys *Design Compiler* with the Mentor Graphics *Capture Station* and *Idea Station*. *Zycad XP* Simulation Accelerators accept Synopsys VHDL, accelerate synthesized structure from RTL, and support full system simulation from VHDL.

Approaching design tool selection from the other end, an ASIC foundry recommends design tools without insisting on adherence to its advice. However its CAD/CAE system is integrated with the recommended third-party tools which its supports, and that recommendation carries considerable weight.

The design flow for the GPS CLA80000 gate array is illustrated in Fig. 11.5, partly in order to show tool deployment and partly to give an example of a data flow schematic compiled by an ASIC manufacturer. Figure 11.6 for the VLSI Technology VSC470 series has been included for the same reason. The description of the design procedure has been for the turnkey route throughout. Many ASIC design flows, although 'turnkey' in nature, have a strong involvement by the customer. This affects the tooling, because compatibility between the tools for simulations and checks carried out by both the ASIC vendor and the customer is essential. GPS support a number of design kits. These contain all the software packages for undertaking the design with recommendations by supporting hardware platforms. GPS also follow the commercial practice of licensing their proprietary PDS design software to the customer for installation on his own DEC VAX or SUN system.

VLSI Technology understandably enough support Compass design tools because of the close link between these companies. The tools include the *Design Assistant*, the *Logic Synthesizer*, and the *Datapath Compiler*. VLSI Technology have also drawn up an ASIC tools matrix giving alternatives available from ten EDA vendors against the headings of the various design processes (schematic entry, synthesis, logic simulation, timing simulation, floorplanning, etc.) Some EDA vendors offer only tools for a few of the design steps. There is however a good choice that demonstrates the growing support of EDA third-party design tools by ASIC foundries.

AMS, like VLSI Technology, list EDA tool suppliers and the names of the tools against the various design processes. There is a good selection with a predominance of tools by Cadence and Mentor Graphics. The tools, part of AMS CAD design systems, have been selected to support mixed-signal designs at different levels of abstraction. The software packages listed for mixed-mode simulation are *Verilog Spectre* (Cadence), *Quicksim II* (Mentor Graphics), *Explorer LSIM* (Mentor Graphics) and *LSIM* (Mentor Graphics). An important hardware item is the tester, especially for mixed-signal testing. AMS use the LTX *Synchromaster* for mixed-mode and digital high speed tests, for which it can go up to 622 MHz. An rf extension to the *Synchromaster* allows tests up to 2.7 GHz. Another tester, the HP82000, is used for digital tests up to 100 MHz. Tools supported by AMS for top level design include the Mentor Graphics *Idea Station* and the Cadence *Verilog-XL* simulator. AMS have assembled some mixed-signal tool kits. The AMS HIT-KITS (High Performance Interface Tool Kits) are based on Mentor Graphics products; a recent version is integrated into the *Falcon Framework*. The Mentor Graphics 8 AMS HIT-KIT enables simulations at different levels of abstraction and includes an analogue option for mixed-signal design covering analogue standard cells. The Mentor 8 KIT is based on the Mentor Graphics *Idea Station* and AMS standard cell libraries.

GPS have chosen Cadence's *Analog Artist Design System* and integrated it with their proprietary *GPS* place and routing software for their mixed signal design. The DA mixed signal family is supported by the GPS *PDM* design linked to the ViewLogic *Workview* CAE tools.

The design methodology for the Micro Linear Tile Arrays is based entirely on Micro Linear's own design tools. It is very much a turnkey approach with Micro Linear drawing extensively on their experience gained with the design of standard ICs and ASSPs.

11.6 FPGA design tools

FPGA design has much in common with the design of MPGAs and cell-based ASICs. What differences there are can be attributed to the nature of the FPGA architecture. The hierarchical top–down approach with behavioural entry and logic synthesis is applied to all ASICs, except where a bottom–up approach with a schematic entry is chosen for chips with a small gate count (up to a few thousand gates). The design technique in such cases resembles the practice for PAL/PLA PLDs outlined in Section 9.4.

The cost effectiveness of an FPGA depends on the elimination of the NRE expenditure and on the customer undertaking the design very much in the mould of a foundry process, but in liaison with the vendor. This section describes the design tools for three FPGA families with different programming technologies, the Actel (antifuse), Lattice (EECMOS) and Xilinx (SRAM) products (Table 9.6). All these FPGAs are described in Chapter 9. The descriptions are based on the data books for these families and permission to quote from these sources is gratefully acknowledged (Actel 1994; Lattice 1994; Xilinx 1994). The configuration program is entered into the FPGA by a hardware programmer for antifuse and EECMOS technologies, or downloaded in a bitstream (BIT) or PROM file for an SRAM FPGA.

Actel, in common with most FPGA vendors, support a number of development systems, giving a choice of software tools and design (hardware) platforms. The use of PCs is becoming widespread. The two in-house systems of Actel are the *Designer* (for up to 2500 gates) and the *Designer Advantage* (for up to 10 000 gates). These systems undertake place and route, timing verification, and draw up the configuration program. They support design systems of Cadence, OrCAD, Mentor Graphics, and ViewLogic. The Actel design systems support schematic entry, PAL descriptions, Boolean equations and state machines (Fig. 9.22). Alternatively, using third-party tools, design descriptions can be entered in VHDL or Verilog. Although place and route is fully automatic, users may choose manual I/O placement. An incremental place and route software allows small local changes to be made whilst freezing the remainder of the logic, and thereby saves much time compared with the traditional global approach. Two programmers are available. The *Activator 2S* programmer programs one device at a time and is intended for low to medium volume applications. Different packages can be programmed using a range of over 20 programming adapters, divided fairly evenly between the ACT1, ACT2, and ACT3 families, and covering package pin ranges of (44–100), (86–172), and (84–257) respectively. The Activator 2 programmer can program up to four devices simultaneously and is intended for medium to high volume production.

Mentor Graphics and Cadence operate with Sun and HP platforms. ViewLogic accepts either a 386/486 PC or a Sun Station. OrCAD operates with a 386/486 PC. These PCs require 8 Mbytes RAM *(Designer)* or 16 Mbytes RAM *(Designer Advantage)* and 30 Mbytes hard disk space. The Sun and HP workstation storage requirements are from 16 Mbytes to 32 Mbytes RAM.

Xilinx have their own *Xilinx Automated CAE Tools (XACT)* design kit, integrated under the *Xilinx Design Manager (XDM)*. The entry can be schematic or, using third-party tools, behavioural in VHDL or Verilog. The XACT is available as a bundled package, but its products can be obtained separately. The individual parts of the *XACT* development system include the *Core Implementation*, the *Parallel Download Cable*, and the *Xchequer Cable*. The *Core Implementation* is the heart of the system. It contains a logic reduction algorithm, automatic and interactive place and route, timing analysis, and generates the configuration program in the format of bitstream (BIT) and PROM files. The *Parallel Download Cable* is for loading the Bit and PROM files into the FPGA directly (BIT file) or into the PROM holding the configuration program (PROM file). The *Xchequer Cable* supports downloading of BIT and PROM files and readback of configuration data and node values (readback was mentioned in Section 9.3.2). Third-party vendors support some other environments. To ease compatibility between diverse entries and simulation packages, Xilinx have produced a standard interface file specification, *XNF*, which simplifies transfer in and out of the XACT development system.

Examples of hardware requirements are:

(i) 386/486 PC supporting ViewLogic Standard System
   8 Mbytes RAM for devices up to XC4006
   16 Mbytes RAM for XC4008 and above
   80–100 Mbytes (minimum) hard-disk space
   3.5 in or 5.25 in floppy disk drive or
   CD-ROM drive.

(ii) HP 700 series supporting Mentor V8 Standard System
   32 Mbytes RAM (minimum)
   (50–100) Mbytes hard-disk space
   CD-ROM drive.

Although the CD-ROM drive in (i) is listed as an alternative, it is strongly recommended in preference to a floppy disk drive for PC platforms in general. It allows much faster loading than is possible with the multiple feed of diskettes into the disk drive. An ASIC vendor can store all the software tools, updates, and modifications on a single CD-ROM.

Lattice have their own stand-alone pLSI and ispLSI Development System *(pDS)*. This they have reinforced by offering the alternative of *pDS + ABEL* software, which augments their own tools with the *ABEL* tools from the Data I/O Corporation, and enhances various processes. The *pDS + ABEL* package extends design entry to HDL. The two languages available are ABEL HDL (AHDL, an acronym which applies to the *pDS + ABEL* software, should not be confused with the more widely known meaning of AHDL, analogue HDL, used in Section 11.5) and VHDL. pDS supports a variety of third-party CAE tools which run on PC and SUN platforms. The PC platforms specified are the 386/486/Pentium PCs. The Pentium PCs are the next generation following on from the 386/486 range. The inclusion of the Pentium microprocessor in the specification can safely be applied to PC platforms for ASIC design in general. Storage demands for the PC and SUN SPARC workstations are the same: 16 Mbytes RAM and 100 Mbytes hard-disk space.

The programming file is generated in JEDEC format, the standard for PAL/PLA PLDs (Section 9.4), and the software includes JEDEC test vectors. All Lattice FPGAs can be programmed with third-party programmers. Seven vendors offering fourteen programmers are available.

11.7 Conclusions

ASICs are becoming the mainstay of electronic systems and engineers for producing electronic hardware will almost certainly become involved in ASIC design. With the gate counts already with us, hierarchical top–down design is clearly the way to go. It gives the best quality and is most efficient for ASICs with about 10 000 equivalent gates or more. A good indicator of VLSI design quality is the mean interconnect length $\overline{L}$, expressed in terms of the *chip edge* L_c. (A square die is assumed. For a rectangular die $L_c = \sqrt{A_c}$, where A_c is the die area). In the early days of micron VLSI, $\overline{L}$ was typically equal to $0.18L_c$ (McGreivy and Pickar 1982). Improvements in design techniques had reduced $\overline{L}$ to $\sim 0.05L_c$, the value quoted in Section 2.4.6, in the late 1980s (Bakoglu 1990). $\overline{L}$ tends to be higher for ASIC than for standard VLSI, because the latter is an in-house design for which more effort and expertise are available. ASIC design has now progressed to yield chips with $\overline{L}$ well below $0.1L_c$. Fujitsu have achieved $\overline{L}$ equal to $0.03L_c$ in a recent complex high-density ASIC and routinely obtain $\overline{L}$ of the order $0.05L_c$ (Harker 1995).

EDA design tools have advanced the techniques of behavioural entry description, decomposed into structural hardware by logic synthesis. Both entry and output of the logic synthesizer are expressed in HDL; VHDL and Verilog are the industry standards for that purpose, with VHDL probably becoming the HDL of choice. The mastery of their syntax is quite an art and the need to acquire the skill points to the urgency of teaching VHDL and/or Verilog on appropriate first degree courses covering electronic hardware and system engineering. VHDL and Verilog must be treated like hardware descriptions and not like programming languages. Their code has to express clearly what the designer wants, and not leave it to the synthesizer to puzzle this out.

Logic synthesizers have made great advances since they became established around 1989 to 1992. Traditional logic synthesis is now being superceded by a combination of logic *with* layout synthesis in which the initial netlist, obtained by purely logic manipulation, is optimized by layout synthesis, primed by the interconnect delays of the layout and interacting with the logic synthesis. This method is superior to the traditional practice. Optional at present - the options offered by three tool vendors were described in Section 11.5 - it is set to become standard practice, and will be indispensable for deep submicron ASICs.

ASIC design relies heavily on simulation. Behavioural simulation prior to logic synthesis is especially vital to confirm the validity of the behavioural HDL description. The reliability of the simulation tools can as a rule be verified only indirectly and complete error detection cannot be guaranteed. The multiplicity of the simulations in the complete design process increases the probability of error detection. Formal verification based on formal methods complements simulation. If simulation cannot be guaranteed to detect every conceivable error, neither can formal verification. However structuring code by using formal methods will by itself aid the simulation process. The development of formal methods for large systems built with ASICs is a very exacting but worthwhile task.

The larger the chip function, the further removed will its description be from a construct which contains an implied hardware architecture. ASICs already contain up to several hundred thousand gates. Carlson (1991, pp.3–4) stresses that in current industrial practice the HDL description of the design for the logic synthesizer is in RTL. Designs are moving towards chips which will contain from ten to hundred million transistors by the turn of the century, and design techniques will have to be modified in order to cope with chips of such magnitude. There will be higher levels of purely behavioural abstraction, and the system will be broken down into independent modules (May *et al.* 1992). Synthesis for such decomposition would be accompanied by behavioural simulation in order to affirm that the modules meet the overall specification. One can visualize logic synthesis on present lines lower down the chain with an RTL HDL input. Logic synthesis, however good, cannot turn a purely behavioural formulation into structural hardware without some data about its architecture, and that raises an issue which has not been mentioned so far, the methods for digital logic design. These the synthesis tool cannot implement unaided. Digital logic design is an essential skill which is called on for standard component and ASIC VLSI design.

The attention given to chips with medium and high densities must not be allowed to detract from the design of ASICs with relatively small gate counts. That spectrum is adequately catered for by design tools whose development has benefitted from high density chip design.

Economics are inevitably a dominant consideration. The findings of a report by McKinsley & Co., quoted in two references in Section 10.4 (Reinertsen 1983; Texas Instruments 1991) not only give a penalty of 34 per cent loss in profit for a six months delay in production, but also estimate the loss in profit to be only 3.5 per cent if the development cost is overrun by 50 per cent *without any production delay*. Time is money, goes an old proverb, and a substantial increase in cost is, if necessary, justified to maintain the development and production schedules. Repeated iterations in the design flow can take up to several days at a time, and in the light of what has just been said the cost of a hardware accelerator is easily justified.

The advances in CAD/CAE ASIC design have in many ways been outstanding. Looking ahead the relentless increase in chip density calls for further efforts to match the accompanying function complexity with adequate design tools. Hardware continues to be ahead of software, technology ahead of design.

References

Actel (1994). *FPGA data book and design guide*. Actel, USA.

Bakoglu, H.B. (1990). *Circuits, interconnections, and packaging for VLSI*, p.196. Addison-Wesley, USA.

Bartee, T.C. (1985). *Digital computer fundamentals*, (6th edn), pp.443–52. McGraw–Hill, USA.

Birtwistle, G. and Subrahmanyam, P.A. (ed) (1988). *VLSI specification, verification, and synthesis*. Kluwer Academic Publishers, USA.

Carlson, S.C. (1991). Introduction to HDL-based design using VHDL. Synopsys, USA.

Fuccio, M.T. and Hinstorff, H.L. (1989). In *VLSI Handbook*, (ed. J. DiGiacomo), pp.22.4–22.5. McGraw-Hill, USA.

Geiger, R.L., Allen, P.E., and Strader, N.R. (1990). *VLSI design techniques for analog and digital circuits*. McGraw-Hill, USA.

Glasser, L.A. and Dobberpuhl, D.W. (1985). *The design and analysis of VLSI circuits*, p.56. Addison-Wesley, USA.

Harker, A.J. (1995). *Personal communication*. Fujitsu Microelectronics, Manchester, England.

Hill, F.J. and Peterson, G.R.P. (1978). *Digital systems: hardware organization and design*, (2nd edn), pp.121–62. Wiley, USA.

Hill, F.J. and Peterson, G.R.P. (1984). *Digital logic and microprocessors*, pp.233–66. Wiley, USA.

Hollis, E.E. (1987). *Design of VLSI gate array ICs*. Prentice Hall, USA.

Huber, J.P. and Rosneck, M.W. (1991). *Successful ASIC design the first time through*. Van Nostrand Reinhold, USA.

IEEE (1987). *IEEE standard VHDL language reference manual 1076—1987*. IEEE, USA.

IEEE (1992). *IEEE standard VHDL reference manual interpretations*. IEEE, USA.

IEEE (1993). *IEEE standard VHDL language reference manual 1076-1993*. IEEE, USA.

Intel (1992). *Intel 486 SX microprocessor data book*. Intel Corporation, USA.

Katsioulas, T. (1994). Designing in deep sub-micron. *Electronic Engineering* **66**, (814), 565–70.

Lattice (1994). *Lattice data book*. Lattice Semiconductor, USA.

Lipsett, R., Schaefer, C., and Ussery, C. (1989). *VHDL: hardware description and design*. Kluwer Academic Publishers, USA.

Maliniak, L. (1994). A beginner's guide to VHDL. *Electronic Design*, **42**, (21), 75–82.

May, D., Barrett, G., and Shepherd, D. (1992). Designing chips that work. *Philosophical Transactions of the Royal Society of London, Series A*, **339**, 3–19.

McGreivy D.J. and Pickar, K.A. (1982). *VLSI technologies through the 80s and beyond*. IEEE Computer Society Press, USA.

Navabi, Z. (1993). *VHDL analysis and modeling of digital systems*. McGraw-Hill, USA.

Ott, D.E. and Wilderotter, T.J. (1994). *A designers guide to VHDL synthesis*. Kluwer Academic Publishers, USA.

Perry,.D.L. (1994). *VHDL*, (2nd edn). McGraw-Hill, USA.

RCA (1986). *Databook, RCA high-speed CMOS logic circuits*. RCA, USA.

Reinertsen, G. (1983). Whodonit? The search for new product killers. *Electronic Business*, **9**, 62–6.

Shankar, R. and Fernandez, E.B. (1989). *VLSI and computer architecture*, pp.217–24. Academic Press, USA.

Sternheim, E., Singh, R., and Trivedi, Y. (1990). *Digital design with Verilog HDL*. Automata Publishing Company, USA.

Texas (1984). *The TTL data book for design engineers, Vol. 1*. Texas Instruments, USA.

Texas (1991). Testability primer. Semiconductor Group, Texas Instruments, USA.

Thomas, D.E. and Moorby, P.R. (1991). *The Verilog hardware description language*. Kluwer Academic Publishers, USA.

Thomas, D.E. and Moorby, P.R. (1995). *The Verilog hardware description language*, (2nd edn). Kluwer Academic Publishers, USA.

Xilinx (1994). *The programmable logic data book*. Xilinx, USA.

Yoeli, M. (ed.) (1990). *Formal verification of hardware design*. IEEE Computer Society Press, USA.

Further reading

Carlson, S. (1991). *Introduction to HDL-based design using VHDL*. Synopsys, USA.

Einspruch, N.G. and Hilbert, J.L. (eds) 1991. *ASIC technology*. Academic Press, USA.

Huber, J.P. and Rosneck, M.W. (1991). *Successful ASIC design the first time through*. Van Nostrand Reinhold, USA.

Lipsett, R., Schaefer, C., and Ussery, C. (1989). *VHDL: hardware description and design*. Kluwer Academic Publishers, USA.

Navabi, Z. (1993). *VHDL analysis and modeling of digital systems*. McGraw-Hill, USA.

Ott, D.E. and Wilderotter, T.J. (1994). *A designers guide to VHDL synthesis*. Kluwer Academic Publishers, USA.

Perry, D.L. (1994). *VHDL*, (2nd edn). McGraw-Hill, USA.

Soin, R.S., Maloberti, F., and Franca, J. (eds)(1991). *Analog-digital ASICs*. Peter Peregrinus, IEE, England.

Thomas, D.E. and Moorby, P.R. (1995). *The Verilog hardware description language*, (2nd edn). Kluwer Academic Publishers, USA.

12
Submicron scaling

12.1 Overview

Submicron features and their continual miniaturization, have been threads woven into the text throughout this book. They highlight the main drives in IC development. The scaling down of components and their interconnections is the dominant pursuit in IC technology. The overwhelming case for going down that route was presented in Chapter 1. Scaling is the driving force behind the growth in IC functionality. Increasing the size of the die plays an important part, but by far the greater proportion of the rise in density is attributable to scaling. The on-going shrinkage is not confined to VLSI. Analogue and digital ICs at lower levels of integration have benefitted similarly from the increased capabilities which have accrued from device shrinkage. SSI and MSI chips supplement VLSI, performing much faster and consuming far less power than did their forerunners with larger geometries.

The growth in chip density has taken place in several phases. The initial phase can be assigned to the era up to about 1970 when the first microprocessor, the Intel 8008 (which contained about 8000 transistors), made its appearance. Until then, the growth in component count was mainly directed towards producing more identical functions on a chip, like for example eight independent flip-flops in place of one. The increase in transistor count (thinking in terms of digital ICs, but bearing in mind that it applies to analogue and mixed-signal ICs as well) followed closely Moore's Law (Chapter 1) according to which the function density FD doubles every year, i.e.

$$FD_n = 2^n.FD_o \qquad (12.1)$$

where FD_o is the original value of FD, which has increased to FD_n after n years. The exponential growth continues to be a characteristic of IC evolution, but the rate of increase has reduced somewhat during the last three decades. The record of IC development during the period 1965 to 1993 has shown that the function density has increased by a factor of 10 every five years (Masaki 1993), leading to what we call Moore's Modified Law, given by

$$FD_n = 1.59^n.FD_o \qquad (12.2)$$

(FD_n/FD_o) is plotted against years in Fig. 12.1. The increase in FD is better assessed in terms of FD_n/FD_o, rather than by taking FD_n and FD_o on their own. What really matters is the growth ratio over a specified interval. The values of FD_o in the years 1959 to 1964, when ICs were mainly in SSI, are not suitable anchor points for calculating FD_n.

426 Submicron scaling

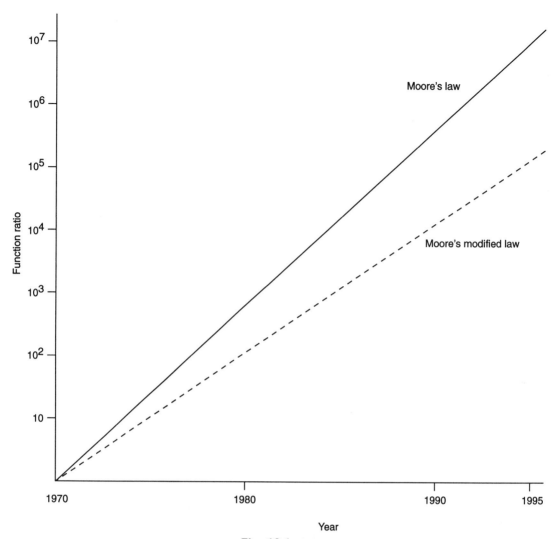

Fig. 12.1 Rate of increase in chip function density

Taking a specific case of growth rate by applying eqn (12.2) to the period 1980–1990, the function density increased by a factor of $1.59^{10} = 103$.

The increase in DRAM capacity follows very closely Moore's Modified Law. Maximum DRAM storage increased by a factor of 1000 from 64 Kb in 1980 to 64 Mb in 1995. Applying eqn (12.2) leads to a factor of 1049. A DRAM with 64 Mb capacity was listed in Table 1.2 with 1991 against the date of its appearance. It has been linked above with 1995 by applying Moore's Modified Law. Experimental DRAMs with 250 Mb and 1 Gb capacity have recently been reported. These and other similar advanced developments, either in respect of capacity or speed, have received little attention in this text, in which illustrations of industrial practice are based on ICs that are in volume production, or are likely to reach that stage within the next few years. Inspite of the announcement of the 64 Mb DRAM around 1991, the mainstay of DRAMs in 1995 has a maximum capacity of 4 Mb, with 16 Mb beginning to emerge in quantity production.

The rise in function density is related to feature size, which decreases by a factor of about 0.7 every three years (Peckerar and Maldonado 1993). Assuming, as for function density, an exponential relationship

$$F_n = 0.89^n . F_o \qquad (12.3)$$

where F_n is the feature size n years after the original feature size F_o. The relationship expressed by eqn (12.3) is a fairly accurate reflection of the reduction in geometry over the years 1985 to 1995. It is plotted in Fig. 12.2 with a postulated geometry of 1.5 μm in 1985 and an extrapolation to the year 2000. The maximum size of die has increased over the years but not, like function density, in a regular exponential manner. The component count is roughly proportional to the square of the geometry, so that a shrinkage from 1.5 μm to 0.5 μm over the decade 1985 to 1995 would be accompanied by a tenfold increase in function density for unchanged die dimensions, increased by the scaled up die area.

Developments in the decade 1970 to 1980 were distinguished from those in the previous ten years by a number of characteristics. First, there was the emergence of MOS, single-channel at the beginning with brief a inaugural spell of pMOS, soon followed by nMOS. Second the microprocessor ushered in the era of a complete system integrated on a chip. Previous expansion was, as has been pointed out, directed chiefly at producing 'more of the same', such as multiple independent flip-flops or counters with higher capacity, on one chip. Third the thrust towards higher and more complex data processing was accompanied by the drive for semiconductor memories with higher storage. It was not long before this drive spearheaded IC development, a situation which still exists and is likely to do so in the future. Once the lithography for obtaining a reduced line width had been established, the other processes were readily adapted. The economic advantages of increased logic power per chip easily justified the initial outlay of setting up a new wafer fabrication for a reduced geometry.

The choice of die area was and is, like the many other facets of IC production, governed by economic considerations. Warwick and Ourmazd (1993) have examined system cost *in terms of die area*. The three components which make up the total cost are:

(i) The cost of wafer production prior to testing the dies: past data has shown this to remain fairly constant.
(ii) The cost of the defective dies: the proportion of defective dies is largely determined by the defect density and its distribution on the die area.
(iii) Assembly cost: this includes the costs of packaging, PCB power supplies, cooling facilities, etc., and PCB production.

The cost of PCB interconnections is several orders of magnitude higher than the cost of wires within the chip. It was explained in Section 2.6.5 that the high cost ratio of PCB to chip wire is due to the far greater length of the PCB wires and that the multichip module (MCM) reduces this dichotomy. The assessment by Warwick and Ourmazd (1993) holds for direct mounting of the ICs on the PCB. The larger

428 Submicron scaling

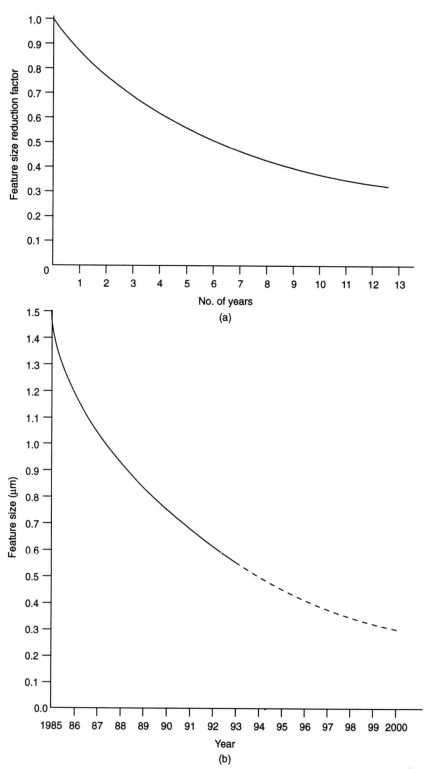

Fig. 12.2 Decrease in feature size with time (a) number of years (b) actual dates

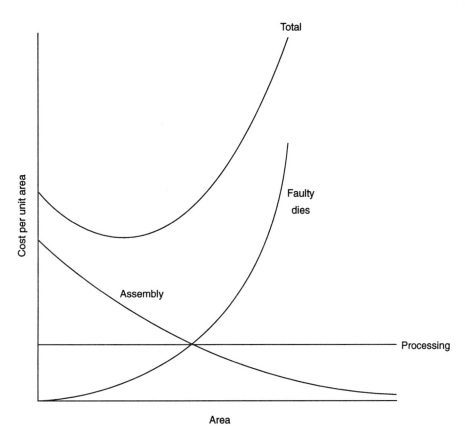

Fig. 12.3 Chip cost vs. die area

the die, the more complex will be its function so that more system interconnections will be within the chip and not off-chip. That saving in cost is counterbalanced by the increase in the cost of good dies, which will constitute a smaller proportion of the total because of defects on the wafer. The total system cost expressed in terms of die cost per unit area has the shape shown in Fig. 12.3. Taking typical 1992 values for the contributions of assembly, processing and faulty die costs, and putting the defect density at ~ 0.3 defects/cm^2, which applied to first class products at that time, the optimum die area comes to 1.3 (1.14 × 1.14) cm^2, and the projected optimum for the year 2010 to 13 (3.6 × 3.6) cm^2, an increase largely due to a much lower postulated defect density of 0.01 defects/cm^2 by then. That estimate is based on a proven reduction in defect density by $\sim$20 per cent per annum over the period 1974–1992. The die area of 1.3 cm^2 is fairly representative of current VLSI practice.

The feature size of current established standard and ASIC VLSI is in the range (0.6–1.2) μm, with 0.8 μm being the digital norm. Analogue and mixed-mode ASICs tend to have larger geometries which are at the upper end of the range, (1.0–1.2) μm being typical. The conditions for down-scaling are very different from those during the first 25 years of the integrated circuit art. A new wafer fabrication facility for a reduced geometry like (0.5–0.6) μm will cost of the order $500 million to $1.5 billion to set up. The cost is likely to be higher for the deep submicron geometries of 0.25 μm and 0.35 μm that will emerge in large–scale production during the next five years. The complexity of production increases with shrinkage: the number of processing steps has gone up from ~80 for a 16 Kb to ~450 for a 16 Mb DRAM (Wieder and Neppl 1992).

The case for continued shrinkage leading to several million transistors per chip has to be justified on economic grounds. The VLSI chips of the future have, like those of today, to satisfy existing or projected functions economically. There is no point in continuing with miniaturization and raising component density for its own sake. It is the deployment of the miniaturized ICs that matters. Miniaturization is the means of producing electronic systems at the smallest possible cost. The continuing shrinkage may recall in some peoples' minds the comment made about the laser at an early stage in its development, when it was sometimes described as a solution in search of a problem. The drive towards the ever decreasing deep submicron geometries is only acceptable if the end justifies the means. Hard-headed economic considerations are the determining factor.

The advantages of miniaturization apply to functions which need a smaller component count and/or operate at a slower speed than the maximum available. The size of the chip is adjusted to accommodate the required component count, making it smaller and less expensive than a chip with a larger geometry. One advantage of a lower operating speed is that delay times in relation to race hazards etc. are less critical.

The move to submicron and deep submicron structures has raised several issues regarding reliability. A supply voltage lower than the entrenched 5 V becomes mandatory for a feature size below 0.8 μm. Many of the SSI, MSI, and VLSI chips listed in this book are powered by a 3.3 V supply. Chips with geometries in the range (0.25–0.35) μm will probably be powered by a 2.2 V supply. The aspects of special importance here are hot electrons in MOS and also—to a lesser extent—in bipolar ICs, ultra thin oxide layers with thicknesses below 10 nm, and electromigration in the interconnections.

With the increase in system complexity, the tasks of designing and testing the chips are becoming as demanding as their fabrication. The previous chapters, Chapter 11 in particular, are highly relevant for appreciating what has to be done. The techniques of ASIC design outlined therein apply in principle to standard VLSI. The only really significant differences are that the designer of standard VLSI has a choice of device dimensions and macro layout, and that the design is likely to be undertaken with more advanced design tools and by a larger design team. The full capability of systems afforded by VLSI only comes about with the support of sophisticated EAD. The CAD/CAE tools are becoming more advanced and expensive with increasing chip density. Whereas scaling down to geometries around (1.5–3) μm was first and foremost a matter of advancing the lithography, backed up by comparatively straightforward adaptations of processing and chip design, the scaling down from ~1 μm to 0.5 μm and beyond calls for not only a completely new wafer fabrication facility, but also for advances in chip design and test techniques to realize their fullest potential. The critical path in the scaling down process is beginning to shift from the fabrication technology to chip design and testing (Nagata 1992, p.471).

Miniaturization is always expected to be accompanied by an increase in speed, but this increase is becoming progessively curtailed by the wires, whose influence has become the dominant factor at submicron geometries. That much will already have become clear from some of the material in Chapter 5 and in the chapters on ASICs. So much depends on the application and the quality of the chip layout that the increase in speed with reduced geometry cannot be specified by a general formula.

This overview would not be complete without discussing the roles of the various technologies in the down-scaling of VLSI from 0.8 μm geometries. CMOS, regardless of scaling, maintains its pre-eminence. Wieder and Neppl (1992) expect its share of the total IC market to increase from 55 per cent in 1992 to 85 per cent by the year 2000. CMOS has always been slower than bipolar logic; BiCMOS has been a means to reduce that gap. With feature sizes of 0.6 μm or less, the role of BiCMOS is no longer so assured. At these dimensions its advantage in system speed is only slight and may be offset by the unquestionably higher cost of production, a direct consequence of the larger number of processing steps for this technology. Table 5.7 highlights the merits of the 0.6 μm HLL and ALVC CMOS families, whose bistables toggle at 350 MHz compared with the 150 MHz bistable of the 0.8 μm BiCMOS family. That gap will be reduced in VLSI, where BiCMOS comes into its own for high capacitive loading. Even so 0.6 μm CMOS may still be faster than 0.8 μm BiCMOS and might be speedwise on a par with 0.6 μm BiCMOS. It could be that BiCMOS will decline when CMOS with 0.6 μm and smaller geometries becomes established. Masaki (1993) estimates in a theoretical study that (0.2 –0.25) μm CMOS may outperform ECL, achieving equal or higher speeds in the various parts of a simulated system. Bipolar ICs have a speed advantage over CMOS for geometries down to 0.35 μm. They will continue to fulfil a vital role in systems, especially at front ends where the required speeds may be beyond the reach of CMOS.

GaAs logic has not yet entered the market place in volume (Section 5.5.3; Chang and Kai 1994, pp.3–4). A paramount attraction for GaAs is for data processing/interfaces will gigahertz optical links. A strong case for GaAs is its superior power performance over equally fast ECL. GaAs DFCL consumes from one-third to one-quarter of the power for an equivalent ECL gate (Section 10.6; Chang and Kai 1994, p.529). With power limitation becoming more pronounced in deep submicron VLSI, GaAs logic could be preferred over ECL for that reason alone, inspite of its higher cost.

MOSFET, BJT, and interconnection scaling are the issues addressed in the sections which follow, and the chapter concludes with a discussion of anticipated developments in the near future.

12.2 MOSFET scaling

The MOSFET scaling rules shown in Table 12.1, which is a slightly modified repeat of Table 5.23, have remained a robust guide for device shrinkage to this day, inspite of being based on an analysis of MOSFETs published over 20 years ago (Dennard et al. 1974). The ideal scaling rules for constant field are shown side by side with the scaling rules for constant voltage. The disadvantage of constant voltage scaling stands out immediately. The differences between the two entries for gate input power and gate input power density sound a warning about the penalty

Table 12.1 MOS scaling rules ($S > 1$)

	Parameters	Scaling factor (Constant field)	Scaling factor (Constant voltage)
Scaling	Dimensions (W, L, t_{ox})	$1/S$	$1/S$
	Substrate doping	S	S
	Voltages (V_{DD}, V_{TH})	$1/S$	1
	Gate area ($A = WL$)	$1/S^2$	$1/S^2$
	Gate capacitance	$1/S$	$1/S$
Effect on performance	Device current (I_{DS})	$1/S$	S
	Gate delay (t_P)	$1/S$	$1/S^2$
	Gate input power (P)	$1/S^2$	S
	Speed-power product ($t_P \times P$)	$1/S^3$	$1/S$
	Gate input power density (P/A)	1	S^3

(©1974 IEEE)

Table 12.2 Selected scaled MOSFET parameters

Feature size (μm)	1.5	0.8	0.6	0.5	0.35	0.25	0.18
V_{DD} (V)	5[†]	5	3.75	3.13	2.19	1.56	1.13
t_{OX} (nm)	30[†]	16	12	10	7	5	3.6
BV_{GS} (V) [‡]	10.5	5.6	4.2	3.5	2.5	1.8	1.3

[†] Values for AC/ACT CMOS (Section 5.5)
[‡] Based on max. electric field = $1/2 \times 700$ V/μm ($1/2$ is a safety factor)

incurred by scaling with constant voltage, a practice adopted for many years. Provided the total power consumption, which scales up with S, is below the maximum chip dissipation, constant voltage scaling can be tolerated. The scaling to deep submicron geometries has changed that. Existing 0.8 μm VLSI chips are often operated at or close to their maximum power rating; an increase in gate power is totally unacceptable. Voltage scaling is necessary for other reasons:

(i) the drain breakdown voltage, which is mainly determined by avalanche breakdown;
(ii) the breakdown voltage of the gate oxide, which scales with S;
(iii) hot carrier effects;
(iv) reducing power consumption for battery operated equipment. The supply voltage may be scaled down by factor larger than S for that purpose.

Scaling down V_{DD} from the standard 5 V, which holds for geometries down to 0.8 μm, leads to the values in Table 12.2, which includes scaling of t_{ox} and BV_{GS}, the breakdown value of V_{GS}. DRAMs with 0.25 μm feature size are at an advanced stage of development, and 0.18 μm 1 Gb memories are expected to be introduced in the year 2001 (Tiwary 1994). It is good engineering practice to operate components below their maximum ratings with a substantial safety factor. This is frequently not observed with submicron ICs; these usually operate at 5 V, a value very close to the maximum permissible. The margin of safety appears however to be adequate. VLSI with 0.8 μm geometry and operated with V_{DD} of 5 V has proved to be reliable, although Kanzaki (1990) estimates that level to be near the drain breakdown voltage of a conventional 1 μm MOSFET.

The breakdown voltage in (ii) above is due to the stress of the electric field across the gate oxide. The maximum value of this field is put by various authors at 1000 V/μm (Woods 1986, p.1725), 700 V/μm (Muller and Kamins 1986, p.496) and 600 V/μm (Tsivides 1987, p.151). The maximum permitted electric field across the oxide remains constant only if the ultra thin gate oxide layers (see t_{ox} in Table 12.2) are free from defects. Such oxides are now available. The PLICE antifuse used in FPGAs and semiconductor memories, where it was first mentioned (Section 7.5), has an oxide–nitrogen–oxide (ONO) dielectric 10 nm thick, and has proved to be very reliable. The value of 700 V/μm has been adopted, combined with a safety factor of two, for calculating BV_{GS} in Table 12.2. The established standard supply voltage of 3.3 V for (0.5–0.6) μm geometries is fairly close to BV_{GS} in Table 12.2 and is acceptable, bearing in mind the safety margin for BV_{GS}. Lower voltages for VLSI below 0.5 μm have not yet been firmly standardized, but we may expect to see 2.2 V for (0.25–0.35) μm feature size and ∼1 V for 0.18 μm chips.

Hot carriers impair the characteristics and reliability of MOSFETs, and their adverse effects become more serious with reduced dimensions. The nature of hot carriers was described in Section 4.3.4. In EPROMs and EEPROMs hot electrons have a positive effect, being used for charging the floating gate (Section 7.5). In MOS logic their influence is negative and must be minimized. The effects of hot carriers are more serious for hot electrons than for hot holes, but both carriers have to be considered. For a fixed supply voltage hot carrier generation increases with reduced channel length on account of the larger electric field. Hot carriers give rise to a substrate current, and occupy trapping states in the gate oxide. Some carriers reach the gate, and their charge causes threshold instability and transconductance degradation.

The high electric field in the drain region is the main cause of hot carrier generation. Its reduction by modified transistor structures is an established technique for reducing the effects of hot carriers. The lightly doped drain (LDD) structure in Fig. 12.4 is a popular formation in which a conventional n^- implantation, self aligned by the gate, is followed by forming sidewalls, either with a CVD deposition of SiO_2 or with a process using polysilicon. A second implantation, masked by these sidewall spacers, produces the bulk n^+ regions for source and drain. The LDD structure reduces the surface electric field near the drain and keeps the boundary of the depletion layer closer to the drain than in a conventional structure. Matsunaga (1990) reports an improvement in the drain breakdown voltage BV_{DS} from ∼7 V for a conventional to ∼10 V for a 1.2 μm LDD MOSFET. The n^- regions of the LDD structure inevitably increase the series source and drain resistances, as does shrinkage. That

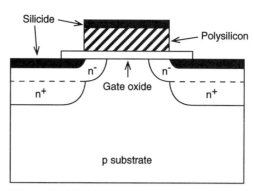

Fig. 12.4 LDD MOSFET

increase lowers the transconductance and reduces switching speed. It is minimized by depositing silicides on the source and drain layers (silicide is also deposited on the polysilicon gate) for contacting these electrodes (Fig. 12.4). Examples of such practice have been given in Figs 3.4 and 3.14 for CMOS and bipolar transistors respectively. The reduction in the series source and drain resistances is brought about by the large electrode contact area obtainable with this method. The LDD structure is proven and favoured for MOSFETs going down to design rules approaching 0.2 μm, but has not completely replaced conventional formation. Chang et al. (1992) and Davari et al. (1992) describe in two parts the design, characterization, and technology of 0.25 μm CMOS with a conventional structure, a 7 nm gate oxide, and operating at 2.5 V. Their project included 0.18 μm devices. The 0.25 μm CMOS gave a 280 ps gate delay in a ring oscillator with a fan-in and fan-out of three, and a load capacitance of 0.2 pF. The gate delay for 0.18 μm CMOS under identical conditions was 185 ps.

The scaling of V_{DD} impacts speed. It is important to quantify the effect on CMOS with deep submicron geometries, and this matter has been investigated by Kakamu and Kinugawa (1990), who have analyzed CMOS inverter transition time in terms of two components in accordance with

$$t_{THL} = \tau_{f1} + \tau_{f2} \qquad (12.4)$$

Figure 12.5 illustrates the composition of t_{THL}, in which τ_{f1} is the component in the saturation region and τ_{f2} the component in the linear region. To a first order approximation τ_{f1} is independent of V_{DD} and $\tau_{f2} \propto 1/V_{DD}$. The transition time t_{THL} is plotted versus V_{DD} in Fig. 12.6, which applies equally to t_{TLH} since the two transition times are equalized in CMOS. V_C is the voltage where τ_{f1} equals τ_{f2}, and is given by

$$V_c = \frac{1.1(E_{cn}L'_n + E_{cp}L'_P)}{2} \qquad (12.5)$$

where E_{cn} and E_{cp} are the critical fields E_c for electrons and holes at which the carrier velocity is saturated. E_c is given by

$$E_c = \frac{2v_{sat}}{\mu_{eff}} \qquad (12.6)$$

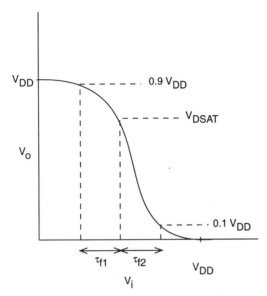

Fig. 12.5 CMOS inverter transition time (©1990 IEEE)

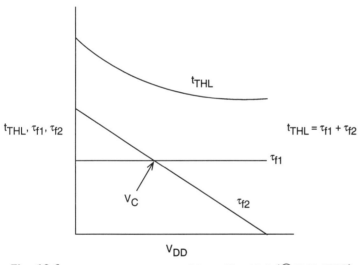

Fig. 12.6 Transition time composition—Fig. 12.5 (©1990 IEEE)

v_{sat} is the saturation velocity (Section 5.5.3; Fig. 5.63) and μ_{eff} is the effective carrier mobility (Section 4.3.1). L'_n and L'_p are the effective channel lengths of the nMOS and pMOS transistors respectively. The computations show V_C to be pretty close to V_{DD} in Table 12.2, witness the values of V_C for the following three geometries: 4.96 V (0.8 μm, t_{ox} = 16 nm), 3.41 V (0.5 μm, t_{ox} = 11 μm), and 2.43 V (0.35 μm, t_{ox} = 7 μm). V_C is not a critical demarcation point, but a convenient boundary. The transition time increases sharply with V_{DD} below V_C, and flattens off when V_{DD} exceeds V_C. The usefulness of V_C for gate delay evaluation was verified by measurements of delay for CMOS inverters fabricated with a wide range of t_{ox} (7 to 45 nm) and for (0.3–2) μm design rules. The measurements of t_P as a function of V_{DD} were based on ring oscillators

Table 12.3 Maximum CMOS gate count and clock frequency (Chip dissipation = 2W)

Feature size (μm)	0.8	0.35	0.25	0.18	0.10
C_L (fF)	175	76.6	54.7	39.4	21.9
V_{DD} (V)	5	2.19	1.56	1.13	0.63
t_P (ps)	150	65.6	46.9	33.8	21.9
$f_C(max)$ (GHz)	1.11	2.54	3.56	4.95	8.93
$N(P)_G$ (max)	2214	11 445	22 551	43 010	97 939

with fan-outs of one and three. They confirmed the expressions given in the paper for τ_{f1} and τ_{f2} and the validity of eqn (12.5). The work of Kakamu and Kinagawa (1990) demonstrates that down-scaling in the deep submicron regime at constant voltage reduces inverter delay time even when V_{DD} is substantially below V_C.

Chips are sometimes operated with V_{DD} in excess of the values recommended in Table 12.2 in order to be logically compatible with larger-geometry CMOS when this is part of the system. One approach adopted for that purpose is to surround the smaller-geometry chip core with a larger-geometry jacket, powered by a high V_{DD} for interfacing with larger-geometry chips.

An estimate of maximum clock frequency $f_C(max)$ and the maximum power limited gate count $N(P)_G(max)$ at $f_C(max)$ is shown in Table 12.3, which is based on the equation

$$P'_T = BN(P)_G(max)f_c(max)(C_L V_{DD}^2) \tag{12.7}$$

where P'_T = chip power available for gates
C_L = gate load capacitance
B = proportion of gates being exercised at any one time.

C_L applies for a fan-out of two and the chip is assumed to consist of simple gates (or array elements) and buffers. The values of V_{DD}, C_L and t_P for the 0.8 μm geometry in Table 12.3 are based on data for the GPS CLA80000 gate array, and fairly represent the performance of 0.8 μm VLSI. The constant field scaling rules in Table 12.1 and the scaled values of V_{DD} in Table 12.2 are used for calculating C_L and V_{DD}. B is put at 0.15 in keeping with Example 10.2, and $P'_T = 0.8 \times 2$ W $= 1.6$ W, allowing for 20 per cent of the power to be consumed by the buffers. All gates are exercised at the clock frequency.

The value of f_c (max) was arrived at by putting the minimum duration of one clock cycle at $6t_p$, made up of $2t_p$ for rise and fall times, and t_p for the flat portions of the high and low states. This duration is similar to the minimum pulse length advocated by Folberth (1981) and leads to

$$f_C(max) = \frac{1}{6t_P} \tag{12.8}$$

$N(P)_G$ was then calculated using eqn (12.7).

Looking at the entries for the 0.8 μm geometry, $N(P)_G(max)$ seems very small, seeing that some of the ASICs in Chapter 10 have up to

260 000 usable gates (Tables 10.6 and 10.7). It must be stressed that Table 12.3 contains the *maximum clock frequency* and the *maximum gate count for that clock frequency*. $N(P)_G$ and f_C can be traded off for a higher gate count at a lower clock frequency in accordance with

$$N(P)_G f_C = N(P)_G(max) f_C(max) \qquad (12.9)$$

$$f_C < f_C(max)$$

The upper limit for N_G is set by the maximum gate count available for the specified die. A simple reverse calculation gives good agreement between N_G in Example 10.2 and N_G calculated by using Table 12.3 and eqn (12.9). N_G at $f = 38.4$ MHz (the frequency obtained in Example 10.2) equals $(2214 \times 1110)/38.4 = 63\,998$. Bearing in mind that the gate utilization in Example 10.2 is 63 per cent, the total gate count comes to $63998/0.63 = 101\,582$, agreeing closely with the figure of 100 048 in the example.

The purpose of Table 12.3 is to quantify the $N(P)_G f_C(max)$ product and to demonstrate that the miniaturization of deep submicron CMOS is leading to clock rates approaching 10 GHz.

12.3 BJT scaling

The vertical structure of and the current flow in the BJT distinguish its operation from the horizontal mode of the MOSFET, and stamp the different approaches to scaling these devices. Scaling the BJT to improve speed requires reducing the base width, a vertical dimension, in order to increase f_T, thereby lowering τ_F (see eqns (4.63) and (4.64)). The lateral dimensions are scaled in order to reduce electrode capacitances. The intrinsic delay times of BJT and MOSFET inverters with submicron dimensions are largely determined by the electrode capacitances; transition times of carriers in the base or the channel are small in comparison. The base width has had submicron dimensions since the late 1960s, when it was about 0.7 μm for the transistors of the faster Motorola MECL families, with base and emitter depths of 3 μm and 2.3 μm respectively. Modern processes yield a base width below 100 nm, a value far below the smallest linewidth of current MOS VLSI. Voltage scaling is less of a problem than it is for MOSFETs, because transistor speed depends only slightly on collector-base voltage, and CML/ECL can operate with a supply voltage down to ~ 1 V (Section 5.2.5). Horizontal scaling has to aim at optimizing power consumption and the speed–power product. There are no scaling rules for the BJT to compare with the universally accepted scaling guidance for the MOSFET in Table 12.1.

The double polysilicon trench-isolated transistor shown in Fig. 3.8 is an established structure for submicron VLSI. Two specially important characteristics are the polysilicon emitter, mentioned in Section 3.3, and the polysilicon base contact. The polysilicon emitter (alternatively polysilicon emitter contact) is an emitter which is covered by a doped layer of polysilicon in place of the usual metal layer. The polysilicon layer should be thicker than 50 nm. The physical operation of the polysilicon emitter is still not fully understood, although it was introduced about twenty years ago (Treitinger and Mura-Mattausch 1988). The polysilicon

Table 12.4 ECL scaling rules ($\lambda > 1$)

Parameter	Scaling factor
Emitter length L_E	$1/\lambda$
Base width W_B	$1/\lambda^{0.8}$
Base doping N_B	$W_B^{-2} = \lambda^{1.6}$
Collector doping N_C	λ^2
Current density J	λ^2
Gate delay t_d	$1/\lambda$

(©1986 IEEE)

emitter gives a much higher current gain, increasing h_{FE} typically by a factor of 10. The polysilicon layer is also the diffusion agent for the emitter. Confining ourselves to npn transistors, the polysilicon layer is first implanted with arsenic, which is subsequently diffused into the substrate by heating. This process makes it possible to form very shallow emitter layers with depths down to 50 nm or even less. Since the base width is the difference between the original base layer and the emitter layer, the formation of such shallow emitter layers enables the production of base widths of 100 nm or less with good accuracy.

The advantages of the very high current gain brought about by the polysilicon emitter are twofold. First current gain decreases when the emitter depth is below ∼200 nm, but it will still be high. Second the base doping which, together with the base dimensions, determines the resistance of the base has to be very much lower than the emitter doping for high emitter injection efficiency and high current gain, and that leads to a high base resistance. The polysilicon emitter allows current gain to be traded for base resistance. With a polysilicon emitter, the base is far more heavily doped, lowering its resistance and yet maintaining a high current gain. The ability to do this is an outstanding improvement, gaining in importance with reduced base width and the increase in base resistance it gives rise to. It is fair to say that the polysilicon emitter has been a tremendous advance which has enabled BJTs to realize their speed potential with scaling by permitting the base resistance to be lowered whilst preserving high current gain.

Base resistance and f_T exert a great influence on speed. The two dominant factors with regard to base resistance are the loss in input drive due to the voltage drop across the extrinsic component, in practice the major proportion of the base resistance, and the $r'_b C_{jc}$ time constant (Fig. 4.11). Great efforts are made to minimize C_{jc}. The polysilicon contact to the base electrode in Fig. 3.8 is a step towards that: it minimizes contact area. The deposition of a silicide, by direct formation or after deposition of the extrinsic base polysilicon layer, is another possible step. This would be an adaptation of a practice developed for deep submicron CMOS (Warnock 1995, p.381).

A set of ECL scaling rules, frequently quoted in the literature, presents a useful guide and is given in Table 12.4 (Ning and Tang 1986; Rosseel and Dutton 1992). The data in Table 12.4 postulates constant transistor

Table 12.5 Selected parameters of submicron SST transistors

Parameter	Transistor 1	Transistor 2
Emitter area (μm^2)	0.35×7	0.3×5
Collector base capacitance (fF)	14.6	10.7
Collector substrate capacitance (fF)	23.5	13.2
Base resistance (Ω)	14.6	96
f_T ($V_{CE} = 1$ V (GHz)	21.1	40.7

Transistor 1 (Ishii *et al.* 1994)
Transistor 2 (Ishii *et al.* 1995)

(©1994, 1995, IEEE)

current. It is the current *density* which is scaled by λ^2 on the assumption that the emitter width scales, like L_E, by $1/\lambda$. The considerations for horizontal scaling are more complex for BJTs than for MOSFETs. The electrode areas are dimensioned in accordance with criteria governed by the physics of transistor operation and cannot be scaled according to simple rules. In general the emitter width exceeds its length by a factor of three or more. The progress achieved in recent times is represented by the 30 GHz f_T submicron BJT described by Yamaguchi *et al.* (1993). Emitter, base, and collector (including isolation) areas are 0.7×1.5 μm^2, 2.3×3.1 μm^2, and 5.4×8.1 μm^2. The base width and emitter junction depths are 82 nm and 50 nm. The emitter current density is 690 $\mu A/\mu m^2$, which is virtually equal to the emitter current since the emitter area is 1.05 μm^2. Selected characteristics of submicron transistors at the forefront of technology, and made with the super self-aligned technology (SST) are contained in Table 12.5 (Ishii *et al.* 1994; Ishii *et al.* 1995).

A comparison of BJT and MOSFET speeds purely in terms of transistor parameters is of limited value. Wire capacitance and wire resistance, which already dominate performance of 0.8 μm geometries and do so increasingly at smaller dimensions, have to be included. Bipolar maintains a speed advantage over CMOS technology because of the BJT's superior drive power for high capacitive loads. The types of transistors described by Yamaguchi *et al.* (1993) and Ishii *et al.* (1994; 1995) spearhead developments, but a substantial engineering effort is called for to turn prototypes of this nature into production (Yamaguchi *et al.* 1994).

12.4 Interconnection scaling

There are two distinct categories of interconnections within the chip, local and global wires. The essential distinction between them is that scaling of length can be applied to local, but not to global wires, whose mean length is proportional to the chip edge (Section 11.7). General scaling rules are in terms of mean local and global lengths and the global length remains the same when transistor dimensions and local wires are scaled without changing the chip edge. (Section 11.7). The aspects considered in this section are electromigration, wire capacitance and resistance, the transient response of wires, and finally MOS interconnection scaling.

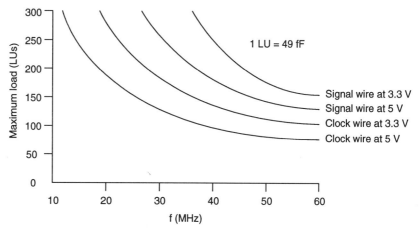

Fig. 12.7 Maximum loading of wires—CLA80000 gate array (Courtesy of GEC Plessey Semiconductors)

Several aspects of transistor reliability were described in the previous section. The major reliability problem for wires is electromigration, an initial coverage of which is contained in Section 2.4.6. Electromigration failure in aluminium wires occurs at current densities in the range $(10^5$–$10^6)$ A/cm^2 ((1–10) mA/μm^2) and is expressed in the mean time to failure (MTTF), which decreases exponentially with temperature (eqn (2.17)), making it important to estimate the MTTF at the higest operating temperature. The steps mentioned in Section 2.4.6 generally reduce the failure rate to acceptable low levels, ~1000 FITS being the target. The alternative of silicide in place of aluminium increases wire resistance and reduces speed, a penalty which may have to be accepted if the current density in aluminium wires exceeds the maximum safe level. Another alternative being pursued is coating aluminium wires with a silicide (Matsunaga 1990, pp.225–6). Electromigration is the subject of much on-going research. Specific attention has to be paid to interconnection scaling in order to avoid an excessive rise in current density with shrinkage.

ASIC vendors sometimes specify the maximum permissible capacitive load to avoid electromigration failure. The capacitance determines the currents demanded for its charge and discharge. The current pulses in signal and clock wires flow in both directions, thereby reducing but not eliminating electromigration. The maximum permissible loads for the GPS CLA80000 gate array are plotted as a function of frequency in Fig. 12.7.

The wire capacitance considered up to now has been confined to the wire–substrate capacitance, expressed for convenience in terms of capacitance per unit length (eqns (5.56) to (5.58)). The effective wire capacitance is modified by the capacitance per unit length between wires, C'_{mm} (Fig. 12.8), to

$$C'_t = C'_W + C'_{mm} \qquad (12.10)$$

C'_t is the total wire capacitance, and C'_{mm} is given by

$$C'_{mm} = \frac{\varepsilon_o e_r T}{W_s} \qquad (12.11)$$

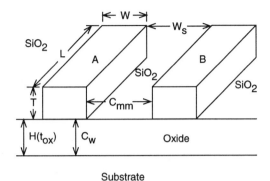

Fig. 12.8 Profile of interconnections

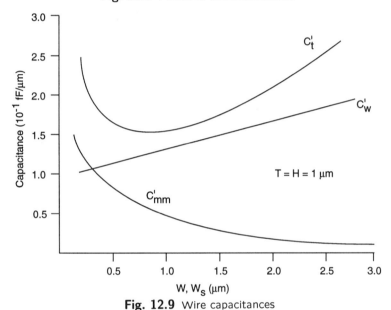

Fig. 12.9 Wire capacitances

e_r being the dielectric constant for SiO$_2$. (H in Fig. 5.60 has been retained for t_{ox} in Fig. 12.8 for the sake of compatibility). The variation of C'_t, C'_w, and C'_{mm} with W and W_s ($W = W_s$) is shown for T and H equal to 1 μm in Fig. 12.9. The minimum of C'_t is fairly flat and the choice of W, W_s is not critical, but the shape of C'_t cautions against too sharp a reduction of W, W_s. The saving in space would be at the cost of increased C'_t and would also raise cross talk. A signal $V(A)$ on wire A (Fig. 12.8) induces a signal $V(B)$ on wire B in accordance with

$$V(B) \simeq \frac{C'_{mm}}{C'_{mm} + C'_w} V(A) \qquad (12.12)$$

and $V(B)$ rapidly rises towards $V(A)$ when W and W_s are reduced below ~0.8 μm.

The interconnection delays are computed from the RC time constant of the wires, obtained by LPE and fed back for analysis via back annotation. This applies to standard and ASIC VLSI design. Depending on the length of wire, a *lumped* (discrete) RC model or a *distributed* RC

Table 12.6 Critical length of wire

Geometry (μm)	0.8	0.5	0.35
t_T (ps)	200	125	87.5
$l_c(w)$ (cm)	0.63	0.39	0.28

(alternatively *transmission line*) model is used for calculating delay time. The software tool decides whether to use the lumped or the distributed model for each wire. That decision depends on the length of wire and on the transition time t_T of the signal driving it. Calculations are usually performed in terms of the leading signal edge, but the expressions given below hold equally for the trailing edge. The standard practice in CMOS is to make t_{THL} and t_{TLH} equal.

The *electrical length* (alternatively *pulse length*) $l(e)$ of a signal is defined by

$$l(e) = \frac{t_T}{t_D} \qquad (12.13)$$

t_D being the delay time *per unit length* of the wire, whose *physical* length is $l(w)$. The lumped RC model is used if

$$l(w) < \frac{l(e)}{K} \qquad (12.14)$$

K being a constant in the range 4 to 6 (Bakoglu 1990, pp.239–40; Johnson and Graham 1993).

Equation (12.14) is alternatively expressed in terms of t_F, the *total signal* propagation delay through the wire, and t_T. Clearly

$$t_F = t_D l(w) \qquad (12.15)$$

Combining eqns (12.13) to (12.15), the lumped RC model is used if

$$t_F < \frac{t_T}{K} \qquad (12.16)$$

t_D is obtained in accordance with (Blood 1980)

$$t_D = 1.017\sqrt{0.475 e_r + 0.67} \text{ ns/ft} \qquad (12.17)$$

and comes to 53.0 ps/cm with SiO_2 for the dielectric. Table 12.6 has been drawn up for the critical length $l_c(w)$ at which $t_F = t_T/6$. The value of t_T for the 0.8 μm geometry is based on some of the simulations in Chapter 5, broadly confirmed by published data for 0.8 μm ASIC 2-input NAND gates. For the other geometries in Table 12.6, t_T has been scaled by $1/S$ (Table 12.1). The mean length of $l(w)$ is, considering global interconnections, $\sim 0.05 L_c$ for standard VLSI (Sections 2.4.6 and 11.7), but will probably be more like $0.1 L_c$ for ASICs on the whole, although a figure of $0.03 L_c$ was reported in Section 11.7. Putting $l(w)$ at $0.1 L_c$, and L_c at 1.6 cm for 0.35 μm VLSI, a significant proportion of wires will exceed $l_c(w)$ for that feature size. Some of the wires for the 0.5 μm and 0.8 μm geometries will also be in excess of $l_c(w)$.

There is no drastic change in delay times calculated either with the lumped or the distributed model in the region of demarcation expressed

Table 12.7 MOS interconnection scaling ($S < 1$; S_c, chip edge scaling factor)

Scaling and effects	Parameter	Scaling factor
Scaling, global and local	Oxide thickness t_{ox}	$1/S$
Scaling, global and local	Width W	$1/S$
Scaling, global and local	Wire separation W_S	$1/S$
Scaling, global and local	Thickness T	$1/S$
Scaling, local	Length L	$1/S$
Scaling, local	Resistance R	S
Scaling, local	Capacitance C	$1/S$
Effects, local	RC product	1
Effects, local	Current density J	S
Effects, local	IR drop	1
Scaling, global	Length L	S_c
Scaling, global	Resistance R	$S^2 S_c$
Scaling, global	Capacitance C	S_c
Effects, global	RC product	$S^2 S_c^2$
Effects, global	Current density J	S
Effects, global	IR drop	$S S_c$

by eqns (12.14) and (12.16). Bakoglu (1990, pp.239–40) suggests a transition region with t_T in the interval $2.5 t_F$ to $5 t_F$, i.e.

$$2.5 t_F < t_T < 5 t_F \qquad (12.18)$$

and points out that the transmission line (distributed RC) model gives the correct answer regardless of wire length. The lumped model has the advantage of making timing computation easier and quicker. For a given RC product t_T is far less for a distributed (transmission line) than for a lumped model; the corresponding values are 0.9 RC and the traditional 2.2 RC (eqn (4.5)) respectively (Bakoglu 1990, p.199). So although t_T—and of course t_P—inevitably increase with wire length, they are far less for long wires than the values deduced from a conventional single pole RC model (Folberth 1981).

Wire delays are greatly reduced by multilayer interconnections. The first metal level on top of the substrate is predominantly taken up by local wires. Global wires are placed as much as possible on the higher levels. The majority of VLSI chips have two or three levels and this number is on the increase for submicron geometries. The values of wire capacitance for the second and third levels are typically $0.3 C'_w$ and $0.2 C'_w$. The wires at these levels are made wider and thicker, reducing their resistance relative to the first-level wires. Furthermore the average length of wire is greatly reduced; Keyes (1982) estimates it to be inversely proportional to the number of levels.

The scaling of MOS interconnections is shown in Table 12.7. The scaling rules therein represent established practice and are varied to suit individual designs. All the wire dimensions, the wire separation and t_{ox}, here the thickness of the field oxide, are scaled by $1/S$ in keeping with the scaling of the gate dimensions and the gate oxide in Table 12.1. The essential difference between local and global scaling is with regard to the (average) wire length which, scaled by 1/S for local interconnections, remains a constant proportion of the chip edge L_c. Dealing with local scaling first, the IR and J entries are based on the reduction of I (I_{DS} in Table 12.1) by 1/S. The RC product remains the same, signifying that wire delay does not improve with scaling. Assuming the gate delay to be reduced by 1/S in accordance with Table 12.1, there will still be some increase in speed lessened, however, by the constancy of the wire delay. The increase in J has to be watched because of electromigration. An alternative set of scaling rules retains 1/S for all horizontal dimensions, and applies $1/\sqrt{S}$ to the vertical dimensions (T and t_{ox}) (Bakoglu, p.197). This reduces RC by $\sim 1/\sqrt{S}$ and J by $\sqrt{S}$, improving speed and immunity to electromigration.

Global performance is far more of a problem. The scaling up of RC by $S^2 S_C^2$ rings an alarm bell. Keeping RC constant in local scaling is bad enough, because the improvement in gate delay should be accompanied by some improvement in wire delay. Instead global scaling increases the RC product by $S^2 S_C^2$! Even if the die size were not altered, RC would still be scaled up by S^2. Moreover enlarging the chip edge is essential for raising the component count. The elongation of L_c predicted by Warwick and Ourmazd (1993) (see Section 12.1) is from ~1.14 cm in 1992 to 3.6 cm in 2010. The scaling factor $S^2 S_C^2$ for RC highlights more forcefully than anything else the limitation on speed imposed by wire. Granted the mitigating effects of multilayer interconnections, the limitations due to wire are abundantly clear, and much research and development is in progress to tackle that problem.

12.5 The immediate future

This section is in the nature of an epilogue. We are now approaching the limits of existing technology. VLSI with 0.35 μm is coming on stream. Achieving 0.25 μm and 0.18 μm feature size cost effectively in volume production will require substantial efforts, but there is every confidence that it can be done; 1 Gb DRAMs with 0.18 μm design rules are targeted to emerge in the year 2001, an estimate arrived at by anchoring the 0.35 μm geometry to 1995 in place of 1997 (as shown in Fig. 12.2(b)) and applying eqn (12.3).

Optical lithography is stretched to its limit at line widths around 0.25 μm. Deep UV optical lithography with an excimer laser (193 nm) reaches a resolution of 0.27 μm, applying eqn (2.19) with k_1 and NA equal to 0.7 and 0.5 respectively. Retaining these values, an F_2 laser (157 nm) improves the resolution to 0.22 μm. Peckerar and Maldonado (1993) expect wafer steppers for those two sources to become available in 1996 (excimer laser) and 1997–8 (F_2 laser). Phase-shifting masks help to extend the resolution, but even so a good reasonably economic yield in production is not likely for 0.25 μm or smaller geometries. The capability of X-ray lithography to achieve feature sizes from 0.25 μm downwards has been established beyond doubt. What needs to be done is bringing it to the stage of reliability possessed by the mature optical lithography and reducing its cost. The weak link in the chain is the mask. Unlike optical reticles, which are typically 5X or 10X the size of the image to be printed, an X-ray mask must have the same size as the image, because there are no X-ray focusing lenses. That rules out projection printing. Instead the image is produced by proximity printing with an uncollimated beam. The absence of diffraction allows this mode to be far more accurate with X-ray than with optical lithography. The UV mask must, like the optical reticle, be completely free from defects. The production by IBM of an 0.35 μm 512 Kb SRAM with X-ray lithography in 1993 gave evidence that this can be done (Peckerar and Maldonado 1993). The resolution limit of X-ray technology based on existing technique is estimated to be $\sim$0.1 μm for proximity-printing with (10–20) μm separation between mask and wafer, which is current practice (Peckerar and Maldonado 1993). Operating costs of a technology at its limit rise sharply. The escalating cost of optical lithography for geometries below 0.35 μm raises a question mark against its economic viability for feature sizes of $\sim$0.25 μm and less, and is an added incentive to intensify the development of X-ray lithography.

Assuming the emergence of 0.18 μm VLSI sometime around the year 2001 and its appearance in volume production within the following quinquennium, let us look at various issues bearing on the shrinkage to $\sim$0.18 μm geometries. The EDA for chip layout has to adapt continually to the increasing component count per chip and to the growing complexity it fulfils. The hierarchy in top-down design will contain one, possibly more new levels at the top. The interlock of synthesis and layout algorithms for optimization will be intensified, and ASIC design tools will become more efficient. Simulation will be increasingly supplemented by formal methods that are not confined to formal verification, but are applied to writing HDL code (May et al. 1992). An acquaintance with formal methods based on an elementary knowledge of mathematics is becoming essential for VLSI design (Cohen 1987 and 1992). Analogue circuits will routinely be incorporated within VLSI: mixed-mode chips will become the design style of choice. These are some predictions based on trends which are already in full swing.

The scaling down to deep submicron dimensions requires advances in behavioural, logic, timing and circuit simulations, DRC, and ERC. Circuit simulation is carried out with SPICE or a similar software package like SCEPTRE, and will have to be based on new, or at least substantially revised transistor models. Instead of being confined to active devices, models will become more and more macro models, greatly reducing simulation time whilst preserving accuracy. Macro models have existed for over 20 years; a well-known case is the macro modelling of op amps. Macro models of op amps, held in the library of most SPICE vendors, are based on simplified circuits and yet gives excellent agreement between full-circuit and macro-based simulations (Boyle et al. 1974). The use of a macro model (Fig. 12.10) reduces the SPICE code from $\sim$50 lines to $\sim$15 lines, and speeds up the simulation. Another vital area is the development of new bulk RC and transmission line models for wires of deep submicron geometries.

Simulation outside CAD/CAE for VLSI design has only received scant mention so far with a fleeting reference in the penultimate paragraph of Section 2.7, because the subject is outside the coverage of this text. Simulations of the various IC fabrication processes like epitaxy, dif-

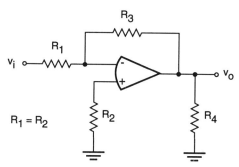

Fig. 12.10 Schematic for SPICE op amp simulation

fusion, implantation, oxidation, lithography and etching are indispensable steps towards shortening the development times for wafer production with new design rules, and for predicting performance, which is optimized by fine tuning. Submicron miniaturization is greatly aided by these supporting tools (Fichtner 1988; Sakurai 1990, pp.267–336). Technology Modeling Associates Inc. (TMA) specialize in semiconductor simulation, covering just about every conceivable process. The brief information about their products, given below by courtesy of TMA, is extracted from various issues of their TCAD Newsletter (now renamed TMA Times) published from 1992 to 1995. The most widely used process simulator, the *SUPREM* package, has become an industry standard. The one-dimensional process simulator *SUPREM-3* is supplemented by the two-dimensional process simulator *TSUPREM - 4*. *SUPREM* predicts one- or two-dimensional structural characteristics like layer thickness and doping profiles resulting from semiconductor processing, evaluates and refines isolation technologies like trench isolation, and helps with the design of MOS and bipolar processes at the forefront of technology. *SUPREM* creates structures for device analysis using the *PISCES - 2B* and *DAVINCI* packages. *DAVINCI* can analyze small geometry MOSFETs and BJTs and more complex structures like DRAM cells. Deep submicron processes to optimize 0.35 μm CMOS can be modelled with the *TSUPREM - 4* and *MEDICI* packages. Several of the packages provide model parameters for established SPICE software like HSPICE (Meta-Software) and PSPICE (Microsim Corporation). *DEPICT - 2*, a tool for simulating the deposition, etching, and photolithographic processes of wafer fabrication, allows the simulation of phase-shift masks. A virtual reality emulating research and development is offered by the *CAESAR 1.1* package, which integrates process and device simulators, parameter extraction, and circuit simulation. These selected extracts from TMA data give an insight into the process simulations available.

The cost of setting up a new wafer fabrication for a scaled-down geometry continues to escalate. Is it all justified on economic grounds, one might ask? The answer is: it has to be. Economic considerations determine whether an engineering project is worth undertaking or not, and pose a potential bar to pursuing miniaturization beyond a certain level. Linked to production costs comes the time to market, a consideration uppermost in the mind of a president or a chief executive. The estimate of McKinsey & Co., contained in references quoted in Sections 10.4 and 11.7 and named in the latter, is still very much in favour with management (Young 1988).

The discussion of scaling has so far been confined to the chip on its own. The intensity of effort to achieve higher chip density has been and is accompanied by a similar effort to improve packaging and chip interconnections on the PCB. In the initial stage of ASICs one of their applications was to mop up *glue* in the redesign of existing equipment. One might be forgiven for thinking that the ASICs of today and tomorrow at the top end of the function density will be similarly used to mop up what can no longer be called *glue*, namely earlier ASICs and standard ICs with much lower component counts, again in the redesign of existing equipment. Very occasionally that may still be the case, but VLSI scaling is directed at the designs of highly complex new systems containing substantial numbers of ICs with greater densities than their predecessors. PCB interconnection density will increase because of the growth in the number of pins on the scaled chips, and the probable growth in the number of ICs which make up the system.

The severe reduction of speed within the chip by the PCB wires was stressed in Section 5.2.4, and the improvement in packaging density and operating speed by the use of multichip modules (MCMs) was described earlier in Section 2.6.5. Significant innovations in packaging and interconnections have been reported in recent years to support the drive for miniaturizing submicron VLSI. MCMs play a large part in this, and one prominent new type of module in the MCM-D category has a thin-film substrate using a copper conductor and a polyimide insulator. It achieves a higher wiring density than previous MCMs (Sudo 1995). (See Table 2.14 for the definition of MCM categories.) The decrease in the number of wires and in the mean wire length due to the deployment of such MCMs raises the proportion of the PCB surface occupied by silicon and pushes up system speed. The efforts to improve packaging and PCB interconnections, which have hitherto been concentrated on MCM development, have widened. The *ball grid array* (BGA), also known by the Motorola trademark of *over molded pad array carrier* (OMPAC) is a surface-mounted package with a maximum total of over 1000 contact points. The chip scale package (CSP) is another new SM package with a similar contact count. A different approach has been adopted with the *few-chip package* (FCP). The FCP is a single-chip package—typically QFPs and PGAs (see Table 2.13) are used for that purpose—which houses a few chips and fulfils a function similar to, but not as efficiently as an MCM. It is an inexpensive development suited to the low end of VLSI by avoiding the far more substantial cost of producing an MCM (Herrell 1993). (The term single chip package (SCP) has been coined to contrast the traditional encapsulation with an FCP). The high end of VLSI is marked by a far larger overall development cost which can more easily absorb the outlay for new MCMs than the costs for the low end, which caters for the mass market of consumer (commodity) electronics. Herrell (1993) feels that the FCP will be a help for introducing the MCM into the commodity market by persuading customers of the cost effectiveness brought about by the combination of several chips in one package. Sudo (1995), acknowledging the competition presented to the MCM by the BGA and the CSP, is convinced that the MCM will find its way into the commodity market, which consists of equipment like portable audiotape machines, cameras, cellular phones, and palmtop computers, to give a few examples. It is quite possible that innovations in MCMs and packaging technology in general will be initially directed at the low end and will later also be used for the high end of VLSI.

The limitations due to wire are now dominating chip performance. The scaling factor $S^2 S_C^2$ for the RC product of global wire in Table 12.7 gives dramatic evidence of that. The limitation it imposes on system speed is, as was mentioned in the previous section, alleviated by multi layer interconnections, stepping up the number of layers from two to three in present structures to three, four, or even more. The effective chip area is thereby increased *without raising the mean interconnection length $\overline{L}$ of the wire*. $\overline{L}$ is a good indicator of design quality. The aim should be to reach a figure like $0.03 L_C$ (Harker 1995).

The overriding limitation on system speed is presented by chip interconnections. These are reduced and the operating speed is enhanced by the MCM and the more recent FCP. An entirely different technique is cryogenic operation at the liquid-nitrogen temperature (77 K), where CMOS performs well at what is a near–optimum temperature for obtaining the highest speed (Kirschmann 1990). Compared with 300 K, mobility and saturation velocity increase as does junction depletion width, leading to lower source and drain capacitances. All these effects raise speed. Equally important is the reduction in wire resistance, typically by factors of six and four for metal (aluminium or copper) and silicide wires respectively. This advantage applies of course to both chip and PCB wires. The intrinsic CMOS gate is between two and three times as fast and CMOS system performance is speeded up by a factor of about two (Bakoglu 1990, pp. 61–3; Masaki 1993, p.1312). One factor which worsens at 77 K is hot carrier degradation, because it has a negative activation energy. Voltage scaling and special processing like the LDD structure are essential to contain hot carrier effects within acceptable levels for cryogenic working and for reliability in general. Low temperature (77 K) CMOS looks like being the ultimate solution for reaching the highest possible speed, and in view of the importance attached to that parameter, cryogenic CMOS may find its way into some mainstream applications.

The huge expenditure of establishing a new foundry for a scaled down process is tending to reduce the number of big players able to keep pace with the inexorable miniaturization. There may be a shift away from the situation reigning hitherto, an established feature size which gives way to a reduced geometry with an interim overlap. It is probable that a pattern will develop with two or even three standard geometry levels. One can visualize two preferred feature sizes, 0.8 μm and 0.6 μm, exemplified by the Intel Pentium microprocessors which are available with 0.8 μm (66 MHz) or a choice of 0.6 μm (75 to 133 MHz) geometries. Mixed-mode chips are set to take a larger share of the VLSI market; in ASIC consumption they are likely to become the norm. Many analogue circuits call for transistors with larger feature size and a 1.2 μm geometry may be retained for mixed-mode chips. The 0.35 μm feature size might be largely reserved for semiconductor memories. Looking further ahead there will no doubt be a move towards 0.1 μm feature size, but whether such chips reach the mass market remains to be seen. Even 0.18 μm VLSI chips will necessitate new approaches to IC engineering (Chatterjee and Larrabee 1993). Solomon (1982), in a detailed and authoritative comparison of semiconductor devices for high-speed logic, estimates the delays of logic inverters based on 0.1 μm geometries to be 6.5 ps, 6.1 ps, and 2.1 ps for the BJT, the MOSFET, and the MESFET respectively. There are other indications that the speed gap between the BJT and the MOSFET is closing (Masaki 1993). The influence of wire will probably reduce the present speed differential between bipolar and GaAs technologies. GaAs is likely to be used not only because of what is becoming a marginal speed advantage, but in order to operate at less power than ECL at much the same speed.

It should be evident by now that the engineering of microelectronic systems has to be undertaken by an interdisciplinary team whose members contribute specialist expertise in subjects like logic design, device processing, device physics, IC production and packaging, CAD/CAE techniques, PCB layout, system assembly and testing, and microelectronic equipment manufacture, embracing system engineering. That list is neither exhaustive nor prescriptive, but purports to portray the variety of expertise needed for producing a microelectronic system. This text is offered to establish a foundation from which the reader can proceed to further study in one or more of these specialized fields.

References

Bakoglu, H.B. (1990). *Circuits, interconnections, and packaging for VLSI*. Addison-Wesley, USA.

Blood, R.W. (1980). MECL system design handbook, (3rd edn), p.39. Motorola Semiconductor Products, USA.

Boyle, G.R., Cohn, B.M., Pederson, D.O., and Solomon, J.E. (1974). Macromodeling of integrated circuit operational amplifiers. *IEEE Journal of Solid-State Circuits*, **9**, 356–63.

Chang, C.Y. and Kai, F. (1994). *GaAs high speed devices*. Wiley, USA.

Chang, W-H., Davari, B., Wordeman, M.R., Taur, Y., Hsu, C.C-H., and Rodriguez, M.D. (1992). A high-performance 0.25 μm CMOS technology: I—design and characterization. *IEEE Transactions on Electron Devices*, **39**, 959–66.

Chatterjee, P.H. and Larrabee, G.B. (1993). Gigabit age microelectronics and their manufacture. *IEEE Transactions on Very Large Scale Integration (VLSI) Systems*, **1**, 7–21.

Cohen, B. (1987). The education of the information systems engineer. *Electronics and Power*, **33**, 203-5.

Cohen, B. (1992). *The inverted curriculum*. National Economic Development Office, England.

Davari, B., Chang, W-H., Petrillo, K.E., Wong, C.Y., Moy, D., Taur, Y., et al. (1992). A high-performance 0.25 μm CMOS technology: II—technology. *IEEE Transactions on Electron Devices*, **39**, 967–75.

Dennard, R.H., Gaensslen, F.H., Yu, H-W., Rideout, V.L., Bassous, E., and LeBlanc, A.R. (1974). Design of ion-planted MOSFETs with very small physical dimensions. *IEEE Journal of Solid-State Circuits*, **9**, 256–66.

Fichtner, W. (1988). *Process Simulation. In VLSI technology, (2nd edn)*, (ed. S.M. Sze), pp.422–65. McGraw-Hill, USA.

Folberth, O.G. (1981). The interdependence of geometrical, thermal, and electrical limitations for VLSI logic. *IEEE Journal of Solid-State Circuits*, **16**, 51–3.

Harker, A.J. (1995). *Personal Communication*. Fujitsu Microelectronics, Manchester, England.

Herrell, D.J. (1993). Addressing the challenges of advanced packaging and interconnection. *IEEE Micro Magazine*, **13**, (2), 10–18.

Ishii, Ichino, H., Togashi, M., Kobayashi, Y., and Yamaguchi, C. (1995). Very high-speed Si bipolar static frequency dividers with new T-type flip-flops. *IEEE Journal of Solid-State Circuits*, **30**, 19–24.

Ishii., Ichino, H., and Yamaguchi, C. (1994). Maximum operating frequency in Si bipolar master-slave toggle flip-flop circuit. *IEEE Journal of Solid-State Circuits*, **29**, 754–60.

Johnson, H.W. and Graham, R. (1993). *High speed digital design*, p.7. Prentice-Hall, USA.

Kakamu, M. and Kinugawa, M. (1990). Power-supply voltage impact on circuit performance for half and lower submicrometer CMOS LSI. *IEEE Transactions on Electron Devices*, **37**, 1900–08.

Kanzaki, K (1990). Foundations and physics of MOS devices. In *Very high speed MOS devices* (ed. S. Kohyama), p.36. Oxford University Press, England.

Keyes, R.W. (1982). The wire-limited logic chip. *IEEE Journal of Solid-State Circuits*, **17**, 1232–3.

Kirschmann, R.K. (1990). Low-temperature electronics. *IEEE Circuits and Devices Magazine*, **6**, (2), 12–24.

Masaki, A. (1993). Possibilities of deep-submicrometer CMOS for very-high speed computer logic. Proceedings of the IEEE, **81**, 1311–24.

Matsunaga, J-I. (1990). Process integration for MOS devices. In *Very high speed MOS devices*, (ed. S. Kohyama), p.203, Oxford University Press, England.

May, D., Barrett, G., and Shepherd, D. (1992). Designing chips that work. *Philosophical Transactions of the Royal Society of London. Series A*, **339**, 3–19.

Muller, R.S. and Kamins, T.I. (1986). *Device electronics for integrated circuits*, (2nd edn), p.496. Wiley, USA.

Nagata, M. (1992). Limitations, innovations, and challenges of circuits and devices into half micrometer and beyond. *IEEE Journal of Solid-State Circuits*, **27**, 465–72.

Ning, T.H. and Tang, D.D. (1986). Bipolar trends. *Proceedings of the IEEE*, **74**, 1669–77.

Peckerar, M.C. and Maldonado, J.R. (1993). X-ray lithography—an overview. *Proceedings of the IEEE*, **81**, 1249–74.

Rosseel, G.P. and Dutton, R.W. (1992). Scaling rules for bipolar transistors in BiCMOS circuits. In *Digital MOS integrated circuits II*, (ed. M.I. Elmasry), pp.157–60. IEEE Press, USA.

Sakurai, T. (1990). Simulation technology. In *Very high speed MOS devices* (ed. S. Kohyama), pp.267–336. Oxford University Press, England.

Solomon, P.M. (1982). A comparison of semiconductor devices for high-speed logic. *Proceedings of the IEEE*, **70**, 489–509.

Sudo, T. (1995). Present and future directions for multichip module technologies. *IEEE Journal of Solid-State Circuits*, **30**, 436–42.

Tiwary, G. (1994). Beyond the half-micrometer mark. *IEEE Spectrum*, **31**, (11), 84–7.

Treitinger, L. and Mura-Mattausch, M. (ed.) (1988). *Ultra-fast silicon bipolar technology*, p.7. Springer, Germany.

Tsivides, Y.P. (1987). *Operation and modeling of the MOS transistor*, p.151. McGraw-Hill, USA.

Warnock, J.D. (1995). Silicon bipolar device structures for digital applications: technology trends and future directions. *IEEE Transactions on Electron Devices*, **42**, 377–87.

Warwick, C.A. and Ourmazd, A. (1993). Trends and limits in monolithic integration by increasing the die area. *IEEE Transactions on Semiconductor Manufacturing*, **6**, 284–9.

Wieder, A.W. and Neppl, F. (1992). MOS technology trends and economics. *IEEE Micro Magazine*, **12**, (4), 10–19.

Woods, M.H. (1986). MOS VLSI reliability and trends. *Proceedings of the IEEE*, **74**, 1715–29.

Yamaguchi, T., Uppili, S., Lee, J.S., Kawamoto, G.H., Dosluoglu, T., and Simpkins, S. (1993). Process and device characterization for a 30-GHz f_T submicrometer double poly-Si bipolar technology using BF_2-implanted base with rapid thermal process. *IEEE Transactions on Electron Devices*, **40**, 1484–95.

Yamaguchi, T., Uppili, S., Lee, J.S. Kawamoto, G.H., and Hanson, R.C. (1994). Process investigations for a 30-GHz f_T submicrometer double poly-Si bipolar technology. *IEEE Transactions on Electron Devices*, **41**, 321–9.

Young, J.A. (1988). Emerging imperatives for engineers. *IEEE Circuits and Devices Magazine*, **4**, (3), 14–17.

Glossary

ABT Bus Interface Logic (BiCMOS ABT) Advanced BiCMOS technology multibyte bus interface logic.

AC/ACT CMOS Advanced high-speed CMOS logic.

Access time The interval between the arrival of an address and a valid data output from a semiconductor memory.

ADC Analogue-to-digital converter.

AHDL Analogue HDL. (This acronym is also used by Altera for their proprietary (digital) HDL).

Aliasing The production of undesired frequencies in the passband of a signal.

ALS TTL Advanced low power Schottky TTL.

ALU Arithmetic logic unit.

ALVC Multibyte version of HLL.

AM Amplitude modulation.

AMD Advanced Micro Devices (a company).

AMS Austria Mikro Systeme International (a company).

Annealing *See* placement.

Antifuse A fuse which is normally open-circuit, but becomes short-circuit when programmed.

APCVD Atmospheric pressure CVD.

Array element (basic cell, core cell, primary cell) The basic uncommitted building block of a gate array, consisting typically of two pMOSFETs and two nMOSFETs.

AS TTL Advanced Schottky TTL.

ASCII American Standard Code for Information Interchange.

ASIC Application specific integrated circuit. A user-specified IC designed in close cooperation between the user and the ASIC vendor.

ASSP Application specific standard product. A vendor-designed IC for a specific application or a range of related applications.

ATE Automatic test equipment.

Back annotation The annotation of the netlist with data obtained from LPE for timing simulation.

Bandgap reference (voltage) Reference voltage based on the bandgap of silicon (about 1.25 V).

Basic cell *See* array element.

BBS Board-level boundary scan. A combination of BIST and BST.

BBU Battery backup.

Bed-of-nails fixture A board with test pins, designed to contact components on a PCB for on-board testing.

BFL Buffered FET logic (GaAs).

BGA (OMPAC, solder grid array) Ball grid array.

BiCMOS Combination of bipolar and CMOS technologies on the same chip.

BiCMOS ABT *See* ABT bus interface logic.

BiMOS Combination of bipolar and MOS technologies on the same chip.

Bird's beak Lateral oxidation, which affects the shape of the boundary between the active area of a transistor and the passivating oxide.

BIST Built-in self test.

BIT Binary digit; also bitstream format (of a program).

BJT Bipolar junction transistor.

Body factor (body-effect coefficient) Modification of MOSFET threshold voltage due to source-to-substrate bias.

BPSG (BP-glass) Borophosphosilicate glass.

BSC Boundary scan cell.

BSR Boundary shift register. A serial register configured by the interconnection of the boundary scan cells.

BST Boundary scan test. A technique for on-board testing of ICs.

BTL Backplane transceiver logic.

Buried register A register (flip-flop), which cannot be accessed for output, within a PLD output cell.

C language (C) A high level language widely used in engineering.

CAD Computer-aided design. The use of computer-based software tools primarily for component design. Many software packages are a combination of CAD and CAE.

CAE Computer-aided engineering. The use of computer-based software tools primarily for engineering a design to be part of an equipment (system). Many software packages are a combination of CAD and CAE.

CAS Column address strobe input to a semiconductor memory.

CCB Controlled-collapse bonding, an alternative expression for solder bonding.

CD Compact disk.

Cell-based ASIC (embedded cell) A hybrid combination of a (usually very large) standard cell and a gate array.

CERDIP Pressed ceramic glass sealed dual-in-line package.

CERQUAD Pressed ceramic glass sealed quad flat pack.

Channelled gate array A gate array in which the array elements are in rows separated by unoccupied spaces for the routing channels. The term gate array on its own is generally understood to signify a channelled structure.

Channelless gate array *See* SOG array

Chip edge The length of a square die, or the square root of the area of a rectangular die.

CML Current mode logic.

CMOS Complementary MOS.

CMRR Common mode rejection ratio.

CMS Current mode switch.

COB Chip on board. The bare attachment of a die to the substrate, usually that of a MCM.

Codec Companding encoder/decoder.

Concurrent engineering (concurrent design, simultaneous engineering) Concurrent engineering is a multidisciplinary approach to the integrated concurrent design of products and their manufacture, achieved with the aid of CAD/CAE.

Core cell *See* array element.

CPGA Ceramic pin grid array.

CPU Central processing unit.

CQFP Ceramic quad flat pack.

CSP Chip scale package. A surface mount package with a very high contact count.

CTE Coefficient of thermal expansion.

Custom ASIC (custom IC, full custom ASIC, handcrafted ASIC) An ASIC designed and processed with a full set of masks. The term custom ASIC is sometimes used to embrace custom and semicustom ASICs.

CVD Chemical vapour deposition.

CZ method Czochralski method of crystal growth.

D flip-flop (D bistable) A flip-flop (bistable) with one data (D) input. The flip-flop can operate in the SR (set-reset) or the T (toggle) mode.

D-MESFET Depletion MESFET.

DAC Digital-to-analogue converter.

Datapath The parts of a data processing system which operate on the data (structured logic), in contrast to the parts which control that operation (random logic).

DCFL Direct coupled FET logic (GaAs).

Decimation The translation by filtering of the bit stream output of a delta-sigma modulator into a multibit word at the Nyquist frequency.

Deep submicron Feature size of 0.5 μm or less.

Design rules (layout rules) The design rules specify the constraints on the mask patterns, specifically with regard to minimum dimensions, to produce an acceptable layout.

Die An unencapsulated chip.

DIP Dual-in-line package.

DMOS Double-diffused MOS.

DoD Department of Defense (US).

DoF Depth of focus.

DOS Disk operating system, held on the hard disk of a PC.

DRAM Dynamic RAM. Data is stored dynamically (i.e. temporarily), and the memory has to be refreshed at regular intervals in order to retain the data.

DRC Design rule check (checker). A CAD/CAE check to ensure that the layout conforms to the design rules.

Dropout voltage The input-output voltage at which a voltage regulator fails to function correctly.

DSP Digital signal processor.

DSW projection (step-and-repeat projection) Direct-step-on-wafer projection. The wafer is moved step-by-step for multiple image formation on the resist.

Dynamic CMOS Dynamic logic implemented with CMOS.

Dynamic logic Logic which relies on temporary storage across an intrinsic nodal (often a transistor electrode) capacitance.

e-beam Electron beam.

E-MESFET Enhancement MESFET.

EAROM Electrically alterable ROM; an obsolete designation for EEPROM.

ECL Emitter coupled logic.

ECMOS ECMOS signifies the programming of a CMOS PLD with an EPROM.

EDA Electronic design automation.

EDIF Electronic design interchange format. An HDL for interchange, capable of expressing a logic schematic, or a circuit description, or a layout in symbolic and graphical form.

EDP Electronic data processing.

EECMOS EECMOS, a trademark of Lattice Semiconductor Corporation, signifies the programming of a CMOS PLD with an EEPROM.

EEPROM (E^2PROM) Electrically programmable and erasable (usually byte-by-byte) CMOS ROM.

EGS First electronic-grade silicon.

EIA Electronics Industries Association (USA).

EIAJ Electronics Industries Association of Japan.

Electromigration Transport of atoms by momentum exchange with conducting electrons.

Embedded cell *See* cell-based ASIC.

EPLD Erasable PLD. Altera use this acronym for their FPGAs.

EPROM Electrically programmable, UV-light erasable CMOS ROM.

ERC Electrical rule check (checker). A CAD/CAE check to ensure that the layout conforms to electrical factors like fan-out, floating I/O pins etc.

ESD Electrostatic discharge.

FAMOS Floating-gate avalanche-injection MOSFET, an Intel trademark.

FAST TTL A fast TTL family. FAST is a trademark of NS.

FC Flip chip.

FC bonding Flip chip bonding.

FCP Few-chip package. A single-chip type

of package (like a QFP or a PGA) which contains several chips.

Feature size (geometry, line width, structure) The smallest dimension which can be achieved by the lithography for a particular process.

FET Field effect transistor.

Field-alterable (field-programmable) ROM *See* user-programmable ROM.

Field oxide (thick oxide) Passivating oxide between the first metal layer and the substrate surface.

Firm macro A macro with fixed placement and flexible routing.

FIT Failure-in-time. One FIT is a failure rate of one in a billion hours.

Flash EEPROM An EEPROM which is erasable by erasing the entire chip or - sometimes - large sections thereof.

FLEX Flexible logic element matrix, a trademark of Altera. FLEX PLDs are Altera SRAM programmable FPGAs.

Floorplan The layout of a die in terms of functional blocks.

FLOTOX Floating gate tunnel oxide technology, an Intel trademark. A structure used in EEPROM MOSFETs.

FM Frequency modulation.

F-N tunnelling *See* Fowler-Nordheim tunnelling.

Footprint The area occupied by a component on the surface on which it is mounted.

Formal methods Formal methods are used to desribe a digital system in terms of logic based on various branches of discrete mathematics.

Formal verification A verification using formal methods to verify the lower-level implementation of a higher-level specification by logic manipulation.

Foundry route An ASIC design route in which the customer undertakes the entire design, and provides the necessary mask patterns to the ASIC manufacturer (the silicon foundry).

Fowler-Nordheim tunnelling (F-N tunnelling) Tunnelling of 'cold' electrons between the drain and the gate of a MOSFET through a very thin oxide layer.

FP Flat pack.

FPGA Field programmable gate array. An array of programmable logic consisting of configurable cells interconnected by a programmable wiring network.

FPLA Field programmable logic array. An earlier, now obsolete designation for PLA.

FQFP Fine pitch quad flat pack.

Full custom ASIC *See* custom ASIC.

GaAs Gallium arsenide.

GAL Generic array logic, a trademark of Lattice Semiconductor Corporation. The GAL is similar to a PAL, but has more powerful logic.

Gate array (ULA) An ASIC made with a prefabricated substrate consisting of array elements, and completed by metallization.

Gate oxide The oxide layer between the substrate and the gate of a MOSFET.

GBP Gain bandwidth product.

GCA Gradual-channel approximation. This approximation applies to a MOSFET and postulates that the gate completely controls the channel.

Geometrical specification language A graphics language for defining mask patterns.

Geometry *See* feature size.

Glue (jelly beans) A widely used jargon for SSI and MSI chips on a PCB.

GPS GEC Plessey Semiconductors (a company).

Graphical interchange format (graphical language) A format (language) for graphical display and often also for mask patterns.

Handcrafted ASIC *See* custom ASIC.

Hard error Repeated setting of a storage element, generally a memory cell, into the wrong logic state due to a structural defect.

Hard macro (hardwired macro) A macro with fixed placement and fixed routing.

Hardware accelerator (hardware engine, hardware simulation accelerator) A special-purpose processor, usually attached as a server to a workstation, which can carry out some simulations much faster than by software.

Hardware platform *See* work station.

Hardwired macro *See* hard macro.

HC/HCT CMOS High-speed CMOS logic.

HDL A language describing the hardware implementation of a design at various levels of abstraction.

HI²L Bipolar GaAs heterojunction I²L.

HLL High speed low-power low-voltage CMOS logic.

Hot carriers (hot electrons, hot holes) Carriers (hot electrons or holes) with energies far above the ambient thermal energy.

I²L Integrated injection logic.

I/O Input/output.

IC Integrated circuit.

ID Intermodulation distortion.

IEEE Std. 1149.1 An industry-accepted IEEE standard (the term Std. is frequently omitted) for the BST.

Intel 286/386/486/Pentium microprocessors. These microprocessors, which are Intel trademarks, are the microprocessors of choice in many workstations.

Intelligent (smart) chip A chip incorporating substantial intelligence to control its function.

Interpolation filters. Filters in a delta-sigma DAC, which change a multibit input at the Nyquist frequency into a bit stream at a much higher frequency.

ispGAL In-system programmable GAL, a trademark of Lattice Semiconductor Corporation. ispGAL is an acronym for Lattice on-board programmable GALs.

ispLSI In-system programmable large scale integration, a trademark of Lattice Semiconductor Corporation. ispLSI is an acronym for Lattice on-board programmable FPGAs.

JEDEC Joint Electron Devices Engineering Council (US).

JEDEC file A file in a standard JEDEC-prescribed format, which defines all fuse links of a PLD for a hardware programmer.

Jelly beans *See* glue.

JFET Junction field effect transistor.

JK flip-flop (JK bistable) A flip-flop which can operate in the SR (set-reset) or T (toggle) mode.

JTAG Joint Test Action Group. A group which established the IEEE Std. 1149.1.

k Boltzmann's constant.

KSPS Kilosamples per second.

L Length of MOSFET channel. (L also stands for Henry, the unit of inductance).

LAN Local area network. A network loosely linking computers within a single building or site.

Latch-up A parasitic effect, causing a CMOS transistor pair to behave like an SCR and to draw a large current, which may destroy the chip.

Layout rules *See* design rules.

Layout synthesis (placement and routing) The transformation of a circuit schematic (netlist) into the physical layout by means of mask patterns.

LCA Logic cell array, a trademark of Xilinx, is the acronym for a Xilinx FPGA.

LCC *See* LLCC.

LCCC Leadless ceramic chip carrier.

LDCC Leaded chip carrier.

LDD Lightly doped drain; a MOSFET structure.

LDO regulator Voltage regulator with a low dropout voltage.

Line width *See* feature size.

LISP List programming language. An HLL designed primarily for simulating high-level behaviour of digital systems.

LLCC (LCC) Leadless chip carrier.

LOCOS Local oxidation of silicon. An oxide isolation process for transistors.

Logic synthesis The transformation of the behavioural description of a system into structured hardware.

LPCVD Low pressure CVD.

LPE Layout parameter extraction. A CAD/CAE tool which extracts the parameters of the chip wires for timing simulation.

LPF Low pass filter.

LS TTL Low power Schottky TTL.

LSB Least significant bit.

LSI Large scale integration. The term applies to ICs with about (1000-100 000) transistors per chip.

LSSD Level sensitive scan design. A technique for the shift register operation in the BST.

LUT Look-up table. LUTs are frequently used as ROM-based function generators in the core blocks of PLDs.

LV-CMOS Low voltage (LV) version of HC/HCT CMOS.

LVC Low voltage (LV) CMOS compatible with TTL Fast and AS families.

LVS Layout-versus-schematic. A CAD/CAE check to ensure that the netlists derived from the schematic and the layout agree.

LVT Low voltage version of ABT Bus Interface Logic.

Macrocell (macro) An interconnection of array elements or a pre-designed component assembly to form a specified circuit like a logic gate, an op amp, or a complex function.

Manhattan style A design style in which all chip wires are orthogenal i.e. horizontal or vertical.

Mask An imaging source, which contains multiple images for lithography.

MAX Multiple array matrix, a trademark of Altera. MAX EPLDs are Altera FPGAs with an advanced architecture.

MBE Molecular beam epitaxy.

MCM Multichip module. A specially designed package containing a number of interconnected chips.

MCR Molded carrier ring.

MDTL Modified diode transistor logic.

MECL Motorola ECL.

Megacell A very large macrocell which fulfils a function like a microcontroller, PLA, RAM etc.

MESFET Metal-semiconductor field effect transistor.

Metallization
The interconnections formed on top of the substrate.

Microcode An instruction set of control signals, held in ROM, for executing cycle-by-cycle a computer program.

Microprogram A program consisting of microcode to execute a computer instruction.

MILSTD Military standard (US).

Minimum-size geometry The smallest MOSFET channel dimensions obtainable with a given technology ($L=W$).

Mixed-mode ASIC (mixed-signal ASIC) An ASIC containing analogue and digital circuits.

MMIC Monolithic microwave IC.

MNOS Metal-nitride-oxide semiconductor.

MOS Metal oxide semiconductor.

MOSFET Metal-oxide-semiconductor field effect transistor.

Motherboard Printed circuit board for a PC.

MPGA Mask programmable gate array. This term is synonymous with gate array. It is used to distinguish between a gate array and an FPGA when these two ASICs are placed in juxtaposition.

MQFP Metric quad flat pack.

MSB Most significant bit.

MSI Medium scale integration. The term applies to ICs with between about (100-1000) transistors per chip.

MSPS Megasamples per second.

MTL Merged transistor logic. An alternative but rarely used term for I^2L.

MTTF Mean time to failure.

Multibyte interface chips *See* widebus (multibyte) interface chips.

MUX Multiplexer. A combinational circuit which selects one of several inputs for transmission on a single output line.

Nanoelectronics Electronic circuits, devices, and systems that employ lateral features of 100 nm or less.

Netlist A list containing all the components and interconnections of a circuit.

nMOS An IC composed of nMOSFETs.

nMOSFET n-channel MOSFET.

NRE Non-recurrent engineering. This term is usually applied to the development cost of an IC, a PCB etc.

NS National Semiconductor (a company).

Nyquist frequency The Nyquist frequency is twice the maximum frequency of the analogue signal being sampled.

OLMC Output logic macrocell. The output cell of a PLD, sometimes just called macrocell, in which case it must be distinguished from the general type of ASIC macrocell (*see* macrocell). It usually contains a flip-flop and routing multiplexers.

OMPAC Over molded pad array carrier. A Motorola trademark for their BGA.

ONO dielectric Oxygen-nitrogen-oxygen dielectric.

OSR Oversampling ratio in ADCs, defined to be the ratio of the oversampling to the Nyquist frequency.

OTA Operational transconductance amplifier. The output current of an OTA is proportional to the input voltage and a control current.

OTP One-time programmable. OTP applies to programmable devices which are UV erasable, but for which such erasure is prevented by a completely opaque package.

OTP EPROM (OTP PROM) An EPROM without the erase facility (*see* OTP).

P-glass *See* PSG.

PAL Programmable array logic. A PLD very similar to a PLA but with more limited logic. PAL is a trademark of AMD.

PALASM PAL assembler, a trademark of AMD. PALASM converts a design description in sum-of-products form into a standard JEDEC file for a hardware programmer (*see* JEDEC file).

PAM Pulse amplitude modulation.

Paracell The GPS designation for megacell.

PC Personal computer, an IBM trademark.

PCB Printed circuit board.

PCBA Printed circuit board assembly.

PCM Pulse code modulation.

PDM Pulse density modulation.

PEVCD Plasma enhanced CVD.

PGA Pin grid array.

PLA Programmable logic array. A combination of AND and OR arrays which generates an output in terms of sum-of-products of the input.

Placement (annealing, place) The location of macros on the die.

Placement and routing (place and route) *See* layout synthesis.

PLCC Plastic leaded chip carrier.

PLD Programmable logic device. A PLD is a completely prefabricated array; the desired function is obtained by programming the interconnections. The PAL, PLA, and FPGA all come into the category of PLDs.

PLICE Programmable low impedance circuit element. A proprietary trademark of Actel for their antifuse.

PLL Phase locked loop.

PLS Programmable logic sequencer. A PLA some or all of whose outputs are registered. (*See* registered output).

pLSI Programmable large scale integration, a trademark of the Lattice Semiconductor Corporation. pLSI is an acronym for Lattice FPGAs.

PML Programmable macro logic, a trademark of Philips Semiconductors. Philips PML devices come into the category of FPGAs.

pMOS An IC composed of pMOSFETs.

pMOSFET p-channel MOSFET.

Polycide Polysilicon-silicon sandwich.

PPGA Plastic pin grid array.

PQFP Plastic quad flat pack.

Primary cell *See* array element.

PROM Bipolar electrically programmable ROM. This type of ROM cannot be erased.

PSG (P-glass) Phosphosilicate glass (phosphorus-doped silicon dioxide).

PSRR Power supply rejection ratio.

PVD Physical vapour deposition.

PWB Printed wire board. US alternative for PCB.

PWM Pulse width modulation.

q Charge of an electron.

QFP Quad flat pack.

QUIP Quad-in-line package.

RAM Random access memory. A semiconductor memory with random read or write. Readout is non-destructive.

RAS Row address strobe input to a semiconductor memory.

Ratio-type (ratio-dependent) logic *see* ratioed logic.

Ratioed logic Logic circuits whose logic levels depend on transistor dimensions.

Ratioless logic Logic circuits whose logic levels are independent of transistor dimensions.

Registered output A PLD output supplied by a register (flip-flop) in contrast to a combinatorial output.

Registered PAL A PAL some whose outputs are registered.

Resist Photosensitive emulsion.

Reticle An imaging source, which contains a single image, or a small number of images for lithography.

RIE Reactive ion etch.

ROM Read only memory. A mask-programmed fixed-content semiconductor memory with non-destructive readout.

Routing (route) The interconnection of macros by metallization.

RTL Register transfer level (alternatively register transfer language). A behavioural description in an HDL at a level with architectural significance.

S TTL Schottky TTL.

S/H amplifier Sample and hold amplifier.

SBD Schottky-barrier diode (Schottky diode).

SBW System bus bandwidth. The number of bytes which can be transferred across the bus per second.

SC Switched capacitor.

Schematic capture A graphical display of structured hardware for inspection and user-interactive modification.

SCP Single-chip package. This term has

been introduced to distinguish traditional packages containing a single chip from the FCP.

SCR Silicon-controlled rectifier.

SDFL Schottky diode FET logic (GaAs).

SEMI Semiconductor Equipment and Materials Institute.

Semicustom ASIC (semicustom IC) An ASIC which is largely (gate array) or completely (PLD) prefabricated, or which is made with a full set of pre-designed masks (standard cell).

SER Soft error rate. The failure rate in terms of incident alpha particles.

Short channel A channel whose length is comparable with or shorter than the maximum depth of the channel depletion layer.

Si Silicon.

Silicide Compound of refractory metals with silicon.

Silicon compiler A silicon compiler transforms a behavioural description, expressed in an HDL, into mask patterns for layout.

Silicon estate The term denotes the surface area of a silicon die (or dies).

Silicon foundry An IC manufacturer. The term is usually applied to ASIC vendors.

Simultaneous engineering *See* concurrent engineering.

SiO$_2$ Silicon dioxide.

SIP Single-in-line package.

Slew rate *See* SR.

SM Surface mount. A mounting style for PCB attachment.

Smart chip *See* intelligent chip. Smart in this context is the US equivalent of intelligent.

SMT Surface mount technology.

SNR Signal-to-noise ratio.

SO (SOIC) Small outline package.

SOA Safe operating area. The range of operating conditions within which the maximum power dissipation of a transistor is not exceeded.

Soft error The setting of a logic cell (generally a memory cell) into the wrong logic state due to radiation, usually of alpha particles, within the IC package.

Soft macro (software macro) A macro consisting of mutiple hard and/or other soft macros, with flexible placement and routing.

Soft saturation The condition of a bipolar transistor with forward emitter base bias, and a forward collector base bias which is much smaller than that for traditional saturation.

Software macro *See* soft macro.

SOG array (SOG, channelless gate array) Sea-of-gates array. A gate array architecture in which the array elements cccupy the entire core of the die.

SOI (SOIC) Silicon-on-insulator structure.

Solder grid array *See* BGA.

SOS Silicon-on-sapphire structure.

SPE Solid phase epitaxy.

SPICE Simulation program with integrated circuit emphasis. An industry standard for simulating semiconductor circuits.

SR Slew rate. The large-signal rate of change of output voltage in op amps with heavy negative feedback.

SRAM Static RAM. There is no restriction on the maximum interval between successive read or write operations.

SRL Shift register latch. An LSSD cell.

SSI Small scale integration. The term applies to ICs with less than about 100 transistors per chip.

SSOP Shrunk small outline package.

Stack capacitor A capacitor formed by a sandwich of conducting plates separated by thin oxide on top of the substrate.

Standard cell An ASIC made with a full set of pre-designed masks and completed by metallization.

Step-and-repeat projecion *See* DSW projection.

Strong inversion Inversion of MOSFET channel for a gate-source voltage in excess of the threshold voltage.

Structure *See* feature size.

Subthreshold conduction Conduction of MOSFET in weak inversion.

T Absolute temperature (degrees Kelvin).

TAB Tape automated bonding.

TAP Test access port for the IEEE Std. 1149.1 BST.

TC Temperature coefficient.

TCK Test clock input connection of TAP.

TDI Test data input connection of TAP.

TDL Tegas design language.

TDO Test data output connection of TAP.

Technology mapping The conversion of a technology-independent netlist into configured circuits of a PLD.

TH Through hole. A mounting style for PCB attachment.

THD Total harmonic distortion.

Thick oxide *See* field oxide.

Thin oxide (thinox) Thin oxide layer, usually the gate oxide of a MOSFET.

Three-state output Output of a logic circuit which can be controlled to function normally or to be in a high impedance (open circuit) state, regardless of the inputs.

TI Texas Instruments (a company).

Tile array An assembly of active and passive components, arranged in various tile patterns on a chip.

TMA Technology Modeling Associates (a company).

TMS Test mode select input connection of TAP.

Totem pole An output stage (usually in TTL) consisting of two transistors, with one on and the other off, in series with the supply line and ground.

t_{PHL} Propagation delay between input and output at half amplitude for a high-to-low output transition.

t_{PLH} Propagation delay between input and output at half amplitude for a low-to-high output transition.

TQFP Thin quad flat pack.

Trench capacitor A capacitor with a trench (isolation) structure.

Trench isolation Isolation between transistors by a trench cut into the substrate and filled with a dielectric.

Tri-state output Widely used proprietary designation of NS for three-state output.

TRST Optional test reset input connection of TAP.

t_{THL} The transition time, high-to-low level, for a change from 90 to 10 per cent of the signal amplitude.

TTL Transistor-transistor logic.

TTL ALS Advanced low-power Schottky TTL.

TTL AS Advanced Schottky TTL

TTL S Schottky TTL.

t_{TLH} The transition time, low-to-high level, for a change from 10 to 90 per cent of the signal amplitude.

Turnkey route An ASIC design route. The customer undertakes the initial system design, and the ASIC vendor then takes over.

UBT Universal bus transceiver.

UHV Ultra high vacuum.

ULA Uncommitted logic array. *See* gate array.

ULSI Ultra large scale integration. This term applies to ICs with over about one million transistors; it is not widely used.

Universal PAL A PAL whose outputs can be arranged to be registered or combinatorial. (*See* registered output).

UNIX An established computer operating system. UNIX is a trademark of AT&T, Inc.

User-programmable (field-alterable, field programmable) ROM A ROM electrically programmed by the user. Most user-programmable ROMs are also user-erasable.

UV Ultra violet

V_t The thermal voltage, which equals kT/q.

V_T Threshold voltage of a MOSFET.

Verilog An HDL which was developed by Cadence and is an established industry standard. Verilog is a trademark of Cadence Design Systems.

VHDL VHSIC hardware description language. An HDL which is an established industry, and also an IEEE standard.

VHSIC Very-high-speed integrated circuits. A term coined for a development program undertaken in the US during the 1980s.

Via Vertical connection through the dielectric which separates horizontal metal layers. The term can be applied to an IC or to a PCB.

ViaLink antifuse A trademark of Quick Logic for their antifuse.

VLSI Very large scale integration. The term applies to ICs with over about 100 000 transistors per chip.

VPE Vapour phase epitaxy.

W Width of MOSFET channel.

Weak inversion Inversion of the MOSFET channel for a range of gate-source voltage below the threshold voltage voltage.

White noise Noise whose power density is independent of frequency.

Widebus (multibyte) interface chips Interface chips which cater for two-byte or wider logic. A trademark of Signetics/Philips Semiconductors.

Wire Metal interconnection within the chip or on a PCB.

Workstation (hardware platform) A stand-alone computer for handling an engineering design with the aid of CAD/CAE.

ZIP Zigzag-in-line package.

Index

ABEL HDL (AHDL), 419
access time, 262
active filters, 6, 226–32
  SC techniques, 227–31
active load, 119, 136, 142–4, 188, 190
ADCs (analogue-to-digital), 181, 224–5, 232–42
  dual-slope integrating, 238–9
  flash, 239
  half flash, 239–40
  ideal converter characteristic, 232–3
  missing codes, 234, 246
  Nyquist frequency, 236
  quantization noise, 232–4
  sampling, 239
  SC charge scaling, 241–2
  signal bandwidth, 235–6
  S/H amplifier, 234–7
  sinc function, 237
  SNR, 233–4, 247–8
  specification and performance, 246–9,
  static, 234
  successive approximation, 238
adder, 402–3
AHDL (analogue hardware description language), 405
  Verilog-A, 405, 415
  VHDL-A, 405, 415
  see also ABEL HDL
aliasing, 236
  anti-aliasing filter, 237, 252
alphanumeric display, 343
alpha radiation, 271, 274–5
aluminium, 2
  addition of silicon, 41
  annealing, 41
  melting point, 41
  in SBD, 81
annealing, 335, 387
  see also aluminium, ion implantation
antifuse
  PLICE, 279–81, 323–4, 331, 433
  ViaLink, 323–4
antilog amplifier, 185
antimony, 32
APCVD (atmospheric pressure chemical vapour deposition), 39
arithmetic logic unit, 385, 390

array element (basic cell, core cell, primary cell), 295
arsenic, 32, 438
ASICs (application specific integrated circuits), 7, 11–12, 293–423
  AMS mixed-mode cells, 373–5, 416–7
  analogue, 294, 298, 363
  analogue arrays, 365
  array element, 295
  basic cell, 295
  BBS, 354–6
  bed-of-nails, 344–5
  BiCMOS, 380–1
  BIST, 353–5, 412
  bottom-up design, 8, 332
  boundary scan test, 345–53
  BR, 346, 348
  BSC, 346–7
  BSR, 346, 349, 353
  Cadence design tools, 414–15
  categories, 293–4
  CD-ROM, 418–19
  channelled gate array, 294–6
  Compass design tools, 411-13
  control logic, 393
  custom (full custom), 11, 294
  datapath, 371, 391, 393, 395, 413, 416
  design flow, 387–98
  design for testability, 340, 344–56, 390
  design styles, 293–302, 360, 362–5
  design tools, 341, 342–4
  DRC, 90, 395, 408, 412
  economics, 356–62
  embedded cell (cell-based ASIC), 294, 371–2, 381
  equivalent gates, 5, 294, 318, 324–5, 365
  ERC, 395, 408, 412
  firm macro, 393
  formal methods and verification, 412, 420, 445
  foundry design route, 340, 395
  FPGA, 318–32
  GaAs, 381
  gate array (ULA), 294–6, 365–70
  GPS CLA80000 family, 366–71, 416, 436, 440
  GPS ULA families, 375–7, 417
  graphical interchange format, 394
  hard macro, 296
  IEEE Standard 1149.1, 345–55
  I/O block, 295

  layout synthesis, 343, 393, 410
  logic synthesis, 391–2, 410, 342
  LOI, 360–1
  LPE, 395, 407
  LVS, 395, 408–9, 412
  macro (macrocell), 296
  Manhattan design style, 394
  megacell, 294, 296–7, 363
  Mentor Graphics design tools, 413–4
  microcell, 366
  Micro Linear Tile Array, 377–80, 417
  migration, 381
  mixed-mode, 298, 363
  mixed-mode arrays, 364–5
  mixed-mode CMOS cells, 364, 372–5
  mixed-signal, 294
  NRE costs, 11, 339, 356–9, 381
  paracell, 366–7
  place and route, 296, 333, 343, 393, 412
  PLDs, 303–36
  primary cell, 295
  primitives, 391, 408
  semicustom, 294
  simulation, 406–9
  SOG, 295–6, 329
  soft macro, 296–7
  standard cell, 294, 296–7
  Synopsys design tools, 415–6
  TAP, 346–53
  technology mapping, 333–5, 391–2
  top-down design, 8, 332, 385–6
  turnkey design, 340, 395, 410, 416
  VLSI Technology VSC470 family, 370–1, 416
ASSP (application specific standard product), 12, 382, 380
ATE (automatic test equipment), 344, 406
automotive temperature range, 202
avalanche breakdown, 210, 281
avalanche breakdown diode, 210
avalanche injection, 26

bandgap voltage, 210–13, 373, 375, 376
bandwidth
  of single pole network, 95
  system bus, 155
basic cell, see array element
battery operated equipment, 157, 217, 432

BBS (board-level boundary scan), 354–6
bed-of-nails, 344–5
BFL (GaAs buffered FET logic), 141
BGA (ball grid array), 447
BiCMOS
  ABT series, 158, 161–2, 155–6
  ASICs, 380–1
  inverter, 138–9
  LVT series, 158, 161, 163
  NAND gate, 139
  radiation resistance, 380–1
  rail-to-rail output, 139–40
  speed compared with CMOS, 141, 431
  structure, 81–2
BiMOS, 194
bird's beak, 28–30
BIST (built-in self test), 353–5, 412
bistables (flip-flops)
  D, 147–9, 304, 307–8, 351–3, 368
  D latch, 306
  edge-triggered, 149
  JK, 146–8, 308
  level-triggered, 148–9
  lock-out logic, 148–50
  master slave, 148
  NAND, 146
  RS, 145
  RS latch, 145
  T, 145, 148, 307, 403–4
BJT (bipolar junction transistor)
  base conductivity modulation (Webster effect), 109
  base resistance, 438–9
  buried collector, 18, 21
  f_T, 79–81, 108–9, 438–9
  Gummel-Poon model, 105, 107–9
  h_{fe}, 79–81, 108–9
  hot electrons, 430
  lateral pnp, 79–80, 136–7, 188, 218
  multiple collector structure, 136
  multiple emitter structure, 132, 134
  planar, 4
  pnp-npn combination, 80–1
  polysilicon emitter, 78–9, 437–8
  production, 4
  profile, 1
  scaling, 437–9
  SPICE parameters, 110–12, 177
  structure, 78–81
  superbeta, 190
  transconductance, 138, 174
  vertical pnp, 79–80
boron, 23, 31–2

bottom-up design, 8, 332
breakdown diodes, 209–10
BSC (boundary scan cell), 346–7
BSR (boundary shift register), 346, 349, 353
buried collector, 18, 21
bus contention, 261, 266–7

C (C language), 109, 404, 410
CAD-CAE (computer aided design-computer aided engineering), 68, 109, 344, 409, 421, 430
capacitors, 30, 86–7, 273, 375, 377–8
CD (compact disk), 254–5, 258, 418–19
CD-ROM, 344, 418–19
cell-based ASIC, see embedded cell
CERQUAD (pressed ceramic glass sealed flat pack), 60, 370
CERDIP (pressed ceramic glass sealed dual-in-line package), 60
charge pump process, 270–1, 290
chip edge, 41, 419, 444
Class AB amplifiers, 188–9
CML (current mode logic), 129–30
  inverter buffer, 129–30
  soft saturation, 129
CMOS
  AC/ACT series, 153–9
  ALVC series, 155–9, 431
  bistable count rate, 153, 155, 159
  cryogenic operation, 448
  domino, 152
  dynamic, 150–2
  dynamic input power, 122–3, 128, 160–2, 228–9
  gate input power, 160–1
  guard rings, 74–5
  HC/HCT series, 153–4
  HLL series, 155–9, 431
  input protection circuit, 77
  inverter, 121–4
  latch up, 30, 73–4, 76
  LV-CMOS series, 154–5, 159–63
  LVC series, 155–9
  NAND gate, 124
  NOR gate, 124
  n-well, 73
  output buffer, 127–8
  p-well, 21
  scaling, 434–7
  SOI, 77
  SOS, 77, 271
  SPICE simulations, 168–9
  SSI/MSI elements, 122–3, 156

  structures, 73–8
  three-state output, 125
  transient response, 123–4, 159, 167–70, 434–6
  transmission gate, 124–5
  trench isolation, 75–6
  twin-tub, 75
  weak inversion, 212, 374
CMRR (common mode rejection ratio), 95
COB (chip on board), 66
codec (companding encoder/decoder), 237
cold electrons, 282
commercial temperature range, 157, 202
concurrent design, 343
contact windows, see windows
control logic, 393
core cell, see array element
counters, 149, 170, 390, 402–3
CPGA (ceramic pin grid array), 60, 372
CPU (control processing unit), 5
CQFP (ceramic quad flat pack), 60
cryogenic operation, 448
crystal oscillator, 373
CSP (chip scale package), 447
CTE (coefficient of thermal expansion), 61, 63
current mirror, 131, 186–8, 190
current-to-voltage converter, 185
custom (full custom) ASICs, 11, 294
CVD (chemical vapour deposition), 30, 39
CZ (Czochralski) crystal growth, 23

DACs (digital-to-analogue converters)
  ideal converter characteristic, 232–3, 242
  monotonicity, 234, 246
  MOSFET switches, 244
  multiplying, 242–3
  performance, 248–9
  SC charge scaling, 242
  thick film resistors, 243
  voltage scaling, 242–3
Darlington stage, 190, 199, 216
database, 343
data converters
  charge scaling, 102
  hybrid ICs in, 10–11
  see also ADCs, DACs
datapath, 371, 391, 393, 395, 413, 416

DCFL (direct coupled FET logic), 141–4, 431
decimation, 252–3
design for testability, 340, 344–56, 390
design rules, 5, 87–90
  lambda-based, 87
D flip-flop, latch, 147–9, 304, 307–8, 351–3, 368
dielectric films
  borophosphosilicate glass, 39
  over-glass, 39
  phosphosilicate glass, 39
differential amplifiers, 181, 184
diffusion, 16, 31–4, 38, 88
  drive-in, 32, 38
  Fick's law, 32
  lateral, 16, 33, 88
  predeposition, 32
digital filters, 252, 254, 258, 391
diode, 81
  see also SBD
DIP (dual-in-line package), 57, 60, 315, 372
DOD (US Department of Defense), 398
DOS (disk operating system), 261
D-MESFET (depletion mode GaAs MESFET), 83
DRAMs (dynamic random access memories), 271–7
  alpha radiation, 271, 274–5
  BBU mode, 277
  burst refresh, 276
  cell size, 90–91, 274
  characteristics, 263, 284–6
  deep submicron, 432
  GaAs, 288
  hidden refresh, 276
  modules, 285–6
  one-transistor cell, 262, 271–3
  readout circuit, 274–5
  self refresh, 277
  sense amplifier, 269, 275
  simulation, 102
  soft errors, 271–3, 287–8
  stack capacitor, 273–4
  testing, 356
  trench capacitor, 30, 273–4
  see also semiconductor memories
DRC (design rule check), 90, 395, 408, 412
drift velocity see GaAs MESFET, MOSFET
dropout voltage, 217
DSP (digital signal processor), 181, 225, 391
DSW (direct-step-on-wafer) projection, 48–9
dynamic CMOS, 150–2

dynamic input power, 118, 160
  see also CMOS

EAROM (electrically alterable ROM), 263, 283
e-beam (electron beam) lithography, 44–5, 47
ECL (emitter coupled logic), 9, 129–32, 431, 437
  bistable count rate, 153, 158
  combined with CML, 131
  current mirror, 131
  ECL 100K series, 157
  input power, 158
  MECL OR/NOR gate, 130–1
  MECL families, 157–8
  propagation delay, 158
  SPICE simulations, 168, 171
  transfer characteristic, 131
  transmission-line output termination, 131, 172
EDA (electronic design automation), 8, 430, 455
  design tools for, 382, 404, 406–7, 410, 414, 419
EDIF (electronic design interchange format), 394, 405, 411
EEPROM (electrically programmable and erasable read only memory), 263, 282–4
effective channel length, 98, 363
EIA (Electronic) Industries Association), 64
EIAJ (Electronic Industries Association of Japan), 65
electrical length, 442
electromigration, 42–3, 430, 439–40, 444
embedded cell (cell-based ASIC), 294, 297–8, 363, 381
E-MESFET (enhancement mode GaAs MESFET), 83
epitaxy, 30–1
  epitaxial layer, 18, 30–1, 76
EPROM (electrically programmable read only memory), 263, 281–2
equivalent gate, 5, 294, 318, 324–5, 365
ERC (electrical rule check), 395, 408, 412
ESD (electrostatic discharge), 76
etching, 52–5
  anisotropic, 54–5
  RIE, 52–4
exclusive - OR/NOR gate, 120, 307

extended temperature range, 202

fabrication (processing), 15–56
  BJT, 18–21
  CMOS, 21–22
  etching, 52–6
  layer formation, 25–44
  lithography, 44–52
  wafer preparation, 22–5
FC (flip-chip) bonding, 61, 66
FCP (few-chip package), 447
feature size, 5
  rate of decrease, 427–8
Fick's law, 32
field oxide, 2, 16, 27
FIFO (first-in first-out register), 322
firm macro, 393
FIT (failure in time), 287–8, 322, 440
flattening, 391
flip-flops, see bistables
FLOTOX (floating gate tunnel oxide technology), 282–4
floorplan, 390–1, 393–5
F-N (Fowler-Nordheim) tunneling, 282–4
footprint, 57–9, 62, 67
formal methods and verification, 412, 420, 445
Fortran, 109
foundry design route, 340, 395
Fowler-Nordheim tunneling see F-N tunneling
FP (flat pack), 60
FPGA (field programmable gate array), 8, 300–1, 318–32, 363, 381
  Actel ACT 1/2/3 series, 319, 323–5, 417–18
  Altera MAX 5000, 7000 and FLEX 800 series, 320, 328–31
  annealing, 335
  coarse grain, 331
  configuration program, 320–1
  design outline, 332–6
  design tools, 417–19
  ECMOS and EECMOS programming, 323, 325–6
  fine grain, 331
  granularity, 331–2
  hardware programmer, 387
  JEDEC file, 335–6, 419
  Lattice pLSI and ispLSI 1000/2000/3000 series
  LUT 318, 320, 322, 331
  macrocells, 318
  migration, 381
  NRE costs, 304, 331

on-board programming, 325
place and route, 333
PLD equivalent gate count, 318, 324
PLICE antifuse, 279–81, 323–4, 331, 433
programming technologies, 320, 323
SRAM programming, 320–3
technology mapping, 333–5
ViaLink antifuse, 323
Xilinx XC2000/3100A/4000E series
FQFP (fine pitch quad flat pack), 60, 370
f_T **(cut off frequency)**, 79–81, 104–5, 108–9, 438–9
full custom ASICs see custom ASICs

GaAs (gallium arsenide), 6, 12
ASICs, 381
compared with Si, 172–5
electron mobility, 172
RAMs, 288
see also GaAs logic, GaAs MESFET
GaAs (gallium arsenide) logic, 141–5
BFL, 141–2
comparison of families, 144–5
DCFL, 141–4, 431
dynamic, 151
HI^2L, 135
noise margins, 144
SDFL, 141
wire capacitance, 166–7
GaAs MESFET (gallium arsenide metal-semiconductor field effect transistor), 82–4, 103–5, 141
characteristics, 103–5
depletion mode, 83
drift velocity, 173
enhancement mode, 83
equivalent circuit, 104
f_T, 104–5
n-channel, 83–4
p-channel, 84
saturation velocity, 104, 173
Schottky gate, 82, 104
SPICE parameters, 108, 112
subthreshold conduction, 104
transconductance, 173
transition time, 174
GAL (generic array logic), 303, 317–18, 325
gate array (ULA), 294–6, 365–70
germanium, 4
alloy, 63
glue (jelly beans), 153, 293, 332, 447

g_m **(transconductance,)**, 94, 138, 173–4
gold, 82–3
GPB (gain bandwidth product), 199, 202–3
granularity, 331–2

hard disk, 261
hard error, 271
hard macro, 296
hardware accelerator, 344, 408
hardware platform, 409
HDL (hardware description language), 343–4, 386–7, 394, 398–406, 410
behavioural level, 387, 399, 401
concurrent execution, 403
EDIF, 394, 405, 411
IEEE 1164 Standard, 399
IEEE VHDL 1076-1987 Standard, 398, 413
IEEE VHDL 1076-1993 Standard, 399
RTL, 390–1, 399–401, 412, 420
structural level, 387, 399, 401
switch level, 390
Verilog, 390, 404, 411, 415, 418–19
Verilog-A, 405, 415
VHDL, 390, 398, 400–4, 411, 414–15, 418
VHDL-A, 405, 415
h_{fe} **(common emitter current gain)**, 79–81, 108–9
HLL (high level language), 395
hot carriers, 103, 281–4, 430, 448
hot electrons, 103, 281, 283–4, 430
hot holes, 103
HSPICE, 109–10
hybrid ICs, 65, 86

I^2L **(integrated injection logic)**, 118, 135–7
HI^2L, 135
lateral pnp transistor, 136–7
MTL, 135
multi-collector transistor, 136–7
Schottky, 137
implantation, see ion implantation
industrial temperature range, 157, 202
INMOS T800 transputer, 413
interactive graphics, 343
intelligent (smart) chips, 163
interconnections, see wires

interpolation, 253–4
ion implantation, 16, 34–9
annealing, 37–8
drive-in diffusion, 38
impurity distribution, 36–7

JEDEC (Joint Electron Devices Engineering Council), 64
file, 335–6, 419
jelly beans, see glue
JFET (junction field effect transistor), 4, 103–4
JK flip-flop, see bistables
JTAG (Joint Action Test Group), 345
junction isolation, 27

kerf, 24

LAN (local area network), 379
latch up, 30, 73–4, 76
layout synthesis (place and route), 343, 393, 410
LDD structure, 433–4, 448
LDCC, (leaded chip carrier), 60, 370
LDO (low dropout voltage) regulators, 217–19
LISP (list programming language), 405
lithography, 44–56
DSW projection, 48–9
e-beam, 44–5, 47
mix-and-match, 48
multiple images, 44, 46
optical, 44–50
overlay error, 49
phase-shifting masks, 444
projection printing, 47, 445
proximity printing, 47, 445
resolution, 49–50, 444
UV light source, 44, 50
X-ray, 50, 444–5
LLCC, LCC (leadless chip carrier), 60
LOCOS (local oxidation of silicon), 27, 29
log amplifier, 185
logic circuits, 117–45
design criteria, 117–18
input power, 122–3, 160–2
inverter transfer characteristic, 117
nMOS, 119–21, 142
noise margin, 144
output buffer, 127–8
positive logic, 120–1
ratioless, 121
ratio-type, 120
see also logic familes
logic families
BiCMOS, 138–41

Index   469

characterization and performance, 157–72
CMOS, 121–9
ECL/CML, 129–320
GaAs, 141–5
I^2L, 135–7
interface conditions, 159
low voltage, 154–7, 161
SSI/MSI elements, 156
TTL, 132–5
logic synthesis, 342, 391–2, 410
LOI (level of integration), 360–1
LPCVD (low pressure chemical vapour disposition)
LPE (layout parameter extraction), 395, 407
LSI (large scale integration), 5, 361
LSSD (level sensitive scan design), 353–4
LUT (look-up table), 261, 318, 320, 322, 351
LVS (layout v. schematic), 395, 408–9, 412

macro (macrocell), 296, 318, 365–6, 370, 393
  firm, 393
  hard, 296, 374, 393
  soft, 296–7, 374, 393
  see also PAL
Manhattan design style, 394
mask, 44
  levels, 16, 48, 68, 80, 88
  phase-shifting, 444
  pattern generation, 343
MBE (molecular beam expitaxy), 30–1
MCM (multichip module), 7, 56–7, 65–7, 285, 410, 427–8
MDTL (modified diode transistor logic), 132, 135
megacell, 294, 296–7, 363
MESFET (metal-semiconductor field effect transistor), see GaAs MESFET
metallization, 3
microprocessors, 390
  Intel 8008, 5, 425
  Intel 386/486, 418–19
  Intel Pentium, 419, 448
microprogram ROM, 385
migration, 381
military temperature range, 157, 202
MILSTD (military standard, USA) 454L, 398
minimum size transistors, 100, 139
MNOS (metal-nitride-oxide semiconductor), 283
Moore's law, 5, 425
  modified, 425–6

MOSFET (metal-oxide-semiconductor field effect transistor)
  body factor (body effect), 98, 270
  capacitances, 100–2
  channel transit time, 103
  depletion mode, 78
  DMOSFET, 216
  drain breakdown voltage, 432–4
  drift velocity, 103, 173
  effective channel length and width, 98
  enhancement mode, 78
  gate oxide breakdown voltage, 433
  hot carriers, 103, 430, 433
  I-V relationships, 96–100
  LDD structure, 433–4, 448
  minimum size geometry, 100
  n-channel, 77–8
  p-channel, 74
  profile, 2
  saturation velocity, 103
  scaling rules, 171–2, 431–2
  self-aligned gate, 22
  SPICE parameters, 106, 111–12
  static characteristics, 96–100
  strong inversion
  submicron scaling, 431–4
  substrate current, 433
  subthreshold conduction, 99
  threshold voltage, 97–8
  transconductance, 94, 138, 173
  weak inversion, 99, 374
motherboard, 1
MPGA (mask-programmable gate array), 304, 318
MQFP (metric quad flat pack), 370
MSI (medium scale integration), 5, 156
MTL (merged transistor logic), 135
MTTF (mean time to failure), 42, 440
multimedia technology, 261, 341
MUX (multiplexer), 170, 300–1, 304, 318–20, 353

netlist, 332, 343, 387, 390, 408
networking, 343
nMOSFET, 77–8, 244
nMOS logic, 418, 427
  inverter, 119–20
  NAND gate, 120
  NOR gate, 120
  output buffer, 127–8
  pass transistor, 121, 30‘
noise
  in avalanche diode, 210
  1/f, 196

in op amps, 263–4
power density, 196
quantization, 232–4, 251–2
spectral density, 196–8
white, 196
NRE (non-recurrent engineering) costs, 11, 304, 311, 339, 356–9, 381
Nyquist frequency, 236, 249, 253–4

ohmic contacts, 21
OMPAC (over moulded pad array carrier), 447
ONO (oxide-introgen-oxide) dielectric, 279, 433
op amps, 181–206
  ac coupled, 183
  active load, 188, 190
  antilog, 185
  BiMOS, 194
  Bode plot, 198
  categories
  Class AB, 188–9
  closed loop gain, 181–2, 198–9
  CMRR, 195
  current mirror, 186–8, 190
  Darlington stage
  differential, 181, 184
  differentiating, 184
  frequency compensation, 198–9
  frequency response, 198–9
  gain bandwidth product, 199, 202–3
  ideal characteristics, 185
  input stage, 183, 187–8, 190, 192
  integrating, 184
  inverting, 181
  JFET input stage, 190, 195
  log, 185
  MOSFET input stage, 190, 192, 195
  noise, 196–8, 203–4
  Nyquist stability criterion, 198
  offset current, 186, 195, 190
  offset voltage, 186, 195
  open loop gain, 181–2, 195, 198
  OTA, 192–4, 205–6, 376
  output protection, 190
  output stage, 188, 192
  overload protection, 190
  PSRR, 196
  rail-to-rail output, 205
  single supply voltage, 204
  SNR, 197, 199
  SR, 199–201
  summing, 184
  superbeta BJT, 190
  V_{BE} multiplier, 189
  virtual earth, 182
  voltage comparator, 206–9

470 Index

voltage follower, 196, 199–200
Widlar current source, 188–9
operating system, 343
optical links, 431
optical lithography, 44–50
OTA (operational transconductance amplifier), 192–4, 205–6, 376
OTP EPROM (one-time programmable EPROM), 263, 282
overlay errors, 49, 88
oversampling sigma-delta converters, 249–58
anti-aliasing filter, 253
bit stream 252–4, 256
CD audio, 254–5, 258
decimation, 252–3
delta demodulation, 249–50
delta modulation, 249–50
digital filters, 252, 254, 258, 391
interpolation, 253–4
multi-level quantizer, 256, 258
Nyquist frequency, 249, 253–4
OSR, 249–50, 252
PCM, 252
PDM, 254
performance, 256–7
PWM, 254
quantization noise, 251–2
SC filters, 254
sigma delta demodulation, 253–4
sigma-delta modulation, 251–3
SNR, 252, 256–7
oxide
barrier to diffusion/implantation, 16, 38
changes in, 26
critical field for breakdown, 76, 432–3
formation, 25
implantation, 25
isolation, 27, 76
LOCOS process, 27, 29
passivation, 16
stress relief, 29
surface passivation, 25
thick (field), 12
thin (thinox), 18, 27
oxide isolation, 27, 76
see also trench isolation

packages, 56–67
fabrication, 57–60
FC bonding, 61, 66
footprint, 57–9
MCM, 56–7, 65–7
soft errors, 62
styles, 58–9
TAB, 61–2, 66

PAL (programmable array logic), 298–300, 303–18
architecture, 299–300
array organization, 315
characteristics, 313–16
combinatorial output, 306
design, 332–6
EEPROM links, 303, 305
EPROM links, 303
folded array structure, 309–11
half-power, 316
macrocell, 307, 311–12
quarter-power, 316
register, 307–9
registered output, 306
universal, 312
zero-power, 14
PAM (pulse amplitude modulation), 234, 236
paracell, 366–7
pass transistor, 121, 301
PC (personal computer), 1, 261, 344, 409, 417–19
switching regulators in, 221
PCB (printed circuit board; US: printed wire board)
CTE, 61, 63
footprint, 67
layout, 410
multilayer, 7, 65, 345
wire, 9, 65–6, 345, 427, 447
wire capacitance, 166
PCM (pulse code modulation), 252
PDM (pulse density modulation), 254
PECVD (plasma enhanced chemical vapour deposition), 39
permittivity, 164–5
PGA (pin grid array) package, 57, 59, 60
phosphorus, 23, 32
PLA (programmable logic array), 298–300, 391
design of, 332–6
place and route, 296, 333, 343, 393, 412
PLCC (plastic leaded chip carrier), 60, 315, 370
PLDs (programmable logic devices), 7, 294, 303–37
AMD MACH family, 311–12
ECMOS and EECMOS programming, 303–4, 312–13
equivalent gate count, 318, 324–5
GAL, 303, 317–18, 325
MUX, 304, 318–20
netlist, 332
PALASM, 335–6
PAL/PLA design, 332–6
place and route, 333

PLS, 311–12
programming hardware, 333–4
security fuse (cell), 312
technology mapping, 333–4
see also FPGA
PLICE (programmable low impedance circuit element) antifuse, 279–81, 323–4, 331, 433
PLS (programmable logic sequencer), 311
pMOSFET, 74
pMOS logic, 118, 427
PPGA (plastic pin grid array), 60
PQFP (plastic quad flat pack package), 60
polycide, 44, 75
polysilicon
emitter, 78–9, 437–8
gate, 22, 43–4, 75
intrinsic, 34
resistor, 77, 86
power dissipation capacitance, 160
precharging of line, 268–9, 275
primary cell, *see* array element
primitives, 391, 408
processing, *see* fabrication
projection printing, 47, 445
PROM (bipolar programmable ROM), 263, 277–81
proximity printing, 47, 445
PSPICE, 110, 446
PSRR (power supply rejection ratio), 196
PVD (physical vapour deposition), 40
PWB, *see* PCB
PWM (pulse width modulation), 254

quantization noise, 232–4, 251–2
QFP (quad flat pack) package, 60
QUIP (quad-in-line package), 60

rail-to-rail output, 139–40, 205
RAMs, *see* DRAMs, SRAMs
raster display, 343
ratioless (ratio-independent) logic, 121
ratio-type (ratioed) logic, 120
refractory metals, 43
regulator (breakdown) diodes, 209–10
Rent's rule, 56–7
resist, 15–16, 150–2
thickness, 38
resistors, 84–6

in BJT technology, 84–5
cermet, 86
characteristics, 86–7
in CMOS technology, 85
pinch, 85
polysilicon, 77, 86
sheet resistance, 84–5
thin film, 85–6, 243
reticle, 44, 374, 444
RIE (reactive ion etch), 52–4
ROMs (read-only memories), 277–84
  access time, 262
  array architecture, 277–8
  avalanche breakdown, 281
  characteristics, 264, 288–90
  charge-pump process, 290
  cold electrons, 282–3
  EAROM, 263, 283
  EEPROM(E^2PROM), 263, 282–4
  EPROM, 263, 281–2
  FAMOS transistor, 281–2
  flash EPROM, 263, 284
  floating gate, 281
  FLOTOX structure, 282, 284
  F-N technology, 282–3
  hot electrons, 281, 283–4
  MNOS transistor, 283
  OTP EPROM, 263, 282
  PROM, 263, 277–81
  UV erasure
RS flip-flop (latch), 145
RTL (register transfer level), 390–1, 399–401, 412, 420

saturation velocity, 103–4, 173
SBD (Schottky barrier diode), 81–3, 112–13, 141
SBW (system bus bandwidth), 155, 285
schematic capture, 343, 391
Schmitt trigger, 208
Schottky clamped transistor, 81, 134
SCP (single chip package), 447
SCR (silicon controlled rectifier), 73
SC (switched capacitor) techniques, 102, 227–32
  active filters, 232
  charge scaling, 241–2
  equivalent resistor, 227–30
  integrator, 231, 237
SDFL (Schottky diode FET logic), 141
security fuse (cell), 312
semiconductor memories, 261–90
  access time, 262
  bus contention, 261, 266–7
  CAS, 275–6
  categories, 261–3

characteristics, 263–4, 284–7
charge-pump circuit, 270–1
DRAMs, 271–7
FITs, 287–8
hard errors, 271
organization and operation, 263–7
overhead circuits, 262
pre-charging of line, 268–9, 275
RAS, 275–6
redundancy, 271
ROMs, 277–84
SER, 287–8
soft errors, 271, 287–8
SRAMs, 267–71
write cycle, 265–6
semicustom ASICs (application-specific integrated circuits), 294
SER (soft error rate), 287–8
S/H (sample-and-hold) amplifier, 234–7
sheet resistance, 84–5
sigma-delta converters, see oversampling sigma–delta converters
silicides, 43, 75, 440, 448
silicon bipolar transistor, see BJT
silicon compiler, 343–4
silicon dioxide (SiO_2) see oxide
silicon nitride (Si_3N_4), 27, 39–40, 83
simulation, 406–9, 445–6
  behavioural, 391, 406
  circuit level, 414
  event-driven, 408
  of fabrication processes, 445–6
  fault, 406
  logic, 406
  mixed-mode, 407
  mixed-signal, 408
  switch-level, 406, 414
  timing, 406–7
  see also SPICE
SIP (single-in-line package), 60
smart chips, 163
SMT (surface mount technology), 57, 60–2, 345, 447
SNR (signal-to-noise ratio), 197, 199, 233–4, 247–8, 252, 256–7
SO, SOIC (small outline, IC) package, 60, 315, 370
soft errors, 62, 271–3, 287–8
soft macro, 296–7
soft saturation, 129, 218
SOG (sea of gates) array, 295–6, 329

SPE (solid phase epitaxy), 37
speed-power product, 158, 160, 172, 432
SPICE (simulation program with integrated circuit emphasis), 93, 106, 109–11, 161, 398, 407, 414, 445
  BJT model, 109–11
  BJT parameters, 110–12, 177
  CMOS inverter simulations, 168–70
  ECL inverter simulations, 171
  GaAs MESFET model, 112
  GaAs MESFET parameters, 108
  HSPICE, 109–10
  IsSPICE, 110
  junction diode parameters, 112–13
  macro models, 445
  MOSFET models, 111–12
  MOSFET parameters, 106, 176–7
  PSPICE, 110, 146
SR (slew rate), 199–201
SRAMs (static random access memories)
  characteristics, 263, 287–8
  dual port architecture, 267
  GaAs 288
  modules, 287
  six-transistor cell, 268
  soft errors, 271–3, 287–8
SSI (small-scale integration), 5, 156
SSOP (shrunk small outline package), 60
SST (super self-aligned technology), 439
standard cell, 294, 296–7
state machine, 351, 417
submicron scaling, 425–444
  BJT, 437–9
  MOSFET, 431–7
  supply voltage, 432, 434–5
  wires, 439–44
subthreshold conduction, 99, 104, 433
switched capacitor filters, 6, 102, 373

TAB (tape automated bonding), 61–2, 66
TAP (test access port), 346–53
technology mapping, 333–5, 391–2
temperature, standard ranges, 157, 202
T flip-flop, 145, 307, 403–4
TH (plated through-hole) mounting, 57
thermal voltage, see V_t
thin film resistor, 7, 85–6, 243

thin films, 41
tile array, 377–9
top-down design, 8, 332, 385–6
totem pole output stage, 133
TQFP (thin quad flat pack), 60, 370
transconductance, see g_m
transition (rise and fall) times
  BJT, 107–8
  CMOS, 123–4, 159, 167–8
  ECL, 160
  GaAs, 174
  MOSFET, 167–8
  single-pole network, 93–6
trench isolation, 27, 29, 40, 75–6, 78, 273
TTL (transistor-transistor logic), 9, 132–5
  characterization and performance, 157–63
  dynamic input power, 160
  FAST series, 153–4
  54/74 ALS, AS, S, and LS series, 153–4, 158, 163
  input protection, 133
  max. output current, 163
  MDTL, 132, 135
  multiple-emitter transistor, 132, 134
  NAND gate, 132–4
  propagation delay, 158
  Schottky, 134–5
  Schottky clamped transistor, 81, 134
  speed-power product, 158
  SSI/MSI elements, 156
  totem pole output stage, 133
tungsten, 75
turnkey design, 340, 395, 410, 416

UBT (universal bus transceiver), 155
ULA (uncommitted logic array) see gate array
ULSI (ultra large scale integration), 5
UNIX, 410, 412
UV (ultra violet) light, 50, 444

Verilog, 390, 404, 411, 415, 418–19
VHDL (VHSIC hardware description language), 390, 398, 400–4, 411, 414–15, 418
VHSIC (very-high-speed integrated circuits), 398
Via, 3
VIPER microprocessor, 412
VLSI (very-large-scale integration)
  designation (V)LSI, 150
  device count, 5
  manufacturing prices, 358
  mean interconnect length, 419
voltage comparators, 206–9
  hysteresis, 208–9
  Schmitt trigger, 208
  transfer characteristic, 207
  transient response, 207
  threshold stability, 206
voltage follower, 196, 199–200
voltage references, 209–15
  avalanche breakdown, 210
  avalanche breakdown diode, 210
  bandgap reference, 210–13, 375–6, 373
  composite diode, 210
  performance, 215
  regulator (breakdown) diode, 209–10, 214
  Widlar current source, 211
  Zener breakdown diode, 209
voltage regulators, 215–24
  Darlington stage, 216
  DMOSFET, 216
  dropout voltage, 217
  efficiency, 220–1, 224
  LDO, 217–19
  overload protection, 216, 218–19, 221
  performance, 216–17, 219, 221, 223–4
  series, 215–19
  switching, 220–4
VPE (vapour phase epitaxy), 30
V_t (thermal voltage), 105, 185, 188, 210–11

wafer
  cost, 358, 427
  defects, 429
  preparation, 22–5
weak inversion, 99, 212, 374
Widlar current source, 188–9, 211
windows (contact windows), 2, 16, 52, 361
Windows (operating system), 261
wires (interconnections)
  capacitance, 164–7, 440–1
  delays, 441–3
  distributed RC model, 441–3
  formation and properties, 40–4
  lumped RC model, 441–2
  mean length in VLSI, 448
  microstrip line, 164–5
  multilayer, 443
  PCB, 73, 126, 164
  scaling, 439–44
workstation, 8, 342–3, 409
WSI (wafer scale integration), 68

X-ray lithography, 50, 444–5

Zener breakdown diode, 209
ZIP (zigzag-in-line package), 60